SpringerWienNewYork

CISM COURSES AND LECTURES

Series Editors:

The Rectors
Giulio Maier - Milan
Jean Salençon - Palaiseau
Wilhelm Schneider - Wien

The Secretary General
Bernhard Schrefler - Padua

Executive Editor
Paolo Serafini - Udine

The series presents lecture notes, monographs, edited works and proceedings in the field of Mechanics, Engineering, Computer Science and Applied Mathematics.
Purpose of the series is to make known in the international scientific and technical community results obtained in some of the activities organized by CISM, the International Centre for Mechanical Sciences.

INTERNATIONAL CENTRE FOR MECHANICAL SCIENCES

COURSES AND LECTURES - No. 488

MECHANICAL VIBRATION: WHERE DO WE STAND?

EDITED BY

ISAAC ELISHAKOFF
FLORIDA ATLANTIC UNIVERSITY,
BOCA RATON, USA

SpringerWienNewYork

This volume contains 134 illustrations

Printed in Italy
SPIN 205641

All contributions have been typeset by the authors.

ISBN-10 3-211-68586-3 SpringerWienNewYork
ISBN-13 978-3-211-68586-0 SpringerWienNewYork

PREFACE

"There is always a delightful sense of movement, vibration and life".

Theodore Robinson (1852-1896)

"I have never solved a major mechanical or interpretive problem at the keyboard. I have always solved it in my mind".

Jorge Bolet (1914-1990)

The idea of this book stems from the realization that scientists, not unlike laymen, should occasionally interrupt their regular work and reflect on the past, to see both the accomplishments and the drawbacks, so as to be able to plan for future research in the "proper" perspective. But an inquisitive reader may ask: Can one really document in any field, let alone mechanical vibrations (whose very name signifies change), "where do we stand"? Did not a Greek philosopher famously claim that one cannot enter a river twice? Another, on an even more sophisticated note, added that actually it is impossible to enter a river even once! For in the process of entering, both entrant and river change. Likewise, one can argue that it is nearly impossible to answer the question posed in the title of this volume.

But experience shows, despite the sage observations of the philosophers, that one does enter a river, lake, sea, or ocean. Likewise, scientists do stop (if not for a minute, for a conference) to reflect on the past, and if not in its detail, then at least in big strokes on various topics presented by the participants; questions by the listeners often change the research direction of the presenter.

The present writer was pleased to locate a short time ago, while searching for references devoted to the topic of statistical linearization, a paper written over 30 years ago under the title "Vibration: Where Do We Stand in 1975" (see the list of papers in the book Random Vibration — Status and Recent Developments: The Stephen Crandall Festschrift, edited by I. Elishakoff and R.H. Lyon, Elsevier, Amsterdam, 1986, page XVI), by a prominent contributor to mechanical vibrations.

The book in front of you inevitably reflects the personal contributions and preferences of the authors. It opens with several contributions of Professor Erasmo Viola of Bologna, Italy, and his co-authors on three-dimensional dynamic problems. In the following chapters Professor Daniel J. Inman of Blacksburg, Virginia, USA, deals with smart structures; Professor Ilya I. Blekhman of St. Petersburg, Russia, offers a new effective approach to the problem of nonlinear vibrations, whereas Professor Leslaw Socha of Warsaw, Poland, treats nonlinear stochastic problems. Dr. Sondipon Adhikari of Bristol, England, deals with problems of validation and verification, and stochastic eigenvalue problems. The contributions by Professor Ivo Caliò of Catania, Italy and the present writer deal with closed-form trigonometric solutions of inhomogeneous structures, preceded by a review of closed-form polynomial solutions of beam and plate problems by the present writer.

It is important to bring to the attention of the readers of this book another undertaking, namely, an extremely important volume titled Structural Dynamics @ 2000: Current Status and Future Directions, edited by D.J. Ewins and D.J. Inman and printed by Research Studies Press Ltd., in Baldock, Hertfordshire, England, 2001.

It is hoped that the readers will enjoy reading, and hopefully studying, the contributions in this volume and those in the above mentioned book. This may shape, albeit partially, their views on the study of mechanical vibrations and its future. We would like to express our sincere appreciation of the excellent atmosphere of the CISM, a true jewel of science and engineering, and its outstanding people — Secretary General Professor Bernhard Schrefler, Resident Rector Professor Giulio Maier, Ms. Elsa Venir, Ms. Carla Toros, Mr. Ezio Cum, and last but not least Ms. Monica del Pin, who had an extensive dedicated correspondence with us. Long before the modern trend of "globalization", CISM became a strong attractor of young and mature scientists and students from all over the world.

It will be rewarding to receive readers' comments by electronic mail elishako@fau.edu or by regular mail.

Isaac Elishakoff
July 2006

CONTENTS

Basic Equations of the Linearized Theory of Elasticity: a Brief Review

Erasmo Viola

Dipartimento di Ingegneria delle Strutture, dei Trasporti, delle Acque, del Rilevamento, del Territorio - University of Bologna, Viale Risorgimento 2, 40136 Bologna, Italy

Abstract The basic relationships of the linearized theory of elasticity of a continuous system are reviewed in different notations. The governing equations are expressed in terms of the displacement field, together with the appropriate initial and boundary conditions. The equations of motion of a few structural members are deduced.

1 Introduction

In order to formulate the mathematical model of a continuous system S undergoing time–dependent deformation, the equations of motion for infinitesimal displacements will be summarized and discussed hereinafter.

The field equations for displacements and deformations refer to a linear elastic solid and the equilibrium under the action of externally applied loads is expressed in the undeformed state.

The coordinates of points of the continuous system in three–dimensional space are denoted by x, y, z or x_1, x_2, x_3, for the Cartesian coordinate systems $Oxyz$ and $Ox_1x_2x_3$ respectively (Fig. 1).

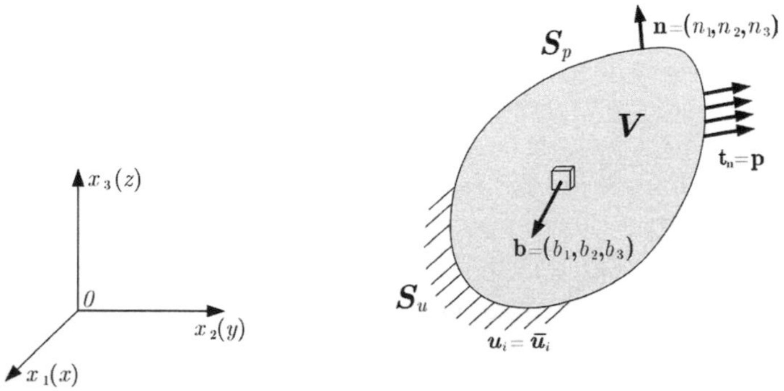

Figure 1. Applied body forces b_i and imposed surface tractions p_i and displacements $\bar{u}_i$ to the equilibrium position of the elastic system in the Cartesian coordinate space $Ox_1x_2x_3$ or $Oxyz$.

The displacement vector **u**, which is defined at every point of the body, is also called the configuration variable and it can be expressed in the extended form

$$\mathbf{u} = \begin{bmatrix} u_1 \\ u_2 \\ u_3 \end{bmatrix} = \begin{bmatrix} u \\ v \\ w \end{bmatrix} \quad \text{or} \quad \mathbf{u}(\mathbf{x}, t) = \begin{bmatrix} u_1(x_1, x_2, x_3, t) \\ u_2(x_1, x_2, x_3, t) \\ u_3(x_1, x_2, x_3, t) \end{bmatrix} = \begin{bmatrix} u(x, y, z, t) \\ v(x, y, z, t) \\ w(x, y, z, t) \end{bmatrix} \tag{1}$$

or in the index notation

$$u_i = u_i(x_1, x_2.x_3.t) \quad \text{or} \quad u_i = u_i(\mathbf{x}, t) \tag{2}$$

where the coordinate vector takes the form

$$\mathbf{x}^{\mathrm{T}} = [x_1\ x_2\ x_3] = [x\ y\ z] \tag{3}$$

The u_i $(i = 1, 2, 3)$ components of the vector **u** refer to the Cartesian coordinate system $Ox_1x_2x_3$, whereas u, v, w denote the displacement components with respect to the coordinate system $Oxyz$.

From above, it should be noted that a vector quantity is denoted by a boldface letter.

The more explicit notation (x, y, z) will be used in the subsequent lectures when it is possible.

The volume occupied by the deformable body is denoted by V and μ indicates the volume density of the material.

The total surface of the body S can be split into two portions S_p and S_u such that $S = S_u \cup S_p$, where $\cup$ denotes the union symbol in set–theoretic notation. On the portion S_u of the surface, the displacements **u** are imposed $(i = 1, 2, 3)$:

$$\mathbf{u} = \bar{\mathbf{u}} \quad \text{or} \quad u_i = \bar{u}_i \tag{4}$$

whereas, the surface traction vector **p**

$$\mathbf{t_n} = \begin{bmatrix} t_{n1} \\ t_{n2} \\ t_{n3} \end{bmatrix} = \begin{bmatrix} p_1 \\ p_2 \\ p_3 \end{bmatrix} = \mathbf{p} \quad \text{or} \quad t_{ni} = p_i \tag{5}$$

is imposed on the S_p surface, namely on the complementary part of the surface S_u.

The boundary conditions (4) are said to be of geometric or essential type, whereas the boundary conditions (5) are referred to as of forced or natural type.

Body forces $\mathbf{b} = (b_1, b_2, b_3)$ act on the unit volume of the body.

The vector of the applied forces **f** may be represented in the vector form

$$\mathbf{f} = \mathbf{b} - \mu\ddot{\mathbf{u}} \quad \text{or} \quad \begin{bmatrix} f_1 \\ f_2 \\ f_3 \end{bmatrix} = \begin{bmatrix} b_1 \\ b_2 \\ b_3 \end{bmatrix} - \mu\frac{\partial^2}{\partial t^2}\begin{bmatrix} u_1 \\ u_2 \\ u_3 \end{bmatrix} \tag{6}$$

as the sum of two terms, namely the vector **b** of the body forces $b_i = b_i(x_1, x_2, x_3, t)$ and the vector $\mu\ddot{\mathbf{u}}$ of the inertia forces $\mu\ddot{u}_i$ $(i = 1, 2, 3)$. A point over the u_i variable stands for derivation of the variable itself with respect to the time t.

In the following, three basic types of equations are mainly considered:

1. kinetic equations or motion equations;
2. kinematic equations of congruence relations;
3. constitutive equations or stress–strain relations.

Kinetic equations describe the static or dynamic equilibrium of a body under the action of applied loads. These equations are also called balance equations.

Kinematic relationships describe the deformation of the elastic body without considering the force causing the deformation itself. Kinematic relations are also called definition equations or congruence relationships.

The constitutive equations describe the constitutive behavior of the body and relate the variables of the kinetic description to the ones of the kinematic description.

These three types of equations can be combined to give the governing equations of the motion of the solid body. These governing equations will also be denoted as fundamental equations. Each basic set of equations will be represented in different notations. The index notation, the extended notation and the vector or matrix notation will be used.

In the above equations four vectors of variables are involved, namely the displacement vector, which describes the configuration of the continuum system; the force vector, which takes the forces applied to the system into account; the strain vector, which represents the strain tensor and the stress vector, which collects the components of the stress tensor.

The strain vector may be put into the extended form

$$\boldsymbol{\varepsilon} = \boldsymbol{\varepsilon}(\mathbf{x}, t) = \begin{bmatrix} \varepsilon_{11} \\ \varepsilon_{22} \\ \varepsilon_{33} \\ 2\varepsilon_{12} \\ 2\varepsilon_{13} \\ 2\varepsilon_{23} \end{bmatrix} = \begin{bmatrix} \varepsilon_{11}(x_1, x_2, x_3, t) \\ \varepsilon_{22}(x_1, x_2, x_3, t) \\ \varepsilon_{33}(x_1, x_2, x_3, t) \\ 2\varepsilon_{12}(x_1, x_2, x_3, t) \\ 2\varepsilon_{13}(x_1, x_2, x_3, t) \\ 2\varepsilon_{23}(x_1, x_2, x_3, t) \end{bmatrix} = \begin{bmatrix} \varepsilon_x(x, y, z, t) \\ \varepsilon_y(x, y, z, t) \\ \varepsilon_z(x, y, z, t) \\ \gamma_{xy}(x, y, z, t) \\ \gamma_{xz}(x, y, z, t) \\ \gamma_{yz}(x, y, z, t) \end{bmatrix} = \begin{bmatrix} \varepsilon_x \\ \varepsilon_y \\ \varepsilon_z \\ \gamma_{xy} \\ \gamma_{xz} \\ \gamma_{yz} \end{bmatrix} \tag{7}$$

or in the index notation

$$\varepsilon_{ij} = \varepsilon_{ij}(x_1, x_2, x_3, t) \quad \text{or} \quad \varepsilon_{ij} = \varepsilon_{ij}(\mathbf{x}, t) \tag{8}$$

In the same way, the stress vector can be expressed in the extended form

$$\boldsymbol{\sigma} = \boldsymbol{\sigma}(\mathbf{x}, t) = \begin{bmatrix} \sigma_{11} \\ \sigma_{22} \\ \sigma_{33} \\ \sigma_{12} \\ \sigma_{13} \\ \sigma_{23} \end{bmatrix} = \begin{bmatrix} \sigma_{11}(x_1, x_2, x_3, t) \\ \sigma_{22}(x_1, x_2, x_3, t) \\ \sigma_{33}(x_1, x_2, x_3, t) \\ \sigma_{12}(x_1, x_2, x_3, t) \\ \sigma_{13}(x_1, x_2, x_3, t) \\ \sigma_{23}(x_1, x_2, x_3, t) \end{bmatrix} = \begin{bmatrix} \sigma_x(x, y, z, t) \\ \sigma_y(x, y, z, t) \\ \sigma_z(x, y, z, t) \\ \sigma_{xy}(x, y, z, t) \\ \sigma_{xz}(x, y, z, t) \\ \sigma_{yz}(x, y, z, t) \end{bmatrix} = \begin{bmatrix} \sigma_x \\ \sigma_y \\ \sigma_z \\ \tau_{xy} \\ \tau_{xz} \\ \tau_{yz} \end{bmatrix} \tag{9}$$

or in the index notation

$$\sigma_{ij} = \sigma_{ij}(x_1, x_2, x_3, t) \quad \text{or} \quad \sigma_{ij} = \sigma_{ij}(\mathbf{x}, t) \tag{10}$$

The state of strain and the state of stress within a body are determined when we know the values at each point of the six components of the vectors (7) and (9), respectively.

2 Dynamic equilibrium equations

In extended notation and in a rectangular Cartesian coordinate system, the equations of motion at time t take the form

$$\begin{aligned}
\frac{\partial \sigma_{11}}{\partial x_1} + \frac{\partial \sigma_{12}}{\partial x_2} + \frac{\partial \sigma_{13}}{\partial x_3} + b_1 &= \mu \frac{\partial^2 u_1}{\partial t^2} \\
\frac{\partial \sigma_{12}}{\partial x_1} + \frac{\partial \sigma_{22}}{\partial x_2} + \frac{\partial \sigma_{23}}{\partial x_3} + b_2 &= \mu \frac{\partial^2 u_2}{\partial t^2} \\
\frac{\partial \sigma_{13}}{\partial x_1} + \frac{\partial \sigma_{23}}{\partial x_2} + \frac{\partial \sigma_{33}}{\partial x_3} + b_3 &= \mu \frac{\partial^2 u_3}{\partial t^2}
\end{aligned} \tag{11}$$

where μ is the density of the material, and the symmetry of the stress tensor is considered too:

$$\sigma_{12} = \sigma_{21} \quad \sigma_{13} = \sigma_{31} \quad \sigma_{23} = \sigma_{32} \tag{12}$$

Equations (11) are also called balance equations.

When no body moments and couples exist, the above system (11) of three equations contains 9 unknowns: 6 stress components σ_{11}, σ_{22}, σ_{33}, σ_{12}, σ_{13}, σ_{23}, and 3 displacements u_1, u_2 and u_3, since b_1, b_2 and b_3 are assumed as given.

The equations of dynamical equilibrium are not sufficient for the determination of the six stress components. Additional equations are needed such as the strain–displacement equations and the constitutive equations.

In index notation, the fundamental stress equations of motion (11) may be put into the form

$$\frac{\partial \sigma_{ij}}{\partial x_j} + f_i = \mu \frac{\partial^2 u_i}{\partial t^2} \quad \text{or} \quad \sigma_{ij,j} + f_i = \mu \ddot{u}_i \tag{13}$$

On the right–hand side of Eq. (13), a comma followed by an index j denotes differentiation with respect to the variable x_j.

Equations of motion or dynamic equilibrium equations (11) can be written in the operatorial form

$$\mathbf{D}^* \boldsymbol{\sigma} = \mathbf{f} \quad \text{or} \quad \mathbf{D}^* \boldsymbol{\sigma} + \mathbf{b} = \mu \frac{\partial^2 \mathbf{u}}{\partial t^2} \tag{14}$$

where $\boldsymbol{\sigma}$ is the vector of the stress tensor and the balance operator $\mathbf{D}^*$, also known as the equilibrium operator, which assumes the aspect:

$$\mathbf{D}^* = -\begin{bmatrix}
\frac{\partial}{\partial x_1} & 0 & 0 & \frac{\partial}{\partial x_2} & \frac{\partial}{\partial x_3} & 0 \\
0 & \frac{\partial}{\partial x_2} & 0 & \frac{\partial}{\partial x_1} & 0 & \frac{\partial}{\partial x_3} \\
0 & 0 & \frac{\partial}{\partial x_3} & 0 & \frac{\partial}{\partial x_1} & \frac{\partial}{\partial x_2}
\end{bmatrix} \tag{15}$$

The equilibrium operator is denoted with the symbol $\mathbf{D}^*$ in order to underline the relationship of adjointness with the congruence operator $\mathbf{D}$, which will be introduced later.

Equations (11) state that at a point inside the body where an infinitesimal cube having unit edges can be supposed to be located, the sum of all the forces acting on the faces of the cube itself as well as at the center of its mass distribution is the vector zero. These equations can be derived using either Newton's second law of motion or the principle of conservation of linear momentum. Equations (11) must be satisfied at every interior point of the body.

2.1 Traction boundary conditions

Let us consider the normal n on a plane at a point B within a body. The components of the unit vector $\mathbf{n}$ of n are n_1, n_2, n_3. It is useful to express the relationship between the stress vector $\mathbf{t}_n$ at the specified point B, on the plane normal to n, and the stress vectors of three mutually perpendicular planes at the same point B, which are parallel to the coordinate planes. In matrix notation, one may write:

$$\begin{bmatrix} t_{n1} \\ t_{n2} \\ t_{n3} \end{bmatrix} = \begin{bmatrix} \sigma_{11} & \sigma_{12} & \sigma_{13} \\ \sigma_{12} & \sigma_{22} & \sigma_{23} \\ \sigma_{13} & \sigma_{23} & \sigma_{33} \end{bmatrix} \begin{bmatrix} n_1 \\ n_2 \\ n_3 \end{bmatrix} \quad \text{or} \quad \mathbf{t}_n = \boldsymbol{\sigma n} \tag{16}$$

in index notation, Eqs. (16) take the form:

$$t_{ni} = \sigma_{ij} n_j \tag{17}$$

for $i = 1, 2, 3$, $j = 1, 2, 3$. Eqs. (16) and (17) are called the Cauchy equations or the Cauchy stress formulas.

It is worth noting that it is customary in mechanics literature to abbreviate a summation of terms by repeating an index, which indicates summation over all the values of that index. This is the well known summation convention, whereby a repeated subscript implies a summation.

The stress components acting across planes parallel to the coordinate planes can also be connected with the external tractions $p_i = p_i(x_1, x_2, x_3, t)$ applied at any point on the bounding surface of a body. Taking account of Eq. (5), Eq. (16) leads to

$$\sigma_{ij} n_j = p_i \tag{18}$$

These relationships may be written in the matrix form

$$\mathbf{B}\boldsymbol{\sigma} = \mathbf{p} \tag{19}$$

where $\boldsymbol{\sigma}$ and $\mathbf{p}$ are the stress vector (9) and the vector of the surface external forces (5) and $\mathbf{B}$ is the matrix

$$\mathbf{B} = \begin{bmatrix} n_1 & 0 & 0 & 0 & n_3 & n_2 \\ 0 & n_2 & 0 & n_3 & 0 & n_1 \\ 0 & 0 & n_3 & n_2 & n_1 & 0 \end{bmatrix} \tag{20}$$

that collects the Cartesian components n_i $(i = 1, 2, 3)$ of the unit normal vector $\mathbf{n}$.

It should be noted that Eq. (19) must be specified at any point of the bounding surface S_p where external loads are applied.

3 Kinematic equations or congruence relations

The deformation of a deformable body in the neighborhood of a given point is characterized by the six components of the symmetrical strain tensor $\boldsymbol{\varepsilon}$, which can be written as a matrix:

$$\boldsymbol{\varepsilon} = \begin{bmatrix} \varepsilon_{11} & \varepsilon_{12} & \varepsilon_{13} \\ \varepsilon_{12} & \varepsilon_{22} & \varepsilon_{23} \\ \varepsilon_{13} & \varepsilon_{23} & \varepsilon_{33} \end{bmatrix} \tag{21}$$

The strain components ε_{11}, ε_{22}, ε_{33} with equal subscripts are denoted by dilatations or normal strains, while the strain components ε_{12}, ε_{13}, ε_{23} with different subscripts are called shearing strains.

In short, ε_{ii} $(i = 1, 2, 3)$ denotes the change in length of a segment originally parallel to the x_i axis, while $\varepsilon_{ij} = \gamma_{ij}/2$ represents half of the change in angle between segments whose original directions were x_i and x_j.

The relationship between the strain components and the displacement components may be written in extended notation

$$\begin{aligned} &\varepsilon_{11} = \frac{\partial u_1}{\partial x_1} \qquad \varepsilon_{22} = \frac{\partial u_2}{\partial x_2} \qquad \varepsilon_{33} = \frac{\partial u_3}{\partial x_3} \\ &\gamma_{12} = 2\varepsilon_{12} = \frac{\partial u_1}{\partial x_2} + \frac{\partial u_2}{\partial x_1} = u_{1,2} + u_{2,1} \\ &\gamma_{13} = 2\varepsilon_{13} = \frac{\partial u_1}{\partial x_3} + \frac{\partial u_3}{\partial x_1} = u_{1,3} + u_{3,1} \\ &\gamma_{23} = 2\varepsilon_{23} = \frac{\partial u_2}{\partial x_3} + \frac{\partial u_3}{\partial x_2} = u_{2,3} + u_{3,2} \end{aligned} \tag{22}$$

In Eq.(22), the derivative notation has been used. That is, differentiation with respect to a variable x_i $(i = 1, 2, 3)$ is indicated by a comma followed by the index i which represents the x_i variable.

Using index notation, the strain–displacement components take the form:

$$\varepsilon_{ij} = \frac{1}{2}\left(u_{i,j} + u_{j,i}\right) = \frac{1}{2}\left(\frac{\partial u_i}{\partial x_j} + \frac{\partial u_j}{\partial x_i}\right) \tag{23}$$

The matrix notation of the congruence equations assumes the aspect

$$\boldsymbol{\varepsilon} = \mathbf{D}\mathbf{u} \tag{24}$$

where

$$\mathbf{D} = \begin{bmatrix} \frac{\partial}{\partial x_1} & 0 & 0 \\ 0 & \frac{\partial}{\partial x_2} & 0 \\ 0 & 0 & \frac{\partial}{\partial x_3} \\ \frac{\partial}{\partial x_2} & \frac{\partial}{\partial x_1} & 0 \\ \frac{\partial}{\partial x_3} & 0 & \frac{\partial}{\partial x_1} \\ 0 & \frac{\partial}{\partial x_3} & \frac{\partial}{\partial x_2} \end{bmatrix} \tag{25}$$

is called the congruence operator or the kinematic operator. Such an operator is also known as the definition operator, because the equations (22) in discussion are known as the definition equations too.

The displacement vector $\mathbf{u}$ and the vector of the strain components are shown by (1) and (7) respectively.

It is worth noting that the congruence operator is the adjoint of the balance operator.

The relationship of adjointness between balance and definition operators can be discovered with respect to the two bilinear functionals

$$< \mathbf{f}, \mathbf{u} >= \int_V f_i u_i \mathrm{d}V \quad , \quad < \boldsymbol{\sigma}, \boldsymbol{\varepsilon} >= \int_V \sigma_{ij} \varepsilon_{ij} \mathrm{d}V \tag{26}$$

3.1 Compatibility equations for strain

When a displacement field $u_i = u_i(x_1, x_2, x_3, t)$ is given, the computation of the strain components of the symmetric small–strain tensor $\boldsymbol{\varepsilon}$ is straightforward. In fact, using Eqs. (23) the strain components ε_{ij} can be expressed in terms of the gradients of the displacement vector $\mathbf{u}$. In addition, the components ω_{ij} of the rotation tensor $\boldsymbol{\omega}$ are defined as

$$\omega_{ij} = \frac{1}{2}(u_{i,j} - u_{j,i}) \tag{27}$$

The last tensor is an antisymmetric tensor, because $\omega_{ij} = -\omega_{ji}$.

However, when the strain components are given, the determination of the displacements is not always possible. The indetermination arises from the system (23) of the six congruence equations which involve only three unknowns u_1, u_2, u_3. Stated in other words, the six components of strain ε_{ij} cannot be given arbitrarily as functions of x_1, x_2, x_3 but must be subject to six differential equations of the type:

$$\varepsilon_{ij,kl} + \varepsilon_{kl,ij} = \varepsilon_{lj,ki} + \varepsilon_{ki,lj} \tag{28}$$

involving the three displacement components.

It should be noted that Eq. (28) provides the necessary and sufficient conditions for the existence of a single–valued displacement field $\mathbf{u} = \mathbf{u}(x_1, x_2, x_3, t)$, when the strains ε_{ij} are given and the region of integration is simply connected. Although Eq. (28) under consideration yields 81 equations, only six of them are different from each other and the remaining ones are either trivial or linear combinations of the above mentioned six independent equations. These 6 essential equations may be put into the operatorial form

$$\boldsymbol{R}\boldsymbol{\varepsilon} = \mathbf{0} \tag{29}$$

where ε is the strain vector (7) and $\mathbf{R}$ is a 6×6 matrix of differential operators of the second order.

Eq. (29) is termed the operatorial form of the compatibility equations for the strain.

4 Constitutive equations or stress–strain relationships

In a linearly elastic isotropic material, the constitutive relations depend only on two elastic constants λ and G, called the Lame's constants:

$$G = \frac{E}{2\,(1+\nu)} \qquad \lambda = \frac{\nu E}{(1+\nu)\,(1-2\nu)} \tag{30}$$

where ν is the Poisson's ratio, E is the Young's modulus or normal modulus, and G is the shear modulus of the material.

The extended notation of the constitutive equations assumes the aspect:

$$\begin{aligned} \sigma_{11} &= 2G\varepsilon_{11} + \lambda I_{1\varepsilon} \quad & \sigma_{22} &= 2G\varepsilon_{22} + \lambda I_{1\varepsilon} \quad & \sigma_{33} &= 2G\varepsilon_{33} + \lambda I_{1\varepsilon} \\ \sigma_{12} &= G\gamma_{12} & \sigma_{13} &= G\gamma_{13} & \sigma_{23} &= G\gamma_{23} \end{aligned} \tag{31}$$

where

$$I_{1\varepsilon} = \varepsilon_{11} + \varepsilon_{22} + \varepsilon_{33} \tag{32}$$

is the cubic dilatation, namely the change in the unit volume of the elastic body.

The first term on the right hand side of the first three Eqs. (31) generates shear deformation, whereas the second one represents the principal stress due to volumetric dilatation.

The stress–strain relationship (31) can be expressed in index notation

$$\sigma_{ij} = 2G\varepsilon_{ij} + \delta_{ij}\lambda I_{1\varepsilon} \tag{33}$$

where δ_{ij} ($\delta_{ij} = 1$ when $i = j$ and $\delta_{ij} = 0$ when $i \neq j$) is the Kronecker delta.

In matrix notation, the relation between the stress components and the strain components takes the form

$$\boldsymbol{\sigma} = \boldsymbol{C}\boldsymbol{\varepsilon} \tag{34}$$

where the stress vector $\boldsymbol{\sigma}$ and the strain vector $\boldsymbol{\varepsilon}$ were indicated above, and

$$\mathbf{C} = \begin{bmatrix} 2G+\lambda & \lambda & \lambda & 0 & 0 & 0 \\ \lambda & 2G+\lambda & \lambda & 0 & 0 & 0 \\ \lambda & \lambda & 2G+\lambda & 0 & 0 & 0 \\ 0 & 0 & 0 & G & 0 & 0 \\ 0 & 0 & 0 & 0 & G & 0 \\ 0 & 0 & 0 & 0 & 0 & G \end{bmatrix} \tag{35}$$

is the constitutive operator, also called matrix of the material rigidity.

It is worth noting that in the case of isotropic material only two elastic constants must be considered. For the general anisotropic case, the elastic constants appearing in the constitutive equations are 21.

5 Fundamental equations

The three basic sets of equations, namely the balance, the kinematic and the constitutive equations may be combined to give the fundamental system of equations, also known as the governing system of equations. Firstly, the fundamental equations are deduced in matrix notation. So, if the strain displacement relations (24) are inserted into the constitutive equations (34), we have the relationships between stresses and displacements:

$$\boldsymbol{\sigma} = \boldsymbol{C\varepsilon} = \mathbf{CDu} \tag{36}$$

When equations (36) are inserted into the equations of motion (14), the fundamental system of equations is derived:

$$\mathbf{D}^*\mathbf{CDu} = \mathbf{f} \tag{37}$$

The equations of motion in terms of displacements take all the three aspects of the problem of the elastic equilibrium into account.

By introducing the fundamental operator, also known as the elasticity operator,

$$\mathbf{L} = \mathbf{D}^*\mathbf{CD} \tag{38}$$

equation (37) can be written as

$$\mathbf{Lu} = \mathbf{f} \quad \text{or} \quad \mathbf{Lu} + \mathbf{b} = \mu\ddot{\mathbf{u}} \tag{39}$$

The fundamental system of equations (39) relates the configuration variable **u** to the source variable **b** of the phenomenon under investigation.

In extended notation, Eq. (39) takes the form:

$$\begin{aligned}
(\lambda + G)\frac{\partial}{\partial x_1}\left(\frac{\partial u_1}{\partial x_1} + \frac{\partial u_2}{\partial x_2} + \frac{\partial u_3}{\partial x_3}\right) + G\nabla^2 u_1 + b_1 = \mu\frac{\partial^2 u_1}{\partial t^2} \\
(\lambda + G)\frac{\partial}{\partial x_2}\left(\frac{\partial u_1}{\partial x_1} + \frac{\partial u_2}{\partial x_2} + \frac{\partial u_3}{\partial x_3}\right) + G\nabla^2 u_2 + b_2 = \mu\frac{\partial^2 u_2}{\partial t^2} \\
(\lambda + G)\frac{\partial}{\partial x_3}\left(\frac{\partial u_1}{\partial x_1} + \frac{\partial u_2}{\partial x_2} + \frac{\partial u_3}{\partial x_3}\right) + G\nabla^2 u_3 + b_3 = \mu\frac{\partial^2 u_3}{\partial t^2}
\end{aligned} \tag{40}$$

Using the index notation, Eqs. (40) assume the aspect

$$(\lambda + G)\, u_{i,ij} + G u_{j,ii} + b_j = \mu\ddot{u}_j \tag{41}$$

Eqs. (40) may also be expressed in the vectorial form:

$$(\lambda + G)\,\nabla I_{1\varepsilon} + G\nabla^2\mathbf{u} + \mathbf{b} = \mu\ddot{\mathbf{u}} \tag{42}$$

where ∇ is the vector operator del (or nabla):

$$\nabla = \mathbf{i_1}\frac{\partial}{\partial x_1} + \mathbf{i_2}\frac{\partial}{\partial x_2} + \mathbf{i_3}\frac{\partial}{\partial x_3} \tag{43}$$

and ∇^2 is the scalar operator called the Laplacian

$$\nabla^2 = \frac{\partial^2}{\partial x_1^2} + \frac{\partial^2}{\partial x_2^2} + \frac{\partial^2}{\partial x_3^2} \tag{44}$$

and

$$I_{1\varepsilon} = \frac{\partial u_1}{\partial x_1} + \frac{\partial u_2}{\partial x_2} + \frac{\partial u_3}{\partial x_3} = \mathrm{div}(\mathbf{u}) = \nabla \cdot \mathbf{u} \tag{45}$$

is the divergence of the displacement field, which denotes the cubical dilatation, that is the change in volume of an infinitesimal rectangular parallelepiped, with sides originally in the coordinate directions, divided by the original volume.

All the basic and fundamental equations, as well as the variables $\mathbf{u}$, $\boldsymbol{\varepsilon}$, $\boldsymbol{\sigma}$ and $\mathbf{b}$ which are involved, are reported in Tonti's diagram (see Appendix A).

The above three equilibrium equations in terms of displacements $u_i(x_1, x_2, x_3, t), i = 1, 2, 3$, are called Navier's equations for the forced vibration. Each form of these equations presents the body forces expressing themselves in different notation, namely as algebraic vector $\mathbf{b} = [b_1 \; b_2 \; b_3]^{\mathrm{T}}$ in (39), by means of the extended notation in (40), in index notation b_j in (41), and as a Cartesian vector $\mathbf{b} = b_1\mathbf{i_1} + b_2\mathbf{i_2} + b_3\mathbf{i_3} = b_i\mathbf{i_i}$ in (42).

It is worth noting that the complete formulation of the dynamic equilibrium problem expressed by the motion equations (40) requires that kinematic boundary conditions (4) on S_u and the forced boundary conditions (18) on S_p must be fixed. In addition, the initial conditions at the $t = 0$ time

$$\mathbf{u}(\mathbf{x}, t) = \mathbf{u}_0(\mathbf{x}, 0) \quad , \quad \left.\frac{\partial \mathbf{u}(\mathbf{x}, t)}{\partial t}\right|_0 = \dot{\mathbf{u}}(\mathbf{x}, 0) = \dot{\mathbf{u}}_0 \tag{46}$$

must be imposed. In short, the considered mixed formulation of the boundary value problem is illustrated in Fig. 1, under the action of a given body force distribution $b_i = b_i(\mathbf{x}, t)$ inside the body, when a displacement field on the S_u surface and boundary forces on the S_p surface are specified.

When Eqs. (40) are solved in conjunction with appropriate initial (46) and boundary conditions, the displacements $u_i = u_i(\mathbf{x}, t)$ can be determined. Then, the application of the congruence relationships (22) and the constitutive relations (31) gives the strains ε_{ij} and the stresses, respectively.

5.1 Dilatational and distortional waves

In the absence of external loading, Navier's equations for the free vibration may be derived. Using the vectorial form (42), for example, we have

$$(\lambda + G)\,\nabla I_{1\varepsilon} + G\nabla^2\mathbf{u} = \mu\ddot{\mathbf{u}} \tag{47}$$

This equation cannot be integrated directly, so a form of solution must be assumed and checked for suitability by differentiation and substitution.

If we assume that the deformation produced by the waves is such that the volume expansion $I_{1\varepsilon}$ is zero, the vector form of the equations of motion becomes:

$$G\nabla^2\mathbf{u} = \mu\ddot{\mathbf{u}} \tag{48}$$

In extended notation, Eq. (48) assumes the aspect:

$$G\nabla^2 u_1 = \mu\frac{\partial^2 u_1}{\partial t^2}, \quad G\nabla^2 u_2 = \mu\frac{\partial^2 u_2}{\partial t^2}, \quad G\nabla^2 u_3 = \mu\frac{\partial^2 u_3}{\partial t^2} \tag{49}$$

These equations for waves are called equivolumic waves, shear waves, waves of distortion, rotational waves or secondary (S) waves. Eqs. (49) may also be put into the form

$$\nabla^2 u_i = \frac{1}{c_T^2}\frac{\partial^2 u_i}{\partial t^2} \tag{50}$$

where

$$c_T = \sqrt{\frac{G}{\mu}} \tag{51}$$

is the velocity of propagation of waves of distortion.

When the deformation produced by the waves is not accompanied by rotation, namely the deformation is irrotational, the components ω_1, ω_2, ω_3 of the rotation vector

$$\omega_1 = \frac{1}{2}\left(\frac{\partial u_3}{\partial x_2} - \frac{\partial u_2}{\partial x_3}\right) \quad \omega_2 = \frac{1}{2}\left(\frac{\partial u_1}{\partial x_3} - \frac{\partial u_3}{\partial x_1}\right) \quad \omega_3 = \frac{1}{2}\left(\frac{\partial u_2}{\partial x_1} - \frac{\partial u_1}{\partial x_2}\right) \tag{52}$$

are zero. Therefore, the displacements $u_i = u_i(x_1, x_2, x_3, t)$ are derivable from a single function $g = g(x_1, x_2, x_3)$ as follows

$$u_1 = \frac{\partial g}{\partial x_1} \quad u_2 = \frac{\partial g}{\partial x_2} \quad u_3 = \frac{\partial g}{\partial x_3} \tag{53}$$

Then, the cubical dilatation and its derivatives take the form

$$I_{1\varepsilon} = \nabla^2 g \quad \text{and} \quad \frac{\partial I_{1\varepsilon}}{\partial x_i} = \nabla^2 u_i \tag{54}$$

Inserting Eqs. (54) into the motion equations (47) leads to the equation for longitudinal waves:

$$(\lambda + 2G)\,\nabla^2\mathbf{u} = \mu\ddot{\mathbf{u}} \tag{55}$$

In different notation we can write ($i = 1, 2, 3$):

$$\nabla^2 u_i = \frac{1}{c_L^2}\frac{\partial^2 u_i}{\partial t^2} \quad \text{or} \quad \nabla^2\mathbf{u} = \frac{1}{c_L^2}\frac{\partial^2\mathbf{u}}{\partial t^2} \tag{56}$$

where

$$c_L = \sqrt{\frac{\lambda + 2G}{\mu}} \tag{57}$$

is the velocity of propagation of the longitudinal waves. A variety of terminology also exists for this wave–type. Longitudinal waves are also called dilatational waves, primary (P) waves ar irrotational waves.

The P and S wave designations have arisen in seismology, where they are also occasionally indicated as the "push" and "shake" waves.

6 Motion equations deduced for structural members

In the following the equations of motion of a few structural elements are derived from the fundamental equations. To this end, a more convenient form of Navier's equations of motion will be used:

$$(\lambda + G) \begin{bmatrix} \frac{\partial}{\partial x_1} \\ \frac{\partial}{\partial x_2} \\ \frac{\partial}{\partial x_3} \end{bmatrix} I_{1\varepsilon} + G\nabla^2 \begin{bmatrix} u_1 \\ u_2 \\ u_3 \end{bmatrix} + \begin{bmatrix} b_1 \\ b_2 \\ b_3 \end{bmatrix} = \mu \frac{\partial^2}{\partial t^2} \begin{bmatrix} u_1 \\ u_2 \\ u_3 \end{bmatrix} \tag{58}$$

The forced and free equations of motion are referred to as both one-dimensional and two-dimensional structural members.

Physical systems such as bars in extension and vibrating strings are included in the category of one–dimensional systems.

The membrane element may be thought of as the two–dimensional analogue of the string.

6.1 Longitudinal vibration of rods

If we assume the one–dimensional displacement field and the vector of the body forces as follows:

$$\mathbf{u}^{\mathrm{T}} = [u_1\; u_2\; u_3] = [u_1(x,t)\; 0\; 0] \quad , \quad \mathbf{b}^{\mathrm{T}} = [b_1(x,t)\; 0\; 0] \tag{59}$$

the cubical dilatation $I_{1\varepsilon} = u_{1,1}$ and its derivative with respect to x_1, x_2, x_3 may be calculated. We have $I_{1\varepsilon,1} = u_{1,11}$, $I_{1\varepsilon,2} = I_{1\varepsilon,3} = 0$.

Using the above positions and results, the first equation of the system (58) of differential equations yields:

$$(\lambda + G) \frac{\partial^2 u_1}{\partial x_1^2} + G\frac{\partial^2 u_1}{\partial x_1^2} + b_1 = \mu \frac{\partial^2 u_1}{\partial t^2} \tag{60}$$

while the second and the third of the equations under consideration are identically verified. Eq. (60) denotes the forced motion equation for a one–dimensional continuous system.

Eq. (60) can be put into the form

$$\frac{\partial^2 u_1}{\partial x_1^2} = \frac{1}{c_L^2} \frac{\partial^2 u_1}{\partial t^2} \tag{61}$$

when the component b_1 of the body force is equal to zero. The velocity of propagation of longitudinal waves c_L can be expressed as

$$c_L = \sqrt{\frac{\lambda + 2G}{\mu}} = \sqrt{\frac{E}{\mu\,(1+\nu)}\left(\frac{1-\nu}{1-2\nu}\right)} \tag{62}$$

It is worth noting that if the Poisson's coefficient is set to zero, one obtains $c_L^2 = E/\mu$ that is the expression of the velocity of propagation of plane wave dilatation deduced by the application of approximate theories.

6.2 Transverse vibration of taut strings

The governing equation for a taut string can be obtained from the fundamental system of equations by assuming the displacement vector $\mathbf{u}$ and the force vector $\mathbf{b}$ as is indicated here:

$$\mathbf{u}^{\mathrm{T}} = [u_1\ u_2\ u_3] = [0\ u_2(x,t)\ 0] \quad , \quad \mathbf{b}^{\mathrm{T}} = [0\ b_2(x,t)\ 0] \tag{63}$$

The evaluation of $I_{1\varepsilon}$ and its derivative leads to $I_{1\varepsilon} = 0$, $I_{1\varepsilon,1} = I_{1\varepsilon,2} = I_{1\varepsilon,3} = 0$ and from the second equation of the system (58) the forced motion of the taut string under the external force $b_2 = b_2(x,t)$ is derived

$$G\frac{\partial^2 u_2}{\partial x_2^2} + b_2 = \mu\frac{\partial^2 u_2}{\partial t^2} \tag{64}$$

When $b_2 = 0$, the equation governing the harmonic motion is written as

$$\frac{\partial^2 u_2}{\partial x_2^2} = \frac{1}{c_T}\frac{\partial^2 u_2}{\partial t^2} \tag{65}$$

where

$$c_T = \sqrt{\frac{G}{\mu}} \tag{66}$$

is the shear wave velocity.

It is worthwhile to note that when an approximate theory is used for the waves in taut strings, the tangential elastic modulus G has to be replaced by the traction force T with which the string is initially taut.

6.3 Governing equation for membranes

In considering the vibration of elastic bodies involving two independent spatial variables, in what follows only the membrane element will be examined.

The equation governing the motion of the membrane can be derived from Eq. (58), by assuming

$$\mathbf{u}^{\mathrm{T}} = [u_1\ u_2\ u_3] = [0\ 0\ u_3(x_1,x_2,t)] \quad , \quad \mathbf{b}^{\mathrm{T}} = [0\ 0\ b_3(x_1,x_2,t)] \tag{67}$$

The assumption (67) leads to the following equation of the forced motion of the membrane:

$$G\left(\frac{\partial^2 u_3}{\partial x_1^2}+\frac{\partial^2 u_3}{\partial x_2^2}\right)+b_3=\mu\frac{\partial^2 u_3}{\partial t^2} \tag{68}$$

Setting $b_3 = 0$ in Eq. (68) the two dimensional form of the wave equation can be derived

$$\frac{\partial^2 u_3}{\partial x_1^2}+\frac{\partial^2 u_3}{\partial x_2^2}=\frac{1}{c_T^2}\frac{\partial^2 u_3}{\partial t^2} \tag{69}$$

7 Concluding remarks

The basic equations of the linearized theory of elasticity were presented and analyzed in a general framework. The equations for dilatational and distortional waves were derived in the three dimensional case and in some specific one and two dimensional structural elements.

Acknowledgements

This topic is one of the subjects of the Centre of Study and Research for the Identification of Materials and Structures (CIMEST) "M. Capurso".

Bibliography

A. E. H. Love. *A Treatise on the Mathematical Theory of Elasticity.* Dover publication, New York, fourth edition, 1944.

S. P. Timoshenko and J. N. Goodier. *Theory of Elasticity. International Student Edition.* McGraw-Hill Kogakusha Ltd, Tokio, 1970.

E. Tonti. The reason for analogies between physical theories. *Applied Mathematical Modelling*, 1:37–50, 1976.

E. Viola. *The Science of Constructions - Theory of Elasticity.* Pitagora Editor, Bologna, 1990. (in Italian).

Appendix A Tonti's diagram

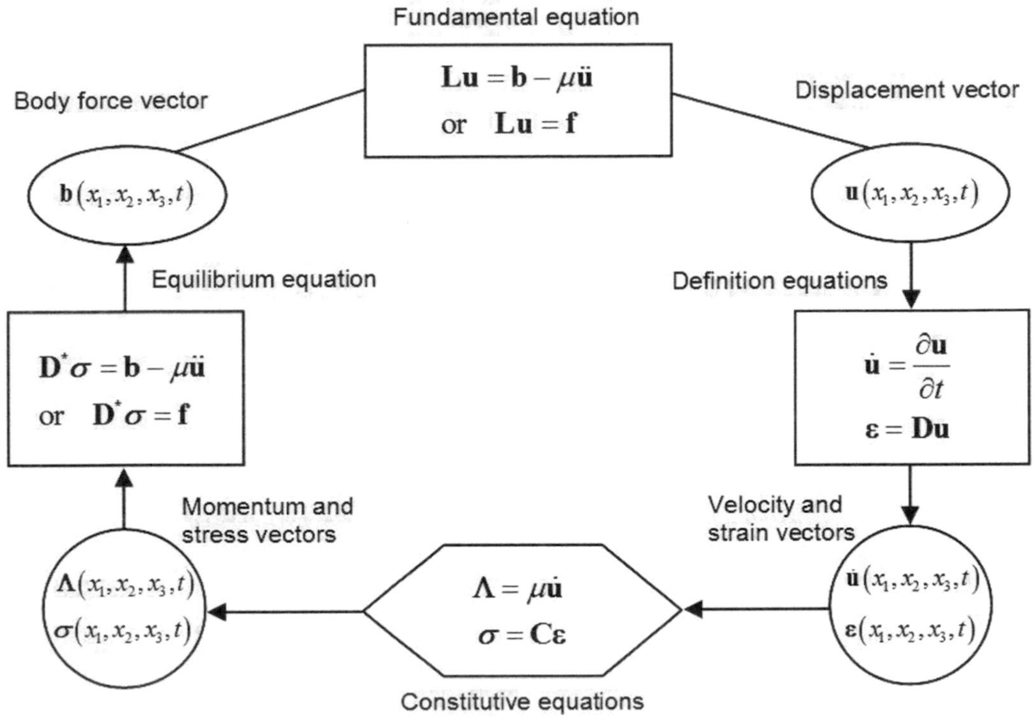
Fundamental equation
$\mathbf{Lu} = \mathbf{b} - \mu\ddot{\mathbf{u}}$
or $\mathbf{Lu} = \mathbf{f}$
Body force vector
Displacement vector
$\mathbf{b}(x_1, x_2, x_3, t)$
$\mathbf{u}(x_1, x_2, x_3, t)$
Equilibrium equation
Definition equations
$\mathbf{D}^*\sigma = \mathbf{b} - \mu\ddot{\mathbf{u}}$
or $\mathbf{D}^*\sigma = \mathbf{f}$
$\dot{\mathbf{u}} = \frac{\partial \mathbf{u}}{\partial t}$
$\boldsymbol{\varepsilon} = \mathbf{Du}$
Momentum and
stress vectors
Velocity and
strain vectors
$\boldsymbol{\Lambda}(x_1, x_2, x_3, t)$
$\sigma(x_1, x_2, x_3, t)$
$\boldsymbol{\Lambda} = \mu\dot{\mathbf{u}}$
$\sigma = \mathbf{C}\boldsymbol{\varepsilon}$
$\dot{\mathbf{u}}(x_1, x_2, x_3, t)$
$\boldsymbol{\varepsilon}(x_1, x_2, x_3, t)$
Constitutive equations

Dynamical Analysis of Spherical Structural Elements Using the First-order Shear Deformation Theory

Erasmo Viola and Francesco Tornabene

Dipartimento di Ingegneria delle Strutture, dei Trasporti, delle Acque, del Rilevamento, del Territorio, University of Bologna, Viale Risorgimento 2, 40136 Bologna, Italy

Abstract. This lecture deals with the dynamical behaviour of hemispherical domes and shell panels. The First-order Shear Deformation Theory (FSDT) is used to analyze the above moderately thick structural elements. The treatment is conducted within the theory of linear elasticity, when the material behaviour is assumed to be homogeneous and isotropic. The governing equations of motion, written in terms of internal resultants, are expressed as functions of five kinematic parameters, by using the constitutive and the congruence relationships. The boundary conditions considered are clamped (C), simply supported (S) and free (F) edge. Numerical solutions have been computed by means of the technique known as the Generalized Differential Quadrature (GDQ) Method. These results, which are based upon the FSDT, are compared with the ones obtained using commercial programs such as Ansys, Femap/Nastran, Straus, Pro/Engineer, which also elaborate a three-dimensional analysis.

1 Introduction

Structures of shell revolution type have been widespread in many fields of engineering, where they give rise to optimum conditions for dynamical behaviour, strength and stability. Pressure vessels, cooling towers, water tanks, dome-shaped structures, dams, turbine engine components and so forth, perform particular functions over different branches of structural engineering.

The purpose of this lecture is to study the dynamic behaviour of structures derived from shells of revolution. The equations given here incorporate the effects of transverse shear deformation and rotary inertia.

The geometric model refers to a moderately thick shell. The analysis will be performed by following two different investigations. In the first one, the solution is obtained by using the numerical technique termed GDQ method, which leads to a generalized eigenvalue problem. The main features of the numerical technique under discussion, as well as its historical development, are illustrated in section 3.

The solution is given in terms of generalized displacement components of the points lying on the middle surface of the shell. At the moment it can only be pointed out that by using the GDQ technique the numerical statement of the problem does not pass through any variational formulation, but deals directly with the governing equations of motion.

Numerical results will also be computed by using commercial programs, which also elaborate three-dimensional analyses.

It should be noted that there are various two-dimensional theories of thin shells. Any two-dimensional theory of shells is an approximation of the real three-dimensional problem. Starting from Love's theory about the thin shells, which dates back to 100 years ago, a lot of contributions on this topic have been made since then. The main purpose has been that of seeking better and better approximations for the exact three-dimensional elasticity solutions for shells.

In the last fifty years refined two-dimensional linear theories of thin shells have developed including important contributions by Sanders (1959), Flügge (1960), Niordson (1985). In these refined shell theories the deformation is based on the Kirchhoff-Love assumption. In other words, this theory assumes that normals to the shell middle-surface remain normal to it during deformations and unstretched in length.

It is worth noting that when the refined theories of thin shells are applied to thick shells, the errors could be quite large. With the increasing use of thick shells in various engineering applications, simple and accurate theories for thick shells have been developed. With respect to the thin shells, the thick shell theories take the transverse shear deformation and rotary inertia into account. The transverse shear deformation has been incorporated into shell theories by following the work of Reissner (1945) for the plate theory.

Several studies have been presented earlier for the vibration analysis of such revolution shells and the most popular numerical tool in carrying out these analyses is currently the finite element method. The generalized collocation method based on the ring element method has also been applied (Viola and Artioli (2004), Artioli, Gould and Viola (2004)). With regard to the latter method each static and kinematic variable is transformed into a theoretically infinite Fourier series of harmonic components, with respect to the circumferential co-ordinates.

In this paper, the governing equations of motion are a set of five bi-dimensional partial differential equations with variable coefficients. These fundamental equations are expressed in terms of kinematic parameters and can be obtained by combining the three basic sets of equations, namely balance, congruence and constitutive equations.

Referring to the formulation of the dynamic equilibrium in terms of harmonic amplitudes of mid-surface displacements and rotations, in this paper the system of second-order linear partial differential equations is solved, without resorting to the one-dimensional formulation of the dynamic equilibrium of the shell. Now, the discretization of the system leads to a standard linear eigenvalue problem, where two independent variables are involved.

In this way it is possible to compute the complete assessment of the modal shapes corresponding to natural frequencies of structures.

2 Basic Governing Equations

2.1 Shell Geometry and Kinematic Equations

The geometry of the shell considered hereafter is a surface of revolution with a circular curved meridian. The notation for the co-ordinates is shown in Figure 1. The total thickness of the shell is represented by h. The distance of each point from the shell mid-surface along the normal is ζ.

The co-ordinate along the meridional and circumferential directions are $\alpha_1 = \alpha_\varphi$ and $\alpha_2 = \alpha_\vartheta$, respectively. The distance of each point from the axis of revolution is $R_0(\varphi)$ and φ is the angle between the normal to the shell surface and the axis of revolution (Figure 2).

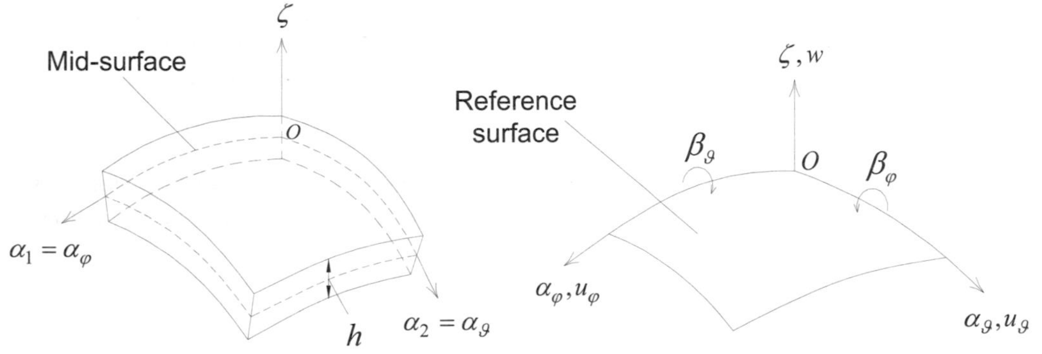

Figure 1. Co-ordinate system of the shell and reference surface.

The position of an arbitrary point within the shell material is known by the co-ordinates φ ($0 \le \varphi \le \pi$), ϑ ($0 \le \vartheta \le 2\pi$) upon the middle surface, and ζ directed along the outward normal and measured from the reference surface ($-h/2 \le \zeta \le h/2$).

R_φ and R_ϑ are, in the general case, the radii of curvature in the meridional and circumferential directions. For a spherical surface R_φ and R_ϑ are constant and equal to the radius of the shell R.

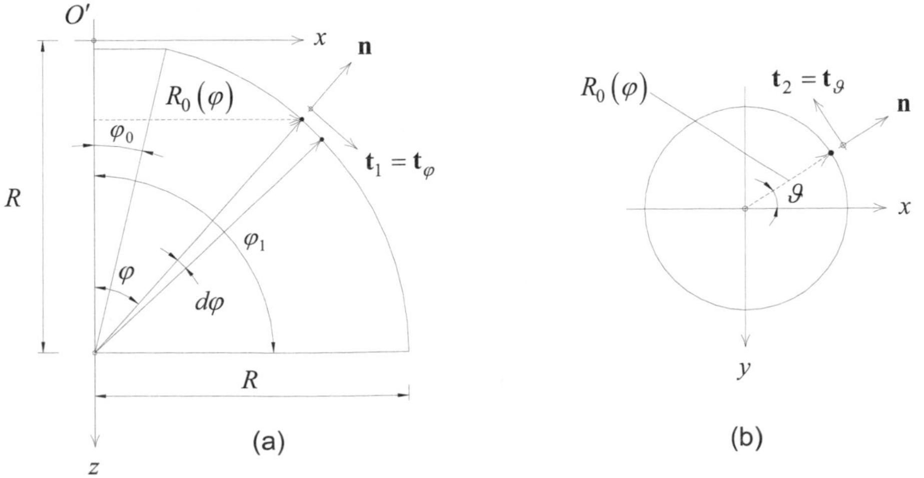

Figure 2. Geometry of hemispherical dome.

The parametric co-ordinates (φ,ϑ) define, respectively, the parallel circles and the meridional curves upon the middle surface of the shell (Figure 2). In developing a moderately thick shell theory we make certain assumptions. They are outlined below:

- The transverse normal is inextensible:

$$\varepsilon_n \approx 0$$

- Normals to the reference surface of the shell before deformation remain straight but not necessarily normal after deformation (a relaxed Kirchhoff-Love hypothesis).
- The transverse normal stress is negligible so that the plane assumption can be invoked:

$$\sigma_n = \sigma_n\left(\alpha_1,\alpha_2,\zeta,t\right) = 0$$

Consistent with the assumptions of a moderately thick shell theory, the displacement field assumed in this study is that of a *First-order Shear Deformation Theory* (FSDT) and can be put in the following form:

$$\begin{cases} U_\varphi\left(\alpha_\varphi,\alpha_\vartheta,\zeta,t\right) = u_\varphi\left(\alpha_\varphi,\alpha_\vartheta,t\right) + \zeta\beta_\varphi\left(\alpha_\varphi,\alpha_\vartheta,t\right) \\ U_\vartheta\left(\alpha_\varphi,\alpha_\vartheta,\zeta,t\right) = u_\vartheta\left(\alpha_\varphi,\alpha_\vartheta,t\right) + \zeta\beta_\vartheta\left(\alpha_\varphi,\alpha_\vartheta,t\right) \\ W\left(\alpha_\varphi,\alpha_\vartheta,\zeta,t\right) = w\left(\alpha_\varphi,\alpha_\vartheta,t\right) \end{cases} \tag{2.1}$$

where u_φ,u_ϑ,w are the displacement components of points lying on the middle surface ($\zeta = 0$) of the shell, along meridional, circumferential and normal directions, respectively. β_φ and β_ϑ are normals-to-mid-surface rotations, respectively.

The kinematics hypothesis expressed by equations (2.1) should be supplemented by the statement that the shell deflections are small and strains are infinitesimal, that is $w\left(\alpha_\varphi,\alpha_\vartheta,t\right) \ll h$.

It is worth noting that in-plane displacements U_φ and U_ϑ vary linearly through the thickness, while W remains independent of ζ. The relationships between strains and displacements along the shell reference (middle) surface $\zeta = 0$ are the following:

$$\begin{gathered} \varepsilon_\varphi = \frac{1}{R}\left(\frac{\partial u_\varphi}{\partial \varphi} + w\right), \quad \varepsilon_\vartheta = \frac{1}{R_0}\left(\frac{\partial u_\vartheta}{\partial \vartheta} + u_\varphi \cos\varphi + w \sin\varphi\right), \quad \gamma_{\varphi\vartheta} = \frac{1}{R}\frac{\partial u_\vartheta}{\partial \varphi} + \frac{1}{R_0}\left(\frac{\partial u_\varphi}{\partial \vartheta} - u_\vartheta \cos\varphi\right) \\ \kappa_\varphi = \frac{1}{R}\frac{\partial \beta_\varphi}{\partial \varphi}, \quad \kappa_\vartheta = \frac{1}{R_0}\left(\frac{\partial \beta_\vartheta}{\partial \vartheta} + \beta_\varphi \cos\varphi\right), \quad \kappa_{\varphi\vartheta} = \frac{1}{R}\frac{\partial \beta_\vartheta}{\partial \varphi} + \frac{1}{R_0}\left(\frac{\partial \beta_\varphi}{\partial \vartheta} - \beta_\vartheta \cos\varphi\right) \\ \gamma_{\varphi n} = \frac{1}{R}\left(\frac{\partial w}{\partial \varphi} - u_\varphi\right) + \beta_\varphi, \quad \gamma_{\vartheta n} = \frac{1}{R_0}\left(\frac{\partial w}{\partial \vartheta} - u_\vartheta \sin\varphi\right) + \beta_\vartheta \end{gathered} \tag{2.2}$$

where $R_0(\varphi) = R\sin\varphi$ is the radius of a generic parallel of the spherical dome.

In the above, the first three strains $\varepsilon_\varphi,\varepsilon_\vartheta,\gamma_{\varphi\vartheta}$ are in-plane meridional, circumferential and shearing components, $\kappa_\varphi,\kappa_\vartheta,\kappa_{\varphi\vartheta}$ are the analogous curvature changes. The last two components are transverse shearing strains.

The matrix notation of the congruence equations assumes the aspect:

$$\boldsymbol{\varepsilon} = \mathbf{D}\mathbf{u} \tag{2.3}$$

where

$$
\mathbf{D}=\begin{bmatrix}
\frac{1}{R}\frac{\partial}{\partial\varphi} & 0 & \frac{1}{R} & 0 & 0 \\
\frac{\cos\varphi}{R_0} & \frac{1}{R_0}\frac{\partial}{\partial\vartheta} & \frac{\sin\varphi}{R_0} & 0 & 0 \\
\frac{1}{R_0}\frac{\partial}{\partial\vartheta} & \frac{1}{R}\frac{\partial}{\partial\varphi}-\frac{\cos\varphi}{R_0} & 0 & 0 & 0 \\
0 & 0 & 0 & \frac{1}{R}\frac{\partial}{\partial\varphi} & 0 \\
0 & 0 & 0 & \frac{\cos\varphi}{R_0} & \frac{1}{R_0}\frac{\partial}{\partial\vartheta} \\
0 & 0 & 0 & \frac{1}{R_0}\frac{\partial}{\partial\vartheta} & \frac{1}{R}\frac{\partial}{\partial\varphi}-\frac{\cos\varphi}{R_0} \\
-\frac{1}{R} & 0 & \frac{1}{R}\frac{\partial}{\partial\varphi} & 1 & 0 \\
0 & -\frac{\sin\varphi}{R_0} & \frac{1}{R_0}\frac{\partial}{\partial\vartheta} & 0 & 1
\end{bmatrix} \tag{2.4}
$$

is called the congruence operator or the kinematic operator and

$$\mathbf{u}(\alpha_\varphi,\alpha_\vartheta,t)=\begin{bmatrix} u_\varphi & u_\vartheta & w & \beta_\varphi & \beta_\vartheta \end{bmatrix}^T \tag{2.5}$$

$$\boldsymbol{\varepsilon}(\alpha_\varphi,\alpha_\vartheta,t)=\begin{bmatrix} \varepsilon_\varphi & \varepsilon_\vartheta & \gamma_{\varphi\vartheta} & \kappa_\varphi & \kappa_\vartheta & \kappa_{\varphi\vartheta} & \gamma_{\varphi n} & \gamma_{\vartheta n} \end{bmatrix}^T \tag{2.6}$$

denote the displacement vector and the generalized strain vector, respectively. The congruence operator is also known as the definition operator, because the equations (2.2) in discussion are known as the definition equations too.

2.2 Constitutive Equations

The shell material assumed in the following is a mono-laminar elastic isotropic one. Accordingly, the following constitutive equations relate internal stress resultants and internal couples with generalized strain components on the middle surface:

$$
\begin{aligned}
&N_\varphi = K(\varepsilon_\varphi+\nu\varepsilon_\vartheta), \quad M_\varphi = D(\kappa_\varphi+\nu\kappa_\vartheta), \quad Q_\varphi = K\frac{(1-\nu)}{2\chi}\gamma_{\varphi n}\\
&N_\vartheta = K(\varepsilon_\vartheta+\nu\varepsilon_\varphi), \quad M_\vartheta = D(\kappa_\vartheta+\nu\kappa_\varphi), \quad Q_\vartheta = K\frac{(1-\nu)}{2\chi}\gamma_{\vartheta n}\\
&N_{\varphi\vartheta} = N_{\vartheta\varphi} = K\frac{(1-\nu)}{2}\gamma_{\varphi\vartheta}, \quad M_{\varphi\vartheta} = M_{\vartheta\varphi} = D\frac{(1-\nu)}{2}\kappa_{\varphi\vartheta}
\end{aligned} \tag{2.7}
$$

where $K = Eh/(1-\nu^2)$, $D = Eh^3/(12(1-\nu^2))$ are the membrane and bending rigidity, respectively. E is the Young modulus, ν is the Poisson ratio and χ is the shear factor which for isotropic materials is usually taken as $\chi = 6/5$. In equations (2.7), the first three components $N_\varphi, N_\vartheta, N_{\varphi\vartheta}$ are the in-plane meridional, circumferential and shearing force resultants, $M_\varphi, M_\vartheta, M_{\varphi\vartheta}$ are the analogous couples, while the last two Q_φ, Q_ϑ are the transverse shears.

In matrix notation, the relation between the generalized stress resultants per unit length and the generalized strain components takes the form:

$$\mathbf{S} = \mathbf{C}\boldsymbol{\varepsilon} \tag{2.8}$$

where

$$\mathbf{C}=\begin{bmatrix} K & \nu K & 0 & 0 & 0 & 0 & 0 & 0 \\ \nu K & K & 0 & 0 & 0 & 0 & 0 & 0 \\ 0 & 0 & K\frac{1-\nu}{2} & 0 & 0 & 0 & 0 & 0 \\ 0 & 0 & 0 & D & \nu D & 0 & 0 & 0 \\ 0 & 0 & 0 & \nu D & D & 0 & 0 & 0 \\ 0 & 0 & 0 & 0 & 0 & D\frac{1-\nu}{2} & 0 & 0 \\ 0 & 0 & 0 & 0 & 0 & 0 & K\frac{1-\nu}{2\chi} & 0 \\ 0 & 0 & 0 & 0 & 0 & 0 & 0 & K\frac{1-\nu}{2\chi} \end{bmatrix} \tag{2.9}$$

is the constitutive operator, also called matrix of the material rigidity and

$$\mathbf{S}(\alpha_\varphi,\alpha_\vartheta,t)=\begin{bmatrix} N_\varphi & N_\vartheta & N_{\varphi\vartheta} & M_\varphi & M_\vartheta & M_{\varphi\vartheta} & Q_\varphi & Q_\vartheta \end{bmatrix}^T \tag{2.10}$$

is the vector of internal stress resultants also termed internal force vector.

2.3 Equations of Motion in terms of internal actions

Following the direct approach or the Hamilton's principle in dynamic version and remembering the Gauss-Codazzi relations for the shells of revolution $dR_0/d\varphi = R_\varphi \cos\varphi = R\cos\varphi$, five equations of dynamic equilibrium in terms of internal actions can be written for the shell element:

$$\begin{aligned}
&\frac{1}{R}\frac{\partial N_\varphi}{\partial \varphi}+\frac{1}{R_0}\frac{\partial N_{\varphi\vartheta}}{\partial \vartheta}+\frac{\cos\varphi}{R_0}\left(N_\varphi-N_\vartheta\right)+\frac{Q_\varphi}{R}+q_\varphi=I_0\ddot{u}_\varphi+I_1\ddot{\beta}_\varphi\\
&\frac{1}{R}\frac{\partial N_{\varphi\vartheta}}{\partial \varphi}+\frac{1}{R_0}\frac{\partial N_\vartheta}{\partial \vartheta}+2\frac{\cos\varphi}{R_0}N_{\varphi\vartheta}+\frac{\sin\varphi}{R_0}Q_\vartheta+q_\vartheta=I_0\ddot{u}_\vartheta+I_1\ddot{\beta}_\vartheta\\
&\frac{1}{R}\frac{\partial Q_\varphi}{\partial \varphi}+\frac{1}{R_0}\frac{\partial Q_\vartheta}{\partial \vartheta}+\frac{\cos\varphi}{R_0}Q_\varphi-\frac{N_\varphi}{R}-\frac{\sin\varphi}{R_0}N_\vartheta+q_n=I_0\ddot{w}\\
&\frac{1}{R}\frac{\partial M_\varphi}{\partial \varphi}+\frac{1}{R_0}\frac{\partial M_{\varphi\vartheta}}{\partial \vartheta}+\frac{\cos\varphi}{R_0}\left(M_\varphi-M_\vartheta\right)-Q_\varphi+m_\varphi=I_1\ddot{u}_\varphi+I_2\ddot{\beta}_\varphi\\
&\frac{1}{R}\frac{\partial M_{\varphi\vartheta}}{\partial \varphi}+\frac{1}{R_0}\frac{\partial M_\vartheta}{\partial \vartheta}+2\frac{\cos\varphi}{R_0}M_{\varphi\vartheta}-Q_\vartheta+m_\vartheta=I_1\ddot{u}_\vartheta+I_2\ddot{\beta}_\vartheta
\end{aligned} \tag{2.11}$$

where

$$I_0=\mu h\left(1+\frac{h^2}{12R^2}\right),\quad I_1=\frac{\mu h^3}{6R},\quad I_2=\mu h^3\left(\frac{1}{12}+\frac{h^2}{80R^2}\right) \tag{2.12}$$

are the mass inertias and μ is the mass density of the material. The first three equations (2.11) represent translational equilibriums along meridional, circumferential and normal directions, while the last two are rotational equilibrium equations about the φ and ϑ directions. Positive sign conventions for external loads per unit area as well as for stress resultants and couples are illustrated in Figure 3.

Equations of motion or dynamic equilibrium equations (2.11) can be written in the operatorial form:

$$\mathbf{D}^*\mathbf{S}=\mathbf{q}-\frac{\partial \Lambda}{\partial t} \quad \text{or} \quad \mathbf{D}^*\mathbf{S}=\mathbf{f} \tag{2.13}$$

where

$$\mathbf{q}\left(\alpha_\varphi,\alpha_\vartheta,t\right)=\begin{bmatrix} q_\varphi & q_\vartheta & q_n & m_\varphi & m_\vartheta \end{bmatrix}^T \tag{2.14}$$

$$\boldsymbol{\Lambda}\left(\alpha_\varphi,\alpha_\vartheta,t\right)=\mathbf{M}\dot{\mathbf{u}} \tag{2.15}$$

is the distributed external load and the momentum vectors, respectively, and

$$\mathbf{M}=\begin{bmatrix} I_0 & 0 & 0 & I_1 & 0 \\ 0 & I_0 & 0 & 0 & I_1 \\ 0 & 0 & I_0 & 0 & 0 \\ I_1 & 0 & 0 & I_2 & 0 \\ 0 & I_1 & 0 & 0 & I_2 \end{bmatrix} \tag{2.16}$$

is the mass matrix, while

$$\dot{\mathbf{u}}\left(\alpha_\varphi,\alpha_\vartheta,t\right)=\frac{\partial}{\partial t}\begin{bmatrix} u_\varphi & u_\vartheta & w & \beta_\varphi & \beta_\vartheta \end{bmatrix}^T \tag{2.17}$$

is the derivative of the displacement vector with respect to the variable t, that is the vector velocity.

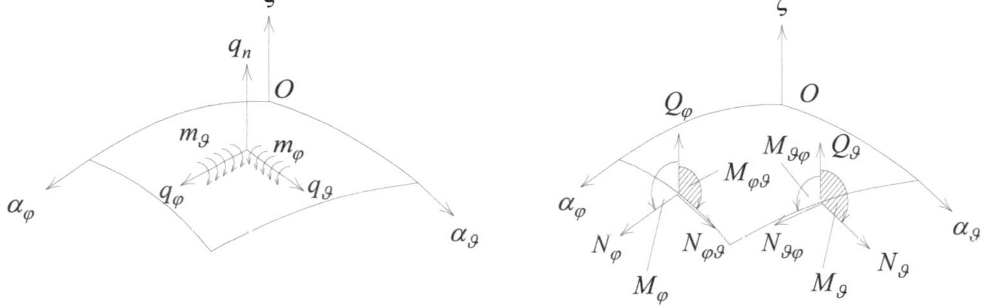

Figure 3. Distributed external loads and generalized stress resultants.

The balance operator, also known as the equilibrium operator, assumes the aspect:

$$\mathbf{D}^*=-\begin{bmatrix} \frac{\cos\varphi}{R_0}+\frac{1}{R}\frac{\partial}{\partial\varphi} & -\frac{\cos\varphi}{R_0} & \frac{1}{R_0}\frac{\partial}{\partial\vartheta} & 0 & 0 & 0 & \frac{1}{R} & 0 \\ 0 & \frac{1}{R_0}\frac{\partial}{\partial\vartheta} & 2\frac{\cos\varphi}{R_0}+\frac{1}{R}\frac{\partial}{\partial\varphi} & 0 & 0 & 0 & 0 & \frac{\sin\varphi}{R_0} \\ -\frac{1}{R} & -\frac{\sin\varphi}{R_0} & 0 & 0 & 0 & 0 & \frac{\cos\varphi}{R_0}+\frac{1}{R}\frac{\partial}{\partial\varphi} & \frac{1}{R_0}\frac{\partial}{\partial\vartheta} \\ 0 & 0 & 0 & \frac{\cos\varphi}{R_0}+\frac{1}{R}\frac{\partial}{\partial\varphi} & -\frac{\cos\varphi}{R_0} & \frac{1}{R_0}\frac{\partial}{\partial\vartheta} & -1 & 0 \\ 0 & 0 & 0 & 0 & \frac{1}{R_0}\frac{\partial}{\partial\vartheta} & 2\frac{\cos\varphi}{R_0}+\frac{1}{R}\frac{\partial}{\partial\varphi} & 0 & -1 \end{bmatrix} \tag{2.18}$$

2.4 Fundamental Equations

The three basic sets of equations, namely the kinematic, the equilibrium and the constitutive equations may be combined to give the fundamental system of equations, also known as the governing system equations. Firstly, the fundamental equations are deducted in the matrix notation. So, if the strain-displacement relations (2.3) are inserted into the constitutive equations (2.8), we have the relationships between stress resultants and the generalized displacement components:

$$\mathbf{S} = \mathbf{C}\boldsymbol{\varepsilon} = \mathbf{CDu} \tag{2.19}$$

When the equations (2.19) are inserted into the equations of motion (2.13), the fundamental system of equations is derived:

$$\mathbf{D}^*\mathbf{CDu} = \mathbf{q} - \frac{\partial \boldsymbol{\Lambda}}{\partial t} \quad \text{or} \quad \mathbf{D}^*\mathbf{CDu} = \mathbf{f} \tag{2.20}$$

The equations of motion in terms of displacements take all the three aspects of the problem of the elastic equilibrium into account.

By introducing the fundamental operator, also known as the elasticity operator,

$$\mathbf{L} = \mathbf{D}^*\mathbf{CDu} \tag{2.21}$$

equation (2.20) can be written as:

$$\mathbf{Lu} = \mathbf{q} - \frac{\partial \boldsymbol{\Lambda}}{\partial t} \quad \text{or} \quad \mathbf{Lu} = \mathbf{f} \tag{2.22}$$

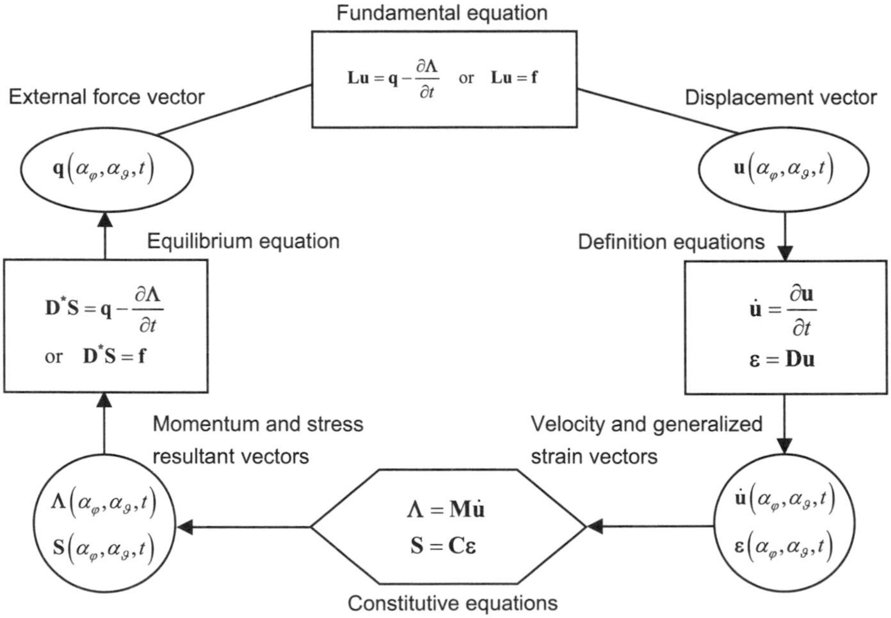

Figure 4. The scheme of the physical theories or the Tonti's diagram.

The fundamental system of equations (2.22) relates the configuration variable $\mathbf{u}$ to the source variable $\mathbf{q}$ of the phenomenon under investigation.

We can summarise all these aspects of each problem of elastic problem of equilibrium into the scheme of the physical theories or Tonti's diagram, which assumes the aspect reported in Figure 4.

Substituting the definition equations (2.2) into the constitutive equations (2.7) and the result of this substitution into the equilibrium equations (2.11), the complete equations of motion in terms of displacements can be written in the extended form as:

$$
\begin{aligned}
&\frac{K}{R^2}\frac{\partial^2 u_\varphi}{\partial\varphi^2}+\frac{(1-\nu)}{2}\frac{K}{R_0^2}\frac{\partial^2 u_\varphi}{\partial\vartheta^2}+\frac{(1+\nu)}{2}\frac{K}{RR_0}\frac{\partial^2 u_\vartheta}{\partial\varphi\partial\vartheta}+K\frac{\cos\varphi}{RR_0}\frac{\partial u_\varphi}{\partial\varphi}+\\
&+\frac{K}{R}\left[\frac{1}{R}\left(1+\frac{(1-\nu)}{2\chi}\right)+\frac{\nu\sin\varphi}{R_0}\right]\frac{\partial w}{\partial\varphi}-K\frac{(3-\nu)}{2}\frac{\cos\varphi}{R_0^2}\frac{\partial u_\vartheta}{\partial\vartheta}+\\
&-K\left[\frac{1}{R_0}\left(\frac{\nu\sin\varphi}{R}+\frac{\cos^2\varphi}{R_0}\right)+\frac{1}{R^2}\frac{(1-\nu)}{2\chi}\right]u_\varphi+K\frac{\cos\varphi}{R_0}\left[\frac{1}{R}-\frac{\sin\varphi}{R_0}\right]w+\\
&+\frac{(1-\nu)}{2\chi}\frac{K}{R}\beta_\varphi+q_\varphi=\mu h\left(1+\frac{h^2}{12R^2}\right)\ddot{u}_\varphi+\frac{\mu h^3}{6R}\ddot{\beta}_\varphi
\end{aligned}
\tag{2.23}
$$

$$
\begin{aligned}
&\frac{(1-\nu)}{2}\frac{K}{R^2}\frac{\partial^2 u_\vartheta}{\partial\varphi^2}+\frac{K}{R_0^2}\frac{\partial^2 u_\vartheta}{\partial\vartheta^2}+\frac{(1+\nu)}{2}\frac{K}{RR_0}\frac{\partial^2 u_\varphi}{\partial\varphi\partial\vartheta}+K\frac{(1-\nu)}{2}\frac{\cos\varphi}{RR_0}\frac{\partial u_\vartheta}{\partial\varphi}+\\
&+K\frac{(3-\nu)}{2}\frac{\cos\varphi}{R_0^2}\frac{\partial u_\varphi}{\partial\vartheta}+\frac{K}{R_0}\left[\frac{\nu}{R}+\frac{\sin\varphi}{R_0}\left(1+\frac{(1-\nu)}{2\chi}\right)\right]\frac{\partial w}{\partial\vartheta}+\\
&+\frac{(1-\nu)}{2}\frac{K}{R_0}\left[\frac{\sin\varphi}{R}-\frac{1}{R_0}\left(\cos^2\varphi+\frac{\sin^2\varphi}{\chi}\right)\right]u_\vartheta+\\
&+K\frac{(1-\nu)}{2\chi}\frac{\sin\varphi}{R_0}\beta_\vartheta+q_\vartheta=\mu h\left(1+\frac{h^2}{12R^2}\right)\ddot{u}_\vartheta+\frac{\mu h^3}{6R}\ddot{\beta}_\vartheta
\end{aligned}
\tag{2.24}
$$

$$
\begin{aligned}
&\frac{(1-\nu)}{2\chi}\frac{K}{R^2}\frac{\partial^2 w}{\partial\varphi^2}+\frac{(1-\nu)}{2\chi}\frac{K}{R_0^2}\frac{\partial^2 w}{\partial\vartheta^2}-\frac{K}{R}\left[\frac{1}{R}\left(1+\frac{(1-\nu)}{2\chi}\right)+\frac{\nu\sin\varphi}{R_0}\right]\frac{\partial u_\varphi}{\partial\varphi}+\\
&+K\frac{(1-\nu)}{2\chi}\frac{\cos\varphi}{RR_0}\frac{\partial w}{\partial\varphi}+\frac{(1-\nu)}{2\chi}\frac{K}{R}\frac{\partial\beta_\varphi}{\partial\varphi}-\frac{K}{R_0}\left[\frac{\nu}{R}+\frac{\sin\varphi}{R_0}\left(1+\frac{(1-\nu)}{2\chi}\right)\right]\frac{\partial u_\vartheta}{\partial\vartheta}+\\
&+\frac{(1-\nu)}{2\chi}\frac{K}{R_0}\frac{\partial\beta_\vartheta}{\partial\vartheta}-K\frac{\cos\varphi}{R_0}\left[\frac{\sin\varphi}{R_0}+\frac{1}{R}\left(\nu+\frac{(1-\nu)}{2\chi}\right)\right]u_\varphi+\\
&-K\left[\frac{1}{R}\left(\frac{1}{R}+\frac{2\nu\sin\varphi}{R_0}\right)+\frac{\sin^2\varphi}{R_0^2}\right]w+K\frac{(1-\nu)}{2\chi}\frac{\cos\varphi}{R_0}\beta_\varphi+q_n=\mu h\left(1+\frac{h^2}{12R^2}\right)\ddot{w}
\end{aligned}
\tag{2.25}
$$

$$
\begin{aligned}
&\frac{D}{R^2}\frac{\partial^2\beta_\varphi}{\partial\varphi^2}+\frac{(1-\nu)}{2}\frac{D}{R_0^2}\frac{\partial^2\beta_\varphi}{\partial\vartheta^2}+\frac{(1+\nu)}{2}\frac{D}{RR_0}\frac{\partial^2\beta_\vartheta}{\partial\varphi\partial\vartheta}+\\
&-\frac{(1-\nu)}{2\chi}\frac{K}{R}\frac{\partial w}{\partial\varphi}+D\frac{\cos\varphi}{RR_0}\frac{\partial\beta_\varphi}{\partial\varphi}-D\frac{(3-\nu)}{2}\frac{\cos\varphi}{R_0^2}\frac{\partial\beta_\vartheta}{\partial\vartheta}+\\
&+\frac{(1-\nu)}{2\chi}\frac{K}{R}u_\varphi-\left[\frac{D}{R_0}\left(\frac{\nu\sin\varphi}{R}+\frac{\cos^2\varphi}{R_0}\right)+K\frac{(1-\nu)}{2\chi}\right]\beta_\varphi+\\
&+m_\varphi=\frac{\mu h^3}{6R}\ddot{u}_\varphi+\mu h^3\left(\frac{1}{12}+\frac{h^2}{80R^2}\right)\ddot{\beta}_\varphi
\end{aligned}
\tag{2.26}
$$

$$
\begin{aligned}
&\frac{(1-\nu)}{2}\frac{D}{R^2}\frac{\partial^2\beta_\vartheta}{\partial\varphi^2}+\frac{D}{R_0^2}\frac{\partial^2\beta_\vartheta}{\partial\vartheta^2}+\frac{(1+\nu)}{2}\frac{D}{RR_0}\frac{\partial^2\beta_\varphi}{\partial\varphi\partial\vartheta}+\\
&+D\frac{(1-\nu)}{2}\frac{\cos\varphi}{RR_0}\frac{\partial\beta_\vartheta}{\partial\varphi}-\frac{(1-\nu)}{2\chi}\frac{K}{R_0}\frac{\partial w}{\partial\vartheta}+D\frac{(3-\nu)}{2}\frac{\cos\varphi}{R_0^2}\frac{\partial\beta_\varphi}{\partial\vartheta}+\\
&+K\frac{(1-\nu)}{2\chi}\frac{\sin\varphi}{R_0}u_\vartheta+\left[\frac{(1-\nu)}{2}\frac{D}{R_0}\left(\frac{\sin\varphi}{R}-\frac{\cos^2\varphi}{R_0}\right)-K\frac{(1-\nu)}{2\chi}\right]\beta_\vartheta+\\
&+m_\vartheta=\frac{\mu h^3}{6R}\ddot{u}_\vartheta+\mu h^3\left(\frac{1}{12}+\frac{h^2}{80R^2}\right)\ddot{\beta}_\vartheta
\end{aligned}
\tag{2.27}
$$

2.5 Boundary and Compatibility Conditions

In the following, three kinds of boundary conditions are considered, namely the fully clamped edge boundary conditions (C), the simply supported edge boundary conditions (S) and the free edge boundary conditions (F). The equations describing the boundary conditions can be written as follows:

Clamped edge boundary condition (C):

$$u_\varphi=u_\vartheta=w=\beta_\varphi=\beta_\vartheta=0 \quad \text{at } \varphi=\varphi_0 \text{ or } \varphi=\varphi_1,\ \ 0\le\vartheta\le\vartheta_0 \tag{2.28}$$

$$u_\varphi=u_\vartheta=w=\beta_\varphi=\beta_\vartheta=0 \quad \text{at } \vartheta=0 \text{ or } \vartheta=\vartheta_0,\ \ \varphi_0\le\varphi\le\varphi_1 \tag{2.29}$$

Simply supported edge boundary condition (S):

$$u_\varphi=u_\vartheta=w=\beta_\vartheta=0,\ \ M_\varphi=0 \quad \text{at } \varphi=\varphi_0 \text{ or } \varphi=\varphi_1,\ \ 0\le\vartheta\le\vartheta_0 \tag{2.30}$$

$$u_\varphi=u_\vartheta=w=\beta_\varphi=0,\ \ M_\vartheta=0 \quad \text{at } \vartheta=0 \text{ or } \vartheta=\vartheta_0,\ \ \varphi_0\le\varphi\le\varphi_1 \tag{2.31}$$

Free edge boundary condition (F):

$$N_\varphi=N_{\varphi\vartheta}=Q_\varphi=M_\varphi=M_{\varphi\vartheta}=0 \quad \text{at } \varphi=\varphi_0 \text{ or } \varphi=\varphi_1,\ \ 0\le\vartheta\le\vartheta_0 \tag{2.32}$$

$$N_\vartheta=N_{\varphi\vartheta}=Q_\vartheta=M_\vartheta=M_{\varphi\vartheta}=0 \quad \text{at } \vartheta=0 \text{ or } \vartheta=\vartheta_0,\ \ \varphi_0\le\varphi\le\varphi_1 \tag{2.33}$$

In addition to the external boundary conditions, the *kinematical* and *physical compatibility* should be satisfied at the common meridian with $\vartheta=0,2\pi$, if we want to consider a complete hemispherical dome of revolution. The kinematical compatibility conditions include the continuity of displacements. The physical compatibility conditions can only be the five continuous conditions for the generalized stress resultants. To consider a complete revolute hemispherical dome characterized by $\vartheta_0 = 2\pi$, it is necessary to implement the kinematical and physical compatibility conditions between the meridians with $\vartheta = 0$ and with $\vartheta_0 = 2\pi$:

Kinematical compatibility conditions:

$$\begin{aligned} &u_\varphi(\varphi,0,t) = u_\varphi(\varphi,2\pi,t), u_\vartheta(\varphi,0,t) = u_\vartheta(\varphi,2\pi,t), w(\varphi,0,t) = w(\varphi,2\pi,t), \\ &\beta_\varphi(\varphi,0,t) = \beta_\varphi(\varphi,2\pi,t), \beta_\vartheta(\varphi,0,t) = \beta_\vartheta(\varphi,2\pi,t) \end{aligned} \qquad \varphi_0 \le \varphi \le \varphi_1 \tag{2.34}$$

Physical compatibility conditions:

$$\begin{aligned} &N_\vartheta(\varphi,0,t) = N_\vartheta(\varphi,2\pi,t), N_{\varphi\vartheta}(\varphi,0,t) = N_{\varphi\vartheta}(\varphi,2\pi,t), Q_\vartheta(\varphi,0,t) = Q_\vartheta(\varphi,2\pi,t), \\ &M_\vartheta(\varphi,0,t) = M_\vartheta(\varphi,2\pi,t), M_{\varphi\vartheta}(\varphi,0,t) = M_{\varphi\vartheta}(\varphi,2\pi,t) \end{aligned} \qquad \varphi_0 \le \varphi \le \varphi_1 \tag{2.35}$$

3 Generalized Differential Quadrature Method

The GDQ method will be used to discretize the derivatives in the governing equations and the boundary conditions. The GDQ approach was developed by Shu and Richards (1992) to improve the Differential Quadrature technique for the computation of weighting coefficients, entering into the linear algebraic system of equations obtained from the discretization of the differential equation system, which can model the physical problem considered. The essence of the differential quadrature method is that the partial derivative of a smooth function with respect to a variable is approximated by a weighted sum of function values at all discrete points in that direction. Its weighting coefficients are not related to any special problem and only depend on the grid points and the derivative order. In this methodology, an arbitrary grid distribution can be chosen without any limitation.

The GDQ method is based on the analysis of a high-order polynomial approximation and the analysis of a linear vector space. For a general problem, it may not be possible to express the solution of the corresponding partial differential equation in a closed form. This solution function can be approximated by the two following types of function approximation: high-order polynomial approximation and Fourier series expansion (harmonic functions). It is well known that a smooth function in a domain can be accurately approximated by a high-order polynomial in accordance with the Weierstrass polynomial approximation theorem. In fact, from the Weierstrass theorem, if $f(x)$ is a real valued continuous function defined in the closed interval $[a,b]$, then there exists a sequence of polynomials $P_r(x)$ which converges to $f(x)$ uniformly as r goes to infinity. In practical applications, a truncated finite polynomial may be used. Thus, if $f(x)$ represents the solution of a partial differential equation, then it can be approximated by a polynomial of a degree less than or equal to $N-1$, for N large enough. The conventional form of this approximation is:

$$f(x) \cong P_N(x) = \sum_{j=1}^{N} d_j p_j(x) \tag{3.1}$$

where d_j is a constant. Then it is easy to show that the polynomial $P_N(x)$ constitutes an N-dimensional linear vector space V_N with respect to the operation of vector addition and scalar multiplication. Obviously, in the linear vector space V_N, $p_j(x)$ is a set of base vectors. It can be seen that, in the linear polynomial vector space, there exist several sets of base polynomials and each set of base polynomials can be expressed uniquely by another set of base polynomials in the space. Using vector space analysis, the method for computing the weighting coefficients can be generalized by a proper choice of base polynomials in a linear vector space. For generality, the Lagrange interpolation polynomials are chosen as the base polynomials. As a result, the weighting coefficients of the first order derivative are computed by a simple algebraic formulation without any restriction on the choice of the grid points, while the weighting coefficients of the second and higher order derivatives are given by a recurrence relationship.

When the Lagrange interpolated polynomials are assumed as a set of vector space base functions, the approximation of the function $f(x)$ can be written as:

$$f(x) \cong \sum_{j=1}^{N} p_j(x) f(x_j) \tag{3.2}$$

where N is the number of grid points in the whole domain, x_j, $j=1,2,...,N$, are the co-ordinates of grid points in the variable domain and $f(x_j)$ are the function values at the grid points. $p_j(x)$ are the Lagrange interpolated polynomials, which can be defined by the following formula:

$$p_j(x) = \frac{\mathcal{L}(x)}{(x-x_j)\mathcal{L}^{(1)}(x_j)}, \quad j=1,2,...,N \tag{3.3}$$

where:

$$\mathcal{L}(x) = \prod_{i=1}^{N}(x-x_i), \quad \mathcal{L}^{(1)}(x_j) = \prod_{i=1, i\neq j}^{N}(x_j - x_i) \tag{3.4}$$

Differentiating equation (3.2) with respect to x and evaluating the first derivative at a certain point of the function domain, it is possible to obtain:

$$f^{(1)}(x_i) \cong \sum_{j=1}^{N} p_j^{(1)}(x_i) f(x_j) = \sum_{j=1}^{N} \varsigma_{ij}^{(1)} f(x_j), \quad i=1,2,...,N \tag{3.5}$$

where $\varsigma_{ij}^{(1)}$ are the GDQ weighting coefficients of the first order derivative and x_i denote the co-ordinates of the grid points. In particular, it is worth noting that the weighting coefficients of the first order derivative can be computed as:

$$p_j^{(1)}(x_i) = \varsigma_{ij}^{(1)} = \frac{\mathcal{L}^{(1)}(x_i)}{(x_i - x_j)\mathcal{L}^{(1)}(x_j)}, \quad i,j=1,2,...,N, \quad i \neq j \tag{3.6}$$

From equation (3.6), $\varsigma_{ij}^{(1)}$ $(i \neq j)$ can be easily computed. However, the calculation of $\varsigma_{ii}^{(1)}$ is not easy to compute. According to the analysis of a linear vector space, one set of base functions can be expressed uniquely by a linear sum of another set of base functions. Thus, if one set of base polynomials satisfy a linear equation like (3.5), so does another set of base polynomials. As a consequence, the equation system for determining $\varsigma_{ij}^{(1)}$ and derived from the Lagrange interpolation polynomials should be equivalent to that derived from another set of base polynomials $p_j(x) = x^{j-1}$, $j=1,2,...,N$. Thus, $\varsigma_{ij}^{(1)}$ satisfies the following equation, which is obtained by the base polynomials $p_j(x) = x^{j-1}$, when $j=1$:

$$\sum_{j=1}^{N} \varsigma_{ij}^{(1)} = 0 \Rightarrow \varsigma_{ii}^{(1)} = -\sum_{j=1, j\neq i}^{N} \varsigma_{ij}^{(1)}, \quad i,j=1,2,...,N \tag{3.7}$$

Equations (3.6) and (3.7) are two formulations to compute the weighting coefficients $\varsigma_{ij}^{(1)}$. It should be noted that, in the development of these two formulations, two sets of base polynomials were used in the linear polynomial vector space V_N. Finally, the nth order derivative of function $f(x)$ with respect to x at grid points x_i, can be approximated by the GDQ approach:

$$\left.\frac{d^n f(x)}{dx^n}\right|_{x=x_i} = \sum_{j=1}^{N} \varsigma_{ij}^{(n)} f(x_j), \quad i = 1,2,...,N \tag{3.8}$$

where $\varsigma_{ij}^{(n)}$ are the weighting coefficients of the nth order derivative. Similar to the first order derivative and according to the polynomial approximation and the analysis of a linear vector space, it is possible to determine a recurrence relationship to compute the second and higher order derivatives. Thus, the weighting coefficients can be generated by the following recurrent formulation:

$$\varsigma_{ij}^{(n)} = n\left(\varsigma_{ii}^{(n-1)}\varsigma_{ij}^{(1)} - \frac{\varsigma_{ij}^{(n-1)}}{x_i - x_j}\right), \quad i \neq j,\ n = 2,3,...,N-1,\ i,j = 1,2,...,N \tag{3.9}$$

$$\sum_{j=1}^{N} \varsigma_{ij}^{(n)} = 0 \Rightarrow \varsigma_{ii}^{(n)} = -\sum_{j=1,\, j\neq i}^{N} \varsigma_{ij}^{(n)}, \quad n = 2,3,...,N-1,\ i,j = 1,2,...,N \tag{3.10}$$

It is obvious from the above equations that the weighting coefficients of the second and higher order derivatives can be determined from those of the first order derivative. Furthermore, it is interesting to note that, the preceding coefficients $\varsigma_{ij}^{(n)}$ are dependent on the derivative order *n*, on the grid point distribution x_j, $j = 1,2,...,N$, and on the specific point x_i, where the derivative is computed. There is no need to obtain the weighting coefficients from a set of algebraic equations which could be ill-conditioned when the number of grid points is large. Furthermore, this set of expressions for the determination of the weighting coefficients is so compact and simple that it is very easy to implement them in formulating and programming, because of the recurrence feature.

Another important point for successful application of the GDQ method is how to distribute the grid points. In fact, the accuracy of this method is usually sensitive to the grid point distribution. The optimal grid point distribution depends on the order of derivatives in the boundary condition and the number of grid points used. The grid point distribution also plays an essential role in determining the convergence speed and stability of the GDQ method. It is demonstrated that non-uniform grid distribution usually yields better results than equally spaced distribution. Quan and Chang (1989) compared numerically the performances of the often-used non-uniform meshes and concluded that the grid points originating from the Chebyshev polynomials of the first kind are optimum in all the cases examined there. The zeros of orthogonal polynomials are the rational basis for the grid points. Shu and Richards (1992) have used other choice which has given better results than the zeros of Chebyshev and Legendre polynomials. Bert and Malik (1996) indicated that the preferred type of grid points changes with problems of interest and recommended the use of Chebyshev-Gauss-Lobatto grid for the structural mechanics computation. With Lagrange interpolating polynomials, the Chebyshev-Gauss-Lobatto sampling point rule proves efficient for numerical reasons [Shu, Chen, Xue and Du (2001)] so that for such a collocation the approximation error of the dependent variables decreases as the number of nodes increases. For the numerical computations presented in this paper, the co-ordinates of grid points (φ_i, ϑ_j) are chosen as:

Chebyshev-Gauss-Lobatto sampling points (C-G-L)

$$
\begin{aligned}
\varphi_i &= \frac{1-\cos\left(\frac{i-1}{N-1}\pi\right)}{2}(\varphi_1-\varphi_0)+\varphi_0, \quad i=1,2,\dots,N, \quad \text{for } \varphi\in[\varphi_0,\varphi_1] \ (\text{with } \varphi_0>0 \text{ and } \varphi_1\le 90^\circ) \\
\vartheta_j &= \frac{1-\cos\left(\frac{j-1}{M-1}\pi\right)}{2}\vartheta_0, \quad j=1,2,\dots,M, \quad \text{for } \vartheta\in[0,\vartheta_0] \ (\text{with } \vartheta_0\le 2\pi)
\end{aligned} \tag{3.11}
$$

where N, M are the total number of sampling points used to discretize a domain in φ and ϑ directions, respectively, of the hemispherical shell.

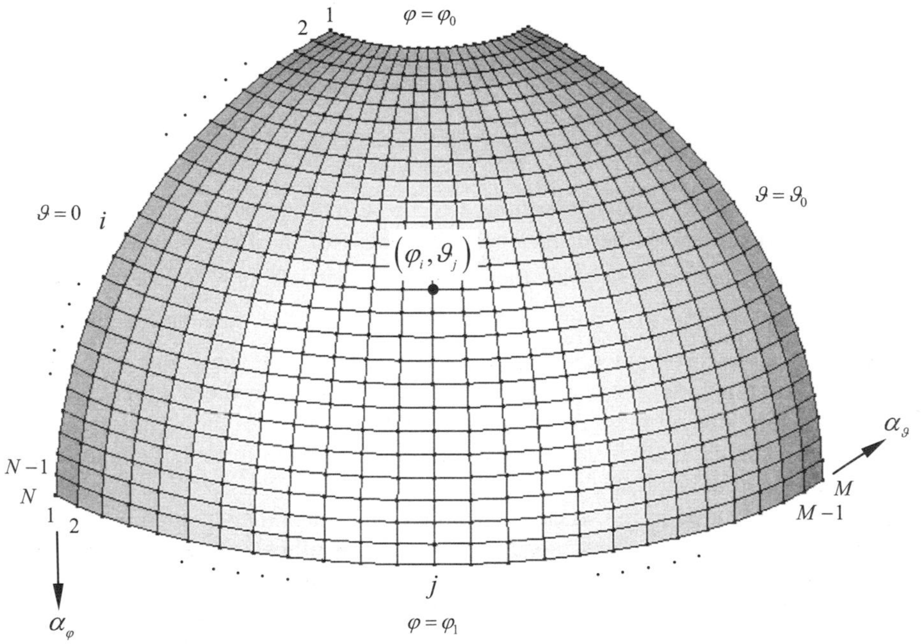

Figure 5. Mesh distribution on a hemispherical panel.

4 Numerical Implementation

A novel approach in numerically solving the governing equations (2.23), (2.24), (2.25), (2.26), and (2.27) is represented by the Generalized Differential Quadrature (GDQ) method. This method, for the problem studied herein, demonstrates its numerical accuracy and extreme coding simplicity.

In the following, only the free vibration of hemispherical dome or panel will be studied. So, setting $\mathbf{q}(\alpha_\varphi,\alpha_\vartheta,t)=\mathbf{0}$ and using the method of variable separation, it is possible to seek solutions that are harmonic in time and whose frequency is ω; then, the displacements and the rotations can be written as follows:

$$\begin{aligned} u_\varphi(\alpha_\varphi,\alpha_\vartheta,t) &= U^\varphi(\alpha_\varphi,\alpha_\vartheta)e^{i\omega t} \\ u_\vartheta(\alpha_\varphi,\alpha_\vartheta,t) &= U^\vartheta(\alpha_\varphi,\alpha_\vartheta)e^{i\omega t} \\ w(\alpha_\varphi,\alpha_\vartheta,t) &= W^\varsigma(\alpha_\varphi,\alpha_\vartheta)e^{i\omega t} \\ \beta_\varphi(\alpha_\varphi,\alpha_\vartheta,t) &= B^\varphi(\alpha_\varphi,\alpha_\vartheta)e^{i\omega t} \\ \beta_\vartheta(\alpha_\varphi,\alpha_\vartheta,t) &= B^\vartheta(\alpha_\varphi,\alpha_\vartheta)e^{i\omega t} \end{aligned} \tag{4.1}$$

where the vibration spatial amplitude values ($U^\varphi(\alpha_\varphi,\alpha_\vartheta)$, $U^\vartheta(\alpha_\varphi,\alpha_\vartheta)$, $W^\varsigma(\alpha_\varphi,\alpha_\vartheta)$, $B^\varphi(\alpha_\varphi,\alpha_\vartheta)$, $B^\vartheta(\alpha_\varphi,\alpha_\vartheta)$) fulfil the fundamental differential system.

The basic steps in the GDQ solution of the free vibration problem of hemispherical dome type structures are as in the following:

- Discretization of independent variables $\varphi \in]0,90°]$, $\vartheta \in [0,\vartheta_0]$ (with $\vartheta_0 \le 2\pi$).
- The spatial derivatives are approximated according to GDQ rule.
- The differential governing systems (2.23), (2.24), (2.25), (2.26), and (2.27) are transformed into linear eigenvalue problems for the natural frequencies. The boundary conditions are imposed in the sampling points corresponding to the boundary. All these relations are imposed pointwise.
- The solution of the previously stated discrete system in terms of natural frequencies and mode shape components is worked out. For each mode, local values of dependent variables are used to obtain the complete assessment of the deformed configuration.

4.1 Discretization of Motion Equations

The simple numerical operations illustrated here, applying the GDQ procedure, enable one to write the equations of motion in discrete form, transforming each space derivative into a weighted sum of node values of dependent variables. Each of the approximated equations is valid in a single sampling point. The governing equations can be discretized as:

$$\begin{aligned} &\frac{K}{R^2}\sum_{k=1}^{N}\varsigma_{ik}^{\varphi(2)}U_{kj}^\varphi+\frac{(1-\nu)}{2}\frac{K}{R_{0i}^2}\sum_{m=1}^{M}\varsigma_{jm}^{\vartheta(2)}U_{mj}^\varphi+\frac{(1+\nu)}{2}\frac{K}{RR_{0i}}\sum_{k=1}^{N}\varsigma_{ik}^{\varphi(1)}\left(\sum_{m=1}^{M}\varsigma_{jm}^{\vartheta(1)}U_{km}^\vartheta\right)+ \\ &+K\frac{\cos\varphi_i}{RR_{0i}}\sum_{k=1}^{N}\varsigma_{ik}^{\varphi(1)}U_{kj}^\varphi+\frac{K}{R}\left[\frac{1}{R}\left(1+\frac{(1-\nu)}{2\chi}\right)+\frac{\nu\sin\varphi_i}{R_{0i}}\right]\sum_{k=1}^{N}\varsigma_{ik}^{\varphi(1)}W_{kj}^\varsigma+ \\ &-K\frac{(3-\nu)}{2}\frac{\cos\varphi_i}{R_{0i}^2}\sum_{m=1}^{M}\varsigma_{jm}^{\vartheta(1)}U_{im}^\vartheta-K\left[\frac{1}{R_{0i}}\left(\frac{\nu\sin\varphi_i}{R}+\frac{\cos^2\varphi_i}{R_{0i}}\right)+\frac{1}{R^2}\frac{(1-\nu)}{2\chi}\right]U_{ij}^\varphi+ \\ &+K\frac{\cos\varphi_i}{R_{0i}}\left[\frac{1}{R}-\frac{\sin\varphi_i}{R_{0i}}\right]W_{ij}^\varsigma+\frac{(1-\nu)}{2\chi}\frac{K}{R}B_{ij}^\varphi=-\omega^2\mu h\left(1+\frac{h^2}{12R^2}\right)U_{ij}^\varphi-\omega^2\frac{\mu h^3}{6R}B_{ij}^\varphi \end{aligned} \tag{4.2}$$

$$\begin{aligned}
&\frac{(1-\nu)}{2}\frac{K}{R^2}\sum_{k=1}^{N}\varsigma_{ik}^{\varphi(2)}U_{kj}^{\vartheta}+\frac{K}{R_{0i}^2}\sum_{m=1}^{M}\varsigma_{jm}^{\vartheta(2)}U_{im}^{\vartheta}+\frac{(1+\nu)}{2}\frac{K}{RR_{0i}}\sum_{k=1}^{N}\varsigma_{ik}^{\varphi(1)}\left(\sum_{m=1}^{M}\varsigma_{jm}^{\vartheta(1)}U_{km}^{\varphi}\right)+\\
&+K\frac{(1-\nu)}{2}\frac{\cos\varphi_i}{RR_{0i}}\sum_{k=1}^{N}\varsigma_{ik}^{\varphi(1)}U_{kj}^{\vartheta}+K\frac{(3-\nu)}{2}\frac{\cos\varphi_i}{R_{0i}^2}\sum_{m=1}^{M}\varsigma_{jm}^{\vartheta(1)}U_{im}^{\varphi}+\\
&+\frac{K}{R_{0i}}\left[\frac{\nu}{R}+\frac{\sin\varphi_i}{R_{0i}}\left(1+\frac{(1-\nu)}{2\chi}\right)\right]\sum_{m=1}^{M}\varsigma_{jm}^{\vartheta(1)}W_{im}^{\varsigma}+K\frac{(1-\nu)}{2\chi}\frac{\sin\varphi_i}{R_{0i}}B_{ij}^{\vartheta}+\\
&+\frac{(1-\nu)}{2}\frac{K}{R_{0i}}\left[\frac{\sin\varphi_i}{R}-\frac{1}{R_{0i}}\left(\cos^2\varphi_i+\frac{\sin^2\varphi_i}{\chi}\right)\right]U_{ij}^{\vartheta}=-\omega^2\mu h\left(1+\frac{h^2}{12R^2}\right)U_{ij}^{\vartheta}-\omega^2\frac{\mu h^3}{6R}B_{ij}^{\vartheta}
\end{aligned}\tag{4.3}$$

$$\begin{aligned}
&\frac{(1-\nu)}{2\chi}\frac{K}{R^2}\sum_{k=1}^{N}\varsigma_{ik}^{\varphi(2)}W_{kj}^{\varsigma}+\frac{(1-\nu)}{2\chi}\frac{K}{R_{0i}^2}\sum_{m=1}^{M}\varsigma_{jm}^{\vartheta(2)}W_{im}^{\varsigma}-\frac{K}{R}\left[\frac{1}{R}\left(1+\frac{(1-\nu)}{2\chi}\right)+\frac{\nu\sin\varphi_i}{R_{0i}}\right]\sum_{k=1}^{N}\varsigma_{ik}^{\varphi(1)}U_{kj}^{\varphi}+\\
&+K\frac{(1-\nu)}{2\chi}\frac{\cos\varphi_i}{RR_{0i}}\sum_{k=1}^{N}\varsigma_{ik}^{\varphi(1)}W_{kj}^{\varsigma}+\frac{(1-\nu)}{2\chi}\frac{K}{R}\sum_{k=1}^{N}\varsigma_{ik}^{\varphi(1)}B_{kj}^{\varphi}-\frac{K}{R_{0i}}\left[\frac{\nu}{R}+\frac{\sin\varphi_i}{R_{0i}}\left(1+\frac{(1-\nu)}{2\chi}\right)\right]\sum_{m=1}^{M}\varsigma_{jm}^{\vartheta(1)}U_{im}^{\vartheta}+\\
&+\frac{(1-\nu)}{2\chi}\frac{K}{R_{0i}}\sum_{m=1}^{M}\varsigma_{jm}^{\vartheta(1)}B_{im}^{\vartheta}-K\frac{\cos\varphi_i}{R_{0i}}\left[\frac{\sin\varphi_i}{R_{0i}}+\frac{1}{R}\left(\nu+\frac{(1-\nu)}{2\chi}\right)\right]U_{ij}^{\varphi}+\\
&-K\left[\frac{1}{R}\left(\frac{1}{R}+\frac{2\nu\sin\varphi_i}{R_{0i}}\right)+\frac{\sin^2\varphi_i}{R_{0i}^2}\right]W_{ij}^{\varsigma}+K\frac{(1-\nu)}{2\chi}\frac{\cos\varphi_i}{R_{0i}}B_{ij}^{\varphi}=-\omega^2\mu h\left(1+\frac{h^2}{12R^2}\right)W_{ij}^{\varsigma}
\end{aligned}\tag{4.4}$$

$$\begin{aligned}
&\frac{D}{R^2}\sum_{k=1}^{N}\varsigma_{ik}^{\varphi(2)}B_{kj}^{\varphi}+\frac{(1-\nu)}{2}\frac{D}{R_{0i}^2}\sum_{m=1}^{M}\varsigma_{jm}^{\vartheta(2)}B_{im}^{\varphi}+\frac{(1+\nu)}{2}\frac{D}{RR_{0i}}\sum_{k=1}^{N}\varsigma_{ik}^{\varphi(1)}\left(\sum_{m=1}^{M}\varsigma_{jm}^{\vartheta(1)}B_{km}^{\vartheta}\right)+\\
&-\frac{(1-\nu)}{2\chi}\frac{K}{R}\sum_{k=1}^{N}\varsigma_{ik}^{\varphi(1)}W_{kj}^{\varsigma}+D\frac{\cos\varphi_i}{RR_{0i}}\sum_{k=1}^{N}\varsigma_{ik}^{\varphi(1)}B_{kj}^{\varphi}-D\frac{(3-\nu)}{2}\frac{\cos\varphi_i}{R_{0i}^2}\sum_{m=1}^{M}\varsigma_{jm}^{\vartheta(1)}B_{im}^{\vartheta}+\\
&+\frac{(1-\nu)}{2\chi}\frac{K}{R}U_{ij}^{\varphi}-\left[\frac{D}{R_{0i}}\left(\frac{\nu\sin\varphi_i}{R}+\frac{\cos^2\varphi_i}{R_{0i}}\right)+K\frac{(1-\nu)}{2\chi}\right]B_{ij}^{\varphi}=\\
&\qquad=-\omega^2\frac{\mu h^3}{6R}U_{ij}^{\varphi}-\omega^2\mu h^3\left(\frac{1}{12}+\frac{h^2}{80R^2}\right)B_{ij}^{\varphi}
\end{aligned}\tag{4.5}$$

$$
\begin{aligned}
&\frac{(1-\nu)}{2}\frac{D}{R^2}\sum_{k=1}^{N}\varsigma_{ik}^{\varphi(2)}B_{kj}^{\vartheta}+\frac{D}{R_{0i}^2}\sum_{m=1}^{M}\varsigma_{jm}^{\vartheta(2)}B_{im}^{\vartheta}+\frac{(1+\nu)}{2}\frac{D}{RR_{0i}}\sum_{k=1}^{N}\varsigma_{ik}^{\varphi(1)}\left(\sum_{m=1}^{M}\varsigma_{jm}^{\vartheta(1)}B_{km}^{\varphi}\right)+\\
&+D\frac{(1-\nu)}{2}\frac{\cos\varphi_i}{RR_{0i}}\sum_{k=1}^{N}\varsigma_{ik}^{\varphi(1)}B_{kj}^{\vartheta}-\frac{(1-\nu)}{2\chi}\frac{K}{R_{0i}}\sum_{m=1}^{M}\varsigma_{jm}^{\vartheta(1)}W_{im}^{\varsigma}+D\frac{(3-\nu)}{2}\frac{\cos\varphi_i}{R_{0i}^2}\sum_{m=1}^{M}\varsigma_{jm}^{\vartheta(1)}B_{im}^{\varphi}+\\
&+K\frac{(1-\nu)}{2\chi}\frac{\sin\varphi_i}{R_{0i}}U_{ij}^{\vartheta}+\left[\frac{(1-\nu)}{2}\frac{D}{R_{0i}}\left(\frac{\sin\varphi_i}{R}-\frac{\cos^2\varphi_i}{R_{0i}}\right)-K\frac{(1-\nu)}{2\chi}\right]B_{ij}^{\vartheta}=\\
&\qquad\qquad=-\omega^2\frac{\mu h^3}{6R}U_{ij}^{\vartheta}-\omega^2\mu h^3\left(\frac{1}{12}+\frac{h^2}{80R^2}\right)B_{ij}^{\vartheta}
\end{aligned}
\tag{4.6}
$$

where $i=2,3,...,N-1$, $j=2,3,...,M-1$ and $\varsigma_{ik}^{\varphi(1)}$, $\varsigma_{jm}^{\vartheta(1)}$, $\varsigma_{ik}^{\varphi(2)}$ and $\varsigma_{jm}^{\vartheta(2)}$ are the weighting coefficients of the first and second derivatives in φ and ϑ directions, respectively. On the other hand, N, M are the total number of grid points in φ and ϑ directions.

4.2 Implementation of Boundary and Compatibility Conditions

Applying the GDQ methodology, the discretized forms of the boundary conditions are given as follows:

Clamped edge boundary condition (C):

$$
\begin{aligned}
&U_{aj}^{\varphi}=U_{aj}^{\vartheta}=W_{aj}^{\varsigma}=B_{aj}^{\varphi}=B_{aj}^{\vartheta}=0 \qquad \text{for } a=1,N \text{ and } j=1,2,...,M\\
&U_{ib}^{\varphi}=U_{ib}^{\vartheta}=W_{ib}^{\varsigma}=B_{ib}^{\varphi}=B_{ib}^{\vartheta}=0 \qquad \text{for } b=1,M \text{ and } i=1,2,...,N
\end{aligned}
\tag{4.7}
$$

Simply supported edge boundary condition (S):

$$
\begin{aligned}
&\begin{cases}
U_{aj}^{\varphi}=U_{aj}^{\vartheta}=W_{aj}^{\varsigma}=B_{aj}^{\vartheta}=0\\
\dfrac{1}{R}\displaystyle\sum_{k=1}^{N}\varsigma_{ak}^{\varphi(1)}B_{kj}^{\varphi}+\frac{\nu}{R_{0a}}\left(\sum_{m=1}^{M}\varsigma_{jm}^{\vartheta(1)}B_{am}^{\vartheta}+B_{aj}^{\varphi}\cos\varphi_a\right)=0
\end{cases} \qquad \text{for } a=1,N \text{ and } j=1,2,...,M\\
&\begin{cases}
U_{ib}^{\varphi}=U_{ib}^{\vartheta}=W_{ib}^{\varsigma}=B_{ib}^{\varphi}=0\\
\dfrac{1}{R_{0i}}\left(\displaystyle\sum_{m=1}^{M}\varsigma_{bm}^{\vartheta(1)}B_{im}^{\vartheta}+B_{ib}^{\varphi}\cos\varphi_i\right)+\frac{\nu}{R}\sum_{k=1}^{N}\varsigma_{ik}^{\varphi(1)}B_{kb}^{\varphi}=0
\end{cases} \qquad \text{for } b=1,M \text{ and } i=1,2,...,N
\end{aligned}
\tag{4.8}
$$

Free edge boundary condition (F):

$$\begin{cases}\frac{1}{R}\left(\sum_{k=1}^{N}\varsigma_{ak}^{\varphi(1)}U_{kj}^{\varphi}+W_{aj}^{\zeta}\right)+\frac{\nu}{R_{0a}}\left(\sum_{m=1}^{M}\varsigma_{jm}^{\vartheta(1)}U_{am}^{\vartheta}+U_{aj}^{\varphi}\cos\varphi_a+W_{aj}^{\zeta}\sin\varphi_a\right)=0\\ \frac{1}{R}\sum_{k=1}^{N}\varsigma_{ak}^{\varphi(1)}U_{kj}^{\vartheta}+\frac{1}{R_{0a}}\left(\sum_{m=1}^{M}\varsigma_{jm}^{\vartheta(1)}U_{am}^{\varphi}-U_{aj}^{\vartheta}\cos\varphi_a\right)=0\\ \frac{1}{R}\left(\sum_{k=1}^{N}\varsigma_{ak}^{\varphi(1)}W_{kj}^{\zeta}-U_{aj}^{\varphi}\right)+B_{aj}^{\varphi}=0 \qquad \text{for } a=1,N \text{ and } j=1,2,...,M\\ \frac{1}{R}\sum_{k=1}^{N}\varsigma_{ak}^{\varphi(1)}B_{kj}^{\varphi}+\frac{\nu}{R_{0a}}\left(\sum_{m=1}^{M}\varsigma_{jm}^{\vartheta(1)}B_{am}^{\vartheta}+B_{aj}^{\varphi}\cos\varphi_a\right)=0\\ \frac{1}{R}\sum_{k=1}^{N}\varsigma_{ak}^{\varphi(1)}B_{kj}^{\vartheta}+\frac{1}{R_{0a}}\left(\sum_{m=1}^{M}\varsigma_{jm}^{\vartheta(1)}B_{am}^{\varphi}-B_{aj}^{\vartheta}\cos\varphi_a\right)=0\end{cases} \tag{4.9}$$

$$\begin{cases}\frac{1}{R_{0i}}\left(\sum_{m=1}^{M}\varsigma_{bm}^{\vartheta(1)}U_{im}^{\vartheta}+U_{ib}^{\varphi}\cos\varphi_i+W_{ib}^{\zeta}\sin\varphi_i\right)+\frac{\nu}{R}\left(\sum_{k=1}^{N}\varsigma_{ik}^{\varphi(1)}U_{kb}^{\varphi}+W_{ib}^{\zeta}\right)=0\\ \frac{1}{R}\sum_{k=1}^{N}\varsigma_{ik}^{\varphi(1)}U_{kb}^{\vartheta}+\frac{1}{R_{0i}}\left(\sum_{m=1}^{M}\varsigma_{bm}^{\vartheta(1)}U_{im}^{\varphi}-U_{ib}^{\vartheta}\cos\varphi_i\right)=0\\ \frac{1}{R_{0i}}\left(\sum_{m=1}^{M}\varsigma_{bm}^{\vartheta(1)}W_{im}^{\zeta}-U_{ib}^{\vartheta}\sin\varphi_i\right)+B_{ib}^{\vartheta}=0 \qquad \text{for } b=1,M \text{ and } i=1,2,...,N\\ \frac{1}{R_{0i}}\left(\sum_{m=1}^{M}\varsigma_{bm}^{\vartheta(1)}B_{im}^{\vartheta}+B_{ib}^{\varphi}\cos\varphi_i\right)+\frac{\nu}{R}\sum_{k=1}^{N}\varsigma_{ik}^{\varphi(1)}B_{kb}^{\varphi}=0\\ \frac{1}{R}\sum_{m=1}^{M}\varsigma_{bm}^{\varphi(1)}B_{im}^{\vartheta}+\frac{1}{R_{0i}}\left(\sum_{k=1}^{N}\varsigma_{ik}^{\vartheta(1)}B_{kb}^{\varphi}-B_{ib}^{\vartheta}\cos\varphi_i\right)=0\end{cases} \tag{4.10}$$

Kinematical and physical compatibility conditions:

$$U_{i1}^{\varphi}=U_{iM}^{\varphi},U_{i1}^{\vartheta}=U_{iM}^{\vartheta},W_{i1}^{\zeta}=W_{iM}^{\zeta},B_{i1}^{\varphi}=B_{iM}^{\varphi},B_{i1}^{\vartheta}=B_{iM}^{\vartheta}$$

$$\begin{cases}\frac{1}{R_{0i}}\left(\sum_{m=1}^{M}\varsigma_{1m}^{\vartheta(1)}U_{im}^{\vartheta}+U_{i1}^{\varphi}\cos\varphi_i+W_{i1}^{\zeta}\sin\varphi_i\right)+\frac{\nu}{R}\left(\sum_{k=1}^{N}\varsigma_{ik}^{\varphi(1)}U_{k1}^{\varphi}+W_{i1}^{\zeta}\right)=\\ \qquad =\frac{1}{R_{0i}}\left(\sum_{m=1}^{M}\varsigma_{bm}^{\vartheta(1)}U_{im}^{\vartheta}+U_{iM}^{\varphi}\cos\varphi_i+W_{iM}^{\zeta}\sin\varphi_i\right)+\frac{\nu}{R}\left(\sum_{k=1}^{N}\varsigma_{ik}^{\varphi(1)}U_{kM}^{\varphi}+W_{iM}^{\zeta}\right)\\ \frac{1}{R}\sum_{k=1}^{N}\varsigma_{ik}^{\varphi(1)}U_{k1}^{\vartheta}+\frac{1}{R_{0i}}\left(\sum_{m=1}^{M}\varsigma_{1m}^{\vartheta(1)}U_{im}^{\varphi}-U_{i1}^{\vartheta}\cos\varphi_i\right)=\frac{1}{R}\sum_{k=1}^{N}\varsigma_{ik}^{\varphi(1)}U_{kM}^{\vartheta}+\frac{1}{R_{0i}}\left(\sum_{m=1}^{M}\varsigma_{Mm}^{\vartheta(1)}U_{im}^{\varphi}-U_{iM}^{\vartheta}\cos\varphi_i\right)\\ \frac{1}{R_{0i}}\left(\sum_{m=1}^{M}\varsigma_{1m}^{\vartheta(1)}W_{im}^{\zeta}-U_{i1}^{\vartheta}\sin\varphi_i\right)+B_{i1}^{\vartheta}=\frac{1}{R_{0i}}\left(\sum_{m=1}^{M}\varsigma_{Mm}^{\vartheta(1)}W_{im}^{\zeta}-U_{iM}^{\vartheta}\sin\varphi_i\right)+B_{iM}^{\vartheta}\\ \frac{1}{R_{0i}}\left(\sum_{m=1}^{M}\varsigma_{1m}^{\vartheta(1)}B_{im}^{\vartheta}+B_{i1}^{\varphi}\cos\varphi_i\right)+\frac{\nu}{R}\sum_{k=1}^{N}\varsigma_{ik}^{\varphi(1)}B_{k1}^{\varphi}=\frac{1}{R_{0i}}\left(\sum_{m=1}^{M}\varsigma_{Mm}^{\vartheta(1)}B_{im}^{\vartheta}+B_{iM}^{\varphi}\cos\varphi_i\right)+\frac{\nu}{R}\sum_{k=1}^{N}\varsigma_{ik}^{\varphi(1)}B_{kM}^{\varphi}\\ \frac{1}{R}\sum_{m=1}^{M}\varsigma_{1m}^{\varphi(1)}B_{im}^{\vartheta}+\frac{1}{R_{0i}}\left(\sum_{k=1}^{N}\varsigma_{ik}^{\vartheta(1)}B_{k1}^{\varphi}-B_{i1}^{\vartheta}\cos\varphi_i\right)=\frac{1}{R}\sum_{m=1}^{M}\varsigma_{Mm}^{\varphi(1)}B_{im}^{\vartheta}+\frac{1}{R_{0i}}\left(\sum_{k=1}^{N}\varsigma_{ik}^{\vartheta(1)}B_{kM}^{\varphi}-B_{iM}^{\vartheta}\cos\varphi_i\right)\end{cases} \tag{4.11}$$

for $i=2,...,N-1$

4.3 Solution Procedure

Applying the differential quadrature procedure, the whole system of differential equations has been discretized and the global assembling leads to the following set of linear algebraic equations:

$$\begin{bmatrix} \mathbf{K}_{bb} & \mathbf{K}_{bd} \\ \mathbf{K}_{db} & \mathbf{K}_{dd} \end{bmatrix} \begin{bmatrix} \boldsymbol{\delta}_b \\ \boldsymbol{\delta}_d \end{bmatrix} = \omega^2 \begin{bmatrix} \mathbf{0} & \mathbf{0} \\ \mathbf{0} & \mathbf{M}_{dd} \end{bmatrix} \begin{bmatrix} \boldsymbol{\delta}_b \\ \boldsymbol{\delta}_d \end{bmatrix} \tag{4.12}$$

In the above matrices and vectors, the partitioning is set forth by subscripts b and d, referring to the system degrees of freedom and standing for boundary and domain, respectively. In order to make the computation more efficient, kinematic condensation of non-domain degrees of freedom is performed:

$$\left(\mathbf{K}_{dd} - \mathbf{K}_{db}\left(\mathbf{K}_{bb}\right)^{-1}\mathbf{K}_{bd}\right)\boldsymbol{\delta}_d = \omega^2 \mathbf{M}_{dd}\boldsymbol{\delta}_d \tag{4.13}$$

The natural frequencies of the structure considered can be determined by making the following determinant vanish:

$$\left|\left(\mathbf{K}_{dd} - \mathbf{K}_{db}\left(\mathbf{K}_{bb}\right)^{-1}\mathbf{K}_{bd}\right) - \omega^2 \mathbf{M}_{dd}\right| = 0 \tag{4.14}$$

5 Applications and Results

Based on the above derivations, in the present paragraph some results and considerations about the free vibration problem of a hemispherical panel and a hemispherical dome are presented. The analysis has been carried out by means of numerical procedures illustrated above. The mechanical characteristics for the considered structures are listed in Table 1. In order to verify the accuracy of the present method, some comparisons have also been performed. The first ten natural frequencies of a hemispherical panel and a hemispherical dome are reported in Tables 2, 3, 4, 5, 6 and 7. The detail regarding the geometry of the considered structures are indicated below:

- *Hemispherical panel*: $R = 1\text{m}$, $h = 0.1\text{m}$, $\varphi_0 = 15°$, $\varphi_1 = 90°$, $\vartheta_0 = 120°$ (Tables 2, 3 and 4).
- *Hemispherical dome*: $R = 1\text{m}$, $h = 0.1\text{m}$, $\varphi_0 = 6°$, $\varphi_1 = 90°$, $\vartheta_0 = 360°$ (Tables 5, 6 and 7).

The geometrical boundary conditions for the hemispherical panel are identified by the following convention. For example, the symbolism C-S-C-F indicates that the edges $\varphi = \varphi_1$, $\vartheta = 0$, $\varphi = \varphi_0$, $\vartheta = \vartheta_0$ are clamped, simply supported, clamped and free, respectively. In particular, we have considered the hemispherical panels characterized by C-F-F-F and S-F-F-F boundary conditions (Tables 2, 3 and 4). For the hemispherical dome, for example, the symbolism C-F indicates that the edges $\varphi = \varphi_1$ and $\varphi = \varphi_0$ are clamped and free, respectively. In this case, the missing boundary conditions are the kinematical and physical compatibility conditions that are applied at the same meridian for $\vartheta = 0$ and $\vartheta = 2\pi$. In this work the hemispherical dome that we have examined is characterized by C-F and S-F boundary conditions (Tables 5, 6 and 7).

One of the aims of this paper is to compare results from the present analysis with those obtained with finite element techniques and based on the same shell theory or 3D element theory. In Tables 2, 4, 5 and 7, we have compared the 2D shell theory results obtained by the GDQ Method with the FEM results obtained by some commercial programs using the same 2D shell theory. On the other hand, the FEM solutions using 3D element theory obtained with the same commercial programs are reported in Tables 3 and 6.

For the GDQM results reported in Tables 2, 4, 5 and 7, we have considered the grid distribution (3.11) with $N = 21$ and $M = 21$. For the commercial programs, we have used shell elements with 4 and 8 nodes in Tables 2, 4, 5 and 7. On the other hand, brick-elements with 20-nodes were used for the 3D element theory in Tables 3 and 6.

It is noteworthy that the results from the present methodology are very close to those obtained by the commercial programs. As can be seen, the numerical results show excellent agreement. Furthermore, it is significant that the computational effort in terms of time and number of grid points is smaller for the GDQ method results than for the finite element method.

In Figure 6, we have reported the first mode shapes for the hemispherical panel characterized by C-F-F-F boundary conditions, while in Figure 7 the mode shapes for the hemispherical dome characterized by C-F boundary conditions are illustrated. In particular, for the hemispherical dome there are some symmetrical mode shapes due to the symmetry of the problem considered in 3D space. In these cases, we have summarized the symmetrical mode shapes in one figure.

6 Conclusions

A Generalized Differential Quadrature Method application to free vibration analysis of spherical shells has been presented to illustrate the versatility and the accuracy of this methodology. The adopted shell theory is a first order shear deformation theory. The dynamic equilibrium equations are discretized and solved with the present method giving a standard linear eigenvalue problem. The vibration results are obtained without the modal expansion methodology. The complete 2D differential system, governing the structural problem, has been solved. Due to the theoretical framework, no approximation (δ-point technique) is needed in modelling the boundary edge conditions.

The GDQ method provides a very simple algebraic formula for determining the weighting coefficients required by the differential quadrature approximation, without restricting in any way the choice of mesh grids. Examples presented show that the generalized differential quadrature method can produce accurate results, utilizing only a small number of sampling points. Furthermore, discretizing and programming procedures are quite easy. Fast convergence and very good stability can be obtained. Furthermore, the computational effort in terms of time and number of grid points is smaller for the GDQ method results than for the finite element method.

Table 1. Physical parameters used in the analysis of free vibrations of the structures considered.

Parameter	Value
Density of mass μ	$7800 \text{ kg} / \text{m}^3$
Young's modulus E	$2.1 \cdot 10^{11}$ Pa
Poisson coefficient ν	0.3
Shear factor χ	$6/5$

Table 2. Shell theory for the hemispherical panel C-F-F-F.

Frequencies [Hz]	GDQ Method	Ansys 8	Femap\Nastran 8.3	Straus 7	Pro\Engineer WildFire 2
f_1	195.35	195.84	196.20	195.11	195.36
f_2	196.01	196.54	196.70	195.85	196.20
f_3	433.95	436.00	435.69	433.48	433.91
f_4	526.27	530.34	531.48	528.90	526.71
f_5	766.43	772.52	771.54	767.01	765.57
f_6	854.57	860.01	860.15	856.80	856.47
f_7	935.20	947.12	943.32	936.96	935.20
f_8	1062.02	1075.60	1065.55	1068.02	1060.61
f_9	1069.86	1084.10	1080.86	1079.32	1069.56
f_{10}	1122.32	1137.80	1137.89	1130.51	1122.62

Table 3. 3D element theory for the hemispherical panel C-F-F-F.

Frequencies [Hz]	Ansys 8	Femap\Nastran 8.3	Straus 7	Pro\Engineer WildFire 2
f_1	195.80	196.52	196.56	196.37
f_2	196.56	197.32	197.40	197.00
f_3	435.02	435.56	435.69	435.77
f_4	527.79	530.55	530.67	529.59
f_5	767.37	766.86	766.93	768.48
f_6	856.50	857.32	857.38	857.35
f_7	937.52	937.12	937.68	940.26
f_8	1060.70	1052.11	1052.27	1063.47
f_9	1069.90	1068.47	1068.61	1072.50
f_{10}	1125.60	1130.15	1130.64	1130.61

Table 4. Shell theory for the hemispherical panel S-F-F-F.

Frequencies [Hz]	f_1	f_2	f_3	f_4	f_5
GDQ Method	159.29	173.05	392.09	453.32	731.42
Ansys 8	158.94	171.90	393.68	454.04	738.00
Femap\Nastran 8.3	159.61	172.40	393.77	456.50	736.96
Straus 7	158.02	171.34	391.20	453.79	733.09
Pro\Engineer WF 2	158.63	171.59	392.07	452.04	730.62

Table 5. Shell theory for the hemispherical dome C-F.

Frequencies [Hz]	GDQ Method	Ansys 8	Femap\Nastran 8.3	Straus 7	Pro\Engineer WildFire 2
f_1	508.93	509.19	509.79	513.14	509.09
f_2	508.93	509.19	509.79	513.14	509.09
f_3	704.71	705.22	705.71	712.14	703.95
f_4	764.88	766.09	767.91	771.44	763.94
f_5	764.88	766.09	767.91	771.44	763.95
f_6	872.77	875.50	876.63	888.37	871.74
f_7	872.77	875.53	876.64	888.37	871.76
f_8	932.50	936.54	931.95	942.37	931.43
f_9	932.50	936.55	931.95	942.37	931.50
f_{10}	1018.30	1024.40	1021.78	1026.28	1016.95

Table 6. 3D element theory for the hemispherical dome C-F.

Frequencies [Hz]	Ansys 8	Femap\Nastran 8.3	Straus 7	Pro\Engineer WildFire 2
f_1	510.09	510.01	509.95	509.32
f_2	510.29	510.05	509.99	509.48
f_3	706.00	705.76	705.46	704.88
f_4	764.35	765.28	764.91	764.23
f_5	764.38	765.39	765.02	764.39
f_6	873.50	873.61	873.12	872.19
f_7	874.10	874.00	873.47	872.74
f_8	932.20	932.11	931.23	931.46
f_9	932.27	932.16	931.30	931.94
f_{10}	1021.10	1020.36	1018.72	1019.64

Table 7. Shell theory for the hemispherical dome S-F.

Frequencies [Hz]	f_1	f_2	f_3	f_4	f_5
GDQ Method	478.49	478.49	666.82	750.93	750.93
Ansys 8	478.71	478.71	667.35	751.65	751.66
Femap\Nastran 8.3	479.59	479.59	668.09	753.39	753.39
Straus 7	479.80	479.80	669.81	755.08	755.08
Pro\Engineer WF 2	478.57	478.57	666.14	749.78	749.78

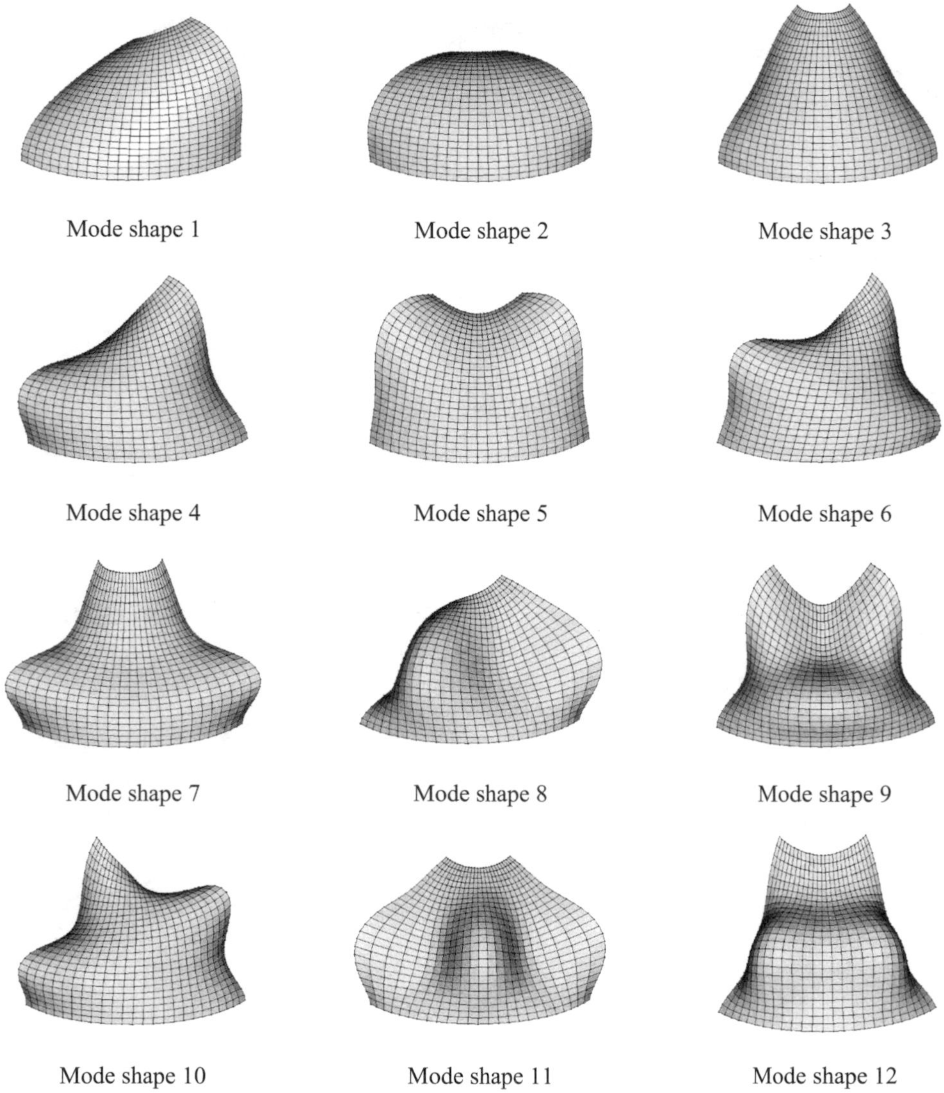

Figure 6. Mode shapes for the hemispherical panel C-F-F-F.

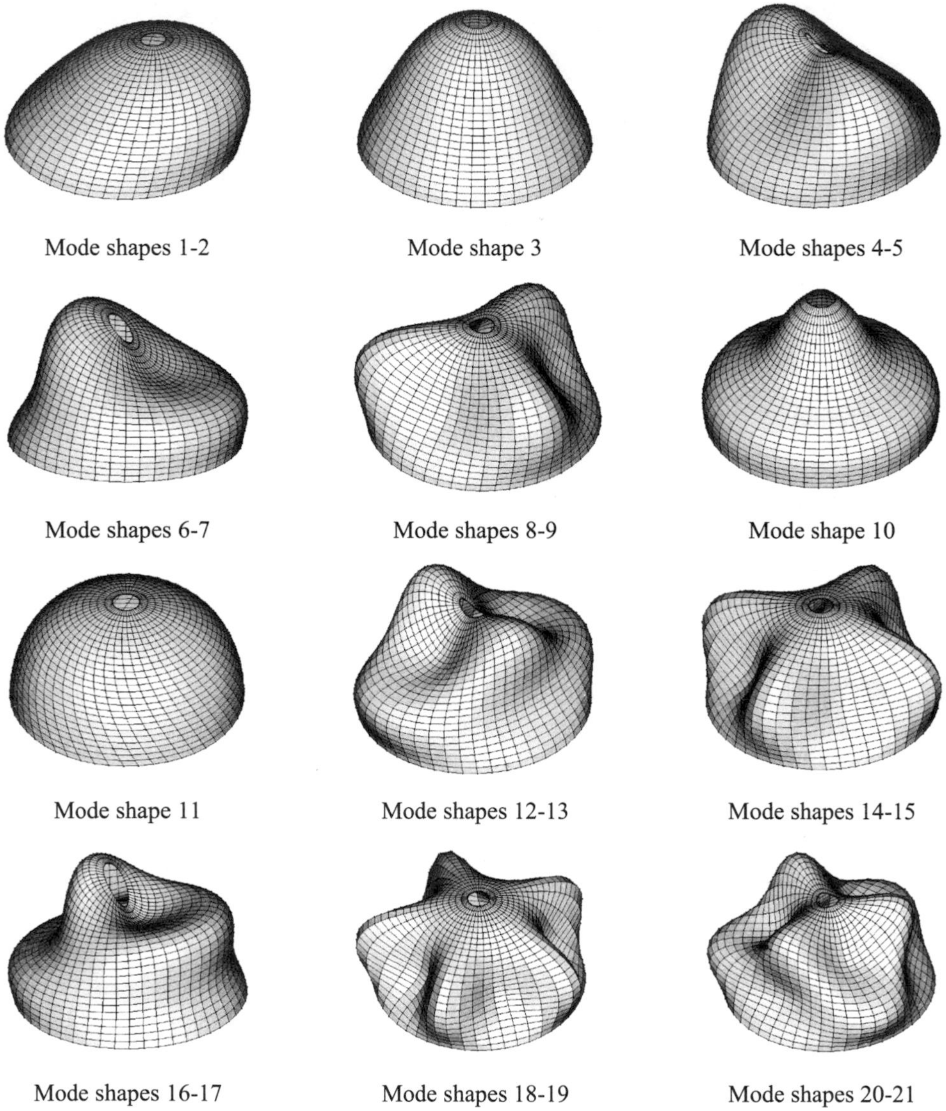

Figure 7. Mode shapes for the hemispherical dome C-F.

Acknowledgment

This research was supported by the Italian Ministry for University and Scientific, Technological Research MIUR (40% and 60%). The research topic is one of the subjects of the Centre of Study and Research for the Identification of Materials and Structures (CIMEST)-"M.Capurso".

References

E. Artioli, P. Gould, E. Viola (2004). Generalized collocation method for rotational shells free vibration analysis. *The Seventh International Conference on Computational Structures Technology*. Lisbon, Portugal, September 7 - 9, 2004.

R. Bellman, J. Casti (1971). Differential quadrature and long-term integration. *Journal of Mathematical Analysis and Applications* 34, 235-238.

R. Bellman, B. G. Kashef, J. Casti (1972). Differential quadrature: a technique for the rapid solution of nonlinear partial differential equations. *Journal of Computational Physic* 10, 40-52, 1972.

C. Bert, M. Malik (1996). Differential quadrature method in computational mechanics. *Applied Mechanical Reviews* 49, 1-27.

C. W. Bert, S. K. Jang, A. G. Striz (1988). Two new approximate methods for analyzing free vibration of structural components. *AIAA Journal* 26, 612-618.

F. Civan, C. M. Sliepcevich (1984). Differential quadrature for multi-dimensional problems. *Journal of Mathematical Analysis and Applications* 101, 423-443.

F. Civan, C. M. Sliepcevich (1985). Application of differential quadrature in solution of pool boiling in cavities. *Proceeding of Oklahoma Academy of Sciences* 65, 73-78.

W. Flügge (1960). Stress in Shells. Springer: New York.

S. Jang, C. Bert, A. G. Striz (1989). Application of differential quadrature to static analysis of structural components. *International Journal of Numerical Methods in Engineering* 28, 561-577.

G. Karami, P. Malekzadeh (2003). Application of a new differential quadrature methodology for free vibration analysis of plates. *International Journal of Numerical Methods in Engineering* 56, 847-868.

F. I. Niordson (1985). Shell Theory. North-Holland: Amsterdam.

J. R. Quan, C. T. Chang (1989). New insights in solving distributed system equations by the quadrature method – I. Analysis. *Computers and Chemical Engineering* 13, 779-788.

J. R. Quan, C. T. Chang (1989). New insights in solving distributed system equations by the quadrature method – II. Numerical experiments. *Computers and Chemical Engineering* 13, 1017-1024.

J.N. Reddy (2003). Mechanics of laminated composite plates and shells. CRC Press, 2nd edition.

E. Reissner (1945). The effect of transverse shear deformation on the bending of elastic plates. *Journal of Applied Mechanics ASME* 12, 66-77.

J. L. Sanders (1959). An improved first approximation theory of thin shells. NASA Report 24.

C. Shu (1991). Generalized differential-integral quadrature and application to the simulation of incompressible viscous flows including parallel computation. *PhD Thesis*. University of Glasgow.

C. Shu (2000). Differential Quadrature and Its Application in Engineering. Springer: Berlin.

C. Shu, W. Chen (1999). On optimal selection of interior points for applying discretized boundary conditions in DQ vibration analysis of beams and plates. *Journal of Sound and Vibration* 222, 239-257.

C. Shu, W. Chen, H. Xue, H. Du (2001). Numerical study of grid distribution effect on accuracy of DQ analysis of beams and plates by error estimation of derivative approximation. *International Journal for Numerical Methods in Engineering* 51, 159-179.

E. Viola, E. Artioli (2004). The G.D.Q. method for the harmonic dynamic analysis of rotational shell structural elements. *Structural Engineering and Mechanics* 17, 789-817.

References

[illegible]

A Comparison of Two Dimensional Structural Theories for Isotropic Plates

Erasmo Viola and Federica Daghia

Dipartimento di Ingegneria delle Strutture, dei Trasporti, delle Acque, del Rilevamento, del Territorio - University of Bologna, Viale Risorgimento 2, 40136 Bologna, Italy

Abstract In the study of thick plates, an accurate description of shear strains is important in order to correctly predict the plate behavior. In the setting of two dimensional theories, improved accuracy with respect to Kirchhoff–Love and Reissner–Mindlin theories can be obtained by introducing higher order terms in the description of the displacement. In this paper, the third order theory proposed by Reddy (1984) for laminated plates is thoroughly evaluated for the isotropic plates. A corresponding finite element model is developed and implemented in a computer code for modal analysis of plates. The results are compared with the three dimensional theory and with the predictions given by traditional two dimensional models.

1 Introduction

The plate structural element can be described in different ways, ranging from three dimensional to two dimensional theories. In the two dimensional environment, the main hypothesis is the choice of the displacement field. This leads to a whole range of theories, each one determining a different representation of the strain and stress fields.

In order to describe the behavior of thick plates, an accurate representation of the shear strains is extremely important, because the significance of shear deformations increases with growing thickness of the plate. Improved accuracy with respect to the classical two dimensional theories can be obtained by introducing higher order terms into the displacement field. In this paper, the third order theory proposed by Reddy (1984) is considered. A finite element formulation is established and implemented in a computer code for modal analysis.

2 Two Dimensional Theories

In two dimensional plate theories, the displacement field defined a priori (kinematic hypothesis) allows us to reduce a 3D problem to a 2D problem. The main difference between the various 2D theories lies in the assumptions made in the choice of the displacement field.

The coordinate system used in this work is shown in Fig. 1. The x and y axes belong to the midplane of the undeformed plate, while the thickness is represented by the z coordinate. The midplane has equation $z = 0$. (u, v, w) are the displacement

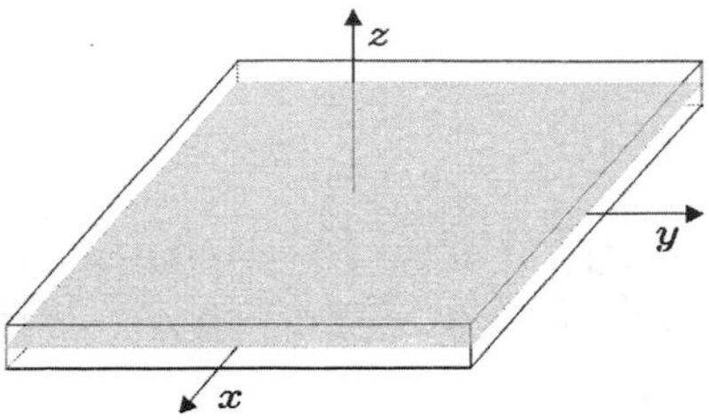

Figure 1. Coordinate system for the plate.

components along the x, y, and z axes respectively, while the variable time is defined as t.

2.1 Kinematic hypotheses

- Classical plate theory (CPT):

$$\begin{aligned} u(x,y,z,t) &= u_0 - z\frac{\partial w_0}{\partial x} \\ v(x,y,z,t) &= v_0 - z\frac{\partial w_0}{\partial y} \\ w(x,y,z,t) &= w_0 \end{aligned} \tag{1}$$

- First order shear deformation theory (FSDT):

$$\begin{aligned} u(x,y,z,t) &= u_0 + z\psi_x \\ v(x,y,z,t) &= v_0 + z\psi_y \\ w(x,y,z,t) &= w_0 \end{aligned} \tag{2}$$

- Higher order shear deformation theory proposed by Reddy (HSDT):

$$\begin{aligned} u(x,y,z,t) &= u_0 + z\psi_x + z^3\left(-\frac{4}{3h^2}\right)\left(\psi_x + \frac{\partial w_0}{\partial x}\right) \\ v(x,y,z,t) &= v_0 + z\psi_y + z^3\left(-\frac{4}{3h^2}\right)\left(\psi_y + \frac{\partial w_0}{\partial y}\right) \\ w(x,y,z,t) &= w_0 \end{aligned} \tag{3}$$

In the previous expressions, $u_0 = u_0(x,y,t)$, $v_0 = v_0(x,y,t)$ and $w_0 = w_0(x,y,t)$ are the displacement components along the (x,y,z) coordinates of a point belonging to the midplane, while $\psi_x = \psi_x(x,y,t)$ and $\psi_y = \psi_y(x,y,t)$ represent the rotations about the y and x axes.

The CPT theory is also known as the Kirchhoff–Love theory. The kinematic hypothesis (1) implies that straight lines normal to the midplane before the deformation remain

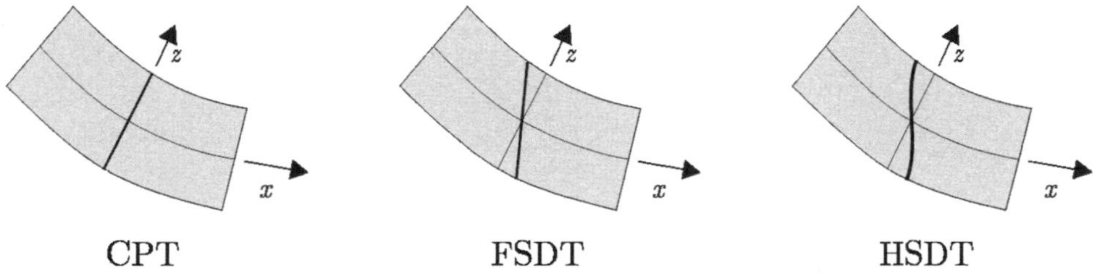

Figure 2. Deformation of a line normal to the midplane in the different theories.

straight and normal to the midplane after the deformation and do not change in length (see Fig. 2). This amounts to neglecting both transverse shear and transverse normal effects, a simplification admissible in the case of thin plates.

The FSDT is also known as the Reissner–Mindlin theory. The introduction of the parameters ψ_x and ψ_y allow constant transverse shear strains across the plate thickness. A straight line normal to the undeformed midplane transforms into a straight line which is no longer normal to the deformed midplane (see Fig. 2). Since the hypothesis of constant shear strains is a simplification, it is possible to correct the model predictions by introducing a shear correction factor k into the constitutive relations (for isotropic plates, $k = 5/6$).

The HSDT theory allows for parabolical shear strains across the plate thickness. The displacement field is obtained by expanding the u and v displacement components up to third order terms in z and imposing the a priori hypothesis of zero shear strains at the top and bottom surfaces of the plate of thickness h ($\gamma_{xz} = \gamma_{yz} = 0$ for $z = \pm\frac{h}{2}$). A straight line normal to the midplane transforms into a cubic line (see Fig. 2). Because of the parabolic distribution of strains, the HSDT does not require a correction factor.

The theory just described was proposed by Reddy in the context of laminated plates, however it can be simplified to fit the description of homogeneous, single layer plates. For a detailed description of two dimensional theories for laminated plates, see Reddy (2003).

2.2 Strain Field

Once the displacement field is defined, it is possible to evaluate the strains associated with each theory. This is done in the hypothesis of small strains in order to study a linear elastic problem. The strain-displacement equations can be written in compact form

$$\mathbf{e} = \mathbf{D}\mathbf{u} \tag{4}$$

where $\mathbf{e}$ is the vector containing the generalized strains, $\mathbf{u}$ is the vector containing the functions on which the displacement field depends and $\mathbf{D}$ is the congruence or kinematic operator. $\mathbf{e}$, $\mathbf{u}$ and $\mathbf{D}$ are different depending on the chosen theory:

- CPT:

$$\begin{bmatrix} \varepsilon_{x0} \\ \varepsilon_{y0} \\ \gamma_{xy0} \\ \chi_x \\ \chi_y \\ \chi_{xy} \end{bmatrix} = \begin{bmatrix} \partial/\partial x & 0 & 0 \\ 0 & \partial/\partial y & 0 \\ \partial/\partial y & \partial/\partial x & 0 \\ 0 & 0 & -\partial^2/\partial x^2 \\ 0 & 0 & -\partial^2/\partial y^2 \\ 0 & 0 & -2\partial^2/\partial x \partial y \end{bmatrix} \begin{bmatrix} u_0 \\ v_0 \\ w_0 \end{bmatrix} \tag{5}$$

- FSDT:

$$\begin{bmatrix} \varepsilon_{x0} \\ \varepsilon_{y0} \\ \gamma_{xy0} \\ \chi_x \\ \chi_y \\ \chi_{xy} \\ \gamma_{yz0} \\ \gamma_{xz0} \end{bmatrix} = \begin{bmatrix} \partial/\partial x & 0 & 0 & 0 & 0 \\ 0 & \partial/\partial y & 0 & 0 & 0 \\ \partial/\partial y & \partial/\partial x & 0 & 0 & 0 \\ 0 & 0 & 0 & \partial/\partial x & 0 \\ 0 & 0 & 0 & 0 & \partial/\partial y \\ 0 & 0 & 0 & \partial/\partial y & \partial/\partial x \\ 0 & 0 & \partial/\partial y & 0 & 1 \\ 0 & 0 & \partial/\partial x & 1 & 0 \end{bmatrix} \begin{bmatrix} u_0 \\ v_0 \\ w_0 \\ \psi_x \\ \psi_y \end{bmatrix} \tag{6}$$

- HSDT:

$$\begin{bmatrix} \varepsilon_{x0} \\ \varepsilon_{y0} \\ \gamma_{xy0} \\ \chi_x \\ \chi_y \\ \chi_{xy} \\ \chi_{xs} \\ \chi_{ys} \\ \chi_{xys} \\ \gamma_{yz0} \\ \gamma_{xz0} \end{bmatrix} = \begin{bmatrix} \partial/\partial x & 0 & 0 & 0 & 0 \\ 0 & \partial/\partial y & 0 & 0 & 0 \\ \partial/\partial y & \partial/\partial x & 0 & 0 & 0 \\ 0 & 0 & 0 & \partial/\partial x & 0 \\ 0 & 0 & 0 & 0 & \partial/\partial y \\ 0 & 0 & 0 & \partial/\partial y & \partial/\partial x \\ 0 & 0 & \partial^2/\partial x^2 & \partial/\partial x & 0 \\ 0 & 0 & \partial^2/\partial y^2 & 0 & \partial/\partial y \\ 0 & 0 & 2\partial^2/\partial x \partial y & \partial/\partial y & \partial/\partial x \\ 0 & 0 & \partial/\partial y & 0 & 1 \\ 0 & 0 & \partial/\partial x & 1 & 0 \end{bmatrix} \begin{bmatrix} u_0 \\ v_0 \\ w_0 \\ \psi_x \\ \psi_y \end{bmatrix} \tag{7}$$

where ε_{x0}, ε_{y0} and γ_{xy0} are the strains due to in plane displacements; χ_x, χ_y and χ_{xy} are the curvatures due to first order terms; γ_{xz0} and γ_{yz0} are the shear strains; χ_{xs}, χ_{ys} and χ_{xys} are the curvatures due to the higher order terms.

In order to understand the physical meaning of the generalized strains in the HSDT, Fig. 3 shows the deformation of a small element under the effect of each generalized strain.

For further calculations it is useful to express the components of the strain tensor in terms of the generalized strain components for the HSDT ($c = -\frac{4}{3h^2}$):

$$\begin{bmatrix} \varepsilon_x \\ \varepsilon_y \\ \gamma_{xy} \\ \gamma_{yz} \\ \gamma_{xz} \end{bmatrix} = \begin{bmatrix} 1 & 0 & 0 & z & 0 & 0 & cz^3 & 0 & 0 & 0 & 0 \\ 0 & 1 & 0 & 0 & z & 0 & 0 & cz^3 & 0 & 0 & 0 \\ 0 & 0 & 1 & 0 & 0 & z & 0 & 0 & cz^3 & 0 & 0 \\ 0 & 0 & 0 & 0 & 0 & 0 & 0 & 0 & 0 & 1+3cz^2 & 0 \\ 0 & 0 & 0 & 0 & 0 & 0 & 0 & 0 & 0 & 0 & 1+3cz^2 \end{bmatrix} \begin{bmatrix} \varepsilon_{x0} \\ \varepsilon_{y0} \\ \gamma_{xy0} \\ \chi_x \\ \chi_y \\ \chi_{xy} \\ \chi_{xs} \\ \chi_{ys} \\ \chi_{xys} \\ \gamma_{yz0} \\ \gamma_{xz0} \end{bmatrix} \tag{8}$$

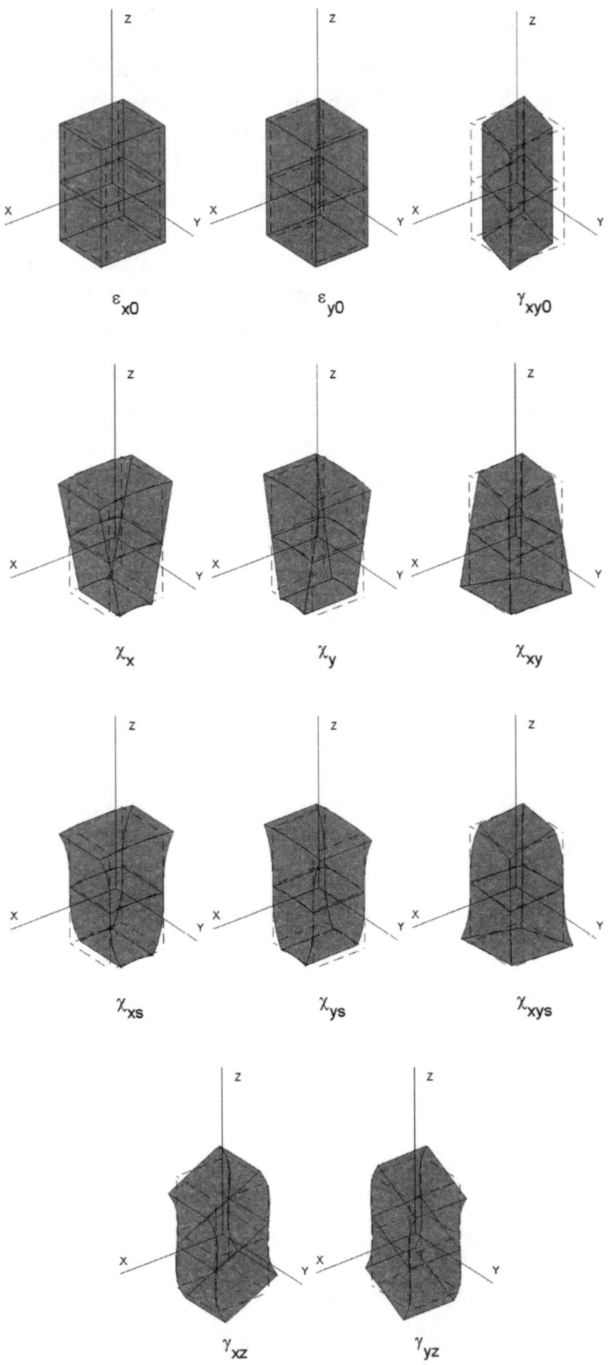

Figure 3. Generalized strain components for the HSDT.

2.3 Constitutive equations

In order to define the constitutive relations for a plate in the two dimensional setting, the first step is to consider the behavior of a three dimensional continuum.

In the following modelling, the materials are assumed to be homogeneous, therefore material properties are independent of position, and isotropic, that is the constitutive relations are the same in every direction. For a more general evaluation of the constitutive relations in the case of laminated composite plates, see Daghia et al. (2005). Note that in this work the definition of the generalized strains and stresses is slightly different, therefore the resulting constitutive relations are not exactly the same. This, however, does not influence the final problem equations in terms of displacement.

In a three dimensional setting, the constitutive behavior is represented by the following relation:

$$\boldsymbol{\sigma} = \mathbf{C}\boldsymbol{\varepsilon} \tag{9}$$

where $\mathbf{C}_{6\times6}$ is the constitutive tensor, $\boldsymbol{\sigma}_{6\times1}$ is the vector of stress components and $\boldsymbol{\varepsilon}_{6\times1}$ is the vector of the strain components. The tensor $\mathbf{C}$ is symmetric and positive definite.

When dealing with a two dimensional problem, equation (9) can be simplified by introducing the hypothesis of plane stress:

$$\sigma_{ij} = \sigma_{ij}(x, y) \quad \text{and} \quad \sigma_{33} = 0 \tag{10}$$

The constitutive equations become:

$$\begin{bmatrix} \sigma_x \\ \sigma_y \\ \sigma_{xy} \\ \sigma_{yz} \\ \sigma_{xz} \end{bmatrix} = \frac{E}{1-\nu^2} \begin{bmatrix} 1 & \nu & 0 & 0 & 0 \\ \nu & 1 & 0 & 0 & 0 \\ 0 & 0 & (1-\nu)/\,2 & 0 & 0 \\ 0 & 0 & 0 & (1-\nu)/\,2 & 0 \\ 0 & 0 & 0 & 0 & (1-\nu)/\,2 \end{bmatrix} \begin{bmatrix} \varepsilon_x \\ \varepsilon_y \\ \gamma_{xy} \\ \gamma_{yz} \\ \gamma_{xz} \end{bmatrix} \tag{11}$$

For the HSDT, the generalized stresses are defined as follows:

$$\begin{aligned}
[N_x \; N_y \; N_{xy}] &= \int_{-\frac{h}{2}}^{\frac{h}{2}} [\sigma_x \; \sigma_y \; \sigma_{xy}] \, \mathrm{d}z \\
[M_x \; M_y \; M_{xy}] &= \int_{-\frac{h}{2}}^{\frac{h}{2}} [\sigma_x \; \sigma_y \; \sigma_{xy}] \, z\mathrm{d}z \\
[M_{xs} \; M_{ys} \; M_{xys}] &= \int_{-\frac{h}{2}}^{\frac{h}{2}} [\sigma_x \; \sigma_y \; \sigma_{xy}] \, cz^3\mathrm{d}z \\
[T_y \; T_x] &= \int_{-\frac{h}{2}}^{\frac{h}{2}} [\sigma_{yz} \; \sigma_{xz}] \left(1 + 3cz^2\right) \mathrm{d}z
\end{aligned} \tag{12}$$

The plate constitutive equations are obtained by substituting (8) and (11) into (12):

$$\begin{bmatrix} N_x \\ N_y \\ N_{xy} \\ \hline M_x \\ M_y \\ M_{xy} \\ \hline M_{xs} \\ M_{ys} \\ M_{xys} \\ \hline T_y \\ T_x \end{bmatrix} = \left[\begin{array}{c|c|c|c} \mathbf{A} & \mathbf{B} & c\mathbf{E} & \mathbf{0} \\ \hline \mathbf{B} & \mathbf{D} & c\mathbf{F} & \mathbf{0} \\ \hline c\mathbf{E} & c\mathbf{F} & c^2\mathbf{H} & \mathbf{0} \\ \hline \mathbf{0} & \mathbf{0} & \mathbf{0} & \mathbf{A}_t + 6c\mathbf{D}_t + 9c^2\mathbf{F}_t \end{array}\right] \begin{bmatrix} \varepsilon_{x0} \\ \varepsilon_{y0} \\ \gamma_{xy0} \\ \hline \chi_x \\ \chi_y \\ \chi_{xy} \\ \hline \chi_{xs} \\ \chi_{ys} \\ \chi_{xys} \\ \hline \gamma_{yz0} \\ \gamma_{xz0} \end{bmatrix} \tag{13}$$

The constitutive matrices are defined as follows:

$$\begin{aligned} \mathbf{A}, \mathbf{B}, \mathbf{D}, \mathbf{E}, \mathbf{F}, \mathbf{H} = \mathbf{X} &= \begin{bmatrix} X_{11} & X_{12} & X_{13} \\ X_{12} & X_{22} & X_{23} \\ X_{13} & X_{23} & X_{33} \end{bmatrix} \\ \mathbf{A}_t, \mathbf{D}_t, \mathbf{F}_t = \mathbf{X}_t &= \begin{bmatrix} X_{44} & X_{45} \\ X_{45} & X_{55} \end{bmatrix} \\ \{A_{ij}, B_{ij}, D_{ij}, E_{ij}, F_{ij}, H_{ij}\} &= \int_{-\frac{h}{2}}^{\frac{h}{2}} C_{ij} \left\{1, z, z^2, z^3, z^4, z^6\right\} \mathrm{d}z \end{aligned} \tag{14}$$

where C_{ij} are the terms of the constitutive matrix in a three dimensional setting, defined in equation (11).

This formulation comes directly from the equations introduced up to now. However, looking at the integrals in (14) it results that, for homogeneous, single layer plate,

$$B_{ij} = E_{ij} = 0 \qquad (\forall \ i, j) \tag{15}$$

thus showing the uncoupling of bending and in plane behavior.

In compact form, equations (13) can be written as

$$\mathbf{s} = \bar{\mathbf{C}}\mathbf{e} \tag{16}$$

In this section, the constitutive equations for a plate in the HSDT theory were obtained. In a similar way it is possible to obtain the constitutive equations for the other two dimensional theories.

2.4 Equations of Motion Through Hamilton's Principle

The equations for the dynamic equilibrium of a plate can be obtained from Hamilton's principle. It states:

$$\delta \int_{t_1}^{t_2} (\mathcal{L} + W)\, dt = 0 \tag{17}$$

where t_1 and t_2 define the time interval, $\mathcal{L}$ is the Lagrangian and W is the external forces potential. The Lagrangian $\mathcal{L}$ is defined as the difference between the strain energy U and the kinetic energy $\mathcal{T}$.

In order to obtain the equations of motion, δU, δT and δW must be written in function of the displacement components and integrated by parts to relieve the terms δu_0, δv_0, δw_0, $\delta \psi_x$ and $\delta \psi_y$. Substituting into equation (17), it is possible to obtain the equations of motion. Written in compact form, they are:

$$\mathbf{D}^* \mathbf{s} + \mathcal{M} \ddot{\mathbf{u}} = \mathbf{f} \tag{18}$$

where $\mathbf{s}$ is the vector of generalized stresses, $\ddot{\mathbf{u}}$ is the vector containing the accelerations of the displacement components, $\mathbf{f}$ is the vector of external distributed loads.

$$\begin{aligned}
\mathbf{s}^{\mathrm{T}} &= \begin{bmatrix} N_x & N_y & N_{xy} & M_x & M_y & M_{xy} & M_{xs} & M_{ys} & M_{xys} & T_y & T_x \end{bmatrix} \\
\ddot{\mathbf{u}}^{\mathrm{T}} &= \begin{bmatrix} \ddot{u}_0 & \ddot{v}_0 & \ddot{w}_0 & \ddot{\psi}_x & \ddot{\psi}_y \end{bmatrix} \\
\mathbf{f}^{\mathrm{T}} &= \begin{bmatrix} q_x & q_y & q_z & m_x & m_y \end{bmatrix}
\end{aligned} \tag{19}$$

The matrices $\mathbf{D}^*$ and $\mathcal{M}$ are given by the following expressions:

$$\mathbf{D}^* = \begin{bmatrix}
-\frac{\partial}{\partial x} & 0 & -\frac{\partial}{\partial y} & 0 & 0 & 0 & 0 & 0 & 0 & 0 & 0 \\
0 & -\frac{\partial}{\partial y} & -\frac{\partial}{\partial x} & 0 & 0 & 0 & 0 & 0 & 0 & 0 & 0 \\
0 & 0 & 0 & 0 & 0 & 0 & \frac{\partial^2}{\partial x^2} & \frac{\partial^2}{\partial y^2} & 2\frac{\partial^2}{\partial x \partial y} & -\frac{\partial}{\partial y} & -\frac{\partial}{\partial x} \\
0 & 0 & 0 & -\frac{\partial}{\partial x} & 0 & -\frac{\partial}{\partial y} & -\frac{\partial}{\partial x} & 0 & -\frac{\partial}{\partial y} & 0 & 1 \\
0 & 0 & 0 & 0 & -\frac{\partial}{\partial y} & -\frac{\partial}{\partial x} & 0 & -\frac{\partial}{\partial y} & -\frac{\partial}{\partial x} & 1 & 0
\end{bmatrix} \tag{20}$$

$$\mathcal{M} = \begin{bmatrix}
I_0 & 0 & I_3 c \frac{\partial}{\partial x} & I_1 + I_3 c & 0 \\
0 & I_0 & I_3 c \frac{\partial}{\partial y} & 0 & I_1 + I_3 c \\
-I_3 c \frac{\partial}{\partial x} & -I_3 c \frac{\partial}{\partial y} & I_0 - I_6 c^2 \left(\frac{\partial^2}{\partial x^2} + \frac{\partial^2}{\partial y^2} \right) & -\left(I_4 c + I_6 c^2 \right) \frac{\partial}{\partial x} & -\left(I_4 c + I_6 c^2 \right) \frac{\partial}{\partial y} \\
I_1 + I_3 c & 0 & \left(I_4 c + I_6 c^2 \right) \frac{\partial}{\partial x} & I_2 + 2 I_4 c + I_6 c^2 & 0 \\
0 & I_1 + I_3 c & \left(I_4 c + I_6 c^2 \right) \frac{\partial}{\partial y} & 0 & I_2 + 2 I_4 c + I_6 c^2
\end{bmatrix}$$

where it results $I_i = \int_{-\frac{h}{2}}^{\frac{h}{2}} \mu z^i dz$ and μ is the density.

The equilibrium operator $\mathbf{D}^*$ is the formal adjoint of the kinematic operator $\mathbf{D}$. Introducing the constitutive relations (16) and the congruence equations (4), the equations of motion can be written in terms of displacement:

$$\mathbf{D}^* \bar{\mathbf{C}} \mathbf{D} \mathbf{u} + \mathcal{M} \ddot{\mathbf{u}} = \mathbf{f} \tag{21}$$

which is the usual form of the equations of motion.

3 Finite Element Formulation

The equations of motion in terms of displacement (21) are partial differential equations. An approximate solution can be obtained through the finite element method. In this work, the finite element formulation for a quadrilateral four–node element is obtained from Hamilton's principle (Zienkiewicz and Taylor, 2002). The master element is shown in Fig. 4.

In order to define the finite element model, it is necessary to choose shape functions that satisfy the continuity requirements imposed by the compatibility operator $\mathbf{D}$. In the

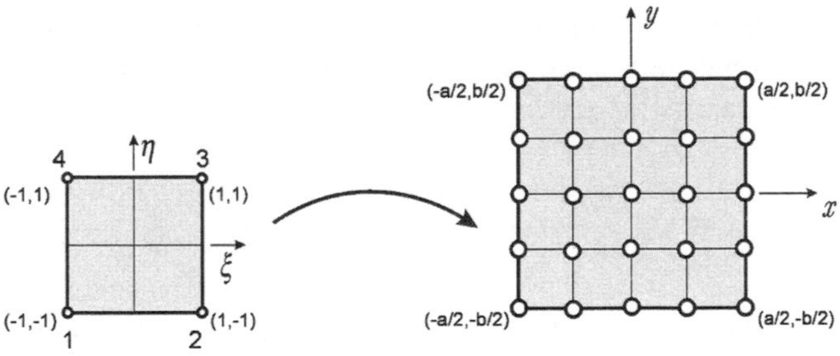

Figure 4. Master element and plate reference system.

HSDT, $\mathbf{D}$ contains second derivatives of the w_0 displacement component (see equation (7)). For this reason, C^1 continuity is needed in the w_0 shape functions. In this work, Lagrange bilinear shape functions were used to represent u_0, v_0, ψ_x and ψ_y, while Hermite bicubic shape functions ensure C^1 continuity to the w_0 displacement component. The same shape functions were used for all theories.

Introducing the shape functions, the displacement field is represented as follows:

$$\mathbf{u} = \mathbf{N}\bar{\mathbf{u}} \tag{22}$$

where $\mathbf{N}$ is the vectors containing the Lagrange and Hermite shape functions. The vector $\bar{\mathbf{u}}$ contains the nodal displacement components.

Substituting the approximate displacement field into the equation (17) describing Hamilton's principle, it is possible to obtain the stiffness and mass matrices.

It is then possible to write the generalized eigenvalue problem:

$$\left(\mathbf{K} - \omega^2\mathbf{M}\right)\mathbf{a} = \mathbf{0} \tag{23}$$

The eigenvalues ω^2 and eigenvectors $\mathbf{a}$ allow us to evaluate the natural frequencies and represent the modal shapes of the plate.

4 Some Numerical Results

In this section, natural frequencies and mode shapes of simply supported square isotropic plates are obtained using the two dimensional theories previously discussed. The results for different side to thickness ratios are compared to the reference 3D solution obtained by Gentilini and Viola (2003).

The plate in its reference system is shown in Fig. 4. All results were obtained using a 32×32 square mesh. The boundary condition is the simply supported *hard*:

$$\begin{aligned} &\text{for } y = \pm\frac{b}{2} \qquad u_0 = v_0 = w_0 = \frac{\partial w_0}{\partial x} = \psi_x = 0 \\ &\text{for } x = \pm\frac{a}{2} \qquad u_0 = v_0 = w_0 = \frac{\partial w_0}{\partial y} = \psi_y = 0 \end{aligned} \tag{24}$$

Figures 5 and 6 show the comparison between the normalized frequencies obtained with the two dimensional and three dimensional theories for different side to thickness ratios. The following quantities are introduced:

$$\bar{\omega} = \omega\sqrt{\frac{\mu h^2}{G}} \quad \text{and} \quad \frac{g^2}{\pi^2} = h^2\left(\frac{m^2}{a^2} + \frac{n^2}{b^2}\right) \tag{25}$$

where $\bar{\omega}$ are the normalized natural frequencies, m and n are the numbers of half waves in the x and y direction of the considered mode shape (bending modes). The only mechanical parameter on which the normalized frequencies depend is ν ($\nu = 0.3$).

Figures 7 and 8 show the first six mode shapes for the CPT, FSDT and HSDT theories for plates with $b/h = 20$ and $b/h = 5$ (both in plane and bending modes are included).

Plate 1 ($b/h = 20$) is a relatively thin specimen, therefore the different theories yield similar results. In particular, the FSDT and HSDT theories follow exactly the 3D solution, while for the higher frequency modes some differences appear in the CPT results. The CPT model is more rigid because no shear deformations are possible.

The difference between the CPT and 3D theories is even more evident in the case of plate 2 ($b/h = 5$). The specimen is thick, therefore the shear deformations are more significant. On the other hand, the FSDT and HSDT theories show once again a very good agreement with the 3D solution.

5 Concluding remarks

Three different 2D theories for isotropic plates were presented in this work. The results in terms of modal analysis were obtained and compared to a 3D reference solution.

As expected, the natural frequencies obtained with the CPT were higher than the corresponding 3D results and the difference increased with increasing thickness of the plate. Both FSDT and HSDT theories yield similar results and compare well to the 3D reference solution.

The agreement between the FSDT and HSDT theories is related to the choice of the shear correction factor k. In the isotropic homogeneous single layer case, k can be accurately determined, while in complicated cases such as laminates the evaluation of k is not obvious. In that case, the difference between the two theories is more evident (see Daghia et al. (2005)).

Acknowledgements

This topic is one of the subjects of the Centre of Study and Research for the Identification of Materials and Structures (CIMEST) "M. Capurso". The numerical tests presented were carried out using the facilities of the Laboratory of Computational Mechanics (LAMC) "A. A. Cannarozzi" of the University of Bologna.

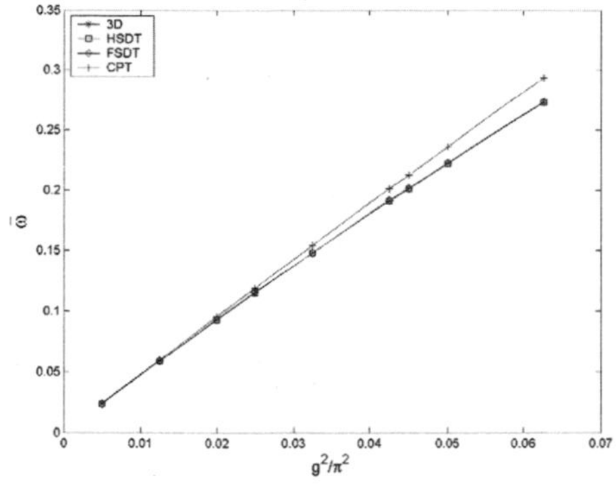

g^2/π^2	3D	HSDT	FSDT	CPT
0.005	0.0238	0.0239	0.0239	0.0240
0.0125	0.0589	0.0589	0.0589	0.0599
0.0125	0.0589	0.0589	0.0589	0.0599
0.02	0.0932	0.0931	0.0931	0.0955
0.025	0.1155	0.1155	0.1157	0.1192
0.025	0.1155	0.1155	0.1157	0.1192
0.0325	0.1484	0.1483	0.1485	0.1545
0.0325	0.1484	0.1483	0.1485	0.1545
0.0425	0.1912	0.1911	0.1918	0.2012
0.0425	0.1912	0.1911	0.1918	0.2012
0.045	0.2017	0.2014	0.2017	0.2128
0.05	0.2226	0.2223	0.2229	0.2360
0.05	0.2226	0.2223	0.2229	0.2360
0.0625	0.2734	0.2729	0.2734	0.2935
0.0625	0.2734	0.2729	0.2734	0.2935

Figure 5. $\bar{\omega}$ for different theories ($b/h = 20$).

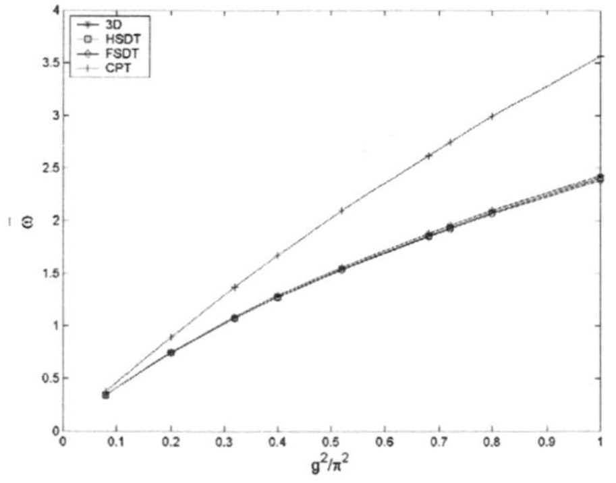

g^2/π^2	3D	HSDT	FSDT	CPT
0.08	0.3420	0.3407	0.3406	0.3732
0.2	0.7511	0.7455	0.7452	0.8926
0.2	0.7511	0.7455	0.7452	0.8926
0.32	1.0888	1.0786	1.0771	1.3712
0.4	1.2881	1.2750	1.2736	1.6710
0.4	1.2881	1.2750	1.2736	1.6710
0.52	1.5589	1.5419	1.5377	2.0959
0.52	1.5589	1.5419	1.5377	2.0959
0.68	1.8808	1.8596	1.8543	2.6226
0.68	1.8808	1.8596	1.8543	2.6226
0.72	1.9557	1.9335	1.9245	2.7480
0.8	2.0999	2.0763	2.0666	2.9921
0.8	2.0999	2.0763	2.0666	2.9921
1.0	2.4332	2.4066	2.3896	3.5674

Figure 6. $\bar{\omega}$ for different theories ($b/h = 5$).

Bibliography

F. Daghia, S. de Miranda, F. Ubertini, and E. Viola. Modal analysis of thick laminated composite plates through higher order theory. Technical Report 160, DISTART, University of Bologna, 2005.

C. Gentilini and E. Viola. On the three dimensional vibration analysis of rectangular plates. In *Problems in structural identification and diagnostics: general aspects and applications*, CISM Courses and Lectures no. 471, pages 149–162. Springer Verlag, Wien-New York, 2003.

J. N. Reddy. A simple higher–order theory for laminated composite plates. *Journal of Applied Mechanics*, 51:745–752, 1984.

J. N. Reddy. *Mechanics of laminated composite plates and shells.* CRC Press, 2nd edition, 2003.

O. C. Zienkiewicz and R. L. Taylor. *The finite element method*, volume 1. Butterworth-Heinemann, 2002.

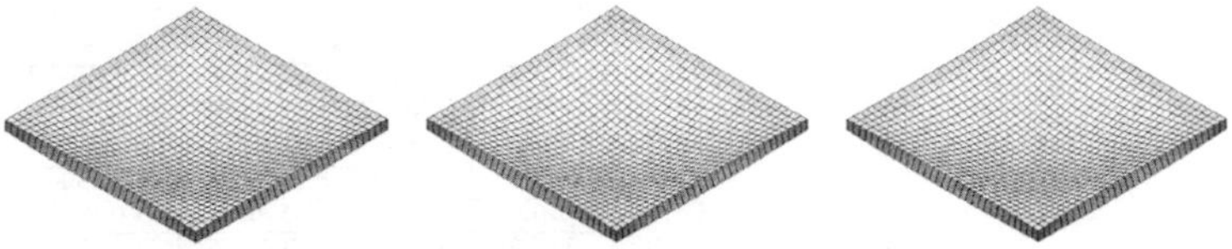

(a) CPT mode 1 (b) FSDT mode 1 (c) HSDT mode 1

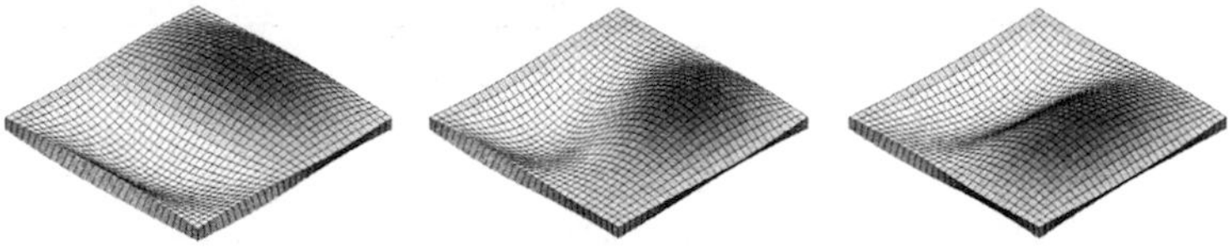

(d) CPT mode 2 (e) FSDT mode 2 (f) HSDT mode 2

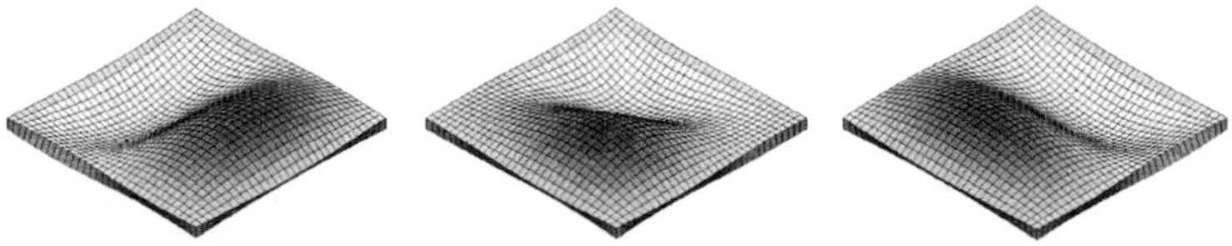

(g) CPT mode 3 (h) FSDT mode 3 (i) HSDT mode 3

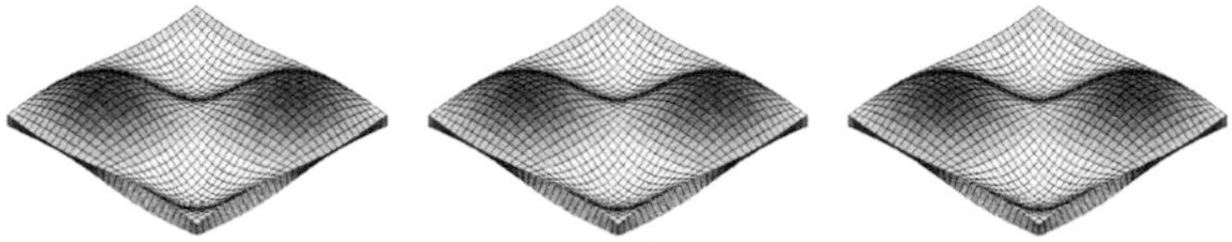

(j) CPT mode 4 (k) FSDT mode 4 (l) HSDT mode 4

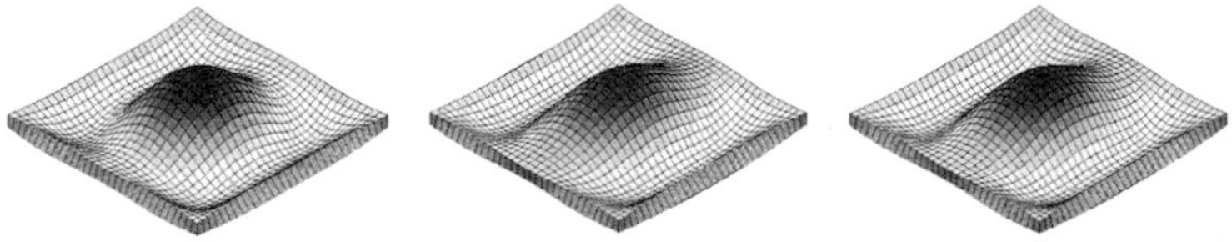

(m) CPT mode 5 (n) FSDT mode 5 (o) HSDT mode 5

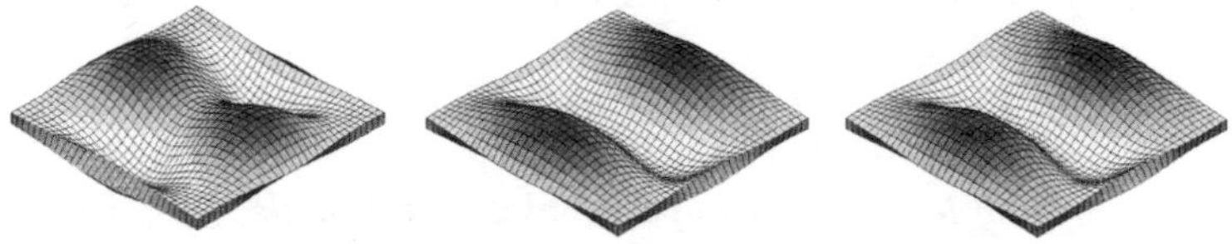

(p) CPT mode 6 (q) FSDT mode 6 (r) HSDT mode 6

Figure 7. Mode shapes ($b/h = 20$).

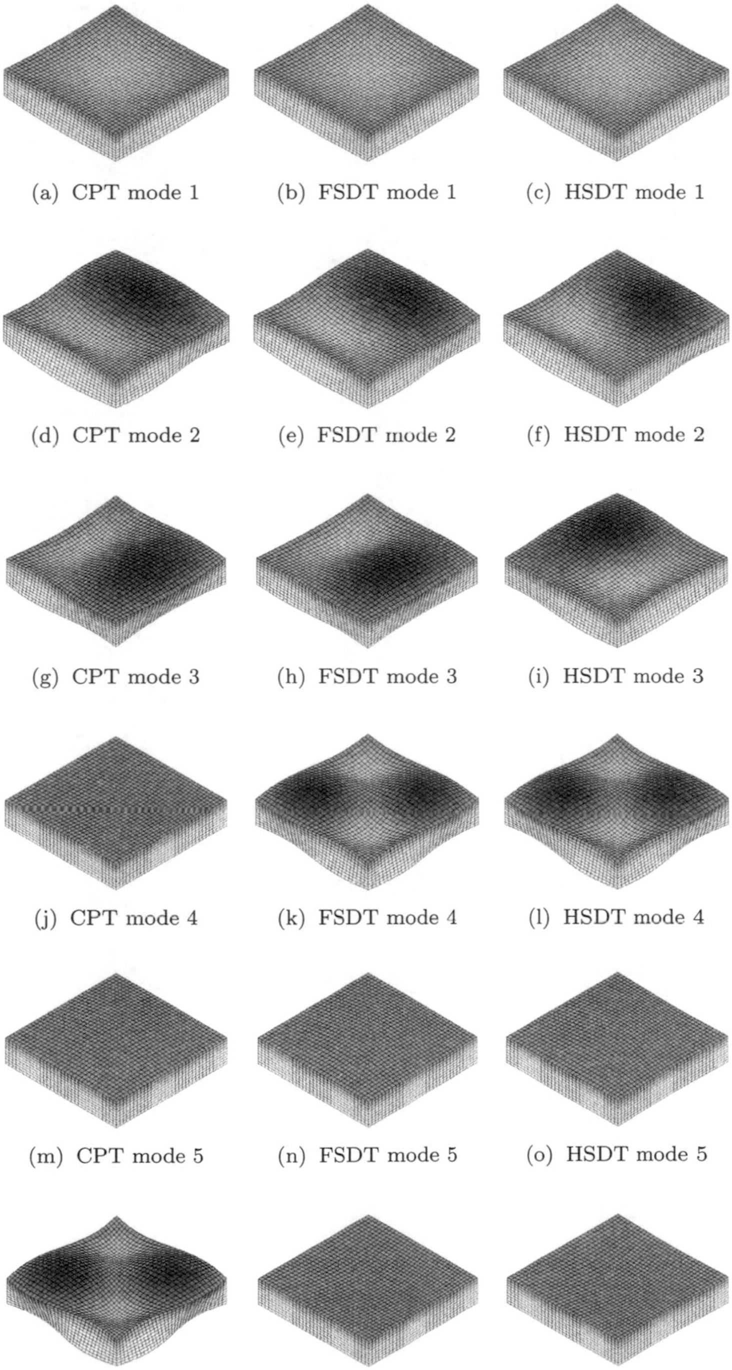

(a) CPT mode 1 (b) FSDT mode 1 (c) HSDT mode 1

(d) CPT mode 2 (e) FSDT mode 2 (f) HSDT mode 2

(g) CPT mode 3 (h) FSDT mode 3 (i) HSDT mode 3

(j) CPT mode 4 (k) FSDT mode 4 (l) HSDT mode 4

(m) CPT mode 5 (n) FSDT mode 5 (o) HSDT mode 5

(p) CPT mode 6 (q) FSDT mode 6 (r) HSDT mode 6

Figure 8. Mode shapes ($b/h = 5$).

Thickness Effect on the Dynamic Behavior of Three-Dimensional Plates by using the Ritz Method

Erasmo Viola and Cristina Gentilini

Dipartimento di Ingegneria delle Strutture, dei Trasporti, delle Acque, del Rilevamento, del Territorio - University of Bologna, Viale Risorgimento 2, 40136 Bologna, Italy

Abstract A continuum three-dimensional Ritz formulation is presented for the free vibration analysis of homogeneous, isotropic, thick rectangular plates. The method is applied to investigate the effects of boundary constraints and thickness ratios on the vibration responses of these plates. The present formulation is based on the linear, three-dimensional, small-strain elasticity theory. The in-plane displacements and deflections are approximated by sets of self generating polynomial shape functions able to satisfy the essential geometric boundary conditions at the outset. Several numerical examples have been computed to demonstrate the accuracy and efficiency of the method. The influence of plate thickness and boundary constraints on the vibratory characteristics of thick plates is also discussed in detail.

1 Introduction

The free vibration analysis of rectangular plates is an important topic in the literature. Two-dimensional theories are widely used to investigate their vibratory characteristics, see Mindlin (1951) and many others. They offer a relatively simple mathematical formulation, reducing the dimensions of the plate problem from three to two and thus the determinant size of the eigenvalue equation. This simplification leads to unreliable results for relatively thick plates. For this reason, much attention has been focused on higher order theories, see Reddy (2003) and the references herein, and three-dimensional analyses (see, for example, Leissa and Zhang (1983), Wittrick (1987) and Liew et al. (1993)).
In particular, three-dimensional vibration analysis, based on linear and small strain elasticity theory, does not rely on any hypotheses involving the kinematics of deformation. Such analysis provides realistic results which cannot otherwise be predicted by the two-dimensional theories and are comparable to results obtained from higher order theories. Herein we employ the Ritz method for the three-dimensional vibration analysis of plates. According to the Ritz method, the spatial displacement components in the three coordinate directions are represented by sets of orthogonal polynomials, obtained as the product of a basic function, chosen to satisfy the essential geometric boundary conditions of the plate, and a mathematically complete one-dimensional polynomial set. These functions are constructed from the Gram-Schmidt recurrence formula, Arfken (1985).
The primary objective of the present work is to investigate the effects of boundary constraints and thickness to width ratios on the vibratory characteristics of isotropic homogeneous rectangular plates. Comparisons are made with the exact theory by Srinivas et al. (1970).

The accurate 3D results presented here may serve as benchmarks against other approximate methods (e.g. finite elements) and 2D plate theories, including first-order and higher order shear deformation theories, may be tested.

2 Description of the problem

2.1 Basic equations

A rectangular plate of length a, width b and uniform thickness h is considered, as represented in Figure 1. The volume of the plate is indicated with $\mathcal{V}$. The plate geometry and dimensions are defined with respect to a Cartesian coordinate system $(O; x, y, z)$, the origin of which is the centre of the plate and the axes are parallel to the edges of the plate. A point of the plate is indicated by the vector $\mathbf{x}$ of its Cartesian coordinates.
In the three-dimensional setting, the generic configuration of the plate is described by the displacement vector $\mathbf{u}(\mathbf{x},t)$, whose independent components are arranged in the following form:

$$\mathbf{u} = \begin{bmatrix} u(\mathbf{x},t) & v(\mathbf{x},t) & w(\mathbf{x},t) \end{bmatrix}^{\mathrm{T}}, \tag{2.1}$$

where t denotes time.

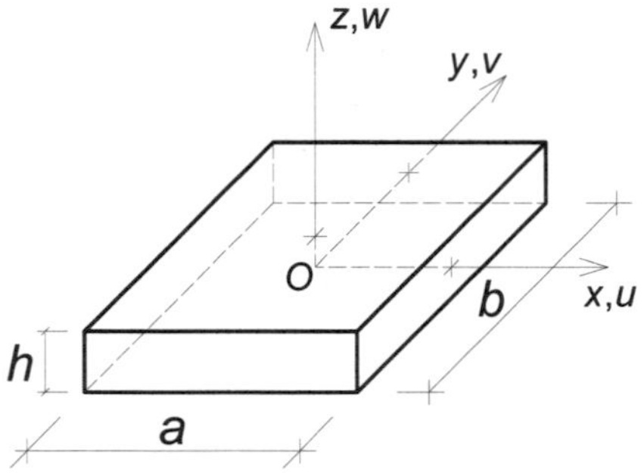

Figure 1. Coordinate system and dimensions of a rectangular plate.

In the hypothesis of small strains, the strain-displacement relations can be written in matrix form as follows:

$$\boldsymbol{\varepsilon}(\mathbf{x},t) = \mathbf{D}\mathbf{u}(\mathbf{x},t) \tag{2.2}$$

where $\boldsymbol{\varepsilon}$ is the strain vector:

$$\boldsymbol{\varepsilon} = \begin{bmatrix} \varepsilon_{xx} & \varepsilon_{yy} & \varepsilon_{zz} & \gamma_{xy} & \gamma_{xz} & \gamma_{yz} \end{bmatrix}^{\mathrm{T}} \tag{2.3}$$

and $\mathbf{D}$ is the compatibility operator defined as:

$$\mathbf{D} = \begin{bmatrix} \partial/\partial x & 0 & 0 \\ 0 & \partial/\partial y & 0 \\ 0 & 0 & \partial/\partial z \\ \partial/\partial y & \partial/\partial x & 0 \\ \partial/\partial z & 0 & \partial/\partial x \\ 0 & \partial/\partial z & \partial/\partial y \end{bmatrix}.$$

Thus, the strain-displacement relations, eq. (2.2), in components are the well-known equations:

$$\varepsilon_{xx} = \frac{\partial u}{\partial x}, \tag{2.4}$$

$$\varepsilon_{yy} = \frac{\partial v}{\partial y}, \tag{2.5}$$

$$\varepsilon_{zz} = \frac{\partial w}{\partial z}, \tag{2.6}$$

$$\gamma_{xy} = \frac{\partial u}{\partial y} + \frac{\partial v}{\partial x}, \tag{2.7}$$

$$\gamma_{xz} = \frac{\partial u}{\partial z} + \frac{\partial w}{\partial x}, \tag{2.8}$$

$$\gamma_{yz} = \frac{\partial v}{\partial z} + \frac{\partial w}{\partial y}. \tag{2.9}$$

Moreover, for a linear elastic body the constitutive relations are given by:

$$\sigma_{xx} = \Gamma(\varepsilon_{xx} + \varepsilon_{yy} + \varepsilon_{zz}) + 2\mu\varepsilon_{xx} \tag{2.10}$$

$$\sigma_{yy} = \Gamma(\varepsilon_{xx} + \varepsilon_{yy} + \varepsilon_{zz}) + 2\mu\varepsilon_{yy}, \tag{2.11}$$

$$\sigma_{zz} = \Gamma(\varepsilon_{xx} + \varepsilon_{yy} + \varepsilon_{zz}) + 2\mu\varepsilon_{zz}, \tag{2.12}$$

$$\tau_{xy} = \mu\gamma_{xy}, \tag{2.13}$$

$$\tau_{xz} = \mu\gamma_{xz}, \tag{2.14}$$

$$\tau_{yz} = \mu\gamma_{yz}, \tag{2.15}$$

where Γ and μ are the Lamè coefficients:

$$\Gamma = \frac{\nu E}{(1+\nu)(1-2\nu)}, \qquad \mu = \frac{E}{2(1+\nu)}. \tag{2.16}$$

with E and ν are Young's modulus and Poisson's ratio, respectively. Equations (2.10)-(2.15) can be written in matrix form:

$$\boldsymbol{\sigma} = \mathbf{H}\boldsymbol{\varepsilon} \tag{2.17}$$

where $\boldsymbol{\sigma}$ is the vector of the stress components:

$$\boldsymbol{\sigma} = \begin{bmatrix} \sigma_{xx} & \sigma_{yy} & \sigma_{zz} & \tau_{xy} & \tau_{xz} & \tau_{yz} \end{bmatrix}^{\mathrm{T}}$$

and $\mathbf{H}$ is the matrix of elastic stiffness moduli:

$$\mathbf{H} = \frac{E(1-\nu)}{(1+\nu)(1-2\nu)} \begin{bmatrix} 1 & \frac{\nu}{1-\nu} & \frac{\nu}{1-\nu} & 0 & 0 & 0 \\ \frac{\nu}{1-\nu} & 1 & \frac{\nu}{1-\nu} & 0 & 0 & 0 \\ \frac{\nu}{1-\nu} & \frac{\nu}{1-\nu} & 1 & 0 & 0 & 0 \\ 0 & 0 & 0 & \frac{1-2\nu}{2(1-\nu)} & 0 & 0 \\ 0 & 0 & 0 & 0 & \frac{1-2\nu}{2(1-\nu)} & 0 \\ 0 & 0 & 0 & 0 & 0 & \frac{1-2\nu}{2(1-\nu)} \end{bmatrix}. \tag{2.18}$$

In the following, the energy terms needed for the free vibration analysis of isotropic elastic plates are introduced.

2.2 Strain energy

The strain energy of a three-dimensional solid in matrix form is given by:

$$\Phi(\mathbf{x},t)=\frac{1}{2}\int_{\mathcal{V}}\boldsymbol{\varepsilon}^{\mathrm{T}}\boldsymbol{\sigma}\mathrm{d}\mathcal{V}. \tag{2.19}$$

Expressing the stress vector $\boldsymbol{\sigma}$ in terms of the strain vector $\boldsymbol{\varepsilon}$ by means of Hooke's law, eq. (2.17), and the strain vector in terms of the displacement vector $\mathbf{u}$ by means of the compatibility equation, eq. (2.2), the strain energy takes the following form:

$$\Phi(\mathbf{x},t)=\frac{1}{2}\int_{\mathcal{V}}(\mathbf{D}\mathbf{u})^{\mathrm{T}}\mathbf{H}(\mathbf{D}\mathbf{u})\,\mathrm{d}\mathcal{V}.$$

Thus, Φ in strain components is:

$$\begin{aligned}\Phi(\mathbf{x},t) = \frac{1}{2}\int_{\mathcal{V}}\Big\{&\frac{E(1-\nu)}{(1+\nu)(1-2\nu)}\left(\varepsilon_{xx}^2+\varepsilon_{yy}^2+\varepsilon_{zz}^2\right)\\ &+2\frac{E\nu}{(1+\nu)(1-2\nu)}\left(\varepsilon_{xx}\varepsilon_{yy}+\varepsilon_{xx}\varepsilon_{zz}+\varepsilon_{yy}\varepsilon_{zz}\right)\\ &+\frac{E}{2(1+\nu)}\left(\gamma_{xy}^2+\gamma_{xz}^2+\gamma_{yz}^2\right)\Big\}\mathrm{d}\mathcal{V},\end{aligned} \tag{2.20}$$

and in terms of displacement components becomes:

$$\begin{aligned}\Phi(\mathbf{x},t) = \tfrac{1}{2}\int_{\mathcal{V}}\Big\{&\tfrac{E(1-\nu)}{(1+\nu)(1-2\nu)}\left[\left(\tfrac{\partial u}{\partial x}\right)^2+\left(\tfrac{\partial v}{\partial y}\right)^2+\left(\tfrac{\partial w}{\partial z}\right)^2\right]\\ &+2\tfrac{E\nu}{(1+\nu)(1-2\nu)}\left[\tfrac{\partial u}{\partial x}\tfrac{\partial v}{\partial y}+\tfrac{\partial u}{\partial x}\tfrac{\partial w}{\partial z}+\tfrac{\partial v}{\partial y}\tfrac{\partial w}{\partial z}\right]\\ &+\tfrac{E}{2(1+\nu)}\Big[\left(\tfrac{\partial u}{\partial y}\right)^2+\left(\tfrac{\partial u}{\partial z}\right)^2+\left(\tfrac{\partial v}{\partial x}\right)^2+\left(\tfrac{\partial v}{\partial z}\right)^2\\ &+\left(\tfrac{\partial w}{\partial x}\right)^2+\left(\tfrac{\partial w}{\partial y}\right)^2+2\tfrac{\partial u}{\partial y}\tfrac{\partial v}{\partial x}+2\tfrac{\partial u}{\partial z}\tfrac{\partial w}{\partial x}+2\tfrac{\partial v}{\partial z}\tfrac{\partial w}{\partial y}\Big]\Big\}\mathrm{d}\mathcal{V}.\end{aligned} \tag{2.21}$$

2.3 Kinetic energy

The kinetic energy of a three-dimensional solid during the vibratory motion in matrix form is given by:

$$\mathcal{T}(\mathbf{x},t)=\frac{\rho}{2}\int_{\mathcal{V}}\dot{\mathbf{u}}^{\mathrm{T}}\dot{\mathbf{u}}\mathrm{d}\mathcal{V} \tag{2.22}$$

where the dot denotes differentiation with respect to time t and ρ is the mass density. In terms of displacement components:

$$\mathcal{T}(\mathbf{x},t)=\frac{\rho}{2}\int_{\mathcal{V}}\left(\dot{u}^2+\dot{v}^2+\dot{w}^2\right)\mathrm{d}\mathcal{V}. \tag{2.23}$$

3 The Ritz method

3.1 Displacement representation

To solve the free vibration problem of the plate, the Ritz method is employed.

For simplicity, the following non-dimensional coordinate system $(O;\xi,\eta,\zeta)$ is introduced:

$$\xi = \frac{x}{a}, \qquad \eta = \frac{y}{b}, \qquad \zeta = \frac{z}{h}. \tag{3.1}$$

The displacement components are written separating the variables in time and space in the non-dimensional coordinate system $(O;\xi,\eta,\zeta)$:

$$\begin{aligned} u(\xi,\eta,\zeta,t) &= U(\xi,\eta,\zeta)e^{i\omega t}, \\ v(\xi,\eta,\zeta,t) &= V(\xi,\eta,\zeta)e^{i\omega t}, \\ w(\xi,\eta,\zeta,t) &= W(\xi,\eta,\zeta)e^{i\omega t}, \end{aligned} \tag{3.2}$$

where ω is the natural frequency and $U(\xi,\eta,\zeta)$, $V(\xi,\eta,\zeta)$ and $W(\xi,\eta,\zeta)$ are the amplitude displacement components expressed in terms of a triple series of orthogonal polynomial functions as follows:

$$U(\xi,\eta,\zeta) = \sum_i^I \sum_j^J \sum_k^K C^u_{ijk}\phi_{ui}(\xi)\psi_{uj}(\eta)\chi_{uk}(\zeta), \tag{3.3}$$

$$V(\xi,\eta,\zeta) = \sum_l^L \sum_m^M \sum_n^N C^v_{lmn}\phi_{vl}(\xi)\psi_{vm}(\eta)\chi_{vn}(\zeta), \tag{3.4}$$

$$W(\xi,\eta,\zeta) = \sum_p^P \sum_q^Q \sum_r^R C^w_{pqr}\phi_{wp}(\xi)\psi_{wq}(\eta)\chi_{wr}(\zeta). \tag{3.5}$$

In eqs. (3.3)-(3.5), the admissible functions $\phi(\xi)$, $\psi(\eta)$ and $\chi(\zeta)$ are polynomial functions so as they satisfy geometric boundary conditions implicitly and C^u_{ijk}, C^v_{lmn}, C^w_{pqr} are unknown coefficients to be determined.
In matrix form, eq. (3.2) can be written as:

$$\mathbf{u}(\boldsymbol{\xi},t) = \mathbf{U}(\boldsymbol{\xi})e^{i\omega t}, \tag{3.6}$$

where $\mathbf{u}(\boldsymbol{\xi},t)$ is the displacement vector and $\mathbf{U}(\boldsymbol{\xi})$ is the displacement amplitude vector. Vector $\mathbf{U}(\boldsymbol{\xi})$ is written in the following form:

$$\mathbf{U}(\boldsymbol{\xi}) = \mathbf{P}(\boldsymbol{\xi})\,\mathbf{C}. \tag{3.7}$$

According to the above displacement representations, eqs. (3.3)-(3.5), matrix $\mathbf{P}(\boldsymbol{\xi})$ in eq. (3.7) of the admissible functions takes the form:

$$\mathbf{P}(\boldsymbol{\xi}) = \begin{bmatrix} P^u_{111}(\boldsymbol{\xi}) & \cdots & P^u_{IJK}(\boldsymbol{\xi}) & 0 & \cdots & 0 & 0 & \cdots & 0 \\ 0 & \cdots & 0 & P^v_{111}(\boldsymbol{\xi}) & \cdots & P^v_{LMN}(\boldsymbol{\xi}) & 0 & \cdots & 0 \\ 0 & \cdots & 0 & 0 & \cdots & 0 & P^w_{111}(\boldsymbol{\xi}) & \cdots & P^w_{PQR}(\boldsymbol{\xi}) \end{bmatrix}, \tag{3.8}$$

where

$$\begin{aligned} P^u_{ijk}(\boldsymbol{\xi}) &= \phi_{ui}(\xi)\psi_{uj}(\eta)\chi_{uk}(\zeta), \\ P^v_{lmn}(\boldsymbol{\xi}) &= \phi_{vl}(\xi)\psi_{vm}(\eta)\chi_{vn}(\zeta), \\ P^w_{pqr}(\boldsymbol{\xi}) &= \phi_{wp}(\xi)\psi_{wq}(\eta)\chi_{wr}(\zeta) \end{aligned}$$

and $\mathbf{C}$ in eq. (3.7) is the following column vector of unknown coefficients:

$$\mathbf{C} = \begin{bmatrix} \mathbf{C}^u \\ \mathbf{C}^v \\ \mathbf{C}^w \end{bmatrix} = \begin{bmatrix} C^u_{111} & .. & C^u_{IJK} & C^v_{111} & .. & C^v_{LMN} & C^w_{111} & .. & C^w_{PQR} \end{bmatrix}^{\mathrm{T}}. \quad (3.9)$$

It is possible to express the partial derivatives of the displacement components with respect to the non-dimensional coordinate system. As example, let us consider the partial derivative of u with respect to x, that is ε_{xx}:

$$\varepsilon_{xx} = \frac{\partial u\,(\mathbf{x},t)}{\partial x} = \frac{\partial u(\boldsymbol{\xi},t)}{\partial \xi}\frac{\partial \xi}{\partial x} = \frac{1}{a}\frac{\partial U(\xi)}{\partial \xi}e^{i\omega t},$$

making use of eqs. (3.1) and (3.2). Moreover, it is possible to obtain the partial derivatives of the amplitude functions in terms of the admissible functions:

$$\frac{\partial U(\boldsymbol{\xi})}{\partial \xi} = \sum_i^I \sum_j^J \sum_k^K C^u_{ijk}\frac{d\phi_{ui}}{d\xi}\psi_{uj}\chi_{uk}.$$

The other strain components are determined by following the same procedure. For the derivative with respect to time, let us consider:

$$\dot{u} = \frac{\partial u(\boldsymbol{\xi},t)}{\partial t} = U(\boldsymbol{\xi})i\omega e^{i\omega t},$$

making use of eq. (3.2). The other velocity components are found by following the same procedure.

Thus, in the non-dimensional coordinate system, the strain energy Φ, eq. (2.21), becomes:

$$\begin{aligned}
\Phi(\boldsymbol{\xi},t) = &\tfrac{1}{2}\left(e^{i\omega t}\right)^2 abh \int\int\int_{-0.5}^{0.5}\left\{\frac{E(1-\nu)}{(1+\nu)(1-2\nu)}\left[\frac{1}{a^2}\left(\frac{\partial U}{\partial \xi}\right)^2 + \frac{1}{b^2}\left(\frac{\partial V}{\partial \eta}\right)^2 + \frac{1}{h^2}\left(\frac{\partial W}{\partial \zeta}\right)^2\right]\right. \\
&+2\frac{E\nu}{(1+\nu)(1-2\nu)}\left[\frac{1}{ab}\frac{\partial U}{\partial \xi}\frac{\partial V}{\partial \eta} + \frac{1}{ah}\frac{\partial U}{\partial \xi}\frac{\partial W}{\partial \zeta} + \frac{1}{bh}\frac{\partial V}{\partial \eta}\frac{\partial W}{\partial \zeta}\right] \\
&+\frac{E}{2(1+\nu)}\left[\frac{1}{b^2}\left(\frac{\partial U}{\partial \eta}\right)^2 + \frac{1}{h^2}\left(\frac{\partial U}{\partial \zeta}\right)^2 + \frac{1}{a^2}\left(\frac{\partial V}{\partial \xi}\right)^2 + \frac{1}{h^2}\left(\frac{\partial V}{\partial \zeta}\right)^2 + \frac{1}{a^2}\left(\frac{\partial W}{\partial \xi}\right)^2 + \frac{1}{b^2}\left(\frac{\partial W}{\partial \eta}\right)^2\right. \\
&\left.\left. + \frac{2}{ab}\frac{\partial U}{\partial \eta}\frac{\partial V}{\partial \xi} + \frac{2}{ah}\frac{\partial U}{\partial \zeta}\frac{\partial W}{\partial \xi} + \frac{2}{bh}\frac{\partial V}{\partial \zeta}\frac{\partial W}{\partial \eta}\right]\right\} d\xi d\eta d\zeta.
\end{aligned} \quad (3.10)$$

Moreover, the kinetic energy in the non-dimensional coordinate system becomes:

$$\mathcal{T}(\boldsymbol{\xi},t) = -\frac{\rho\omega^2\left(e^{i\omega t}\right)^2}{2}abh\int\int\int_{-0.5}^{0.5}(U^2 + V^2 + W^2)d\xi d\eta d\zeta. \quad (3.11)$$

In free vibration analysis, the maximum energy functional Π_{max} is defined as:

$$\Pi_{\mathrm{max}} = \Phi_{\mathrm{max}} - \mathcal{T}_{\mathrm{max}} \quad (3.12)$$

where Φ_{max} and $\mathcal{T}_{\mathrm{max}}$ are the maximum values of the strain energy and of the kinetic energy, obtained from eqs. (3.10) and (3.11) setting $\left(e^{i\omega t}\right)^2 = 1$, respectively. Now, inserting the expressions of the displacement amplitude components, eqs. (3.3)-(3.5), into the functional, eq. (3.12), the following expression is obtained:

$$
\begin{aligned}
&\Pi_{\max} = abh \int\int\int_{-0.5}^{0.5} \left\{ \frac{E(1-\nu)}{2(1+\nu)(1-2\nu)} \left[\frac{1}{a^2} \left(\sum^I \sum^J \sum^K C^u_{ijk} \frac{\mathrm{d}\phi_{ui}}{\mathrm{d}\xi} \psi_{uj} \chi_{uk} \right)^2 \right.\right. \\
&+ \frac{1}{b^2} \left(\sum^L \sum^M \sum^N C^v_{lmn} \phi_{vl} \frac{\mathrm{d}\psi_{vm}}{\mathrm{d}\eta} \chi_{vn} \right)^2 + \frac{1}{h^2} \left(\sum^P \sum^Q \sum^R C^w_{pqr} \phi_{wp} \psi_{wq} \frac{\mathrm{d}\chi_{wr}}{\mathrm{d}\zeta} \right)^2 \Bigg] \\
&\frac{E\nu}{(1+\nu)(1-2\nu)} \left[\frac{1}{ab} \left(\sum^I \sum^J \sum^K C^u_{ijk} \frac{\mathrm{d}\phi_{ui}}{\mathrm{d}\xi} \psi_{uj} \chi_{uk} \right) \left(\sum^L \sum^M \sum^N C^v_{lmn} \phi_{vl} \frac{\mathrm{d}\psi_{vm}}{\mathrm{d}\eta} \chi_{vn} \right) \right. \\
&+ \frac{1}{ah} \left(\sum^I \sum^J \sum^K C^u_{ijk} \frac{\mathrm{d}\phi_{ui}}{\mathrm{d}\xi} \psi_{uj} \chi_{uk} \right) \left(\sum^P \sum^Q \sum^R C^w_{pqr} \phi_{wp} \psi_{wq} \frac{\mathrm{d}\chi_{wr}}{\mathrm{d}\zeta} \right) \\
&+ \frac{1}{bh} \left(\sum^L \sum^M \sum^N C^v_{lmn} \phi_{vl} \frac{\mathrm{d}\psi_{vm}}{\mathrm{d}\eta} \chi_{vn} \right) \left(\sum^P \sum^Q \sum^R C^w_{pqr} \phi_{wp} \psi_{wq} \frac{\mathrm{d}\chi_{wr}}{\mathrm{d}\zeta} \right) \Bigg] \\
&+ \frac{E}{4(1+\nu)} \left\{ \frac{1}{a^2} \left(\sum^L \sum^M \sum^N C^v_{lmn} \frac{\mathrm{d}\phi_{vl}}{\mathrm{d}\xi} \psi_{vm} \chi_{vn} \right)^2 + \frac{1}{a^2} \left(\sum^P \sum^Q \sum^R C^w_{pqr} \frac{\mathrm{d}\phi_{wp}}{\mathrm{d}\xi} \psi_{wq} \chi_{wr} \right)^2 \right. \\
&+ \frac{1}{b^2} \left(\sum^I \sum^J \sum^K C^u_{ijk} \phi_{ui} \frac{\mathrm{d}\psi_{uj}}{\mathrm{d}\eta} \chi_{uk} \right)^2 + \frac{1}{b^2} \left(\sum^P \sum^Q \sum^R C^w_{pqr} \phi_{wp} \frac{\mathrm{d}\psi_{wq}}{\mathrm{d}\eta} \chi_{wr} \right)^2 \\
&+ \frac{1}{h^2} \left(\sum^I \sum^J \sum^K C^u_{ijk} \phi_{ui} \psi_{uj} \frac{\mathrm{d}\chi_{uk}}{\mathrm{d}\zeta} \right)^2 + \frac{1}{h^2} \left(\sum^L \sum^M \sum^N C^v_{lmn} \phi_{vl} \psi_{vm} \frac{\mathrm{d}\chi_{vn}}{\mathrm{d}\zeta} \right)^2 \\
&+ \frac{2}{ab} \left(\sum^I \sum^J \sum^K C^u_{ijk} \phi_{ui} \frac{\mathrm{d}\psi_{uj}}{\mathrm{d}\eta} \chi_{uk} \right) \left(\sum^L \sum^M \sum^N C^v_{lmn} \frac{\mathrm{d}\phi_{vl}}{\mathrm{d}\xi} \psi_{vm} \chi_{vn} \right) \\
&+ \frac{2}{ah} \left(\sum^I \sum^J \sum^K C^u_{ijk} \phi_{ui} \psi_{uj} \frac{\mathrm{d}\chi_{uk}}{\mathrm{d}\zeta} \right) \left(\sum^P \sum^Q \sum^R C^w_{pqr} \frac{\mathrm{d}\phi_{wp}}{\mathrm{d}\xi} \psi_{wq} \chi_{wr} \right) \\
&+ \frac{2}{bh} \left(\sum^L \sum^M \sum^N C^v_{lmn} \phi_{vl} \psi_{vm} \frac{\mathrm{d}\chi_{vn}}{\mathrm{d}\zeta} \right) \left(\sum^P \sum^Q \sum^R C^w_{pqr} \phi_{wp} \frac{\mathrm{d}\psi_{wq}}{\mathrm{d}\eta} \chi_{wr} \right) \Bigg] \\
&- \frac{\rho\omega^2}{2} \left[\left(\sum^I \sum^J \sum^K C^u_{ijk} \phi_{ui} \psi_{uj} \chi_{uk} \right)^2 + \left(\sum^L \sum^M \sum^N C^v_{lmn} \phi_{vl} \psi_{vm} \chi_{vn} \right)^2 \right. \\
&\left.\left. + \left(\sum^P \sum^Q \sum^R C^w_{pqr} \phi_{wp} \psi_{wq} \chi_{wr} \right)^2 \right] \right\} \mathrm{d}\xi \mathrm{d}\eta \mathrm{d}\zeta.
\end{aligned}
\tag{3.13}
$$

The Ritz method requires $\Pi_{\max}$ in eq. (3.13) to attain minimum with respect to the unknown coefficients C^u_{ijk}, C^v_{lmn} and C^w_{pqr}, so that:

$$
\left\{
\begin{aligned}
\frac{\partial \Pi_{\max}}{\partial C^u_{ijk}} &= 0 \\
\frac{\partial \Pi_{\max}}{\partial C^v_{lmn}} &= 0 \\
\frac{\partial \Pi_{\max}}{\partial C^w_{pqr}} &= 0
\end{aligned}
\right.
$$

which leads to the following governing equation:

$$
\left(\mathbf{K} - \overline{\lambda}^2 \mathbf{M} \right) \mathbf{C} = \mathbf{0} \tag{3.14}
$$

where $\overline{\lambda}$ is a frequency parameter, $\mathbf{C}$ is the vector given in eq. (3.9) and

$$\mathbf{K} = \begin{bmatrix} \mathbf{K}_{uu} & \mathbf{K}_{uv} & \mathbf{K}_{uw} \\ \mathbf{K}_{uv}^{\mathrm{T}} & \mathbf{K}_{vv} & \mathbf{K}_{vw} \\ \mathbf{K}_{uw}^{\mathrm{T}} & \mathbf{K}_{vw}^{\mathrm{T}} & \mathbf{K}_{ww} \end{bmatrix}, \tag{3.15}$$

$$\mathbf{M} = \begin{bmatrix} \mathbf{M}_{uu} & \mathbf{0} & \mathbf{0} \\ \mathbf{0} & \mathbf{M}_{vv} & \mathbf{0} \\ \mathbf{0} & \mathbf{0} & \mathbf{M}_{ww} \end{bmatrix}. \tag{3.16}$$

In Appendix A, some simple examples of how the derivatives of the functional $\Pi_{\max}$ with respect to the unknown coefficients have been derived are reported.

In eqs. (3.14)-(3.16), $\mathbf{K}$ and $\mathbf{M}$ are the stiffness and mass matrices, respectively. The explicit form of the elements in the stiffness sub-matrices $\mathbf{K}_{ij}$ with $i, j = u, v, w$ in eq. (3.15), is given by:

$$\begin{aligned}
\mathbf{K}_{uu}|_{ijk\overline{ijk}} &= \frac{(1-\nu)}{(1-2\nu)} E^{11}_{ui u\bar{i}} F^{00}_{uju\bar{j}} G^{00}_{uku\bar{k}} + \frac{1}{2}(\frac{a}{b})^2 E^{00}_{uiu\bar{i}} F^{11}_{uju\bar{j}} G^{00}_{uku\bar{k}} \\
&+ \frac{1}{2}(\frac{a}{h})^2 E^{00}_{uiu\bar{i}} F^{00}_{uju\bar{j}} G^{11}_{uku\bar{k}}, \\
\mathbf{K}_{uv}|_{ijklmn} &= (\frac{a}{b}) \left(\frac{\nu}{(1-2\nu)} E^{10}_{uivl} F^{01}_{ujvm} G^{00}_{ukvn} + \frac{1}{2} E^{01}_{uivl} F^{10}_{ujvm} G^{00}_{ukvn} \right), \\
\mathbf{K}_{uw}|_{ijkpqr} &= (\frac{a}{h}) \left(\frac{\nu}{(1-2\nu)} E^{10}_{uiwp} F^{00}_{ujwq} G^{01}_{ukwr} + \frac{1}{2} E^{01}_{uiwp} F^{00}_{ujwq} G^{10}_{ukwr} \right), \\
\mathbf{K}_{vv}|_{lmn\overline{lmn}} &= (\frac{a}{b})^2 \frac{(1-\nu)}{(1-2\nu)} E^{00}_{vlv\bar{l}} F^{11}_{vmv\bar{m}} G^{00}_{vnv\bar{n}} + \frac{1}{2}(\frac{a}{h})^2 E^{00}_{vlv\bar{l}} F^{00}_{vmv\bar{m}} G^{11}_{vnv\bar{n}} \\
&+ \frac{1}{2} E^{11}_{vlv\bar{l}} F^{00}_{vmv\bar{m}} G^{00}_{vnv\bar{n}}, \\
\mathbf{K}_{vw}|_{lmnpqr} &= (\frac{a^2}{bh}) \left(\frac{\nu}{(1-2\nu)} E^{00}_{vlwp} F^{10}_{vmwq} G^{01}_{vnwr} + \frac{1}{2} E^{00}_{vlwp} F^{01}_{wmwq} G^{10}_{vnwr} \right), \\
\mathbf{K}_{ww}|_{pqr\overline{pqr}} &= (\frac{a}{h})^2 \frac{(1-\nu)}{(1-2\nu)} E^{00}_{wpw\bar{p}} F^{00}_{wqw\bar{q}} G^{11}_{wrw\bar{r}} + \frac{1}{2}(\frac{a}{b})^2 E^{00}_{wpw\bar{p}} F^{11}_{wqw\bar{q}} G^{00}_{wrw\bar{r}} \\
&+ \frac{1}{2} E^{11}_{wpw\bar{p}} F^{00}_{wqw\bar{q}} G^{00}_{wrw\bar{r}}.
\end{aligned} \tag{3.17}$$

The elements in the mass sub-matrices $\mathbf{M}_{ij}$, with $i, j = u, v, w$ in eq. (3.16), are given by:

$$\begin{aligned}
\mathbf{M}_{uu}|_{ijk\overline{ijk}} &= (1+\nu) E^{00}_{uiu\bar{i}} F^{00}_{uju\bar{j}} G^{00}_{uku\bar{k}}, \\
\mathbf{M}_{vv}|_{lmn\overline{lmn}} &= (1+\nu) E^{00}_{vlv\bar{l}} F^{00}_{vmv\bar{m}} G^{00}_{vnv\bar{n}}, \\
\mathbf{M}_{ww}|_{pqr\overline{pqr}} &= (1+\nu) E^{00}_{wpw\bar{p}} F^{00}_{wqw\bar{q}} G^{00}_{wrw\bar{r}},
\end{aligned}$$

where

$$E^{rs}_{\alpha l \beta i} = \int_{-0.5}^{0.5} \left(\frac{\mathrm{d}^r \phi_{\alpha l}(\xi)}{\mathrm{d}\xi^r} \frac{\mathrm{d}^s \phi_{\beta i}(\xi)}{\mathrm{d}\xi^s} \right) \mathrm{d}\xi, \tag{3.18}$$

$$F^{rs}_{\alpha m \beta j} = \int_{-0.5}^{0.5} \left(\frac{\mathrm{d}^r \psi_{\alpha m}(\eta)}{\mathrm{d}\eta^r} \frac{\mathrm{d}^s \psi_{\beta j}(\eta)}{\mathrm{d}\eta^s} \right) \mathrm{d}\eta, \tag{3.19}$$

$$G^{rs}_{\alpha n \beta k} = \int_{-0.5}^{0.5} \left(\frac{\mathrm{d}^r \chi_{\alpha n}(\zeta)}{\mathrm{d}\zeta^r} \frac{\mathrm{d}^s \chi_{\beta k}(\zeta)}{\mathrm{d}\zeta^s} \right) \mathrm{d}\zeta \tag{3.20}$$

with $r, s = 0, 1$ and the subscripts α and β indicate the corresponding displacement components u, v and w.
A non-trivial solution of eq. (3.14) is obtained by setting the determinant of the coefficient matrix equal to zero. Roots of the determinant are the square of the frequency parameter $\overline{\lambda}$:

$$\overline{\lambda} = \omega a \frac{\rho}{E}. \tag{3.21}$$

All the calculations that lead to the governing equation (3.14) for plates that have material properties functionally graded in the thickness direction are reported in Gentilini (2005).

3.2 Admissible displacement functions

In the expression of the displacement components, eqs. (3.3)-(3.5), sets of polynomial functions have been employed. The one-dimensional polynomial displacement function is obtained as the product of a basic function chosen to satisfy the essential geometric boundary conditions of the plate and a mathematically complete one-dimensional polynomial set.
The generic function $\phi_{k+1}(\xi)$ is constructed from the Gram-Schmidt recurrence formula, Arfken (1985):

$$\phi_{k+1}(\xi) = (g(\xi) - \theta_k)\, \phi_k(\xi) - \Xi_k \phi_{k-1}(\xi) \qquad k = 1, 2, 3...$$

where

$$\theta_k = \frac{\int_{-0.5}^{0.5} g(\xi)\phi_k^2(\xi)\mathrm{d}\xi}{\int_{-0.5}^{0.5} \phi_k^2(\xi)\mathrm{d}\xi}, \quad \Xi_k = \frac{\int_{-0.5}^{0.5} \phi_k^2(\xi)\mathrm{d}\xi}{\int_{-0.5}^{0.5} \phi_{k-1}^2(\xi)\mathrm{d}\xi}.$$

The polynomial $\phi_0(\xi)$ is defined as zero and the generating function $g(\xi)$ is chosen to ensure that the higher order polynomial terms satisfy the geometric boundary constraints:

$$g(\xi) = \xi^2.$$

In addition, this set of polynomials satisfies the orthogonality condition:

$$\int_{-0.5}^{0.5} \phi_i(\xi)\phi_j(\xi)\mathrm{d}\xi = n_{ij}\delta_{ij},$$

where δ_{ij} is the Kronecker delta and the value of n_{ij} depends on the normalization used. For example, if we consider as basic function $\phi_1(\xi) = 1$, the first five $\phi_i(\xi)$ polynomials are given below:

$$
\begin{aligned}
\phi_0(\xi) &= 0 \\
\phi_1(\xi) &= 1 \\
\phi_2(\xi) &= \xi^2 - \frac{1}{12} \\
\phi_3(\xi) &= \xi^4 - \frac{3}{14}\xi^2 + 0.00556 \\
\phi_4(\xi) &= \xi^6 - \frac{15}{44}\xi^4 + \frac{5}{176}\xi^2 - 0.00034.
\end{aligned}
$$

Polynomials $\psi_k(\eta)$ and $\chi_k(\zeta)$ in the η and ζ–directions, respectively, can be constructed by following the same procedure. The generating function $g(\eta)$ is η^2, while $g(\zeta)$ is the linear function ζ. The basic function $\chi_1(\zeta)$ is always chosen constant and equal to 1, since it satisfies the essential geometric boundary requirements of the stress free surfaces

$$\sigma_z = \tau_{xz} = \tau_{yz} = 0 \quad \text{at } \zeta = \pm 0.5. \tag{3.22}$$

The basic functions in the ξ and η–directions depend on the boundary conditions and symmetry classes of vibration. In the following, plates with different boundary conditions are considered and the expressions of the displacement functions are illustrated.

3.3 Boundary conditions: all-round simply-supported plate (SSSS plate)

As presented in the previous section, in the Ritz formulation it is sufficient to define admissible functions that satisfy the essential geometric boundary constraints.
In this analysis, the boundary conditions are chosen such that the $x-z$ and $y-z$ planes are the symmetry planes. This symmetry consideration permits the classification of the vibration modes into four distinct symmetry classes: double symmetry modes (SS), symmetry-antisymmetry modes (SA), antisymmetry-symmetry modes (AS) and double antisymmetry modes (AA), about the $x-z$ and $y-z$ planes. The computational advantage of this classification is that it reduces the determinant size of the resulting eigenvalue equation, eq (3.14).
If an all-round simply-supported plate is considered, the boundary conditions are:

$$
\begin{aligned}
&\sigma_x = 0, \quad v = 0, \quad w = 0 \quad \text{at } \xi = \pm 0.5, \\
&\sigma_y = 0, \quad u = 0, \quad w = 0 \quad \text{at } \eta = \pm 0.5.
\end{aligned} \tag{3.23}
$$

Now, we want to investigate how the basic interpolating functions ($\phi_1(\xi)$ and $\psi_1(\eta)$) are chosen in order the displacement components satisfy the boundary conditions. As example, consider the amplitude displacement function in the ξ–direction, $U(\xi, \eta, \zeta)$, eq. (3.3), here reported:

$$U(\xi,\eta,\zeta) = \sum_i^I \sum_j^J \sum_k^K C^u_{ijk} \phi_{ui}(\xi) \psi_{uj}(\eta) \chi_{uk}(\zeta). \tag{3.24}$$

The aim is to investigate which functions are chosen as basic functions (i.e. when $i = j = k = 1$) for the interpolation of the amplitude displacement U. Disregarding C^u_{111} and since $\chi_{u1}(\zeta)$ is constant and equal to one, U reduces to:

$$U(\xi,\eta) = \phi_{u1}(\xi)\psi_{u1}(\eta).$$

Since an all-round simply-supported plate is considered, at $\eta = \pm 0.5$ the displacement component U has to be zero, eq. (3.23). Thus, the basic interpolating function $\psi_{u1}(\eta)$ can be chosen between $\eta^2 - 0.25$ and $\eta^3 - 0.25\eta$, that are zero at $\eta = \pm 0.5$. Analogously, since the displacement component U does not have to satisfy any geometric constraints at $\xi = \pm 0.5$, the basic interpolating function $\phi_{u1}(\xi)$ can be chosen between the constant unitary function and the ξ function. The basic interpolating functions are listed in Table 1.

Table 1. Basic interpolating functions in the ξ and η–directions

ξ–direction $\phi_{u1}(\xi)$	η–direction $\psi_{u1}(\eta)$
1	1
ξ	η
$\xi^2 - 0.25$	$\eta^2 - 0.25$
$\xi^3 - 0.25\xi$	$\eta^3 - 0.25\eta$

As a consequence, the possible combination of starting polynomials for U that satisfy the geometric conditions for a SSSS plate are four and are listed in Table 2.

Table 2. Basic interpolating functions for the displacement component U for a SSSS plate

$\phi_{u1}(\xi)\psi_{u1}(\eta)$	Symmetry class	
$1\,(\eta^2 - 0.25)$	SS	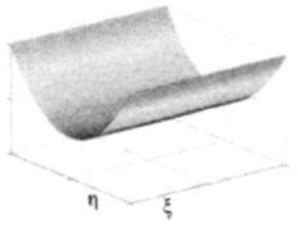
$1\,(\eta^3 - 0.25\eta)$	SA	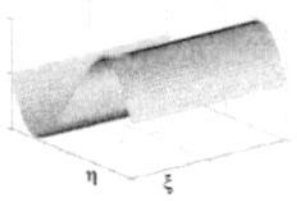
$\xi\,(\eta^2 - 0.25)$	AS	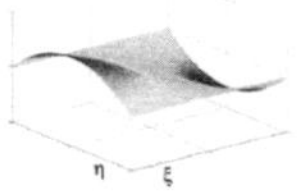
$\xi\,(\eta^3 - 0.25\eta)$	AA	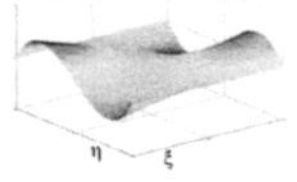

These displacement classes origin vibration modes that have the same characteristic of symmetry and antisymmetry in which the displacement components are subdivided. In Appendix B, see Table 15, the interpolating functions for the displacement components that have been employed in the ξ, η and ζ–directions, are listed and subdivided into symmetry classes.

3.4 Boundary conditions: simply-supported clamped plate (SCSC plate)

If a simply-supported clamped plate is considered, the boundary conditions are:

$$\begin{aligned} &u = 0, \quad v = 0, \quad w = 0 \quad \text{at } \xi = \pm 0.5, \\ &\sigma_y = 0, \quad u = 0, \quad w = 0 \quad \text{at } \eta = \pm 0.5. \end{aligned} \tag{3.25}$$

As example, consider the amplitude displacement function in the η direction, $V(\xi, \eta, \zeta)$, eq. (3.4), here reported:

$$V(\xi, \eta, \zeta) = \sum_{l}^{L} \sum_{m}^{M} \sum_{n}^{N} C^{v}_{lmn} \phi_{vl}(\xi) \psi_{vm}(\eta) \chi_{vn}(\zeta).$$

As before, disregarding C^{v}_{111} and since $\chi_{v1}(\zeta)$ is constant and equal to one, V reduces to:

$$V(\xi, \eta) = \phi_{v1}(\xi) \psi_{v1}(\eta).$$

The basic interpolating functions are listed in Table 1. Since a simply-supported clamped plate is considered, the displacement component V has to be zero at $\xi = \pm 0.5$, eq. (3.25). Thus, the basic interpolating function $\phi_{v1}(\xi)$ can be chosen between $\xi^2 - 0.25$ and $\xi^3 - 0.25\xi$, that are zero at $\xi = \pm 0.5$. Analogously, since the displacement component V does not have to satisfy geometric constraints at $\eta = \pm 0.5$, the basic interpolating function $\psi_{v1}(\eta)$ can be chosen between the constant unitary function and the η function. As a consequence, the possible combination for V of polynomials that satisfy the geometric conditions are four and are listed in Table 3. In Appendix B, see Table 16, the interpolating functions for the displacement components that have been employed in the ξ, η and ζ−directions, are listed and subdivided into symmetry classes.

Table 3. Basic interpolating functions for the displacement component V for a SCSC plate

$\phi_{v1}(\xi)\psi_{v1}(\eta)$	Symmetry class	
$(\xi^2 - 0.25)1$	SS	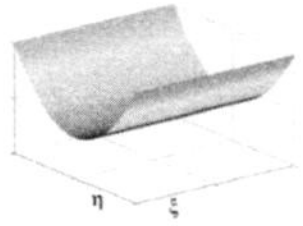
$(\xi^3 - 0.25\xi)1$	AS	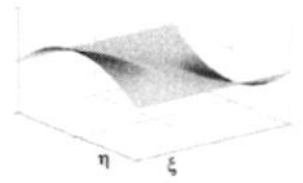
$(\xi^2 - 0.25)\eta$	SA	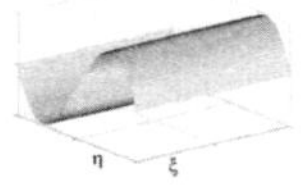
$(\xi^3 - 0.25\xi)\eta$	AA	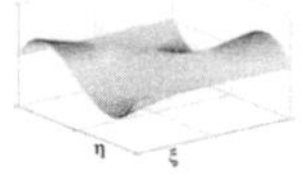

3.5 Boundary conditions: all-round clamped plate (CCCC plate)

If an all-round clamped plate is considered, the boundary conditions are:

$$\begin{array}{llll} u=0, & v=0, & w=0 & \text{at } \xi=\pm 0.5, \\ u=0, & v=0, & w=0 & \text{at } \eta=\pm 0.5. \end{array} \tag{3.26}$$

As example, consider the amplitude displacement function in the ζ direction, $W(\xi,\eta,\zeta)$, eq. (3.5), here reported:

$$W(\xi,\eta,\zeta)=\sum_{p}^{P}\sum_{q}^{Q}\sum_{r}^{R} C^{w}_{pqr}\phi_{wp}(\xi)\psi_{wq}(\eta)\chi_{wr}(\zeta).$$

As before, disregarding C^{w}_{111} and since $\chi_{w1}(\zeta)$ is constant and equal to one, W reduces to:

$$W(\xi,\eta)=\phi_{w1}(\xi)\psi_{w1}(\eta)$$

The basic interpolating functions are listed in Table 1. Since an all round-clamped plate is considered, the displacement component W has to be zero at $\xi=\pm 0.5$, eq. (3.26). Thus, the basic interpolating function $\phi_{w1}(\xi)$ can be chosen between $\xi^2-0.25$ and $\xi^3-0.25\xi$, that are zero at $\xi=\pm 0.5$. Analogously, since the displacement component W has to be zero also at $\eta=\pm 0.5$, the basic interpolating function $\psi_{w1}(\eta)$ can be chosen between $\eta^2-0.25$ and $\eta^3-0.25\eta$, that are zero at $\eta=\pm 0.5$. As a consequence, the possible combination of polynomials for W that satisfy the geometric conditions are four and are listed in Table 4. In Appendix B, see Table 17, the interpolating functions for the displacement components that have been employed in the ξ,η and ζ−directions, are listed and subdivided into symmetry classes.

Table 4. Basic interpolating functions for the displacement component W for a CCCC plate

$\phi_{w1}(\xi)\psi_{w1}(\eta)$	Symmetry class	
$(\xi^2-0.25)\,(\eta^2-0.25)$	SS	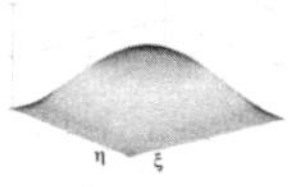
$(\xi^2-0.25)\,(\eta^3-0.25\eta)$	SA	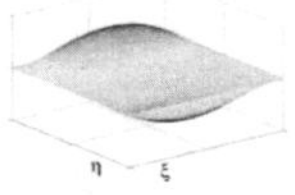
$(\xi^3-0.25\xi)\,(\eta^2-0.25)$	AS	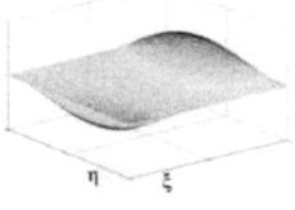
$(\xi^3-0.25\xi)\,(\eta^3-0.25\eta)$	AA	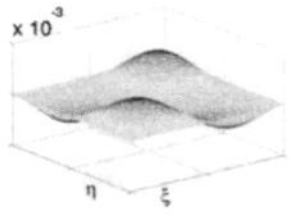

4 Numerical results

The continuum three-dimensional approach presented in the previous sections has been applied to study the free vibration of square ($a/b = 1$) plates with different boundary conditions. A wide range of thickness to width ratios is considered, from thin ($h/b = 0.05$), moderately thick ($h/b = 0.1$ and 0.2) to very thick plates ($h/b = 0.3 \div 1.2$). The influence of thickness ratio and boundary conditions on the vibration frequencies is examined in detail.
The non-dimensional frequency parameter λ is introduced:

$$\lambda = \bar{\lambda}\frac{b^2}{ah\pi^2}\sqrt{12(1-\nu^2)}$$

where $\bar{\lambda}$ is given in eq. (3.21). In the calculation, Poisson's ratio ν is considered constant equal to 0.3. The first non-dimensional frequency parameters λ are calculated. Repeated frequencies are not reported in the tables. The integrations in eqs. (3.18)-(3.20) are performed by using numerical integration schemes, as Gaussian quadrature, when analytical solutions are not available.
To demonstrate the accuracy of the present approach, convergence tests with respect to the exact theory of Srinivas et al. (1970) are carried out. The objective of these studies is to determine the number of terms in x, y and $z-$directions in the series of the displacements, eqs. (3.3)-(3.5), that are needed to get the first frequency parameters accurate up to four significant figures. In general, in the representation of the displacement components, eqs. (3.3)-(3.5), the number of terms in x, y and $z-$directions can be different. In practice, in the following calculations the same number of terms, indicated with N_x, N_y and N_z, in the x, y and $z-$directions are considered ($I = L = P = N_x$, $J = M = Q = N_y$, $K = N = R = N_z$).
The error ε is calculated as

$$\varepsilon = \frac{\lambda_3 - \lambda_{sr}}{\lambda_{sr}}100$$

where λ_3 corresponds to the frequency parameter found with a particular number of terms in the series and λ_{sr} correspond to the frequency parameter calculated with the exact analysis conducted by Srinivas et al. (1970) and reported in Gentilini and Viola (2003).

4.1 SSSS plate

Numerical results are listed in Tables 5-12 for all-round simply-supported plates. In each table, eingenvalues are subdivided into symmetry classes (SS, SA, AS, AA). A careful scrutiny of the tables reveals that convergence rate is very rapid and in almost all the cases frequency parameters monotonically decrease with the increase of the number of terms of admissible functions. Furthermore, as the thickness ratio increases, the number of terms needed for the polynomial function in the thickness direction ($z-$direction) increases. For a relatively thin simply-supported plate ($h/a = 0.05$), see Table 5, 3 terms in the $z-$direction are sufficient to give accurate results. However, it needs as high as 10 terms for a plate with thickness ratio $h/a = 1.2$, see Table 12, to achieve the same degree of accuracy. On the other hand, the number of terms needed in the admissible functions along the plate surface (x and $y-$directions) reduces from 5×5 to 3×3 terms as the plate thickness dimension becomes significant. In general, few terms are needed to get a good convergence also when the plate is very thick. In fact, for the simply-supported plate with $h/a = 1.2$, only $3 \times 3 \times 10$ terms in the three directions are sufficient to get a good accuracy (i.e. low error ε) for the first frequencies.

Table 5. Frequency parameters for SSSS plate ($a/b = 1$, $h/b = 0.05$)

Symmetry class	λ_1 (3×3×3)	λ_2 (4×4×3)	λ_3 (5×5×3)	λ_{sr}	ε
SS	0.0238	0.0238	0.0238	**0.0238**	0.000
	0.1176	0.1159	0.1159	**0.1155**	0.350
	0.2053	0.2031	0.2030	**0.2017**	0.640
	0.2221	0.2221	0.2221	**0.2221**	0.000
	0.3753	0.3062	0.2870	**0.2834**	1.270
SA	0.0592	0.0590	0.0590	**0.0589**	0.170
	0.1506	0.1491	0.1491	**0.1484**	0.472
	0.3091	0.1994	0.1926	**0.1912**	0.914
	0.3518	0.2811	0.2759	**0.2734**	1.278
AS	0.0592	0.0590	0.0590	**0.0589**	0.170
	0.1056	0.1491	0.1491	**0.1484**	0.472
	0.3091	0.1994	0.1926	**0.1912**	0.732
	0.3518	0.2811	0.2759	**0.2734**	0.914
	0.3673	0.3350	0.3170	**0.3130**	1.278
AA	0.0936	0.0934	0.0934	**0.0932**	0.214
	0.3141	0.2304	0.2243	**0.2226**	0.764

Table 6. Frequency parameters for SSSS plate ($a/b = 1$, $h/b = 0.1$)

Symmetry class	λ_1 (3×3×3)	λ_2 (4×4×3)	λ_3 (5×5×3)	λ_{sr}	ε
SS	0.0934	0.0934	0.0934	**0.0932**	0.214
	0.4275	0.4222	0.4221	**0.4171**	1.199
	0.4442	0.4442	0.4442	**0.4443**	-0.020
	0.7075	0.7012	0.7011	**0.6889**	1.771
	0.7498	0.7498	0.7498	**0.7498**	0.000
SA	0.2247	0.2241	0.2226	**0.2226**	0.000
	0.3147	0.3141	0.3141	**0.3142**	0.032
	0.5359	0.5315	0.5240	**0.5239**	0.019
	0.6648	0.6321	0.6289	**0.6247**	0.672
	0.7037	0.6881	0.6579	**0.6571**	0.122
AS	0.2247	0.2241	0.2226	**0.2226**	0.000
	0.3147	0.3141	0.3141	**0.3142**	0.032
	0.5359	0.5315	0.5240	**0.5239**	0.019
	0.6648	0.6321	0.6289	**0.6247**	0.672
	0.7037	0.6881	0.6579	**0.6571**	0.122
AA	0.3463	0.3455	0.3455	**0.3421**	0.994
	0.6283	0.6283	0.6283	**0.6283**	0.000
	0.8930	0.6681	0.6670	**0.6571**	1.507
	1.0001	0.7819	0.7618	**0.7511**	1.424

Table 7. Frequency parameters for SSSS plate ($a/b = 1$, $h/b = 0.2$)

Symmetry class	λ_1 (3×3×3)	λ_2 (4×4×3)	λ_3 (5×5×3)	λ_{sr}	ε
SS	0.3455	0.3455	0.3421	**0.3420**	0.029
	0.8885	0.8885	0.8885	**0.8885**	0.000
	1.3354	1.3230	1.2898	**1.2881**	0.132
	1.4925	1.4992	1.4922	**1.4922**	0.000
	1.9972	1.9871	1.9614	**1.9557**	0.291
	1.9972	1.9870	1.9869	**1.9869**	0.000
SA	0.6283	0.6283	0.6283	**0.6283**	0.000
	0.7669	0.7652	0.7514	**0.7511**	0.040
	1.4075	1.4049	1.4049	**1.4050**	0.000
	1.6155	1.6061	1.5619	**1.5589**	0.192
	1.8886	1.8849	1.8849	**1.8808**	0.218
AS	0.6283	0.6283	0.6283	**0.6283**	0.000
	0.7669	0.7652	0.7514	**0.7511**	0.040
	1.4075	1.4049	1.4049	**1.4050**	0.000
	1.6155	1.6061	1.5619	**1.5589**	0.192
	1.8886	1.8849	1.8849	**1.8808**	0.218
AA	1.1170	1.1153	1.0899	**1.0888**	0.101
	1.2566	1.2566	1.2566	**1.2566**	0.000
	1.7860	1.7771	1.7771	**1.7772**	0.000

Table 8. Frequency parameters for SSSS plate ($a/b = 1$, $h/b = 0.3$)

Symmetry class	λ_1 (3×3×3)	λ_2 (4×4×3)	λ_3 (5×5×3)	λ_{sr}	ε
SS	0.7011	0.7011	0.7011	**0.6889**	1.771
	1.3328	1.3328	1.3328	**1.3329**	-0.007
	2.2193	2.2193	2.2193	**2.2171**	0.099
	2.3716	2.3547	2.3547	**2.2710**	3.686
	2.9961	2.9806	2.9806	**2.9804**	0.007
SA	0.9424	0.9424	0.9424	**0.8985**	4.886
	1.4469	1.4445	1.4068	**1.4046**	0.157
	2.1114	2.1074	2.1074	**2.1073**	0.005
	2.8108	2.7986	2.7053	**2.6915**	0.513
	2.8329	2.8274	2.8274	**2.8273**	0.003
AS	0.9424	0.9424	0.9424	**0.8985**	4.886
	1.4469	1.4445	1.4068	**1.4046**	0.157
	2.1114	2.1074	2.1074	**2.1073**	0.005
	2.8108	2.7986	2.7053	**2.6915**	0.513
AA	1.8849	1.8849	1.8849	**1.8849**	0.000
	2.0246	2.0225	1.9614	**1.9557**	0.291
	2.6792	2.6657	2.6657	**2.6657**	0.000

Table 9. Frequency parameters for SSSS plate ($a/b = 1$, $h/b = 0.4$)

Symmetry class	λ_1 (3×3×4)	λ_2 (3×3×5)	λ_3 (3×3×6)	λ_{sr}	ε
SS	1.0899	1.0889	1.0888	**1.0888**	0.005
	1.7771	1.7771	1.7771	**1.7772**	-0.003
	2.9073	2.9070	2.9070	**2.9070**	0.000
	3.3125	3.2895	3.2884	**3.2711**	0.529
SA	1.2566	1.2566	1.2566	**1.2566**	0.003
	2.1094	2.1025	2.1024	**2.0999**	0.119
	2.8153	2.8153	2.8153	**2.8099**	0.191
	3.3844	3.3844	3.3836	**3.3836**	0.000
	3.7773	3.7773	3.7773	**3.7698**	0.198
	3.8789	3.8444	3.8426	**3.8300**	0.328
AS	1.2566	1.2566	1.2566	**1.2566**	0.003
	2.1094	2.1025	2.1024	**2.0999**	0.119
	2.8153	2.8153	2.8153	**2.8099**	0.191
	3.3844	3.3844	3.3836	**3.3836**	0.000
	3.8789	3.8444	3.8426	**3.8300**	0.328
AA	2.5133	2.5133	2.5133	**2.5133**	0.001
	2.8671	2.8513	2.8509	**2.8488**	0.072
	3.5721	3.5722	3.5721	**3.5543**	9.203

Table 10. Frequency parameters for SSSS plate ($a/b = 1$, $h/b = 0.6$)

Symmetry class	λ_1 (3×3×4)	λ_2 (3×3×6)	λ_3 (3×3×8)	λ_{sr}	ε
SS	1.9614	1.9557	1.9557	**1.9557**	0.002
	2.6657	2.6657	2.6657	**2.6657**	0.001
	4.0471	4.0423	4.0423	**4.0423**	0.001
	4.1208	4.1202	4.1202	**4.1201**	0.001
SA	1.8849	1.8849	1.8849	**1.8849**	0.003
	3.5446	3.5160	3.5159	**3.5133**	0.073
	3.6644	3.6637	3.6637	**3.6637**	0.000
	4.2232	4.2231	4.2232	**4.2147**	0.201
	5.2632	5.2627	5.2627	**5.2505**	0.233
AS	1.8849	1.8849	1.8849	**1.8849**	0.003
	3.5446	3.5160	3.5159	**3.5133**	0.073
	3.6644	3.6637	3.6637	**3.6637**	0.000
	4.2232	4.2231	4.2232	**4.2147**	0.201
	5.0813	5.0802	5.0802	**5.0732**	0.138
AA	3.7700	3.7700	3.7699	**3.7698**	0.005
	4.6894	4.6304	4.6299	**4.6279**	0.000
	4.9079	4.9074	4.9074	**4.9073**	0.138

Table 11. Frequency parameters for SSSS plate ($a/b = 1$, $h/b = 1.0$)

Symmetry class	λ_1 (3×3×4)	λ_2 (3×3×6)	λ_3 (3×3×9)	λ_{sr}	ε
SS	3.7758	3.7420	3.7419	**3.7419**	0.000
	4.4429	4.4429	4.4429	**4.4428**	0.002
	5.2015	5.2013	5.2013	**5.2013**	0.001
	5.4419	5.4414	5.4414	**5.4414**	0.000
SA	3.1416	3.1416	3.1416	**3.1416**	0.000
	4.4435	4.4429	4.4429	**4.4428**	0.002
	6.4252	6.2996	6.2971	**6.2943**	0.043
	7.0330	6.9611	6.9577	**6.9519**	0.123
	7.0439	7.0395	7.0258	**7.0248**	0.000
AS	3.1416	3.1416	3.1416	**3.1416**	0.000
	4.4435	4.4429	4.4429	**4.4428**	0.002
	6.4252	6.2996	6.2971	**6.2943**	0.043
	7.0330	6.9611	6.9577	**6.9519**	0.123
	7.0439	7.0395	7.0258	**7.0248**	0.000
AA	6.2833	6.2833	6.2833	**6.2830**	0.003
	7.0253	7.0249	7.0249	**7.0248**	0.000
	8.3289	8.1097	8.0996	**8.0966**	0.032
	8.6803	8.4825	8.4833	**8.4708**	0.839
	8.9325	8.9019	8.8866	**8.8855**	0.003

This is due to the fact that as the thickness grows, flexural modes, i.e. vibration out of plane of the plate, becomes more difficult, while modes in the plane of the plate acquire more importance. It should be remarked that the presence of surface parallel, thickness modes in the lower vibration spectrum restrict the applicability of the refined plate theory such as Mindlin theory to the analysis of moderately thick plates. Since the first order Mindlin theory only considers flexural modes and in this study, it is found that the thickness modes often precede some of the flexural modes in thick plate vibrations. Although the method has the capability of analyzing accurately very thick plates, which two dimensional theories cannot, it can also be applied to thin plates, thereby determining conclusively the accuracies of the plate theories. In fact, the in-plane modes of vibration are not predicted by a plate theory, such as the classical one, that neglects the in-plane displacements u and v.
It is found that the results from the present formulation are in good agreement with the three-dimensional study by Srinivas et al. (1970) and Gentilini and Viola (2003).
Moreover, observing the results it is important to underline other important aspects. As it is well-known, Ritz's method gives accurate solutions, Liew et al. (1993), Liew et al. (1994) and Liew et al. (1995). However, its efficiency is greatly influenced by the choice of the admissible functions. Natural frequencies obtained from the Ritz method converge as upper bounds to the exact values. This upper limit can be improved increasing the number of polynomial functions employed in the summations, so in theory, every accuracy can be obtained. In reality, a practical limit to the numbers of terms exists due to the capacity of computers. In particular, in the three-dimensional vibration analysis of an elastic solid, numerical instability can occur with a high number of admissible functions. From the analysis of the data listed in the mentioned tables, it can be inferred that the frequencies converge from the top to the solution, but in some cases, even if in a limited number, this does not happen. This behavior can be due to the kind of *admissible functions* that are employed in the approximation for the displacement representation. In the present study, orthogonal polynomial functions have been employed. These polynomials, when the degree is increasing, independently from the thickness, may give numerical problems and thus, stiffness and mass matrices result ill-conditioned. For

Table 12. Frequency parameters for SSSS plate ($a/b = 1$, $h/b = 1.2$)

Symmetry class	λ_1 (3×3×5)	λ_2 (3×3×6)	λ_3 (3×3×10)	λ_{sr}	ε
SS	4.6872	4.6284	4.6280	**4.6279**	0.002
	5.3315	5.3315	5.3315	**5.3314**	0.002
	5.7483	5.7402	5.7402	**5.7401**	0.002
	6.1887	6.1882	6.1882	**6.1882**	0.000
	8.8640	8.1933	8.1761	**8.1760**	0.001
SA	3.7699	3.7699	3.7699	**3.7698**	0.003
	4.9079	4.9073	4.9073	**4.9073**	0.001
	7.3469	7.3469	7.3274	**7.3273**	0.001
	8.1497	8.0989	8.0514	**8.0865**	-0.434
	8.4471	8.4471	8.4482	**8.4297**	0.219
AS	3.7699	3.7699	3.7699	**3.7698**	0.003
	4.9079	4.9073	4.9073	**4.9073**	0.001
	7.3469	7.3469	7.3274	**7.3273**	0.001
	7.6949	7.6683	7.6604	**7.6574**	0.040
	8.1497	8.0989	8.0514	**8.0865**	-0.434
	8.4471	8.4471	8.4482	**8.4297**	0.219
AA	7.5399	7.5399	7.5398	**7.5398**	0.000
	8.1686	8.1682	8.1681	**8.1681**	0.001

this reason, the authors investigated in Gentilini et al. (2003) the employment of other kind of functions such as Chebyshev polynomials, in order to solve numerical problems and, consequently, to improve the solution.

4.2 SCSC and CCCC plates

In the previous section it has been shown as, for the all-round simply-supported plate, the Ritz method gives good results and the convergence to the exact solution is fast. The method is applied to the calculation of frequency parameters of plates with different thickness side ratios and boundary conditions. In particular, plates with simply-supported and clamped edges and with all-round clamped edges are considered.
In Tables 13 and 14 frequency parameters of SCSC and CCCC plates are listed, respectively. In this analysis, the nomenclature SCSC, as was already specified, denotes a plate with simply supported edges at $x = \pm a/2$ and clamped at $y = \pm b/2$. Plates with thickness to width ratios h/b equal to 0.1, 0.2, 0.3 and 0.4 have been considered. The eigenvalues are subdivided into symmetry classes and determined with $3 \times 3 \times 3$ and $4 \times 4 \times 4$ polynomials in the x, y and $z-$directions, respectively.

5 Conclusions

A three-dimensional elasticity solution to the problem of free vibrations of rectangular homogeneous isotropic plates has been presented. The solution is obtained by the Ritz method. Sets of polynomials orthogonalized by means of Gram-Schmidt procedure have been employed as admissible functions. Numerical results are supported by appropriate comparisons with results available in the literature for the case of all-round simply-supported plates. Parametric studies have been performed for different boundary conditions and thickness side ratios. The convergence of vibration frequencies has been examined and frequency parameters have been presented for different thickness ratios (h/b). Since no simplifying assumptions have been made on the displacement fields, the

Table 13. Frequency parameters for SCSC plates with different thickness to width ratios

$N_x \times N_y \times N_z$	SS			SA			AS			AA		
				h/b=0.1								
$3 \times 3 \times 3$	2.758	9.086	10.871	6.167	11.526	11.750	5.072	6.523	12.630	8.204	13.047	15.51[illegible]
$4 \times 4 \times 3$	2.749	8.973	10.624	6.155	11.521	11.658	5.054	6.523	12.455	8.186	13.047	15.49[illegible]
				h/b=0.2								
$3 \times 3 \times 3$	2.350	6.134	7.011	4.746	5.769	8.591	3.262	4.185	8.829	6.217	6.523	7.757
$4 \times 4 \times 3$	2.297	6.133	6.768	4.609	5.767	8.297	3.262	4.087	8.521	6.042	6.523	7.748
				h/b=0.3								
$3 \times 3 \times 3$	1.968	4.083	5.505	3.700	3.850	5.795	2.174	3.439	6.536	4.349	4.841	5.173
$4 \times 4 \times 3$	1.912	4.081	5.291	3.581	3.848	5.788	2.174	3.338	6.387	4.349	4.688	5.167
				h/b=0.4								
$3 \times 3 \times 3$	1.667	3.054	4.469	2.890	2.990	4.329	1.631	2.877	4.782	3.262	3.881	3.921
$4 \times 4 \times 4$	1.615	3.050	4.292	2.888	2.897	4.315	1.631	2.784	4.393	3.262	3.796	3.876

Table 14. Frequency parameters for CCCC plates with different thickness to width ratios

$N_x \times N_y \times N_z$	SS			SA			AS			AA		
				h/b=0.1								
$3 \times 3 \times 3$	3.382	10.987	11.101	6.477	12.543	13.203	6.477	12.543	13.203	9.106	14.873	18.280
$4 \times 4 \times 3$	3.331	10.528	10.629	6.362	12.529	12.740	6.362	12.529	12.740	9.087	14.871	17.761
				h/b=0.2								
$3 \times 3 \times 3$	2.809	7.643	7.763	4.939	6.283	9.064	4.939	6.283	9.064	6.650	7.438	9.148
$4 \times 4 \times 3$	2.734	7.354	7.457	4.789	6.278	8.736	4.789	6.278	8.736	6.442	7.437	9.140
				h/b=0.3								
$3 \times 3 \times 3$	2.304	5.687	5.789	3.821	4.196	6.705	3.821	4.196	6.705	4.960	5.045	6.101
$4 \times 4 \times 3$	2.226	5.495	5.579	3.691	4.192	6.479	3.691	4.192	6.479	4.960	5.045	6.100
				h/b=0.4								
$3 \times 3 \times 3$	1.917	4.493	4.576	3.063	3.152	5.175	3.063	3.152	5.175	3.720	4.013	4.574
$4 \times 4 \times 4$	1.845	4.363	4.429	2.962	3.147	5.068	2.962	3.147	5.068	3.720	3.869	4.561

method is capable of providing accurate frequency solutions for the vibration analysis of thick plates. Numerical results that are presented can serve as a reference for establishing the validity of approximate theories as well as checking numerical solutions. The formulation is readily extendible to other kinds of boundary conditions.

Acknowledgement

This topic is one of the subjects of the Centre of Study and Research for the Identification of Materials and Structures (CIMEST) "M. Capurso". The numerical tests presented were carried out using the facilities of the Laboratory of Computational Mechanics

(LAMC) "A. A. Cannarozzi" of the University of Bologna.

Bibliography

G. Arfken. *Mathematical Methods for Physicists.* Academic Press, 1985.

C. Gentilini, E. Artioli and E. Viola. On the use of polynomial series in the three-dimensional vibration analysis of rectangular plates. In *Proocedings of the 16th AIMETA Congress of Theoretical and Applied Mechanics-AIMETA'03*, pages 1-10, 2003.

C. Gentilini. *Modelling of the static and dynamic behaviour of structures with variable parameters.* PhD Dissertation, 2005.

C. Gentilini and E. Viola. On the three-dimensional vibration analysis of rectangular plates. In *Problems in Structural Identification and Diagnostics: General Aspects and Applications*, C. Davini and E. Viola editors, pages 149-162. Springer Verlag, 2003.

K.M. Liew, K.C. Hung and M.K. Lim. A continuum three-dimensional vibration analysis of thick rectangular plates. *International Journal of Solids and Structures*, 30:3357-3379, 1993.

K.M. Liew, K.C. Hung and M.K. Lim. Three-dimensional vibration of rectangular plates: variance of simple support conditions and influence of in-plane inertia. *International Journal of Solids and Structures*, 31:3233-3247, 1994.

K.M. Liew, K.C. Hung and M.K. Lim. Three-dimensional vibration of rectangular plates: effects of the thickness and edge constraints. *Journal of Sound and Vibration*, 182: 709-727, 1995.

A.W. Leissa and Z.D. Zhang. On the three-dimensional vibrations of the cantilevered rectangular parallelepiped. *Journal of Acoustical Society of America*, 73:2013-2021, 1983.

R.D. Mindlin. Influence of rotatory inertia and shear on flexural motion of isotropic elastic plates. *ASME Journal of Applied Mechanics*, 18:123-130, 1951.

J.N. Reddy. *Mechanics of Laminated Composite Plates and Shells: Theory and Analysis.* CRC Press, 2003.

S. Srinivas, C.V. Joga Rao and K. Rao. An exact analysis for vibration of simple-supported homogeneous and laminated thick rectangular plates. *Journal of Sound and Vibration*, 12:187-199, 1970.

W.H. Wittrick. Analytical three-dimensional elasticity solutions to some plate problems, and some observations on Mindlin's plate theory. *International Journal of Solids and Structures*, 23:441-464, 1987.

Appendix A

Let us consider the case when $I=2,\ J=2,\ K=2$ and consider a term of the functional eq. (3.13):

$$\begin{aligned} H &= \sum_i^2\sum_j^2\sum_k^2 C^u_{ijk}\phi_{ui}\psi_{uj}\chi_{uk} = C^u_{111}\phi_{u1}\psi_{u1}\chi_{u1} + C^u_{112}\phi_{u1}\psi_{u1}\chi_{u2} + \\ &\quad + C^u_{121}\phi_{u1}\psi_{u2}\chi_{u1} + C^u_{122}\phi_{u1}\psi_{u2}\chi_{u2} + \\ &\quad + C^u_{211}\phi_{u2}\psi_{u1}\chi_{u1} + C^u_{212}\phi_{u2}\psi_{u1}\chi_{u2} + \\ &\quad + C^u_{221}\phi_{u2}\psi_{u2}\chi_{u1} + C^u_{222}\phi_{u2}\psi_{u2}\chi_{u2}. \end{aligned}$$

The derivative of H with respect to C^u_{111} is:

$$\partial H/\partial C^u_{111} = \phi_{u1}\psi_{u1}\chi_{u1}.$$

If $I = 1,\ J = 1$ and $K = 2$, H is:

$$\begin{aligned} H &= \left(\sum_{i}^{1}\sum_{j}^{1}\sum_{k}^{2} C_{ijk}^{u}\phi_{ui}\psi_{uj}\chi_{uk}\right)^2 = \left(C_{111}^{u}\phi_{u1}\psi_{u1}\chi_{u1} + C_{112}^{u}\phi_{u1}\psi_{u1}\chi_{u2}\right)^2 = \\ &= \left(C_{111}^{u}\phi_{u1}\psi_{u1}\chi_{u1}\right)^2 + \left(C_{112}^{u}\phi_{u1}\psi_{u1}\chi_{u2}\right)^2 + 2C_{111}^{u}\phi_{u1}\psi_{u1}\chi_{u1}C_{112}^{u}\phi_{u1}\psi_{u1}\chi_{u2}. \end{aligned}$$

Thus, the derivative of H with respect to C_{111}^{u} is:

$$\begin{aligned} \partial H/\partial C_{111}^{u} &= 2C_{111}^{u}\phi_{u1}\psi_{u1}\chi_{u1}\phi_{u1}\psi_{u1}\chi_{u1} + 2\phi_{u1}\psi_{u1}\chi_{u1}C_{112}^{u}\phi_{u1}\psi_{u1}\chi_{u2} = \\ &= 2\phi_{u1}\psi_{u1}\chi_{u1}\left(C_{111}^{u}\phi_{u1}\psi_{u1}\chi_{u1} + C_{112}^{u}\phi_{u1}\psi_{u1}\chi_{u2}\right) = \\ &= 2\phi_{u1}\psi_{u1}\chi_{u1}\sum_{\bar{i}}^{1}\sum_{\bar{j}}^{1}\sum_{\bar{k}}^{2} C_{\bar{i}\bar{j}\bar{k}}^{u}\phi_{u\bar{i}}\psi_{u\bar{j}}\chi_{u\bar{k}}. \end{aligned}$$

Let us consider a further example. Let $I = 1,\ J = 1$ and $K = 3, H$ is:

$$\begin{aligned} &H = \left(\sum_{i}^{1}\sum_{j}^{1}\sum_{k}^{3} C_{ijk}^{u}\phi_{ui}\psi_{uj}\chi_{uk}\right)^2 = \\ &= \left(C_{111}^{u}\phi_{u1}\psi_{u1}\chi_{u1} + C_{112}^{u}\phi_{u1}\psi_{u1}\chi_{u2} + C_{113}^{u}\phi_{u1}\psi_{u1}\chi_{u3}\right)^2 = \\ &= \left(C_{111}^{u}\phi_{u1}\psi_{u1}\chi_{u1}\right)^2 + \left(C_{112}^{u}\phi_{u1}\psi_{u1}\chi_{u2}\right)^2 + \left(C_{113}^{u}\phi_{u1}\psi_{u1}\chi_{u3}\right)^2 + \\ &+2C_{111}^{u}\phi_{u1}\psi_{u1}\chi_{u1}C_{112}^{u}\phi_{u1}\psi_{u1}\chi_{u2} + 2C_{111}^{u}\phi_{u1}\psi_{u1}\chi_{u1}C_{113}^{u}\phi_{u1}\psi_{u1}\chi_{u3} \\ &+2C_{112}^{u}\phi_{u1}\psi_{u1}\chi_{u2}C_{113}^{u}\phi_{u1}\psi_{u1}\chi_{u3}. \end{aligned}$$

The derivative of H with respect to C_{111}^{u} is:

$$\begin{aligned} \partial H/\partial C_{111}^{u} &= 2C_{111}^{u}\phi_{u1}\psi_{u1}\chi_{u1}\phi_{u1}\psi_{u1}\chi_{u1} + 2\phi_{u1}\psi_{u1}\chi_{u1}C_{112}^{u}\phi_{u1}\psi_{u1}\chi_{u2} \\ &\quad +2\phi_{u1}\psi_{u1}\chi_{u1}C_{113}^{u}\phi_{u1}\psi_{u1}\chi_{u3} = \\ &= 2\phi_{u1}\psi_{u1}\chi_{u1}\left(C_{111}^{u}\phi_{u1}\psi_{u1}\chi_{u1} + C_{112}^{u}\phi_{u1}\psi_{u1}\chi_{u2} + C_{113}^{u}\phi_{u1}\psi_{u1}\chi_{u3}\right) = \\ &= 2\phi_{u1}\psi_{u1}\chi_{u1}\sum_{\bar{i}}^{1}\sum_{\bar{j}}^{1}\sum_{\bar{k}}^{3} C_{\bar{i}\bar{j}\bar{k}}^{u}\phi_{u\bar{i}}\psi_{u\bar{j}}\chi_{u\bar{k}}. \end{aligned}$$

Thus, if H has the expression:

$$H = \sum_{i}^{I}\sum_{j}^{L}\sum_{k}^{K} C_{ijk}^{u}\phi_{ui}\psi_{uj}\chi_{uk},$$

then, its derivative with respect to the generic C_{ijk}^{u} is:

$$\partial H/\partial C_{ijk}^{u} = \phi_{ui}\psi_{uj}\chi_{uk}.$$

If H has the expression:

$$H = \left(\sum_{i}^{I}\sum_{j}^{L}\sum_{k}^{K} C_{ijk}^{u}\phi_{ui}\psi_{uj}\chi_{uk}\right)^2,$$

then, its derivative with respect to the generic C_{ijk}^{u} is:

$$\partial H/\partial C_{ijk}^{u} = 2\phi_{ui}\psi_{uj}\chi_{uk}\sum_{\bar{i}}\sum_{\bar{j}}\sum_{\bar{k}} C_{\bar{i}\bar{j}\bar{k}}^{u}\phi_{u\bar{i}}\psi_{u\bar{j}}\chi_{u\bar{k}}.$$

Appendix B

Table 15. SSSS plate: basic interpolating functions for the amplitude displacements.

Symmetry class	Generating functions $\begin{cases} U(\xi,\eta,\zeta) = C_{u111}\phi_{u1}(\xi)\psi_{u1}(\eta)\chi_{u1}(\zeta) \\ V(\xi,\eta,\zeta) = C_{v111}\phi_{v1}(\xi)\psi_{v1}(\eta)\chi_{v1}(\zeta), \\ W(\xi,\eta,\zeta) = C_{w111}\phi_{w1}(\xi)\psi_{w1}(\eta)\chi_{w1}(\zeta) \end{cases}$
SS	$U = (\eta^2 - 0.25)$ where $\begin{cases} \phi_{u1}(\xi) = 1 \\ \psi_{u1}(\eta) = \eta^2 - 0.25 \\ \chi_{u1}(\zeta) = 1 \end{cases}$
	$V = (\xi^2 - 0.25)$ where $\begin{cases} \phi_{v1}(\xi) = \xi^2 - 0.25 \\ \psi_{v1}(\eta) = 1 \\ \chi_{v1}(\zeta) = 1 \end{cases}$
	$W = (\xi^2 - 0.25)(\eta^2 - 0.25)$ where $\begin{cases} \phi_{w1}(\xi) = \xi^2 - 0.25 \\ \psi_{w1}(\eta) = \eta^2 - 0.25 \\ \chi_{w1}(\zeta) = 1 \end{cases}$
SA	$U = \left(\eta^3 - 0.25\eta\right)$ where $\begin{cases} \phi_{u1}(\xi) = 1 \\ \psi_{u1}(\eta) = \eta^3 - 0.25\eta \\ \chi_{u1}(\zeta) = 1 \end{cases}$
	$V = (\xi^2 - 0.25)\eta$ where $\begin{cases} \phi_{v1}(\xi) = \xi^3 - 0.25\xi \\ \psi_{v1}(\eta) = \eta \\ \chi_{v1}(\zeta) = 1 \end{cases}$
	$W = (\xi^2 - 0.25)(\eta^3 - 0.25\eta)$ where $\begin{cases} \phi_{w1}(\xi) = \xi^2 - 0.25 \\ \psi_{w1}(\eta) = \eta^3 - 0.25\eta \\ \chi_{w1}(\zeta) = 1 \end{cases}$
AS	$U = \xi(\eta^2 - 0.25)$ where $\begin{cases} \phi_{u1}(\zeta) = \xi \\ \psi_{u1}(\eta) = \eta^2 - 0.25 \\ \chi_{u1}(\zeta) = 1 \end{cases}$
	$V = (\xi^3 - 0.25\xi)$ where $\begin{cases} \phi_{v1}(\xi) = \xi^3 - 0.25\xi \\ \psi_{v1}(\eta) = 1 \\ \chi_{v1}(\zeta) = 1 \end{cases}$
	$W = (\xi^3 - 0.25\xi)(\eta^2 - 0.25)$ where $\begin{cases} \phi_{w1}(\xi) = \xi^3 - 0.25\xi \\ \psi_{w1}(\eta) = \eta^2 - 0.25 \\ \chi_{w1}(\zeta) = 1 \end{cases}$
AA	$U = \xi(\eta^3 - 0.25\eta)$ where $\begin{cases} \phi_{u1}(\xi) = \xi \\ \psi_{u1}(\eta) = \eta^3 - 0.25\eta \\ \chi_{u1}(\zeta) = 1 \end{cases}$
	$V = (\xi^3 - 0.25\xi)\eta$ where $\begin{cases} \phi_{v1}(\xi) = \xi^3 - 0.25\xi \\ \psi_{v1}(\eta) = \eta \\ \chi_{v1}(\zeta) = 1 \end{cases}$
	$W = (\xi^3 - 0.25\xi)(\eta^3 - 0.25\eta)$ where $\begin{cases} \phi_{w1}(\xi) = \xi^3 - 0.25\xi \\ \psi_{w1}(\eta) = \eta^3 - 0.25\eta \\ \chi_{w1}(\zeta) = 1 \end{cases}$

Table 16. SCSC plate: basic interpolating functions for the amplitude displacements.

Symmetry class	Amplitude functions $\begin{cases} U(\xi,\eta,\zeta) = C_{u111}\phi_{u1}(\xi)\psi_{u1}(\eta)\chi_{u1}(\zeta) \\ V(\xi,\eta,\zeta) = C_{v111}\phi_{v1}(\xi)\psi_{v1}(\eta)\chi_{v1}(\zeta), \\ W(\xi,\eta,\zeta) = C_{w111}\phi_{w1}(\xi)\psi_{w1}(\eta)\chi_{w1}(\zeta) \end{cases}$
SS	$U = (\xi^2 - 0.25)(\eta^2 - 0.25)$ where $\begin{cases} \phi_{u1}(\xi) = \xi^2 - 0.25 \\ \psi_{u1}(\eta) = \eta^2 - 0.25 \\ \chi_{u1}(\zeta) = 1 \end{cases}$ $V = (\xi^2 - 0.25)$ where $\begin{cases} \phi_{v1}(\xi) = \xi^2 - 0.25 \\ \psi_{v1}(\eta) = 1 \\ \chi_{v1}(\zeta) = 1 \end{cases}$ $W = (\xi^2 - 0.25)(\eta^2 - 0.25)$ where $\begin{cases} \phi_{w1}(\xi) = \xi^2 - 0.25 \\ \psi_{w1}(\eta) = \eta^2 - 0.25 \\ \chi_{w1}(\zeta) = 1 \end{cases}$
SA	$U = (\xi^2 - 0.25)(\eta^3 - 0.25\eta)$ where $\begin{cases} \phi_{u1}(\xi) = \xi^2 - 0.25 \\ \psi_{u1}(\eta) = \eta^3 - 0.25\eta \\ \chi_{u1}(\zeta) = 1 \end{cases}$ $V = (\xi^3 - 0.25\xi)\eta$ where $\begin{cases} \phi_{v1}(\xi) = \xi^3 - 0.25\xi \\ \psi_{v1}(\eta) = \eta \\ \chi_{v1}(\zeta) = 1 \end{cases}$ $W = (\xi^2 - 0.25)(\eta^3 - 0.25\eta)$ where $\begin{cases} \phi_{w1}(\xi) = \xi^2 - 0.25 \\ \psi_{w1}(\eta) = \eta^3 - 0.25\eta \\ \chi_{w1}(\zeta) = 1 \end{cases}$
AS	$U = (\xi^3 - 0.25\xi)(\eta^2 - 0.25)$ where $\begin{cases} \phi_{u1}(\xi) = \xi^3 - 0.25\xi \\ \psi_{u1}(\eta) = \eta^2 - 0.25 \\ \chi_{u1}(\zeta) = 1 \end{cases}$ $V = (\xi^3 - 0.25\xi)$ where $\begin{cases} \phi_{v1}(\xi) = \xi^3 - 0.25\xi \\ \psi_{v1}(\eta) = 1 \\ \chi_{v1}(\zeta) = 1 \end{cases}$ $W = (\xi^3 - 0.25\xi)(\eta^2 - 0.25)$ where $\begin{cases} \phi_{w1}(\xi) = \xi^3 - 0.25\xi \\ \psi_{w1}(\eta) = \eta^2 - 0.25 \\ \chi_{w1}(\zeta) = 1 \end{cases}$
AA	$U = (\xi^3 - 0.25\xi)(\eta^3 - 0.25\eta)$ where $\begin{cases} \phi_{u1}(\xi) = \xi^3 - 0.25\xi \\ \psi_{u1}(\eta) = \eta^3 - 0.25\eta \\ \chi_{u1}(\zeta) = 1 \end{cases}$ $V = (\xi^3 - 0.25\xi)$ where $\begin{cases} \phi_{v1}(\xi) = \xi^3 - 0.25\xi \\ \psi_{v1}(\eta) = 1 \\ \chi_{v1}(\zeta) = 1 \end{cases}$ $W = (\xi^3 - 0.25\xi)(\eta^3 - 0.25\eta)$ where $\begin{cases} \phi_{w1}(\xi) = \xi^3 - 0.25\xi \\ \psi_{w1}(\eta) = \eta^3 - 0.25\eta \\ \chi_{w1}(\zeta) = 1 \end{cases}$

Table 17. CCCC plate: basic interpolating functions for the amplitude displacements.

Symmetry class	Amplitude functions $\begin{cases} U(\xi,\eta,\zeta) = C_{u111}\phi_{u1}(\xi)\psi_{u1}(\eta)\chi_{u1}(\zeta) \\ V(\xi,\eta,\zeta) = C_{v111}\phi_{v1}(\xi)\psi_{v1}(\eta)\chi_{v1}(\zeta), \\ W(\xi,\eta,\zeta) = C_{w111}\phi_{w1}(\xi)\psi_{w1}(\eta)\chi_{w1}(\zeta) \end{cases}$
SS	$U = (\xi^2 - 0.25)(\eta^2 - 0.25)$ where $\begin{cases} \phi_{u1}(\xi) = \xi^2 - 0.25 \\ \psi_{u1}(\eta) = \eta^2 - 0.25 \\ \chi_{u1}(\zeta) = 1 \end{cases}$
	$V = (\xi^2 - 0.25)(\eta^2 - 0.25)$ where $\begin{cases} \phi_{v1}(\xi) = \xi^2 - 0.25 \\ \psi_{v1}(\eta) = \eta^2 - 0.25 \\ \chi_{v1}(\zeta) = 1 \end{cases}$
	$W = (\xi^2 - 0.25)(\eta^2 - 0.25)$ where $\begin{cases} \phi_{w1}(\xi) = \xi^2 - 0.25 \\ \psi_{w1}(\eta) = \eta^2 - 0.25 \\ \chi_{w1}(\zeta) = 1 \end{cases}$
SA	$U = (\xi^2 - 0.25)(\eta^3 - 0.25\eta)$ where $\begin{cases} \phi_{u1}(\xi) = \xi^2 - 0.25 \\ \psi_{u1}(\eta) = \eta^3 - 0.25\eta \\ \chi_{u1}(\zeta) = 1 \end{cases}$
	$V = (\xi^2 - 0.25)\left(\eta^3 - 0.25\eta\right)$ where $\begin{cases} \phi_{v1}(\xi) = \xi^2 - 0.25 \\ \psi_{v1}(\eta) = \eta^3 - 0.25\eta \\ \chi_{v1}(\zeta) = 1 \end{cases}$
	$W = (\xi^2 - 0.25)(\eta^3 - 0.25\eta)$ where $\begin{cases} \phi_{w1}(\xi) = \xi^2 - 0.25 \\ \psi_{w1}(\eta) = \eta^3 - 0.25\eta \\ \chi_{w1}(\zeta) = 1 \end{cases}$
AS	$U = (\xi^3 - 0.25\xi)(\eta^2 - 0.25)$ where $\begin{cases} \phi_{u1}(\xi) = \xi^3 - 0.25\xi \\ \psi_{u1}(\eta) = \eta^2 - 0.25 \\ \chi_{u1}(\zeta) = 1 \end{cases}$
	$V = (\xi^3 - 0.25\xi)(\eta^2 - 0.25)$ where $\begin{cases} \phi_{v1}(\xi) = \xi^3 - 0.25\xi \\ \psi_{v1}(\eta) = \eta^2 - 0.25 \\ \chi_{v1}(\zeta) = 1 \end{cases}$
	$W = (\xi^3 - 0.25\xi)(\eta^2 - 0.25)$ where $\begin{cases} \phi_{w1}(\xi) = \xi^3 - 0.25\xi \\ \psi_{w1}(\eta) = \eta^2 - 0.25 \\ \chi_{w1}(\zeta) = 1 \end{cases}$
AA	$U = (\xi^3 - 0.25\xi)(\eta^3 - 0.25\eta)$ where $\begin{cases} \phi_{u1}(\xi) = \xi^3 - 0.25\xi \\ \psi_{u1}(\eta) = \eta^3 - 0.25\eta \\ \chi_{u1}(\zeta) = 1 \end{cases}$
	$V = (\xi^3 - 0.25\xi)(\eta^3 - 0.25\eta)$ where $\begin{cases} \phi_{u1}(\xi) = \xi^3 - 0.25\xi \\ \psi_{u1}(\eta) = \eta^3 - 0.25\eta \\ \chi_{u1}(\zeta) = 1 \end{cases}$
	$W = (\xi^3 - 0.25\xi)(\eta^3 - 0.25\eta)$ where $\begin{cases} \phi_{u1}(\xi) = \xi^3 - 0.25\xi \\ \psi_{u1}(\eta) = \eta^3 - 0.25\eta \\ \chi_{u1}(\zeta) = 1 \end{cases}$

Exact Analysis of Wave Motions in Rods and Hollow Cylinders

Erasmo Viola and Alessandro Marzani

Dipartimento di Ingegneria delle Strutture, dei Trasporti, delle Acque, del Rilevamento, del Territorio - University of Bologna, Viale Risorgimento 2, 40136 Bologna, Italy

Abstract In a deformable isotropic infinitely long cylinder a discrete number of propagating guided modes regularly exists in a limited interval of frequency (f) and wavenumber (ξ). The calculation of the guided modes is best done via Helmholtz's method, where the Bessel functions are used to scale the scalar and wave potentials. Solving the three-dimensional wave equations, leads to displacement and stress componenets in terms of potential to be found. By imposing the stress free boundary conditions for the inner and outer surface of the cylinder, the dispersion equation can be obtained. The dispersion equation shows how the phase velocity, $c_p = 2\pi f/\xi$, change with the frequency. The group velocity, i.e. the speed of the propagating guided modes along the cylinder, can be obtained as $c_g = \partial(2\pi f)/\partial\xi$.

1 Introduction

This work describes the development of the governing equations for ultrasonic elastic wave propagation in cylindrical systems like rods and hollow cylinders. Based on the geometry and material properties of the waveguide, the equations determine what resonances can exist in order to satisfy the bulk wave propagation and the boundary conditions for the system. First, Navier's equations of motion in cylindrical coordinates are decoupled in a scalar and vector wave equations via Helmholtz potentials. The displacement and stress fields due to the bulk wave propagation are obtained in terms of Bessel functions. Then, the boundary conditions on displacements and stresses must be satisfied so that constructive combinations of these bulk waves, i.e. guided modes, can exist. This places restrictions on the bulk waves that must have the same frequency and the same wavenumber in the direction of propagation. They therefore have the same velocity component in this direction, as will have any wave which is described by the superposition of these waves. Physically, this means that for given geometry and material properties, suitable values of the frequency (time varying component) and the wavenumber (spatially varying component) combine to form a guided wave which propagates along the structure. These valid combinations can be found by solving the system of equations for its modal response, finding the zeros of the characteristic function. Solutions must be found iteratively by varying these two parameters until a valid root is converged upon. The solutions to the guided wave problem lie on continuous lines named dispersion curves that show how the velocity of the guided waves changes with the frequency.

The velocities which are of interest are the phase velocity and the group velocity. The phase velocity of a particular wave for a given frequency is the velocity at which the wavefront or crest travels and it can be expressed as $c_p = \omega/k$, where ω is the angular frequency of the wave and k is its wavenumber. The group velocity $c_g = \partial\omega/\partial k$ is the velocity at which the waves carry their potential energy along the structure. This latter velocity is of particular interest in the use of guided waves for non-destructive testing of structures.

An enormous amount of work has contributed to our current understanding of cylindrical wave propagation. The problem was first studied analytically in the late nineteenth century. Pochhammer and Chree (Chree, 1889) were the first researchers to mathematically investigate the propagation of free time-harmonic waves in an infinitely long cylinder and their names are still associated with the equation that describes some of the modes of a solid cylinder. However, most of the applications of cylindrical wave propagation have occurred much more recently. In the mid twentieth century, a good deal of research was performed on the modes of solid bars. In 1941, Bancroft (Bancroft, 1941) proposed a preliminary numerical study on the propagation of longitudinal waves in cylindrical bars. Based on the work of Bancroft, in 1943, Hudson (Hudson, 1943) attempted to study the dispersive nature of the fundamental flexural modes in a solid cylinder. A more detailed study on the longitudinal modes of a bar was given by Davies in 1948 (Davies, 1948). Later work by researchers such as Pao, Mindlin, Meeker and Meitzler (Pao and Mindlin, 1960)-(Pao, 1962)-(Meeker and Meitzler, 1972) fully developed all of the branches of the complete three dimensional problem of a solid circular cylinder in vacuum. Similar waves in hollow circular cylinders have been investigated, under the restriction of axial symmetry of motion, by McFadden (McFadden, 1954) and Hermann and Mirsky (Herrmann and Mirsky, 1956). The whole spectrum solution for hollow cylinders, i.e. axisymmetric and non-axisymmetric propagating modes, was given by Gazis (Gazis, 1959a)-(Gazis, 1959b). Nowadays, a full treatment of wave propagation in cylindrical structures can be found collected in classical textbooks, see for example: Achenbach (1973) and Auld (1990).

Based on these works, now we will consider the complete analytical derivation of the dispersion equation, for elastic rods and hollow cylinders. The result is that there are three families of modes that propagate along the structure with different behavior. The solution will be characterized according to the different mode families and some fundamental results will be given.

2 Equations of elasticity in cylindrical coordinates

In developing the cylindrical wave propagation model, certain assumptions will be made. This model assumes that the cylindrical systems, rods or hollow cylinders, are axisymmetric and infinitely long. The systems are considered immersed in vacuum and made of linear elastic homogeneous isotropic materials. These assumptions imply that no energy leakage in the surrounding media will be considered as well as no material attenuation. Although the geometry of the system is considered axisymmetric, it will be shown that non axisymmetric wave propagation can exist.

2.1 Equations of motion

In cylindrical coordinates (r, θ, z), the equations of motion in terms of stress components $(\sigma_r, \sigma_\theta, \sigma_z, \sigma_{r\theta}, \sigma_{rz}, \sigma_{\theta z})$ are given by:

$$\frac{\partial \sigma_r}{\partial r} + \frac{1}{r}\frac{\partial \sigma_{r\theta}}{\partial \theta} + \frac{\partial \sigma_{rz}}{\partial z} + \frac{1}{r}(\sigma_r - \sigma_\theta) + b_r = \mu \ddot{u}_r \quad (2.1)$$

$$\frac{\partial \sigma_{r\theta}}{\partial r} + \frac{1}{r}\frac{\partial \sigma_\theta}{\partial \theta} + \frac{\partial \sigma_{\theta z}}{\partial z} + \frac{2}{r}\sigma_{r\theta} + b_\theta = \mu \ddot{u}_\theta \quad (2.2)$$

$$\frac{\partial \sigma_{rz}}{\partial r} + \frac{1}{r}\frac{\partial \sigma_{\theta z}}{\partial \theta} + \frac{\partial \sigma_z}{\partial z} + \frac{1}{r}\sigma_{rz} + b_z = \mu \ddot{u}_z \quad (2.3)$$

where μ is the density, (b_r, b_θ, b_z) are the body forces per unit volume and $(\ddot{u}_r, \ddot{u}_\theta, \ddot{u}_z)$ are the accelerations, in the r, θ and z-direction, respectively. In addition, equilibrium involving the moments requires the symmetry conditions $\sigma_{ij} = \sigma_{ji}$. Hooke's law defines the stress-strain relations for an isotropic solid as $\sigma_{ij} = \lambda I_{1\varepsilon} + 2G\varepsilon_{ij}$, where λ and G are the Lamè constants, $I_{1\varepsilon}$ is the cubical dilatation $(I_{1\varepsilon} = \varepsilon_r + \varepsilon_\theta + \varepsilon_z)$ and $(\varepsilon_r, \varepsilon_\theta, \varepsilon_z, \varepsilon_{r\theta}, \varepsilon_{rz}, \varepsilon_{\theta z})$ are the strain components which are also symmetric, i.e. $\varepsilon_{ij} = \varepsilon_{ji}$. The strain-displacement relations for small strain are:

$$\varepsilon_r = \frac{\partial u_r}{\partial r}, \qquad \varepsilon_{r\theta} = \frac{1}{2}\left(\frac{1}{r}\frac{\partial u_r}{\partial \theta} + \frac{\partial u_\theta}{\partial r} - \frac{u_\theta}{r}\right) \quad (2.4)$$

$$\varepsilon_\theta = \frac{1}{r}\frac{\partial u_\theta}{\partial \theta} + \frac{u_r}{r}, \qquad \varepsilon_{rz} = \frac{1}{2}\left(\frac{\partial u_r}{\partial z} + \frac{\partial u_z}{\partial r}\right) \quad (2.5)$$

$$\varepsilon_z = \frac{\partial u_z}{\partial z}, \qquad \varepsilon_{\theta z} = \frac{1}{2}\left(\frac{\partial u_\theta}{\partial z} + \frac{1}{r}\frac{\partial u_z}{\partial \theta}\right) \quad (2.6)$$

Using eqs.(2.4)-(2.6), the stresses can be written in terms of displacement components:

$$\sigma_r = \lambda\left(\frac{\partial u_r}{\partial r} + \frac{1}{r}\frac{\partial u_\theta}{\partial \theta} + \frac{u_r}{r} + \frac{\partial u_z}{\partial z}\right) + 2G\frac{\partial u_r}{\partial r} \quad (2.7)$$

$$\sigma_\theta = \lambda\left(\frac{\partial u_r}{\partial r} + \frac{1}{r}\frac{\partial u_\theta}{\partial \theta} + \frac{u_r}{r} + \frac{\partial u_z}{\partial z}\right) + 2G\left(\frac{1}{r}\frac{\partial u_\theta}{\partial \theta} + \frac{u_r}{r}\right) \quad (2.8)$$

$$\sigma_z = \lambda\left(\frac{\partial u_r}{\partial r} + \frac{1}{r}\frac{\partial u_\theta}{\partial \theta} + \frac{u_r}{r} + \frac{\partial u_z}{\partial z}\right) + 2G\frac{\partial u_z}{\partial z} \quad (2.9)$$

$$\sigma_{r\theta} = G\left(\frac{1}{r}\frac{\partial u_r}{\partial \theta} + \frac{\partial u_\theta}{\partial r} - \frac{u_\theta}{r}\right) \quad (2.10)$$

$$\sigma_{rz} = G\left(\frac{\partial u_r}{\partial z} + \frac{\partial u_z}{\partial r}\right) \quad (2.11)$$

$$\sigma_{\theta z} = G\left(\frac{\partial u_\theta}{\partial z} + \frac{1}{r}\frac{\partial u_z}{\partial \theta}\right) \quad (2.12)$$

in addition to the cubical dilatation:

$$I_{1\varepsilon} = \frac{\partial u_r}{\partial r} + \frac{1}{r}\frac{\partial u_\theta}{\partial \theta} + \frac{u_r}{r} + \frac{\partial u_z}{\partial z} \quad (2.13)$$

Replacing in eqs.(2.1)-(2.3) the stresses in terms of displacements (2.7)-(2.12), leads to the Navier equations in terms of the displacement components:

$$G\left(\nabla^2 u_r - \frac{u_r}{r^2} - \frac{2}{r^2}\frac{\partial u_\theta}{\partial \theta}\right) + (\lambda + G)\frac{\partial I_{1\varepsilon}}{\partial r} + b_r = \mu \ddot{u}_r \tag{2.14}$$

$$G\left(\nabla^2 u_\theta - \frac{u_\theta}{r^2} + \frac{2}{r^2}\frac{\partial u_r}{\partial \theta}\right) + (\lambda + G)\frac{1}{r}\frac{\partial I_{1\varepsilon}}{\partial \theta} + b_\theta = \mu \ddot{u}_\theta \tag{2.15}$$

$$G\nabla^2 u_z + (\lambda + G)\frac{\partial I_{1\varepsilon}}{\partial z} + b_z = \mu \ddot{u}_z \tag{2.16}$$

where ∇^2 is the Laplacian operator. Navier equations can be rewritten in synthetic vectorial form:

$$G\nabla^2 \mathbf{u} + (\lambda + G)\nabla\left(\nabla \cdot \mathbf{u}\right) = \mu \frac{\partial^2 \mathbf{u}}{\partial t^2} \tag{2.17}$$

where the body forces have been neglected and $\mathbf{u} = \left[u_r, u_\theta, u_z\right]^{\mathrm{T}}$ is the displacement vector. The differential operators in cylindrical coordinates are given in Appendix A. The resulting Navier equations constitute a system of three, second-order partial differential equations. The equations are coupled and, except for special cases, have a structure more complex than the one of the standard wave equation.

2.2 Helmholtz decomposition

Helmholtz's theorem states that any vector can be decomposed into the sum of the gradient of a scalar potential plus the curl of a vector potential. For the displacement vector $\mathbf{u} = \left[u_r, u_\theta, u_z\right]^{\mathrm{T}}$, which is a function of the position and time, t, and is finite, uniform, continuous, and vanishes at infinity, we can accordingly write:

$$\mathbf{u} = \nabla f + \nabla \times \mathbf{h} = \begin{bmatrix} \partial f/\partial r \\ \frac{1}{r}\partial f/\partial \theta \\ \partial f/\partial z \end{bmatrix} + \frac{1}{r}\begin{bmatrix} \hat{e}_r & r\hat{e}_\theta & \hat{e}_z \\ \partial/\partial r & \partial/\partial \theta & \partial/\partial z \\ h_r & r h_\theta & h_z \end{bmatrix} \tag{2.18}$$

where f is a compressional scalar potential and $\mathbf{h} = (h_r, h_\theta, h_z)$ an equivoluminal vector potential. In general $\nabla \cdot \mathbf{h} = g(\mathbf{r}, t)$ where g is a function of the coordinate vector, $\mathbf{r} = (r, \theta, z)$, and the time, t. The function g can be chosen arbitrarily due to the *gauge invariance* of the field transformations (Gazis, 1959a). This means that the potentials are not unique, but we can always select them so that $\nabla \cdot \mathbf{h} = 0$. Making the equivoluminal vector potential a zero-divergence vector implies that the field is solenoidal (i.e. there are no sources or sinks of energy within the region) and provides the necessary additional condition to uniquely determine the three components of $\mathbf{u}$ from the four components of the two Helmholtz potentials. By using Helmholtz potentials, the Navier's equation of motion (2.17) yields:

$$\nabla\left[(\lambda + 2G)\nabla^2 f - \mu\left(\partial^2 f/\partial t^2\right)\right] + \nabla \times \left[G\nabla^2 \mathbf{h} - \mu\left(\partial^2 \mathbf{h}/\partial t^2\right)\right] = 0 \tag{2.19}$$

A sufficient condition for this equation to hold is that both terms vanish, which leads to the standard scalar and vector decoupled wave equations:

$$\nabla^2 f = \frac{1}{c_L^2}\frac{\partial^2 f}{\partial t^2}, \qquad \nabla^2 \mathbf{h} = \frac{1}{c_T^2}\frac{\partial^2 \mathbf{h}}{\partial t^2} \tag{2.20}$$

where $c_L = \left[(\lambda + 2G)/\mu\right]^{1/2}$ and $c_T = (G/\mu)^{1/2}$ are the longitudinal and shear bulk wave material velocities, respectively. Exploiting the Laplacian, the vector wave equation in (2.20) yields to three scalar equations in the r, θ and z-direction:

$$\nabla^2 h_r - \frac{h_r}{r^2} - \frac{2}{r^2}\frac{\partial h_\theta}{\partial \theta} = \frac{1}{c_T^2}\frac{\partial^2 h_r}{\partial t^2} \quad (2.21)$$

$$\nabla^2 h_\theta - \frac{h_\theta}{r^2} + \frac{2}{r^2}\frac{\partial h_r}{\partial \theta} = \frac{1}{c_T^2}\frac{\partial^2 h_\theta}{\partial t^2} \quad (2.22)$$

$$\nabla^2 h_z = \frac{1}{c_T^2}\frac{\partial^2 h_z}{\partial t^2} \quad (2.23)$$

Navier equation of motion (2.17) is thus decoupled in four scalar equations (2.20)-(2.23). It is obvious, and can be shown by direct substitution, that if the potentials satisfy the four scalar wave equations, the Navier equation of motion (2.17) is identically satisfied. The introduction of the potentials f and $\mathbf{h}$ has simplified the equation of motion to the degree that two of the potentials (f and h_z) satisfy the uncoupled scalar wave equations (2.20) and (2.23). However, the remaining two potentials (h_r and h_θ) must still satisfy the more complex coupled equations (2.21) and (2.22). Equation (2.18) gives:

$$u_r = \frac{\partial f}{\partial r} + \frac{1}{r}\frac{\partial h_z}{\partial \theta} - \frac{\partial h_\theta}{\partial z} \quad (2.24)$$

$$u_\theta = \frac{1}{r}\frac{\partial f}{\partial \theta} + \frac{\partial h_r}{\partial z} - \frac{\partial h_z}{\partial r} \quad (2.25)$$

$$u_z = \frac{\partial f}{\partial z} + \frac{1}{r}\frac{\partial (r h_\theta)}{\partial r} - \frac{1}{r}\frac{\partial h_r}{\partial \theta} \quad (2.26)$$

At last, inserting eqs.(2.24)-(2.26) into eqs.(2.7)-(2.12), the stresses in terms of potentials can be obtained:

$$\sigma_r = \lambda \nabla^2 f + 2G\left(\frac{\partial^2 f}{\partial r^2} - \frac{\partial^2 h_\theta}{\partial r \partial z} - \frac{1}{r^2}\frac{\partial h_z}{\partial \theta} + \frac{1}{r}\frac{\partial^2 h_z}{\partial r \partial \theta}\right) \quad (2.27)$$

$$\sigma_\theta = \lambda \nabla^2 f + 2G\left(\frac{\partial f}{r\partial r} + \frac{\partial^2 f}{r^2 \partial \theta^2} + \frac{1}{r}\frac{\partial^2 h_r}{\partial \theta \partial z} - \frac{\partial h_\theta}{r \partial z} + \frac{\partial h_z}{r^2 \partial \theta} - \frac{1}{r}\frac{\partial^2 h_z}{\partial r \partial \theta}\right) \quad (2.28)$$

$$\sigma_z = \lambda \nabla^2 f + 2G\left(\frac{\partial^2 f}{\partial z^2} + \frac{1}{r}\frac{\partial h_\theta}{\partial z} + \frac{\partial^2 h_\theta}{\partial r \partial z} - \frac{1}{r}\frac{\partial^2 h_r}{\partial \theta \partial z}\right) \quad (2.29)$$

$$\sigma_{r\theta} = G\left(\frac{2}{r}\frac{\partial^2 f}{\partial r \partial \theta} - \frac{2}{r^2}\frac{\partial f}{\partial \theta} - \frac{\partial h_r}{r \partial z} + \frac{\partial^2 h_r}{\partial r \partial z} - \frac{\partial^2 h_\theta}{r \partial \theta \partial z} + \frac{\partial^2 h_z}{r^2 \partial \theta^2} - \frac{\partial^2 h_z}{\partial r^2} + \frac{\partial h_z}{r \partial r}\right) \quad (2.30)$$

$$\sigma_{rz} = G\left(2\frac{\partial^2 f}{\partial r \partial z} + \frac{\partial h_r}{r^2 \partial \theta} - \frac{1}{r}\frac{\partial^2 h_r}{\partial r \partial \theta} + \frac{\partial h_\theta}{r \partial r} + \frac{\partial^2 h_\theta}{\partial r^2} - \frac{h_\theta}{r^2} - \frac{\partial^2 h_\theta}{\partial z^2} + \frac{1}{r}\frac{\partial^2 h_z}{\partial \theta \partial z}\right) \quad (2.31)$$

$$\sigma_{\theta z} = G\left(\frac{2}{r}\frac{\partial^2 f}{\partial \theta \partial z} + \frac{\partial^2 h_r}{\partial z^2} - \frac{1}{r^2}\frac{\partial^2 h_r}{\partial \theta^2} + \frac{1}{r^2}\frac{\partial h_\theta}{\partial \theta} + \frac{1}{r}\frac{\partial^2 h_\theta}{\partial r \partial \theta} - \frac{\partial^2 h_z}{\partial r \partial z}\right) \quad (2.32)$$

It is clear that although the form of the partial differential equations of motion has been simplified, the expressions for the displacement and stress components have become more complex.

3 Wave propagation in the axial z-direction

A general form of the potentials representing a wave propagating in unbounded media assumes the aspect:

$$f, h_r, h_\theta, h_z = R(r)\Theta(\theta)Z(z)e^{i(\mathbf{k}\cdot\mathbf{r}-\omega t)} = R(r)\Theta(\theta)Z(z)e^{i(k_r r + k_\theta \theta + k_z z - \omega t)} \tag{3.1}$$

where $\mathbf{k}=(k_r,\ k_\theta,\ k_z)$ is the wavenumber, ω the circular frequency, $R(r)$, $\Theta(\theta)$ and $Z(z)$ describe how the field varies in each spatial coordinate. It can be assumed that the wave does not propagate in the r-direction and that the displacement field does not vary in the θ or z-direction except for the harmonic oscillation described by the wavenumber. Based on these assumptions, the potentials (3.1) can be rewritten as:

$$f, h_r, h_\theta, h_z = R(r)e^{i(k_\theta \theta + \xi z - \omega t)} \tag{3.2}$$

where k_θ is the angular wavenumber component and $\xi = k_z$ is the component of the wavenumber vector in the propagation z-direction, which will henceforth be referred to simply as the wavenumber. For wave propagating in the z-direction, due to the radial symmetry of the system, the solution at θ and $\theta+2\pi$ has to be unique. Therefore, the angular wavenumber component, k_θ, must be an integer, n. Phase angles may be arbitrarily added to each of the potential functions, since the added phases will be compensated by the coefficients of the functions themselves. The Helmholtz potentials components (3.2) for cylindrical waves propagating in the z-direction can be assumed as (Achenbach, 1973):

$$f = F(r)\cos(n\theta + \theta_0)e^{i(\xi z - \omega t)} \tag{3.3}$$
$$h_r = H_r(r)\sin(n\theta + \theta_0)e^{i(\xi z - \omega t)} \tag{3.4}$$
$$h_\theta = H_\theta(r)\cos(n\theta + \theta_0)e^{i(\xi z - \omega t)} \tag{3.5}$$
$$h_z = H_z(r)\sin(n\theta + \theta_0)e^{i(\xi z - \omega t)} \tag{3.6}$$

The solution results in sines and cosines for the θ dependence. The constant θ_0 is introduced to allow us to use the same general displacement potentials for both torsional and flexural motion. As mentioned before, the three displacement components are related to four scalar potential functions: the scalar potential f and the three components of the vector potential $\mathbf{h}$. The property of the *gauge invariance* tell us that one of the three potentials h_i (i=r,θ,z) can be set equal to zero without loss of generality in the solution (Gazis, 1959a). Physically this implies that the displacement field corresponding to an equivoluminal potential h_i can also be derived by a combination of the other two equivoluminal potentials. Analytically, this allows us to eliminate one of the scalar equations because it is linearly dependent of the others. For the problem considered here, it is convenient to require that:

$$H_\theta(r) = -H_r(r) \tag{3.7}$$

Under this hypothesis, the four scalar differential equations reduce to the following three independent scalar equations:

$$\frac{\partial^2 F}{\partial r^2} + \frac{1}{r}\frac{\partial F}{\partial r} + \left(p^2 - \frac{n^2}{r^2}\right) F = 0 \tag{3.8}$$

$$\frac{d^2 H_r}{dr^2} + \frac{1}{r}\frac{dH_r}{dr} + \left(q^2 - \frac{(n+1)^2}{r^2}\right) H_r = 0 \tag{3.9}$$

$$\frac{d^2 H_z}{dr^2} + \frac{1}{r}\frac{dH_z}{dr} + \left(q^2 - \frac{n^2}{r^2}\right) H_z = 0 \tag{3.10}$$

where:

$$p^2 = \frac{\omega^2}{c_L^2} - \xi^2 = \xi^2\left(\frac{c_p^2}{c_L^2} - 1\right), \quad q^2 = \frac{\omega^2}{c_T^2} - \xi^2 = \xi^2\left(\frac{c_p^2}{c_T^2} - 1\right) \tag{3.11}$$

and $c_p = \omega/\xi$ is the phase velocity of waves propagating in the axial z-direction. The differential operator:

$$B_{n,r} = \left[\frac{\partial^2}{\partial r^2} + \frac{1}{r}\frac{\partial}{\partial r} + \left(1 - \frac{n^2}{r^2}\right)\right] \tag{3.12}$$

where, in general, n is an arbitrary complex number, is known as Bessel's operator. By the aid of the Bessel operator, eqs.(3.8)-(3.10), can be written in the synthetic form:

$$B_{n,pr}[F] = 0 \tag{3.13}$$

$$B_{n+1,qr}[H_r] = 0 \tag{3.14}$$

$$B_{n,qr}[H_z] = 0 \tag{3.15}$$

The series solution of these differential equations are the Bessel functions, also termed cylindrical functions. There are several combinations of Bessel functions available to produce the solutions of eqs.(3.13)-(3.15):

- Bessel functions of the first and second kind, $\mathbf{J}_n$ and $\mathbf{Y}_n$, respectively,
- modified Bessel functions of the first and second kind, $\mathbf{I}_n$ and $\mathbf{K}_n$, respectively,
- Hankel functions named also Bessel functions of the third kind, $\mathbf{H}_{1n} = \mathbf{J}_n + i\mathbf{Y}_n$ and $\mathbf{H}_{2n} = \mathbf{J}_n - i\mathbf{Y}_n$,

A classical representation of some low-order Bessel functions, for real argument x, is given in Figure 1. From here it is obvious that the ordinary Bessel functions are oscillatory in nature and that the modified Bessel functions tend to look more like decaying and growing exponentials (this is a rough description only).

3.1 Considerations on the choice of the Bessel functions

In order to fully satisfy the Bessel operator, a pair of linearly independent Bessel functions is required, which physically correspond to inward and outward propagating waves. Thus, each of the three independent functions, $F(r)$, $H_r(r)$, and $H_z(r)$, and a

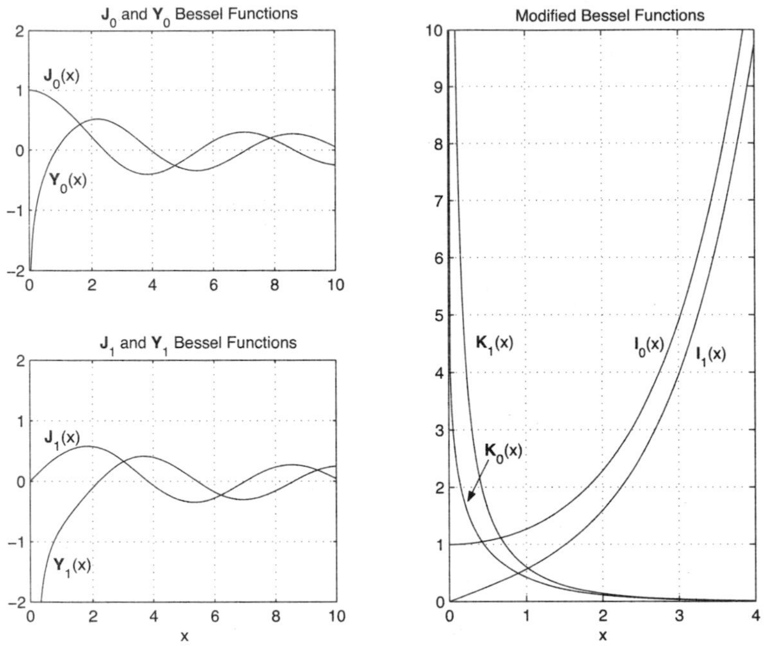

Figure 1. Classical representation of the first and second kind of the unmodified and modified Bessel functions

linear combination of them satisfy the Bessel differential operator. The "standard" pair of functions that satisfy the Bessel differential operator are the Bessel function of the first kind ($\mathbf{J}_n$) and the second kind ($\mathbf{Y}_n$). These functions represent a pair of standing waves. The modified Bessel functions, $\mathbf{I}_n$ and $\mathbf{K}_n$, which represent exponential waves, may also be used. The third valid combination of functions are the Hankel functions of the first and second kind ($\mathbf{H}_n^{(1)}$ and $\mathbf{H}_n^{(2)}$), which represent waves that propagate toward or away from the origin (radial directions). Since we are studying guided wave propagation in axial direction, standing or exponential waves best represent the solution. Here, the solution is given in terms of Bessel's function of the first $\mathbf{J}_i$ and second $\mathbf{Y}_i$ kind of order n for $F(r)$ and $H_z(r)$, and order $n+1$ for $H_r(r)$:

$$F(r) = A_1\mathbf{J}_n(pr) + A_2\mathbf{Y}_n(pr) \tag{3.16}$$

$$H_r(r) = B_1\mathbf{J}_{n+1}(qr) + B_2\mathbf{Y}_{n+1}(qr) \tag{3.17}$$

$$H_z(r) = C_1\mathbf{J}_n(qr) + C_2\mathbf{Y}_n(qr) \tag{3.18}$$

where constants A_i, B_i, and C_i are to be determined by the boundary conditions of the given problem. It is worth observing that the choice between the unmodified or modified Bessel functions is mainly based on the argument of the Bessel functions themselves. The arguments of the Bessel functions, pr and qr, will be real or purely imaginary. The Bessel functions $\mathbf{J}_n$ and $\mathbf{Y}_n$ can handle complex arguments, however for some problems it is more stable and elegant to use for imaginary arguments the modified Bessel functions, $\mathbf{I}_n$ and $\mathbf{K}_n$. In fact, since both of the unmodified Bessel functions ($\mathbf{J}_n$ and $\mathbf{Y}_n$) increase exponentially along the imaginary axis, the effect of the inward and outward propagating waves cannot be clearly separated. This, in turning, may lead to an unstable solution as the arguments become large. This problem is analogous to the "*large fd*" problem that is discussed for plates and is especially important for large radius pipes. In this note, regarding the problem we are dealing with, the choice of the unmodified Bessel function ensures a correct solution. For future reference, the limiting values for the zero and first order Bessel functions on the interval of the argument $[0 \leq x \leq \infty]$ are summarized in Table 1, where the symbol $\sim$ means an oscillating solution.

Table 1. Limiting values of the low-order integer Bessel functions.

	$\mathbf{J}_0(x)$	$\mathbf{J}_1(x)$	$\mathbf{Y}_0(x)$	$\mathbf{Y}_1(x)$	$\mathbf{I}_0(x)$	$\mathbf{I}_1(x)$	$\mathbf{K}_0(x)$	$\mathbf{K}_1(x)$
$x \to 0$	1	0	$-\infty$	$-\infty$	1	0	∞	∞
$x \to \infty$	$\sim$	$\sim$	$\sim$	$\sim$	∞	∞	0	0

3.2 Wave propagation displacement field

We are now ready to apply the information that we know about the solution in order to calculate the displacements at an arbitrary location in the medium. By means of the solution (3.16)-(3.18) and the position (3.7), the potentials that satisfy the scalar equations of motion (3.8)-(3.10) are:

$$f = [A_1\mathbf{J}_n(pr) + A_2\mathbf{Y}_n(pr)]\cos(n\theta+\theta_0)e^{i(\xi z-\omega t)} \tag{3.19}$$

$$h_r = [B_1\mathbf{J}_{n+1}(qr) + B_2\mathbf{Y}_{n+1}(qr)]\sin(n\theta+\theta_0)e^{i(\xi z-\omega t)} \tag{3.20}$$

$$h_\theta = -[B_1\mathbf{J}_{n+1}(qr) + B_2\mathbf{Y}_{n+1}(qr)]\cos(n\theta+\theta_0)e^{i(\xi z-\omega t)} \tag{3.21}$$

$$h_z = [C_1\mathbf{J}_n(qr) + C_2\mathbf{Y}_n(qr)]\sin(n\theta+\theta_0)e^{i(\xi z-\omega t)} \tag{3.22}$$

Inserting the potentials (3.19)-(3.22) into eqs.(2.24)-(2.26) after using positions (3.16)-(3.18), leads to the displacement field in the form:

$$u_r = \left(F' + \frac{n}{r}H_z + i\xi H_r\right)\cos(n\theta+\theta_0)e^{i(\xi z-\omega t)} \tag{3.23}$$

$$u_\theta = \left(-\frac{n}{r}F + i\xi H_r - H_z'\right)\sin(n\theta+\theta_0)e^{i(\xi z-\omega t)} \tag{3.24}$$

$$u_z = \left(i\xi F - \frac{n+1}{r}H_r - H_r'\right)\cos(n\theta+\theta_0)e^{i(\xi z-\omega t)} \tag{3.25}$$

where the prime notation indicates the derivative with respect to the radius, r.

3.3 Wave propagation stress field

Once the displacement field in terms of potentials is found the stresses can also be defined in terms of potentials. For the problem we are studying, not all the stress components are necessary for the solution. In fact, for a free cylindrical rod, the boundary conditions can be uniquely determined by imposing a stress-free condition on the outer surface of the system. Therefore, only the σ_r, $\sigma_{r\theta}$ and σ_{rz} are sufficient. Substituting the expressions of the displacements in terms of potentials (3.23)-(3.25), into eqs.(2.7), (2.10) and (2.11), yields:

$$\sigma_r = S_r \cos(n\theta + \theta_0) e^{i(\xi z - \omega t)} \tag{3.26}$$

$$\sigma_{r\theta} = S_{r\theta} \sin(n\theta + \theta_0) e^{i(\xi z - \omega t)} \tag{3.27}$$

$$\sigma_{rz} = S_{rz} \cos(n\theta + \theta_0) e^{i(\xi z - \omega t)} \tag{3.28}$$

where:

$$S_r = \lambda \left(F'' + \frac{F'}{r} - \xi^2 F - \frac{n^2}{r^2} F \right) + 2G \left(F'' + i\xi H_r' - \frac{n}{r^2} H_z + \frac{n}{r} H_z' \right) \tag{3.29}$$

$$S_{r\theta} = G \left(\frac{2n}{r^2} F - \frac{2n}{r} F' - i\xi \frac{n+1}{r} H_r + i\xi H_r' - \frac{n^2}{r^2} H_z + \frac{1}{r} H_z' - H_z'' \right) \tag{3.30}$$

$$S_{rz} = G \left(2i\xi F' - H_r'' - \frac{1+n}{r} H_r' - \xi^2 H_r + \frac{1+n}{r^2} H_r + i\xi \frac{n}{r} H_z \right) \tag{3.31}$$

Before concluding this section, we remark that the foregoing potentials depend upon eight constants, A_i, B_i, C_i (for $i = 1, 2$), and n and θ_0. Regardless of the values of these constants, the displacements and stresses derived from these potentials satisfy the equations of motion. In what follows, we will assign particular values for these constants in order to investigate some specific motions of practical interest of wave propagation in cylindrical waveguide.

4 Axial waves in solid cylinder

Consider a solid cylinder of circular cross section of radius b. Since the wave motion is finite at $r = 0$, the coefficients A_2, B_2, C_2 of the Bessel functions of the second kind, $\mathbf{Y}_n$, must be zero in the general solution. The potentials (3.19)-(3.22) turn into the following:

$$f = A_1 \mathbf{J}_n(pr) \cos(n\theta + \theta_0) e^{i(\xi z - \omega t)} \tag{4.1}$$

$$h_r = B_1 \mathbf{J}_{n+1}(qr) \sin(n\theta + \theta_0) e^{i(\xi z - \omega t)} \tag{4.2}$$

$$h_\theta = -B_1 \mathbf{J}_{n+1}(qr) \cos(n\theta + \theta_0) e^{i(\xi z - \omega t)} \tag{4.3}$$

$$h_z = C_1 \mathbf{J}_n(qr) \sin(n\theta + \theta_0) e^{i(\xi z - \omega t)} \tag{4.4}$$

and therefore the amplitudes result as:

$$F(r) = A_1 \mathbf{J}_n(pr) \tag{4.5}$$

$$H_r(r) = B_1 \mathbf{J}_{n+1}(qr) \tag{4.6}$$

$$H_z(r) = C_1 \mathbf{J}_n(qr) \tag{4.7}$$

The above solution forms a complete set of modes for time-harmonic waves propagating in the axial direction of a solid cylinder of circular cross-section. For a given boundary value problem, the solution can be obtained as a linear superposition of these modes.

In what follows, the dispersion equation will be derived; this dispersion equation relates the frequency ω, the axial wavenumber ξ and the circumferential order n. The stresses can be derived in terms of the potentials by using the amplitudes (4.5)-(4.7) to express eqs.(3.26)-(3.28):

$$\begin{bmatrix} \sigma_r \\ \sigma_{rz} \\ \sigma_{r\theta} \end{bmatrix} = G \begin{bmatrix} d_{11}^J(n,r) & d_{12}^J(n,r) & d_{13}^J(n,r) \\ d_{21}^J(n,r) & d_{22}^J(n,r) & d_{23}^J(n,r) \\ d_{31}^J(n,r) & d_{32}^J(n,r) & d_{33}^J(n,r) \end{bmatrix} \begin{bmatrix} A_1 \\ B_1 \\ C_1 \end{bmatrix} \tag{4.8}$$

Note that $d_{ij}^J(n,r)$ are functions of the circumferential order n, the frequency, ω, the axial wavenumber, ξ, as well as of the radius r. The upper script J denotes the fact that the coefficient depends on the $\mathbf{J}_n(x)$ Bessel functions of the first kind. These coefficients are given in Appendix B and can be obtained by applying the relationships among Bessel functions and their derivatives. For a cylinder immersed in vacuum, the free-stress conditions for $r = b$ result as:

$$\sigma_r|_{r=b} = 0, \qquad \sigma_{r\theta}|_{r=b} = 0, \qquad \sigma_{rz}|_{r=b} = 0 \tag{4.9}$$

Applying the conditions (4.9), eq.(4.8) yields:

$$\begin{bmatrix} d_{11}^J(n,b) & d_{12}^J(n,b) & d_{13}^J(n,b) \\ d_{21}^J(n,b) & d_{22}^J(n,b) & d_{23}^J(n,b) \\ d_{31}^J(n,b) & d_{32}^J(n,b) & d_{33}^J(n,b) \end{bmatrix} \begin{bmatrix} A_1 \\ B_1 \\ C_1 \end{bmatrix} = \begin{bmatrix} 0 \\ 0 \\ 0 \end{bmatrix} \tag{4.10}$$

or in a more synthetic form:

$$\mathbf{D}(n,\xi,\omega)\mathbf{A} = 0 \tag{4.11}$$

where $\mathbf{A} = [A_1, B_1, C_1]^{\mathrm{T}}$ and:

$$\mathbf{D}(n,\xi,\omega) = \begin{bmatrix} d_{11}^J(n,b) & d_{12}^J(n,b) & d_{13}^J(n,b) \\ d_{21}^J(n,b) & d_{22}^J(n,b) & d_{23}^J(n,b) \\ d_{31}^J(n,b) & d_{32}^J(n,b) & d_{33}^J(n,b) \end{bmatrix} \tag{4.12}$$

The dependence of D_{ij} on n and r is explicitly indicated in the coefficients matrix $\mathbf{D}$, while it is implicit for ξ and ω. For nontrivial solutions, the determinant of the coefficients matrix must vanish:

$$|\mathbf{D}(n,\xi,\omega)| = 0 \tag{4.13}$$

The above equation, for assigned radius and circumferential order, provides the desired dispersion relationship between the wavenumber ξ and the frequency ω. For each given ξ value, eq.(4.13) may have many roots for ω. Solutions to this equation yield a family of dispersion curves. Each one is called a branch, which corresponds to a wave mode propagating in the axial direction of the cylinder. Once a root is obtained, it can be substituted back into eq.(4.11) to determine the corresponding eigenvector, $\mathbf{A}$. Obviously, $\mathbf{A}$ can only be determined up to an arbitrary multiplier. For a given ξ and a given value of n, let the ith root of eq.(4.13) be denoted by ω_i and let the corresponding eigenvector be $\mathbf{A}_i = [\mathbf{A}_1^i, \mathbf{B}_1^i, \mathbf{C}_1^i]^{\mathrm{T}}$. It then follows from substituting (4.5)-(4.7) into the expression for the displacement (3.23)-(3.25), that the corresponding mode can be written as:

$$\begin{bmatrix} u_r^{(i)} \\ u_\theta^{(i)} \\ u_z^{(i)} \end{bmatrix} = \begin{bmatrix} U_r^{(i)}(r)\cos(n\theta+\theta_0) \\ U_\theta^{(i)}(r)\sin(n\theta+\theta_0) \\ U_z^{(i)}(r)\cos(n\theta+\theta_0) \end{bmatrix} e^{i(\xi z-\omega t)} \tag{4.14}$$

with:

$$\begin{bmatrix} U_r^{(i)} \\ U_\theta^{(i)} \\ U_z^{(i)} \end{bmatrix} = \begin{bmatrix} \mathbf{J}_n'(pr) & i\xi\mathbf{J}_{n+1}(qr) & \frac{n\mathbf{J}_n(qr)}{r} \\ -\frac{n\mathbf{J}_n(pr)}{r} & i\xi\mathbf{J}_{n+1}(qr) & -q\mathbf{J}_n'(qr) \\ i\xi\mathbf{J}_n(pr) & -q\mathbf{J}_n(qr) & 0 \end{bmatrix} \mathbf{A}_i \tag{4.15}$$

and where the prime still holds the meaning of derivative with respect to the radius. It should be noted that the waves described by the solutions (4.1)-(4.4) are rather complicated. So, to gain some insight about the wave motion, let us consider some special cases of the above general solutions.

4.1 Torsional waves in solid cylinder

In a pure torsional motion, u_θ should be the only nonzero displacement component. It should also be independent of θ. These conditions are met by setting, in the general solution (4.1)-(4.4), $A_1 = B_1 = n = 0$, and $\theta_0 = \pi/2$. In fact, under these assumptions the displacement field follows from (3.23)-(3.25):

$$u_r = 0 \tag{4.16}$$

$$u_\theta = [-C_1\mathbf{J}_0'(qr)]\, e^{i(\xi z-\omega t)} = [C_1 q\mathbf{J}_1(qr)]\, e^{i(\xi z-\omega t)} \tag{4.17}$$

$$u_z = 0 \tag{4.18}$$

where we have used the property of Bessel functions $\mathbf{J}_0'(x) = -\mathbf{J}_1(x)$. Some recurrence formulas for the Bessel functions are given in Appendix C. The corresponding stresses follow from eqs.(3.26)-(3.28):

$$\sigma_r = 0 \tag{4.19}$$

$$\sigma_{r\theta} = GC_1\left[q^2\mathbf{J}_0(qr) - \frac{2q}{r}\mathbf{J}_1(qr)\right] e^{i(\xi z-\omega t)} \tag{4.20}$$

$$\sigma_{rz} = 0 \tag{4.21}$$

The traction-free boundary condition at $r = b$, yields:

$$qb\mathbf{J}_0(qb) - 2\mathbf{J}_1(qb) = 0 \tag{4.22}$$

This is the dispersion equation for the torsional waves propagating in the axial direction of a solid cylinder of radius b. The left hand side of eq.(4.22) describes an oscillating function around the zero value for increasing argument qb. Thus, an infinite number of roots exist. Each of them represents a particular wave mode. Since, for very small argument x, the Bessel's function $\mathbf{J}_n$ for $n = 0, 1$ gives:

$$\mathbf{J}_0(x) \simeq 1, \qquad \mathbf{J}_1(x) \simeq x/2 \tag{4.23}$$

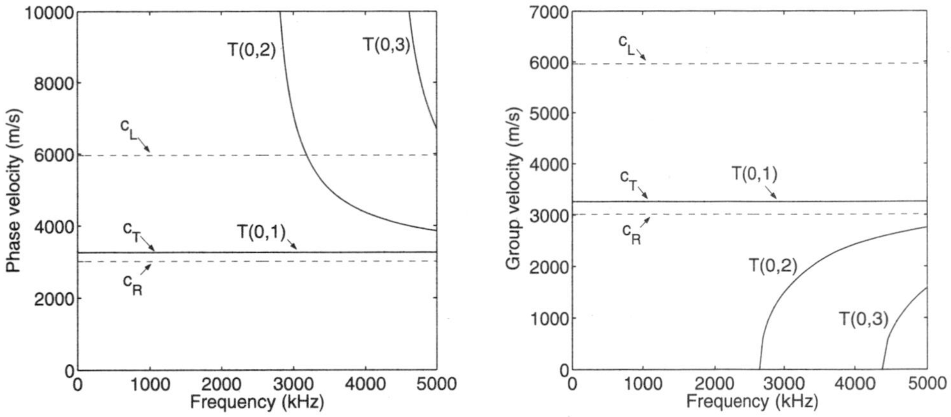

Figure 2. Torsional modes for a 1-mm radius steel bar: $c_L = 5960 \ m/s$, $c_T = 3260 \ m/s$ and $\mu = 7800 \ \ kg/m^3$

one can easily see that $q = 0$ is a root of eq.(4.22). Using eqs.(4.23) in eq.(4.17) for the $q = 0$ mode, we have:

$$u_\theta = Ure^{i(\xi z - \omega t)} \tag{4.24}$$

This displacement represents the well-known lowest torsional mode, named $T(0, 1)$. In this lowest mode, the displacement is proportional to the radius, and the motion consists of a rigid rotation of each cross section of the cylinder about its center. Note that, since $q = (\omega^2/c_T^2 - \xi^2) = 0$ implies that the phase velocity equals the shear bulk velocity, $c_p = c_T$, this mode is not dispersive. The other roots of eq.(4.22) must be obtained numerically. Higher torsional modes are dispersive with frequencies which follow from the definition of q as:

$$\left(\frac{\omega b}{c_T}\right)^2 = (q_i b)^2 + (\xi b)^2 \tag{4.25}$$

where $q_i b$ are the roots of eq.(4.22). The first five roots are:

$$q_1 b = 0, \quad q_2 b \simeq 5.14, \quad q_3 b \simeq 8.42, \quad q_4 b \simeq 11.62, \quad q_5 b \simeq 14.79 \tag{4.26}$$

It is noted that for a given real-valued frequency, the wavenumber may be real-valued or imaginary. This is similar to the case of shear waves in a layer, where the branches are hyperboles for real-valued ξ and circles for imaginary values of the wavenumber. Higher modes start to propagate at their cut-on frequency that can be obtained from eq.(4.25) for a zero wavenumber. The cut-on frequencies for the first five modes in $[Hz - m]$ are:

$$\omega_1 b = 0, \quad \omega_2 b \simeq 1.67, \quad \omega_3 b \simeq 2.74, \quad \omega_4 b \simeq 3.79, \quad \omega_5 b \simeq 4.82 \tag{4.27}$$

Below the cut-on frequency a mode does not propagate and can exist only as an evanescent wave. From the behavior of Bessel functions it follows that eq.(4.22) has no roots for $q^2 < 0$. This means that the phase velocity of the torsional waves is always greater than or equal to the shear wave speed c_T. The torsional modes for a 1-mm radius steel bar ($c_L = 5960 \quad m/s$, $c_T = 3260 \quad m/s$ and $\mu = 7800 \quad kg/m^3$) are shown in Figure 2 in terms of phase and group velocity. In the same Figure 2, the dashed lines have been used to indicate the bulk shear velocity c_T, the bulk longitudinal velocity c_L as long as the Rayleigh velocity c_T. It can be seen that up to 5 Mhz, only three modes are present: the non-dispersive $T(0,1)$ mode that corresponds to the $q_0 b$ solution and the modes $T(0,2)$ and $T(0,3)$ for the solutions $q_2 b$ and $q_3 b$, respectively.

4.2 Longitudinal waves in solid cylinder

Longitudinal waves refer to the axially symmetrical motion in the cylinder. They are characterized by the presence of displacement components in the radial and axial directions, but none in the θ-direction. Furthermore, these nonzero displacement components must be independent of θ because of the axisymmetric requirement. It is seen that these conditions are met by setting $C_1 = n = 0$, and $\theta_0 = 0$ in the general solution given by eqs.(4.1)-(4.4). In this case, it follows from eq.(3.23) through eq.(3.25) that:

$$u_r = \left[-A_1 p \mathbf{J}_1(pr) + i\xi B_1 \mathbf{J}_1(qr)\right] e^{i(\xi z - \omega t)} \tag{4.28}$$

$$u_\theta = 0 \tag{4.29}$$

$$u_z = \left[i\xi A_1 \mathbf{J}_0(pr) - C_1 q \mathbf{J}_0(qr)\right] e^{i(\xi z - \omega t)} \tag{4.30}$$

where the relation $\mathbf{J}'_n(x) = \mathbf{J}_{n-1}(x) - \frac{n}{x}\mathbf{J}_n(x)$ has been used. From eqs.(3.26)-(3.28) the corresponding pertinent stresses are found:

$$\begin{bmatrix} \sigma_r \\ \sigma_{rz} \end{bmatrix} = G \begin{bmatrix} d_{11}^J(0,r) & d_{21}^J(0,r) \\ d_{12}^J(0,r) & d_{22}^J(0,r) \end{bmatrix} \begin{bmatrix} A_1 \\ B_1 \end{bmatrix} \tag{4.31}$$

where once again the functions $d_{ij}^J(n,r)$ are given in Appendix B. As mentioned earlier, they are functions of the circumferential mode number n, and the radial coordinate r, the frequency ω and wavenumber ξ. The traction-free boundary conditions on the surface $r = b$ yields:

$$\begin{bmatrix} d_{11}^J(0,b) & d_{12}^J(0,b) \\ d_{21}^J(0,b) & d_{22}^J(0,b) \end{bmatrix} \begin{bmatrix} A_1 \\ B_1 \end{bmatrix} = \begin{bmatrix} 0 \\ 0 \end{bmatrix} \tag{4.32}$$

The requirement that the determinant of the coefficients matrix must vanish leads to the following dispersion equation:

$$\left[\frac{2p}{b}(q^2+\xi^2)\mathbf{J}_1(pb) - (q^2-\xi^2)\mathbf{J}_0(pb)\right]\mathbf{J}_1(qb) - 4\xi^2 pq \mathbf{J}_1(pb)\mathbf{J}_0(qb) = 0 \tag{4.33}$$

This is the well-known Pochhammer (1876) dispersion equation for the longitudinal modes. Solutions of this equation are depicted in Figure 3 for a 1-mm steel bar up to 5 Mhz. Once the roots of the dispersion equation are found, the corresponding eigenvectors can be obtained from eq.(4.32). Thus the displacements are also known from eqs.(4.28)-(4.30). For practical applications, low order modes are the most important. To find the lower modes, one may assume $\hat{\xi} = \xi b \ll 1$. In this case, the Bessel functions

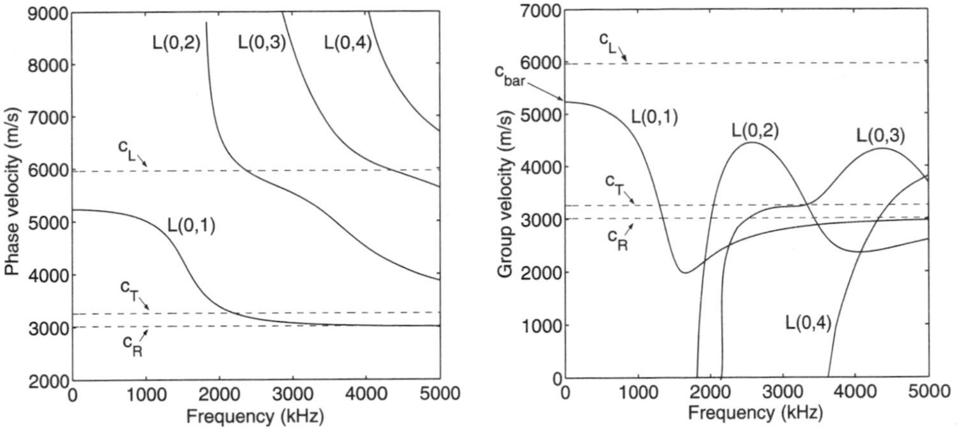

Figure 3. Longitudinal modes for a 1-mm radius steel bar: $c_L = 5960 \ \ m/s$, $c_T = 3260 \ \ m/s$ and $\mu = 7800 \ \ kg/m^3$

in eq.(4.33) can be approximated by their asymptotic representations for small argument (Stakgold, 1979):

$$\mathbf{J}_0(x) \simeq 1 - \frac{x^2}{4}, \qquad \mathbf{J}_1(x) \simeq \frac{x}{2} - \frac{x^3}{16}, \qquad \text{for } x \to 0 \tag{4.34}$$

Using the above approximations in eq.(4.33), we obtain an asymptotic representation of the dispersion equation valid for $\hat{\xi} \ll 1$:

$$c = \sqrt{\frac{E}{\mu}} \left(1 - \frac{v^2}{4}\hat{\xi}^2\right) + O(\hat{\xi}^4), \quad \text{with} \quad E = G\frac{(3\lambda + 2G)}{(\lambda + G)} = \mu c_T^2 \left(\frac{3c_L^2 - 4c_T^2}{cL - c_T^2}\right) \tag{4.35}$$

where E is the Young's modulus of the material. In the limit when $\hat{\xi} \to 0$, the wave velocity becomes equal to $\sqrt{E/\mu}$. This is called the bar velocity, c_{bar}, and is the value found from the simplest theory of thin rods. To investigate the behavior of eq.(4.33) for high frequency (or large diameter), consider the asymptotic expansions of the Bessel functions for $|x| \to \infty$ (Stakgold, 1979):

$$\mathbf{J}_n(x) = \left\{ \begin{array}{ll} \sqrt{\frac{2}{\pi x}} \cos\left(x - \frac{n\pi}{4} - \frac{\pi}{4}\right) + O(\hat{\xi}^4) & \text{when} \quad Im(x) = 0 \\ \frac{i^n e^{-ix}}{\sqrt{-2\pi i x}} & \text{when} \quad Re(x) = 0 \end{array} \right. \tag{4.36}$$

For the case of $c_p < c_T$, both p and q are purely imaginary. In this case, when $\hat{\xi} \to \infty$, by the aid of eq.(4.36) it follows from eq.(4.33) that:

$$\left(2 - \frac{c_p^2}{c_T^2}\right) - 4\sqrt{1 - \frac{c_p^2}{c_L^2}}\sqrt{1 - \frac{c_p^2}{c_T^2}} = 0 \tag{4.37}$$

This is identical to the Rayleigh wave equation. This result is expected because large $\hat{\xi}$ means either a very thick cylinder or very high frequency. In both cases, the waves propagate as if they were in a semi-infinite space that supports only the Rayleigh surface wave. Because the Rayleigh equation has only one root, this result also indicates that as the frequency increases, all modes whose velocity is less than c_T will asymptotically approach the Rayleigh wave velocity c_R. For modes with $c_L > c_p > c_T$, p is still imaginary, but q becomes real. It thus follows from eq.(4.36) that the dispersion equation (4.33) can be written asymptotically for large $\hat{\xi}$ as:

$$\frac{\tan\left[\hat{\xi}\left(\hat{c}^2-1\right)^{1/2}\right]}{1+\tan\left[\hat{\xi}\left(\hat{c}^2-1\right)^{1/2}\right]}-\frac{4(\hat{c}^2-1)^{1/2}(1-\hat{c}/\kappa^2)^{1/2}}{(\hat{c}^2-2)^2}=0 \tag{4.38}$$

where $\hat{c}=c_p/c_T$ and $\kappa=c_L/c_T$. Clearly, one of the roots to the above equation is $\hat{c}=1$ or $c_p=c_T$. All the modes whose velocity is in between c_T and c_L gravitate to the shear wave speed c_T for large $\hat{\xi}$. When $c_p > c_L$, both p and q are real. The dispersion equation can therefore be written asymptotically for large $\hat{\xi}$ as:

$$\frac{\tan\left[\hat{\xi}\left(\hat{c}^2-1\right)^{1/2}-\pi/4\right]}{\tan\left[\hat{\xi}\left(\hat{c}^2/\kappa^2-1\right)^{1/2}-\pi/4\right]}+\frac{4(\hat{c}^2-1)^{1/2}(\hat{c}^2/\kappa^2-1)^{1/2}}{(\hat{c}^2-2)^2}=0 \tag{4.39}$$

The close resemblance between this and the Rayleigh-Lamb dispersion equation exhibits the fact that the frequency spectra for a cylinder and a plate, for axially symmetric and symmetric waves, respectively, are very similar.

4.3 Flexural waves in solid cylinder

Now consider the solution (4.1)-(4.4) for $n = 1$ and $\theta_0 = 0$. The corresponding displacement components follow from eqs.(3.23)-(3.25):

$$u_r = \left[-A_1\mathbf{J}_1'(pr)+\frac{C_1}{r}\mathbf{J}_1(qr)+i\xi B_1\mathbf{J}_2(qr)\right]\cos(\theta)e^{i(\xi z-\omega t)} \tag{4.40}$$

$$u_\theta = \left[-\frac{A_1}{r}\mathbf{J}_1(pr)+i\xi B_1\mathbf{J}_2(qr)-C_1\mathbf{J}_1'(qr)\right]\sin(\theta)e^{i(\xi z-\omega t)} \tag{4.41}$$

$$u_z = \left[i\xi A_1\mathbf{J}_1(pr)-B_1 q\mathbf{J}_1(qr)\right]\cos(\theta)e^{i(\xi z-\omega t)} \tag{4.42}$$

By these displacement components, it can be noticed that the section of the cylinder at $y=0$ (which is the same as $\theta=0,\pi$) remains in the plane, because $u_\theta=0$ on this plane. The material particles that lie in this plane can only move in the x or the z-direction within the xz-plane. Furthermore, all the material particles that lie in the vertical plane $x=0$ (which is the same as $\theta=\pm\pi/2$) can move only in the x-direction, because both $u_r=u_z=0$. Looking from the top (down from the positive y-axis), the overall motion of the cylinder resembles the motion of a snake moving on the ground (the xz-plane). These observations suggest the terminology flexural waves for the motion (Rose, 1999). The pertinent stresses of the flexural waves are:

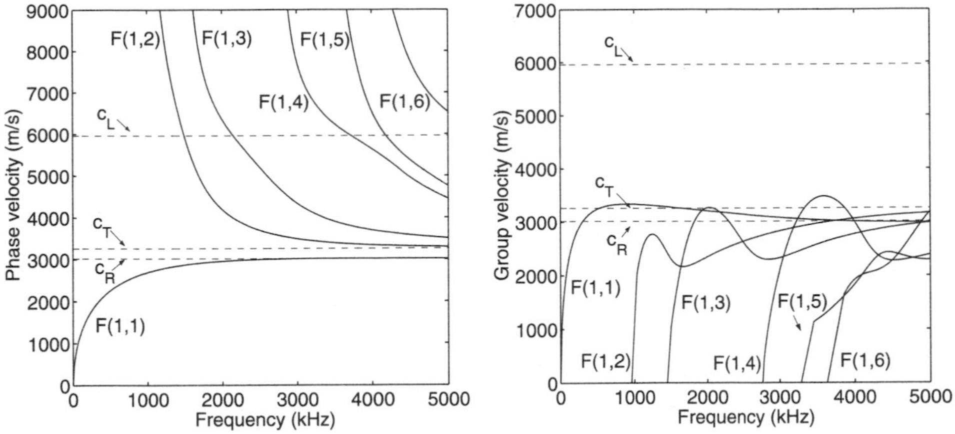

Figure 4. Flexural modes for a 1-mm radius steel bar: $c_L = 5960\ m/s$, $c_T = 3260\ m/s$ and $\mu = 7800\ kg/m^3$

$$\begin{bmatrix} \sigma_r \\ \sigma_{r\theta} \\ \sigma_{rz} \end{bmatrix} = G \begin{bmatrix} d^J_{11}(1,r) & d^J_{12}(1,r) & d^J_{13}(1,r) \\ d^J_{21}(1,r) & d^J_{22}(1,r) & d^J_{23}(1,r) \\ d^J_{31}(1,r) & d^J_{32}(1,r) & d^J_{33}(1,r) \end{bmatrix} \begin{bmatrix} A_1 \\ B_1 \\ C_1 \end{bmatrix} \tag{4.43}$$

where the functions $d^J_{ij}(n,r)$ are given in Appendix B. By imposing the traction-free boundary conditions on the surface $r = b$ yields:

$$\begin{bmatrix} d^J_{11}(1,b) & d^J_{12}(1,b) & d^J_{13}(1,b) \\ d^J_{21}(1,b) & d^J_{22}(1,b) & d^J_{23}(1,b) \\ d^J_{31}(1,b) & d^J_{32}(1,b) & d^J_{33}(1,b) \end{bmatrix} \begin{bmatrix} A_1 \\ B_1 \\ C_1 \end{bmatrix} = \begin{bmatrix} 0 \\ 0 \\ 0 \end{bmatrix} \tag{4.44}$$

For nontrivial solutions, the determinant of the coefficient matrix of this system of equations must vanish, meaning:

$$|\mathbf{D}(n = 1, \xi, \omega)| = 0 \tag{4.45}$$

This is the desired dispersion equation for flexural waves propagating in the axial direction of the cylinder of radius b. Solutions are represented in Figure 4 in terms of phase and group velocity for the steel bar of 1-mm radius. The asymptotic behavior of the frequency spectrum was examined in considerable detail by Pao and Mindlin (Pao and Mindlin, 1960). Similar to the analysis of the axially symmetric waves, they have shown that, as $\hat{\xi} \to \infty$, the velocities of all propagating modes will asymptotically approach one of the three speeds, Rayleigh wave speed c_R, the shear wave speed c_T, or the dilatational wave speed c_L. In the low frequency range, it can be shown (Achenbach (1973), p. 248) that:

$$c_p = \frac{1}{2}\sqrt{\frac{E}{\mu}}\hat{\xi}, \qquad \text{for } \hat{\xi} \to 0$$

5 Axial waves in hollow cylinders

Consider a hollow cylinder of inner radius a and outer radius b. It is assumed that both the inner and outer surfaces are traction free, i.e.:

$$\sigma_r = 0, \qquad \sigma_{r\theta} = 0, \qquad \sigma_{rz} = 0 \quad \text{for } r = a, b \tag{5.1}$$

The general solution to the waves propagating in the axial direction of this hollow cylinder is given by using the potential functions in their general form (3.19)-(3.22). The Bessel functions of the second kind have to be retained in this case, because the presence of both incoming (toward the center) and outgoing (away from the center) waves is necessary to satisfy the traction-free boundary conditions on both inner and outer surfaces. The displacement components can be computed by eqs.(3.23)-(3.25) by using the proper potential functions. The corresponding stress field is defined by eqs.(3.26)-(3.28):

$$\begin{bmatrix} \sigma_r \\ \sigma_{r\theta} \\ \sigma_{rz} \end{bmatrix} = G \begin{bmatrix} d^J_{11}(n,r) & d^J_{12}(n,r) & d^J_{13}(n,r) & d^Y_{11}(n,r) & d^Y_{12}(n,r) & d^Y_{13}(n,r) \\ d^J_{21}(n,r) & d^J_{22}(n,r) & d^J_{23}(n,r) & d^Y_{21}(n,r) & d^Y_{22}(n,r) & d^Y_{23}(n,r) \\ d^J_{31}(n,r) & d^J_{32}(n,r) & d^J_{33}(n,r) & d^Y_{31}(n,r) & d^Y_{32}(n,r) & d^Y_{33}(n,r) \end{bmatrix} \begin{bmatrix} A_1 \\ B_1 \\ C_1 \end{bmatrix} \tag{5.2}$$

where the elements $d^Z_{ij}(n,r)$ are given in Appendix B. The superscript Z in $d^Z_{ij}(n,r)$ indicates which of the first kind Bessel functions, $\mathbf{J}$, $\mathbf{Y}$, has to be used. The traction-free condition (5.1) on both surfaces $r = a, b$ results in the following eigenvalue problem:

$$\begin{bmatrix} d^J_{11}(n,b) & d^J_{12}(n,b) & d^J_{13}(n,b) & d^Y_{11}(n,b) & d^Y_{12}(n,b) & d^Y_{13}(n,b) \\ d^J_{21}(n,b) & d^J_{22}(n,b) & d^J_{23}(n,b) & d^Y_{21}(n,b) & d^Y_{22}(n,b) & d^Y_{23}(n,b) \\ d^J_{31}(n,b) & d^J_{32}(n,b) & d^J_{33}(n,b) & d^Y_{31}(n,b) & d^Y_{32}(n,b) & d^Y_{33}(n,b) \\ d^J_{11}(n,a) & d^J_{12}(n,a) & d^J_{13}(n,a) & d^Y_{11}(n,a) & d^Y_{12}(n,a) & d^Y_{13}(n,a) \\ d^J_{21}(n,a) & d^J_{22}(n,a) & d^J_{23}(n,a) & d^Y_{21}(n,a) & d^Y_{22}(n,a) & d^Y_{23}(n,a) \\ d^J_{31}(n,a) & d^J_{32}(n,a) & d^J_{33}(n,a) & d^Y_{31}(n,a) & d^Y_{32}(n,a) & d^Y_{33}(n,a) \end{bmatrix} \begin{bmatrix} A_1 \\ B_1 \\ C_1 \\ A_2 \\ B_2 \\ C_2 \end{bmatrix} \tag{5.3}$$

that in synthetic form can be viewed as eq.(4.10), where $\mathbf{A} = [A_1, B_1, C_1, A_2, B_2, C_2]^{\mathrm{T}}$. By setting the determinant of the matrix $\mathbf{D}$ to zero, we obtain the dispersion equation for waves propagating in the axial direction of a hollow cylinder of inner radius a and outer radius b. Similar to the solid cylinder case, let us consider some specific wave motions of practical interest.

5.1 Torsional waves in hollow cylinders

In a pure torsional motion, u_θ should be the only nonzero displacement component. Furthermore, it should be independent of θ. These conditions are met by setting $A_i = B_i = n = 0$ and $\theta = \pi/2$ in the general solution. The corresponding dispersion equation is thus deduced from eq.(5.3):

$$\begin{bmatrix} d^J_{33}(0,b) & d^Y_{33}(0,b) \\ d^J_{33}(0,a) & d^Y_{33}(0,a) \end{bmatrix} \begin{bmatrix} C_1 \\ C_2 \end{bmatrix} = \begin{bmatrix} 0 \\ 0 \end{bmatrix} \tag{5.4}$$

The vanishing of the determinant (5.4) yields $q = 0$, or:

$$\mathbf{J}_2(qb)\mathbf{Y}_2(qa) - \mathbf{J}_2(qa)\mathbf{Y}_2(qb) = 0 \tag{5.5}$$

Clearly, the modes determined from eq.(5.5) are all dispersive. The lowest mode corresponding to $q = 0$, however, is non-dispersive. Its displacement field is similar to eq.(4.24) for the solid bar. In fact, one can easily show that in the limit $a \to 0$, eq.(5.5) reduces to eq.(4.22), as expected. For the case $b > a \to \infty$, one can use the asymptotic expressions of the Bessel functions in eq.(5.5), which results in:

$$\sin(qh) = 0 \tag{5.6}$$

where $h = b - a$ is the wall thickness of the hollow cylinder. This is identical to the dispersion equation for the symmetric horizontally polarized shear waves in a plate of thickness $2h$ (Achenbach (1973), p.206). Similar to the solid cylinder case, it can be shown (Gazis, 1959a) that eq.(5.5) does not have any real roots for $q^2 < 0$. Thus, it can be ascertained that the phase velocity of the torsional waves in a hollow cylinder is always greater than or equal to the shear velocity c_T.

5.2 Longitudinal waves in hollow cylinders

Longitudinal waves refer to the axially symmetrical motion in the cylinder. They are characterized by the presence of displacement components in the radial and axial directions, but none in the θ-direction. Furthermore, these nonzero displacement components must be independent of θ because of the axisymmetric requirement. It is seen that these conditions are met by setting $C_i = n = 0$ and $\theta_0 = 0$ in the general solution. The corresponding eigenvalue problem is thus obtained from eq.(5.3):

$$\begin{bmatrix} d_{11}^J(0,b) & d_{12}^J(0,b) & d_{11}^Y(0,b) & d_{12}^Y(0,b) \\ d_{21}^J(0,b) & d_{22}^J(0,b) & d_{21}^Y(0,b) & d_{22}^Y(0,b) \\ d_{11}^J(0,a) & d_{12}^J(0,a) & d_{11}^Y(0,a) & d_{12}^Y(0,a) \\ d_{21}^J(0,a) & d_{22}^J(0,a) & d_{21}^Y(0,a) & d_{22}^Y(0,a) \end{bmatrix} \begin{bmatrix} A_1 \\ B_1 \\ A_2 \\ B_2 \end{bmatrix} = \begin{bmatrix} 0 \\ 0 \\ 0 \\ 0 \end{bmatrix} \tag{5.7}$$

The vanishing of the determinant yields the dispersion equation.

5.3 Flexural waves in hollow cylinders

As in the solid cylinder case, flexural waves in a hollow cylinder are defined by setting $n = 1$, and $\theta_0 = 0$ in the general solution. The corresponding eigenvalue problem is thus obtained from eq.(5.3).

6 Conclusions

This lecture deals with wave propagation in structures with cylindrical symmetry, namely rods and tubes. As we know, propagation in such waveguides is restricted to discrete modes, instead of the free propagation existing in bulk media. Furthermore, those modes show dispersion, i.e., dependence of the phase speed on the frequency, and cut-off frequencies, below which the waves cannot propagate and exist only as evanescent waves. There are three families of modes, named *torsional* $T(0,i)$, *longitudinal* $L(0,i)$ and *flexural* $F(n,i)$, each containing an infinite number of modes.

The *Torsional modes* $T(0, i)$ are the axisymmetric rotational modes whose displacement is primarily in the θ direction and is equal to zero in the z-direction. The fundamental torsional mode $T(0, 1)$ is non-dispersive for the whole frequency range. The *Longitudinal modes* $L(0, i)$ are the longitudinal axially symmetric modes, with predominant displacement in the propagation z-direction. The fundamental longitudinal mode $L(0, 1)$ has a purely extensional displacement in the axial direction at zero frequency. The *Flexural modes* $F(n, i)$ are the non-axially symmetric bending modes that activate all the displacement components. The fundamental flexural mode $F(1, 1)$ is a pure bending mode at zero frequency. The index n is termed as the circumferential order, and it specifies the integer number of wavelengths around the circumference, i.e. the order of symmetry around the z-axis (the modes with $n = 0$ are called axisymmetric). For example, the displacements for zero order modes are constant with angle and the displacements on the top and bottom of the cylinder are 180 degrees out of phase for first order modes. Therefore, all longitudinal modes are of circumferential order 0, and all of the flexural modes have a circumferential order greater than or equal to 1. The second index i is a counter variable and it is used to sort the modes for a given family, into ascending phase velocity. The fundamental modes (those modes that can propagate at zero frequency) are given the value $i = 1$ and the higher order modes are numbered consecutively. This index roughly reflects the modes of vibration within the wall of the cylinder. For example, using this convention, the third flexural mode of circumferential order 1 would be named $F(1, 3)$. The results are given in terms of dispersion curves. Dispersion means, that the velocity of a wave is a function of its frequency or wavelength. Dispersion diagrams are an effective means for visualizing the dispersive behavior of the structure and to anticipate what modes to expect at a particular frequency. The modes depend on material properties and characteristic geometrical parameters of the waveguide, such as radius and wall thickness in the case of hollow cylinders.

Acknowledgement

This topic is one of the subjects of the Centre of Study and Research for the Identification of Materials and Structures (CIMEST) "M. Capurso". The numerical tests presented were carried out using the facilities of the Laboratory of Computational Mechanics (LAMC) "A. A. Cannarozzi" of the University of Bologna.

Bibliography

J.D. Achenbach. *Wave Propagation in Elastic Solids.* North Holland, 1973.

B.A. Auld. *Acoustic Fields and Waves in Solids.* Krieger, 1990.

D. Bancroft. The velocity of longitudinal waves in cylindrical bars. *Physical Review*, 59: 588–593, 1941.

C. Chree. The equations on an isotropic elastic solid in polar and cylindrical coordinates, their solutions, an applications. *Trans. Cambridge Philos. Soc.*, 14:250–369, 1889.

R. M. Davies. A critical study of the hopkinson pressure bar. *Phil. Trans. Roy. Soc. London A*, 240:375–457, 1948.

D. Gazis. Three dimensional investigation of the propagation of waves in hollow circular cylinders. I Analitical foundation. *Journal of the Acoustical Society of America*, 31(5): 568–573, 1959a.

D. Gazis. Three dimensional investigation of the propagation of waves in hollow circular cylinders. II Numerical results. *Journal of the Acoustical Society of America*, 31(5): 573–578, 1959b.

G. Herrmann and I. Mirsky. Three-dimensional and shell-theory analysis of axially symmetric motions of cylinders. *Trans. ASME J. Appl. Mech.*, 78:563568, 1956.

G. E. Hudson. Dispersion of elastic waves in solid circular cylinders. *Physical Review*, 63:46–51, 1943.

J.A. McFadden. Radial vibrations of thick-walled hollow cylinders. *Journal of the Acoustical Society of America*, 26:714–715, 1954.

T. R. Meeker and A. H. Meitzler. Guided wave propagation in elongated cylinders and plates. In W. P. Mason and R. N. Thurston, editors, *Physical Acoustics, Principles and Methods*, volume 1A, pages 111–167. Academic Press, 1972.

Y. H. Pao. The dispersion of flexural waves in an elastic, circular cylinder. Part 2. *Journal of Applied Mechanics*, 29:61–64, 1962.

Y. H. Pao and R. Mindlin. Dispersion of flexural waves in an elastic, circular cylinder. *Journal of Applied Mechanics*, 27:513–520, 1960.

J.L. Rose. *Ultrasonic waves in solid media.* Cambridge Press, 1999.

I. Stakgold. *Green's Functions and Boundary Value Problems.* John Wiley, 1979.

Appendix A: Common differential operators on scalar and vector fields in cylindrical coordinates

$$
\begin{aligned}
\mathbf{r} &= (r, \theta, z) \\
\nabla f &= \frac{\partial f}{\partial r}\hat{e}_r + \frac{1}{r}\frac{\partial f}{\partial \theta}\hat{e}_\theta + \frac{\partial f}{\partial z}\hat{e}_z \\
\nabla^2 f &= \frac{\partial^2 f}{\partial r^2} + \frac{1}{r}\frac{\partial f}{\partial r} + \frac{1}{r^2}\frac{\partial^2 f}{\partial \theta^2} + \frac{\partial^2 f}{\partial z^2} \\
\nabla \cdot \mathbf{A} &= \frac{1}{r}\frac{\partial}{\partial r}(rA_r) + \frac{1}{r}\frac{\partial A_\theta}{\partial \theta} + \frac{\partial A_z}{\partial z} \\
\nabla \times \mathbf{A} &= \left(\frac{1}{r}\frac{\partial A_z}{\partial \theta} - \frac{\partial A_\theta}{\partial z}\right)\hat{e}_r + \left(\frac{\partial A_r}{\partial z} - \frac{\partial A_z}{\partial r}\right)\hat{e}_\theta + \left(\frac{1}{r}\frac{\partial}{\partial r}(rA_\theta) - \frac{1}{r}\frac{\partial A_r}{\partial \theta}\right)\hat{e}_z \\
\nabla^2 \mathbf{A} &= \left(\nabla^2 A_r - \frac{A_r}{r^2} - \frac{2}{r^2}\frac{\partial A_\theta}{\partial \theta}\right)\hat{e}_r + \left(\nabla^2 A_\theta - \frac{A_\theta}{r^2} + \frac{2}{r^2}\frac{\partial A_r}{\partial \theta}\right)\hat{e}_\theta + (\nabla^2 A_z)\,\hat{e}_z
\end{aligned}
$$

Appendix B: d_{ij}^Z coefficients

$$
\begin{aligned}
d_{11}^Z(n,r) &= \left[\frac{2n(n-1)}{r^2} + \xi^2 - q^2\right] Z_n(pr) + \frac{2p}{r} Z_{n+1}(pr) \\
d_{12}^Z(n,r) &= 2i\xi \left[qZ_n(pr) - \frac{(n+1)}{r} Z_{n+1}(qr)\right] \\
d_{13}^Z(n,r) &= 2n \left[\frac{(n-1)}{r^2} Z_n(pr) - \frac{q}{r} Z_{n+1}(qr)\right] \\
d_{21}^Z(n,r) &= 2i\xi \left[\frac{n}{r} Z_n(pr) - pZ_{n+1}(pr)\right] \\
d_{22}^Z(n,r) &= -\frac{qn}{r} Z_n(qr) + (q^2 - \xi^2) Z_{n+1}(qr) \\
d_{23}^Z(n,r) &= \frac{i\xi n}{r} Z_n(qr) \\
d_{31}^Z(n,r) &= -2n \left[\frac{(n-1)}{r^2} Z_n(pr) - \frac{p}{r} Z_{n+1}(pr)\right] \\
d_{32}^Z(n,r) &= i\xi \left[qZ_n(qr) - \frac{2(n+1)}{r} Z_{n+1}(qr)\right] \\
d_{33}^Z(n,r) &= \left[q^2 - \frac{2n(n-1)}{r^2}\right] Z_n(qr) - \frac{2q}{r} Z_{n+1}(qr)
\end{aligned}
$$

where $\mathbf{Z}_n(\bullet)$ can be either Bessel function of the first kind, $\mathbf{J}_n(\bullet)$, or the second kind, $\mathbf{Y}_n(\bullet)$. The superscript Z in $d_{ij}^Z(n,r)$ is to indicate which Bessel function has to be used. For example:

$$
d_{23}^J(n,r) = \frac{i\xi n}{r} J_n(qr), \quad \text{while} \quad d_{23}^Y(n,r) = \frac{i\xi n}{r} Y_n(qr)
$$

Appendix C: Recurrence formulas for the Bessel equations

(where we have only considered the functional dependence on x)

$$
\begin{aligned}
\mathbf{J}_{n+1}(x) &= \frac{2n}{x}\mathbf{J}_n(x) - \mathbf{J}_{n-1}(x) \\
\mathbf{Y}_{n+1}(x) &= \frac{2n}{x}\mathbf{Y}_n(x) - \mathbf{Y}_{n-1}(x) \\
\mathbf{J}'_n(x) &= \mathbf{J}_{n-1}(x) - \frac{n}{x}\mathbf{J}_n(x) = -\mathbf{J}_{n+1}(x) + \frac{n}{x}\mathbf{J}_n(x) \\
\mathbf{Y}'_n(x) &= \mathbf{Y}_{n-1}(x) - \frac{n}{x}\mathbf{Y}_n(x) = -\mathbf{Y}_{n+1}(x) + \frac{n}{x}\mathbf{Y}_n(x)
\end{aligned}
$$

Semi-analytical Formulation for Guided Wave Propagation

Erasmo Viola, Alessandro Marzani and Ivan Bartoli
Dipartimento di Ingegneria delle Strutture, dei Trasporti, delle Acque, del Rilevamento, del Territorio - University of Bologna, Viale Risorgimento 2, 40136 Bologna, Italy

Abstract Guided wave is a type of wave propagation in which the waves are guided in plates, rods, pipes or elongated structures such as rails and I-beams. In order to extract the guided wave velocities and wavestructures, for waveguides of arbitrary cross section, a theoretical framework is developed. Here, a semi-analytical finite element (SAFE) method is used for the calculation of the wave propagation characteristics in elastic waveguides immersed in vacuum. The method couples an approximate displacement field over the cross-section of the waveguide and assumes time harmonic representation of the propagating waves along the length of the guide. The Hamilton's principle and the finite element discretization lead to a discrete weak form of the energy balance equation. The wave propagation problem reduces to a system of algebraic equations, from which the dispersive equation can be obtained. The solution, which depends on both time t and propagation coordinate z, i.e. $e^{i(\xi z-\omega t)}$, results in a two-parameter eigensystem. By specifying a real axial wavenumber ξ, the eigenproblem permits real frequencies of propagating modes to be determined. Giving instead real frequency ω, both real and complex axial wavenumbers can be extracted, where real values pertain to propagating modes and the complex ones to the evanescent modes. The method allows us to model a generic cross-section of solid waveguide and it is well suited for computing the phase velocity, the group velocity and the wavestructure or cross-sectional mode shape.

1 Introduction

Guided acoustic waves provide a highly efficient method for inspecting certain types of structures. Problems associated with using guided waves are well known and include the multiple modes and dispersion. Guided waves are complex vibrational waves that travel through the entire thickness of a material. Guided waves behavior in plate, bar or pipe structures is described by the dispersion curves, that show how the velocities of the waves change with the frequency, and by the wavestructures that represent the cross sectional displacement of the waveguide. For elastic waveguides immersed in vacuum, phenomena of leakage and material attenuation are not present. In this context, the velocities which are of interest are the phase and the group velocity. The phase velocity of a particular wave is the velocity at which the wave fronts or crests travel and it can be expressed as $c_p = \omega/\xi$, where ω is the angular frequency of the wave and ξ is its wavenumber in the direction of propagation. The group velocity c_g is the velocity at which waves carry their potential energy and kinetic energy along the structure. The group velocity is the derivative of the frequency with respect to the wavenumber $c_g = \partial\omega/\partial\xi$, and it can be calculated for real wavenumbers once the relation $\omega = \omega(\xi)$ is known. It is worth reminding that for waveguide that leaks energy into the surrounding media or when material damping is considered, the group velocity does not represent anymore the speed at which the energy is traveling along the structure (Bernard et al., 2001). It is well established in this case that the energy velocity is the meaningful parameter. However, in this lecture non-absorbing waveguides surrounded by vacuum are considered, therefore the group velocity represents the velocity of the traveling energy.

In literature, superposition of partial bulk waves (SPBW) approaches to model guided wave propagation phenomena, are well known. The exact solution of guided wave propagation in plates, rods and hollow cylinders can be obtained in the case of elastic isotropic materials. Moreover, matrices techniques like the Transfer Matrix Method (TMM) and Global Matrix Method (GMM), have been used to obtain the solution for multilayer structures made up of flat (Lowe, 1995) or cylindrical multiple layers (Pavlakovic, 1998). In SPBW methods, the characteristic equation that describes wave propagation takes the form of a nonlinear transcendental eigenvalue problem. Numerical root-finding algorithms are then usually employed to determine the dispersion properties of the system. From a numerical point of view the approach under consideration often lacks robustness. Moreover, careful attention needs to be paid to the step sampling in the searching algorithm in order to ensure that no wave types have been missed. This is especially true for roots that are closely spaced or repeated. It is worth noting that a sampling too coarse results in missed roots whereas sampling too fine results in excessive computational expense. In addition the superposition of partial bulk waves, allows to describe the wave propagation only for systems with a "standard" geometry, where the behavior of bulk wave propagation is known. For example, when the waveguide cross-section is generic, such for I-beam or rail section, the superposition methods become unfeasible. The solution can still be found by using numerical techniques such as the finite difference method (FDM), finite element method (FEM) and boundary element method (BEM) (see e.g. Brebbia and Walker (1980), Banerjee and Kobayashi (1990), Zienkiewicz and Taylor (2000), respectively). It is significant that in guided wave inspections, waves propagating with a very short wavelength are often used. The conventional techniques of FDM, FEM and BEM generally require a minimum of four/six nodes per wavelength to express the waveform accurately. Thus, to express guided wave propagation in large structures, a large number of nodes in the propagating direction is necessary. Generally, problems of long calculation times and insufficient computational memory are common when using these conventional techniques for modeling guided waves. These types of problems can be overcome by using a Semi-Analytical Finite Element (SAFE) method. This hybrid technique combines a discrete two-dimensional finite element displacement field over the cross section of the waveguide with a series of orthogonal harmonic functions $\exp(i\xi z\text{-}i\omega t)$ to express the distribution of the displacement field in the propagation direction. The technique reduces a three-dimensional propagation problem to a bi-dimensional one for all kinds of bar-like structures, including plates and pipes, with constant material properties and geometrical structure in the propagation direction with a severe reduction of

computational time and memory needs. The method can be used for the computation of waves with even very short wavelengths, since polynomial approximation of the displacement field along the waveguide is avoided. Apparently, it was employed for the first time in 1973 by two different authors, Lagasse (1973) and Aalami (1973). Both works contained a finite-element technique, based on a variational formulation of the guided-wave problem. Using this technique, a decade later, Huang and Dong (1984) obtained the complete set of dispersion curves for circular anisotropic cylinders including the evanescent branches, i.e. those which pertain to complex wavenumbers. Evanescent modes are also known in literature as non propagative modes. It should be noted that propagative modes are traveling waves with energy transport capabilities, while the evanescent modes are standing vibrations which, in contrast, do not transport any energy along the structure. Even though the evanescent modes are not able to carry energy along the structure, they become essential from the point of view of the solution, when a traction-free end condition has to be satisfied. Propagation phenomena in thin-walled and rail structures have been studied by Gavrić (1994, 1995), respectively. The technique has been extensively applied to the study of wave propagation in wedges by Hladky-Hennion (1996). In these latter works, a particular displacement field is adopted: the displacement along the wave propagation direction was assumed to be shifted by a phase of $\pi/2$ with respect to the in-plane displacement over the cross section. This shift leads to an eigenproblem with real symmetric matrices with practical benefits in representing the solution.
Studies on the effect of twisted waveguides on the behavior of propagating and evanescent modes can be found in Onipede and Dong (1996). In Taweel et al. (2000) a study on the wave reflection phenomena from the end of a waveguide can be found. In Volovoi et al. (1998) the semi analytical finite element technique for characterizing the dispersion relation for a non-homogeneous anisotropic beam is applied. Composite plates were studied using this method for the first time by Dong and Huang (1985). Layered composite plates of both finite and infinite width, have been extensively studied by Mukdadi and Datta (Mukdadi et al., 2002; Mukdadi and Datta, 2003). In all these studies is clear and well explained how to compute the solution in terms of frequency-wavenumber, i.e. the phase velocity information for propagative and evanescent modes. However, few approaches concerning the group velocity information have been found in literature. The work proposed by Hayashi et al. (2003), extracts the group velocity information as a function of the modal solution of the system and the chosen step in the frequency domain. This means that when a coarse step in the frequency domain is chosen, poor group velocity accuracy regardless the quality of the modal solution will be produced. In others works, see e.g. Finnveden (2004) and Han et al. (2002), the calculation of the group velocity is based on the derivation of the Rayleigh quotient respect to the wavenumber. In these works the group velocity is extracted by algebra manipulation as a function of the modal solution of the system only, removing the step frequency dependency.

2 Governing equations modeling

The mathematical model is presented here for the case of a linear waveguide immersed in vacuum, which is uniform and infinite in the z-direction, and whose arbitrary cross-section, is set in the $x-y$ plane. In the most general case, the waveguide is composed of possible distinct anisotropic viscoelastic materials that are perfectly bonded together along its entire length. Any type of elastic material in deriving the present model was considered as homogeneous at the wavelength scale and behaving in a linear elastic range. Let's consider that an acoustic wave, characterized by its wavenumber ξ in the propagation z-direction and circular frequency ω, is propagating along the waveguide. The energy and momentum carried by the traveling wave displace the material point of the waveguide. The point defined by the coordinates x-y-z undergoes a state of harmonic motion with frequency ω. The mechanical variables at a point, the displacement $\mathbf{u}$, the stress $\boldsymbol{\sigma}$ and the strain $\boldsymbol{\varepsilon}$, can be expressed in a vector form by their components along the coordinate axes:

$$\mathbf{u}(x,y,z,t) = [u,\ v,\ w]^{\mathrm{T}} \tag{2.1}$$

$$\boldsymbol{\varepsilon}(x,y,z,t) = [\varepsilon_x,\ \varepsilon_y,\ \varepsilon_z,\ \gamma_{yz},\ \gamma_{xz},\ \gamma_{xy}]^{\mathrm{T}} \tag{2.2}$$

$$\boldsymbol{\sigma}(x,y,z,t) = [\sigma_x,\ \sigma_y,\ \sigma_z,\ \sigma_{yz},\ \sigma_{xz},\ \sigma_{xy}]^{\mathrm{T}} \tag{2.3}$$

In the most general case, the stress-strain relations at a point is given by $\boldsymbol{\sigma} = \mathbf{C}\boldsymbol{\varepsilon}$, where $\mathbf{C}$ is the constitutive elastic stiffness tensor expressed in Voight notation. The strain can be represented by the strain-displacements relation:

$$\boldsymbol{\varepsilon} = \begin{bmatrix} \varepsilon_x \\ \varepsilon_y \\ \varepsilon_z \\ \gamma_{xy} \\ \gamma_{xz} \\ \gamma_{yz} \end{bmatrix} = \begin{bmatrix} \partial u/\partial x \\ \partial v/\partial y \\ \partial w/\partial z \\ \partial v/\partial z + \partial w/\partial y \\ \partial u/\partial z + \partial w/\partial x \\ \partial u/\partial y + \partial v/\partial x \end{bmatrix} \tag{2.4}$$

or in a more synthetic form eq.(2.4) can also be expressed as:

$$\boldsymbol{\varepsilon} = [\mathbf{L}_x \frac{\partial}{\partial x} + \mathbf{L}_y \frac{\partial}{\partial y} + \mathbf{L}_z \frac{\partial}{\partial z}]\mathbf{u} \tag{2.5}$$

where:

$$\mathbf{L}_x = \begin{bmatrix} 1 & 0 & 0 \\ 0 & 0 & 0 \\ 0 & 0 & 0 \\ 0 & 0 & 0 \\ 0 & 0 & 1 \\ 0 & 1 & 0 \end{bmatrix}, \quad \mathbf{L}_y = \begin{bmatrix} 0 & 0 & 0 \\ 0 & 1 & 0 \\ 0 & 0 & 0 \\ 0 & 0 & 1 \\ 0 & 0 & 0 \\ 1 & 0 & 0 \end{bmatrix}, \quad \mathbf{L}_z = \begin{bmatrix} 0 & 0 & 0 \\ 0 & 0 & 0 \\ 0 & 0 & 1 \\ 0 & 1 & 0 \\ 1 & 0 & 0 \\ 0 & 0 & 0 \end{bmatrix} \tag{2.6}$$

2.1 Hamilton principle

The equations of motion can be obtained by using the first variation of the Hamiltonian $\mathcal{A}$ of the waveguide:

$$\delta\mathcal{A} = \int_{t_1}^{t_2} \delta(\mathcal{U} - \mathcal{T})dt = 0 \tag{2.7}$$

where $\mathcal{U}$ is the strain energy of the system and $\mathcal{T}$ is its kinetic energy. The strain energy $\mathcal{U}$ is given by:

$$\mathcal{U} = \frac{1}{2}\int_V \boldsymbol{\varepsilon}^{\mathrm{T}}\mathbf{C}\boldsymbol{\varepsilon}dV \tag{2.8}$$

where the upper script T stands for transpose operator and V is the volume of the unit length waveguide. The kinetic energy in terms of the velocity vector $\dot{\mathbf{u}}$ and unit mass density μ can be expressed as:

$$\mathcal{T} = \frac{1}{2}\int_V \dot{\mathbf{u}}^{\mathrm{T}}\boldsymbol{\mu}\dot{\mathbf{u}}dV \qquad \text{with } \boldsymbol{\mu} = \begin{bmatrix} \mu & 0 & 0 \\ 0 & \mu & 0 \\ 0 & 0 & \mu \end{bmatrix} \tag{2.9}$$

where the overdots represent time derivatives.

2.2 Calculus of variations

The variations of the energy and work terms are:

$$\delta\mathcal{U} = \int_V \delta\left(\boldsymbol{\varepsilon}^{\mathrm{T}}\right)\mathbf{C}\boldsymbol{\varepsilon}dV \tag{2.10}$$

$$\delta\mathcal{T} = \int_V \delta\left(\dot{\mathbf{u}}^{\mathrm{T}}\right)\boldsymbol{\mu}\dot{\mathbf{u}}dV \tag{2.11}$$

Inserting the variations (2.10) and (2.11) into the Hamilton's principle (2.7), yields:

$$\delta\mathcal{A} = \int_{t_1}^{t_2}\left[\int_V \delta\left(\boldsymbol{\varepsilon}^{\mathrm{T}}\right)\mathbf{C}\boldsymbol{\varepsilon}dV - \int_V \delta\left(\dot{\mathbf{u}}^{\mathrm{T}}\right)\boldsymbol{\mu}\dot{\mathbf{u}}dV\right]dt = 0 \tag{2.12}$$

Integrating by parts the variation of the kinetic term (2.11):

$$\int_V\left[\int_{t_1}^{t_2}\delta\left(\dot{\mathbf{u}}^{\mathrm{T}}\right)\boldsymbol{\mu}\dot{\mathbf{u}}dt\right]dV = \underbrace{\int_V\left[\delta\left(\mathbf{u}^{\mathrm{T}}\right)\boldsymbol{\mu}\dot{\mathbf{u}}\right]_{t_1}^{t_2}dV}_{=0} - \int_V\int_{t_1}^{t_2}\delta\left(\mathbf{u}^{\mathrm{T}}\right)\boldsymbol{\mu}\ddot{\mathbf{u}}dtdV \tag{2.13}$$

In eq.(2.13) the change of order in the integration is possible via Fubini's theorem since the volume V does not change over the time. The element configuration is known to be $\mathbf{u}_1$ at time t_1, and at some other configuration $\mathbf{u}_2$ at a later time t_2, i.e. the configuration element is known at both the time instants t_1 and t_2, $\delta\mathbf{u}(t_1) = \delta\mathbf{u}(t_2) = 0$. In the case of guided waves this consideration is always possible since before and after the transit of energy the waveguide can be considered in a steady situation. Thus, eq.(2.13) yields:

$$\int_V\left[\int_{t_1}^{t_2}\delta\left(\dot{\mathbf{u}}^{\mathrm{T}}\right)\boldsymbol{\mu}\dot{\mathbf{u}}dt\right]dV = -\int_V\left[\int_{t_1}^{t_2}\delta\left(\mathbf{u}^{\mathrm{T}}\right)\boldsymbol{\mu}\ddot{\mathbf{u}}dt\right]dV \tag{2.14}$$

By the calculus of the variations and integrating by parts the kinetic term $\mathcal{T}$, allows us to write the Hamilton principle as:

$$\delta\mathcal{A} = \int_{t_1}^{t_2}\left[\int_V \delta\left(\boldsymbol{\varepsilon}^{\mathrm{T}}\right)\mathbf{C}\boldsymbol{\varepsilon}dV + \int_V \delta\left(\mathbf{u}^{\mathrm{T}}\right)\boldsymbol{\mu}\ddot{\mathbf{u}}dV\right]dt = 0 \tag{2.15}$$

2.3 Semi-analytical displacement field

In view of the nature of the problem concerned and upon the recognition of constant physical properties of the waveguide along the propagation direction, a specific displacement field $\mathbf{u}$, characteristic for wave motion, is considered:

$$\mathbf{u}(x,y,z,t) = \begin{bmatrix} u(x,y,z,t) \\ v(x,y,z,t) \\ w(x,y,z,t) \end{bmatrix} = \begin{bmatrix} U(x,y) \\ V(x,y) \\ W(x,y) \end{bmatrix} e^{i(\xi z - \omega t)} \tag{2.16}$$

Here $i = \sqrt{-1}$ is the imaginary unit. The spatial functions $U(x,y)$, $V(x,y)$ and $W(x,y)$ describe the amplitudes of the displacements of the waveguide cross-section in the x, y and z-directions, respectively. The solution consists in finding the displacement fields $U(x,y)$, $V(x,y)$, $W(x,y)$ and the scalars ξ and ω which satisfy the condition $\delta\mathcal{A} = 0$. It can be seen that by assuming a dependence for displacement field in (z,t) by exponential harmonic term, leaves the (x,y) dependence to be modeled by interpolation functions via discrete technique.

2.4 Finite element discretization

The cross section domain (Ω) is subdivided into a system of planar elements of finite domain (Ω_e) as it can be seen from Figure 1. Even if in Figure 1 a quadrilateral finite

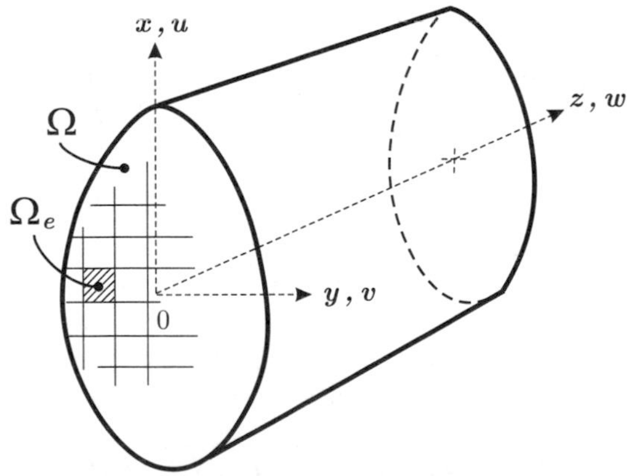

Figure 1. Waveguide discretization

element is represented, the following formulation is general and it can be applied also for beam and triangular elements. Leaving the axial z and time t dependences as arbitrary, the displacement vector (2.16) over the element domain (Ω_e) can be approximated with the aid of shape functions $N_k(x,y)$ and nodal unknown displacements (u_k, v_k, w_k):

$$\mathbf{u}^{(e)}(x,y,z,t) = \begin{bmatrix} \sum_{k=1}^{n} N_k(x,y)u_k \\ \sum_{k=1}^{n} N_k(x,y)v_k \\ \sum_{k=1}^{n} N_k(x,y)w_k \end{bmatrix}^{(e)} e^{i(\xi z-\omega t)} \tag{2.17}$$

or in a more shyntetic form as:

$$\mathbf{u}^{(e)}(x,y,z,t) = \mathbf{N}(x,y)\mathbf{q}^{(e)} e^{i(\xi z-\omega t)} \tag{2.18}$$

where in eq.(2.18):

$$\mathbf{N}(x,y) = \begin{bmatrix} N_1 & & & N_2 & & & \ddots & & & N_n & & \\ & N_1 & & & N_2 & & & \ddots & & & N_n & \\ & & N_1 & & & N_2 & & & \ddots & & & N_n \end{bmatrix} \tag{2.19}$$

$$\mathbf{q}^{(e)} = \begin{bmatrix} u_1 & v_1 & w_1 & u_2 & v_2 & w_2 & \cdots & \cdots & u_n & v_n & w_n \end{bmatrix}^{\mathrm{T}} \tag{2.20}$$

and n denotes the number of nodes per element. The strain for a point in the element can be represented in vector form as function of the nodal displacements. Combining eq.(2.5) with eq.(2.18) yields:

$$\begin{aligned} \boldsymbol{\varepsilon}^{(e)} &= [\mathbf{L}_x \frac{\partial}{\partial x} + \mathbf{L}_y \frac{\partial}{\partial y} + \mathbf{L}_z \frac{\partial}{\partial z}]\mathbf{N}(x,y)\mathbf{q}^{(e)} e^{i(\xi z-\omega t)} = \\ &= \left(\mathbf{L}_x \mathbf{N}_{,x} + \mathbf{L}_y \mathbf{N}_{,y} + i\xi \mathbf{L}_z \mathbf{N}\right) \mathbf{q}^{(e)} e^{i(\xi z-\omega t)} \end{aligned} \tag{2.21}$$

where $\mathbf{N}_{,x}$ and $\mathbf{N}_{,y}$ are the differentiations of the shape functions matrix (2.19) with respect to x and y, respectively. More concisely:

$$\boldsymbol{\varepsilon}^{(e)} = (\mathbf{B}_1 + i\xi \mathbf{B}_2)\, \mathbf{q}^{(e)} e^{i(\xi z-\omega t)} \tag{2.22}$$

where $\mathbf{B}_1 = (\mathbf{L}_x \mathbf{N}_{,x} + \mathbf{L}_y \mathbf{N}_{,y})$ and $\mathbf{B}_2 = \mathbf{L}_z \mathbf{N}$.

2.5 Discrete equation of motion

The balance eq.(2.15) can now be written as a sum of the energy terms for all the elements of the cross-section. Supposing that n_{el} is the total number of finite elements used to represent the cross-section, the discrete form of the Hamilton formulation becomes:

$$\int_{t_1}^{t_2} \left\{ \bigcup_{e=1}^{n_{el}} \left[\int_{V_e} \delta\left(\boldsymbol{\varepsilon}^{(e)\mathrm{T}}\right) \mathbf{C}_e \boldsymbol{\varepsilon}^{(e)} dV_e + \int_{V_e} \delta\left(\mathbf{u}^{(e)\mathrm{T}}\right) \boldsymbol{\mu}_e \ddot{\mathbf{u}}^{(e)} dV_e \right] \right\} dt = 0 \tag{2.23}$$

where $\mathbf{C}_e$ and $\boldsymbol{\mu}_e$ are the elastic stiffness tensor and the density for the generic element, respectively. Now, the first terms under the integral in eq.(2.23), that is the elastic energy contribute for the generic finite element, can be written as:

$$\begin{aligned}\delta\mathcal{U}^{(e)} &= \int_{V_e}\delta\left(\boldsymbol{\varepsilon}^{(e)\mathrm{T}}\right)\mathbf{C}_e\boldsymbol{\varepsilon}^{(e)}dV_e = \int_{\Omega_e}\int_z\delta\left(\boldsymbol{\varepsilon}^{(e)\mathrm{T}}\right)\mathbf{C}_e\boldsymbol{\varepsilon}^{(e)}dzd\Omega_e = \\ &= \int_{\Omega_e}\int_z\delta\left[\left(\mathbf{B}_1+i\xi\mathbf{B}_2\right)\mathbf{q}^{(e)}e^{i(\xi z-\omega t)}\right]^{\mathrm{T}}\mathbf{C}_e\left[\left(\mathbf{B}_1+i\xi\mathbf{B}_2\right)\mathbf{q}^{(e)}e^{i(\xi z-\omega t)}\right]dzd\Omega_e = \\ &= \int_{\Omega_e}\int_z\delta\left[\mathbf{q}^{(e)\mathrm{T}}\left(\mathbf{B}_1^{\mathrm{T}}-i\xi\mathbf{B}_2^{\mathrm{T}}\right)\left(e^{i(\xi z-\omega t)}\right)^*\right]\mathbf{C}_e\left(\mathbf{B}_1+i\xi\mathbf{B}_2\right)\mathbf{q}^{(e)}e^{i(\xi z-\omega t)}dzd\Omega_e\end{aligned} \tag{2.24}$$

remembering that $i^{\mathrm{T}} = -i$. The $e^{i(\xi z-\omega t)}$ terms can be simplified because of the complex conjugate. Moreover, the integration relative to the z variable introduces only the unit length (Hladky-Hennion et al., 1997). So, eq.(2.24) turn into

$$\begin{aligned}&\int_{\Omega_e}\delta\left[\mathbf{q}^{(e)\mathrm{T}}\left(\mathbf{B}_1^{\mathrm{T}}-i\xi\mathbf{B}_2^{\mathrm{T}}\right)\right]\mathbf{C}_e\left(\mathbf{B}_1+i\xi\mathbf{B}_2\right)\mathbf{q}^{(e)}d\Omega_e = \\ &= \delta\mathbf{q}^{(e)\mathrm{T}}\int_x\int_y\left(\mathbf{B}_1^{\mathrm{T}}\mathbf{C}_e\mathbf{B}_1+i\xi\mathbf{B}_1^{\mathrm{T}}\mathbf{C}_e\mathbf{B}_2-i\xi\mathbf{B}_2^{\mathrm{T}}\mathbf{C}_e\mathbf{B}_1+\xi^2\mathbf{B}_2^{\mathrm{T}}\mathbf{C}_e\mathbf{B}_2\right)dydx\mathbf{q}^{(e)}\end{aligned} \tag{2.25}$$

where the integration on the element surface (Ω^e) is splitted over the x and y variables. The kinetic energy contribution in eq.(2.23) for the generic element can be written as:

$$\begin{aligned}\delta\mathcal{T}^{(e)} &= \int_{V_e}\delta\left(\mathbf{u}^{(e)\mathrm{T}}\right)\boldsymbol{\mu}_e\ddot{\mathbf{u}}^{(e)}dV_e = \int_{\Omega_e}\int_z\delta\left(\mathbf{u}^{(e)\mathrm{T}}\right)\boldsymbol{\mu}_e\ddot{\mathbf{u}}^{(e)}dzd\Omega_e = \\ &= \int_{\Omega_e}\int_z\delta\left[\mathbf{N}\mathbf{q}^{(e)}e^{i(\xi z-\omega t)}\right]^{\mathrm{T}}\boldsymbol{\mu}_e\frac{\partial^2}{\partial t^2}\left[\mathbf{N}\mathbf{q}^{(e)}e^{i(\xi z-\omega t)}\right]dzd\Omega_e = \\ &= \int_{\Omega_e}\int_z\delta\left(\mathbf{q}^{(e)\mathrm{T}}\mathbf{N}^{\mathrm{T}}\left[e^{i(\xi z-\omega t)}\right]^*\right)\boldsymbol{\mu}_e\left(i\omega\right)^2\mathbf{N}\mathbf{q}^{(e)}e^{i(\xi z-\omega t)}dzd\Omega_e = \\ &= \delta\mathbf{q}^{(e)\mathrm{T}}\int_{\Omega_e}\mathbf{N}^{\mathrm{T}}\boldsymbol{\mu}_e\left(i\omega\right)^2\mathbf{N}d\Omega_e\mathbf{q}^{(e)} = \\ &= -\omega^2\delta\mathbf{q}^{(e)\mathrm{T}}\int_x\int_y\mathbf{N}^{\mathrm{T}}\boldsymbol{\mu}_e\mathbf{N}dydx\mathbf{q}^{(e)}\end{aligned} \tag{2.26}$$

It is here worth remarking that, despite the displacement field is z dependent, a bidimensional mesh, depending on x and y, is sufficient to calculate the elastic and kinetic contribute for the generic element of the waveguide. Substituting eqs.(2.25)-(2.26) into the energy balance equation (2.23), yields:

$$\int_{t_1}^{t_2}\left\{\bigcup_{e=1}^{n_{el}}\delta\mathbf{q}^{(e)\mathrm{T}}\int_x\int_y\left(\begin{array}{c}\mathbf{B}_1^{\mathrm{T}}\mathbf{C}_e\mathbf{B}_1+i\xi\mathbf{B}_1^{\mathrm{T}}\mathbf{C}_e\mathbf{B}_2-i\xi\mathbf{B}_2^{\mathrm{T}}\mathbf{C}_e\mathbf{B}_1+ \\ +\,\xi^2\mathbf{B}_2^{\mathrm{T}}\mathbf{C}_e\mathbf{B}_2-\omega^2\mathbf{N}^{\mathrm{T}}\boldsymbol{\mu}_e\mathbf{N}\end{array}\right)dydx\mathbf{q}^{(e)}\right\}dt = 0 \tag{2.27}$$

Indicating with:

$$\mathbf{m}^{(e)} = \int_x \int_y \mathbf{N}^{\mathrm{T}} \boldsymbol{\mu}_e \mathbf{N} dy dx \tag{2.28}$$

$$\mathbf{k}_1^{(e)} = \int_x \int_y \mathbf{B}_1^{\mathrm{T}} \mathbf{C}_e \mathbf{B}_1 dy dx \tag{2.29}$$

$$\mathbf{k}_2^{(e)} = \int_x \int_y \left(\mathbf{B}_1^{\mathrm{T}} \mathbf{C}_e \mathbf{B}_2 - \mathbf{B}_2^{\mathrm{T}} \mathbf{C}_e \mathbf{B}_1\right) dy dx \tag{2.30}$$

$$\mathbf{k}_3^{(e)} = \int_x \int_y \mathbf{B}_2^{\mathrm{T}} \mathbf{C}_e \mathbf{B}_2 dy dx \tag{2.31}$$

eq.(2.27) can be rewritten as:

$$\int_{t_1}^{t_2} \left\{ \bigcup_{e=1}^{n_{el}} \delta \mathbf{q}^{(e)\mathrm{T}} \left(\mathbf{k}_1^{(e)} + i\xi \mathbf{k}_2^{(e)} + \xi^2 \mathbf{k}_3^{(e)} - \omega^2 \mathbf{m}^{(e)} \right) \mathbf{q}^{(e)} \right\} dt = 0 \tag{2.32}$$

By the standard finite element assembling procedure, eq.(2.32) leads to:

$$\int_{t_1}^{t_2} \left\{ \delta \mathbf{U}^{\mathrm{T}} \left(\mathbf{K}_1 + i\xi \mathbf{K}_2 + \xi^2 \mathbf{K}_3 - \omega^2 \mathbf{M} \right) \mathbf{U} \right\} dt = 0 \tag{2.33}$$

where in eq.(2.33) $\mathbf{U}$ is the global vector of unknown nodal displacements and:

$$\mathbf{M} = \bigcup_{e=1}^{n_{el}} \mathbf{m}^{(e)}, \quad \mathbf{K}_1 = \bigcup_{e=1}^{n_{el}} \mathbf{k}_1^{(e)}, \quad \mathbf{K}_2 = \bigcup_{e=1}^{n_{el}} \mathbf{k}_2^{(e)}, \quad \mathbf{K}_3 = \bigcup_{e=1}^{n_{el}} \mathbf{k}_3^{(e)} \tag{2.34}$$

Due to the arbitrary of $\delta \mathbf{U}$ and the time interval $t_1 - t_2$ in eq.(2.33), the application of the waveguide-SAFE yields the following homogeneous discrete wave equation:

$$\left[\mathbf{K}_1 + i\xi \mathbf{K}_2 + \xi^2 \mathbf{K}_3 - \omega^2 \mathbf{M} \right]_{n_{dof}} \mathbf{U} = 0 \tag{2.35}$$

where the subscript n_{dof} is the number of total degrees of freedom in the problem and represents the system dimension.

2.6 Properties of the matrices

The matrices defined above have some properties that will be stated in the following. The matrices $\mathbf{M}$, $\mathbf{K}_1$ and $\mathbf{K}_3$ are real symmetric matrices and positive definite while the matrix $\mathbf{K}_2$ is a real skew symmetric matrix. The stiffness matrix $\mathbf{K}_1$ is related with the strain-transformation matrix $\mathbf{B}_1$ that pertains to generalized planar deformations, and thus describes the generalized plane strain behavior or essentially the cross-sectional twisting. The stiffness matrix $\mathbf{K}_3$ models out-of-plane deformations behavior since is characterized by the strain-transformation matrix $\mathbf{B}_2$. $\mathbf{K}_2$ containing both $\mathbf{B}_1$ and $\mathbf{B}_2$ is the agent coupling the cross-sectional warpage with the out-of-plane effects. Without loss of generality, in order to lose the imaginary unit in eq.(2.35), a $n_{dof} \times n_{dof}$ diagonal transformation matrix $\mathbf{T}$ is introduced, with each diagonal element equal to 1, except every third (corresponding to the displacement in the z-direction) being equal to i:

$$\mathbf{T} = \begin{bmatrix} 1 & & & & & & \\ & 1 & & & & & \\ & & i & & & & \\ & & & \ddots & & & \\ & & & & 1 & & \\ & & & & & 1 & \\ & & & & & & i \end{bmatrix} \tag{2.36}$$

The matrix $\mathbf{T}$ has the property that $\mathbf{T}^{\mathrm{T}} = \mathbf{T}^*$ and $\mathbf{T}^*\mathbf{T} = \mathbf{T}\mathbf{T}^* = \mathbf{I}$, where here the asterisk (*) denotes the complex conjugate. The terms in eq.(2.35) are pre-multiplied by $\mathbf{T}^{\mathrm{T}}$ and post multiplied by $\mathbf{T}$. The symmetric matrices $\mathbf{K}_1$, $\mathbf{K}_3$ and $\mathbf{M}$ do not mix the w component of displacement with u and v. Therefore, this algebraic manipulation leaves them in the same form as before:

$$\mathbf{T}^{\mathrm{T}}\mathbf{K}_1\mathbf{T} = \mathbf{K}_1 \tag{2.37}$$
$$\mathbf{T}^{\mathrm{T}}\mathbf{K}_3\mathbf{T} = \mathbf{K}_3 \tag{2.38}$$
$$\mathbf{T}^{\mathrm{T}}\mathbf{M}\mathbf{T} = \mathbf{M} \tag{2.39}$$

On the other hand, the skew symmetric matrix $\mathbf{K}_2$, mixes the displacement u and v with w but it does not mix the displacements u and v with each other. It follows that:

$$\mathbf{T}^{\mathrm{T}}\mathbf{K}_2\mathbf{T} = \tilde{\mathbf{K}}_2 = -i\hat{\mathbf{K}}_2 \tag{2.40}$$

where now the matrix $\tilde{\mathbf{K}}_2$ is a symmetric matrix and $\hat{\mathbf{K}}_2$ is also a real matrix. The introduction of the operator $\mathbf{T}$ is equivalent to using a specific displacement field as assumed by other researchers (Gavrić, 1994; Volovoi et al., 1998; Hladky-Hennion, 1996), where to assure the quadrature between the two in-plane displacements and the displacement in the propagation z-direction (normal to the plane), the w displacement component is multiplied by the imaginary unit i:

$$\mathbf{u}(x,y,z,t) = \begin{bmatrix} u(x,y)e^{i(\xi z-\omega t)} \\ v(x,y)e^{i(\xi z-\omega t)} \\ w(x,y)e^{i(\xi z-\omega t+\pi/2)} \end{bmatrix} = \begin{bmatrix} u(x,y) \\ v(x,y) \\ iw(x,y) \end{bmatrix} e^{i(\xi z-\omega t)} \tag{2.41}$$

An equivalent form of the wave equation (2.35) is:

$$\left[\mathbf{K}_1 + \xi\hat{\mathbf{K}}_2 + \xi^2\mathbf{K}_3 - \omega^2\mathbf{M}\right]_{n_{dof}} \hat{\mathbf{U}} = 0 \tag{2.42}$$

where $\hat{\mathbf{U}}$ is a new nodal displacement vector.

2.7 Eigenvalue problem solution

Non trivial solutions of eq.(2.42) can be found by solving the twin parameter generalized eigenproblem in ξ and ω:

$$det\left[\mathbf{K}_1 + \xi\hat{\mathbf{K}}_2 + \xi^2\mathbf{K}_3 - \omega^2\mathbf{M}\right]_{n_{dof}} = 0 \tag{2.43}$$

The eigenvalue problem (2.43) can either, be solved as a linear generalized eigenvalue problem in $\omega(\xi)$ for a given real wavenumber ξ, or for a given frequency, ω, as a second order polynomial eigenvalue problem in $\xi(\omega)$ form. In order to solve the latter case, $\xi(\omega)$, eq.(2.42) has to be recasted as a first-order eigensystem in the wavenumber:

$$(\mathbf{A}-\xi\mathbf{B})\,\mathbf{Q}=\mathbf{0} \tag{2.44}$$

where $\mathbf{A}$ and $\mathbf{B}$ are real symmetric matrices:

$$\mathbf{A}=\begin{bmatrix} \mathbf{0} & \mathbf{K}_1-\omega^2\mathbf{M} \\ \mathbf{K}_1-\omega^2\mathbf{M} & \hat{\mathbf{K}}_2 \end{bmatrix}, \qquad \mathbf{B}=\begin{bmatrix} \mathbf{K}_1-\omega^2\mathbf{M} & \mathbf{0} \\ \mathbf{0} & -\mathbf{K}_3 \end{bmatrix}, \qquad \mathbf{Q}=\begin{bmatrix} \hat{\mathbf{U}} \\ \xi\hat{\mathbf{U}} \end{bmatrix} \tag{2.45}$$

A generic eigensolution of eq.(2.43) or eq.(2.45) consists in finding the spatial and temporal wave propagation parameters, frequency ω and wavenumber ξ, as well as in the calculation of the wavestructure or cross-sectional mode shape $\hat{\mathbf{U}}$, that can be obtained back substituting the solution ω-ξ in eq.(2.42).

3 Guided wave general solution

The solutions $\omega(\xi)$ of (2.43) represent the couples of wavenumber and ferquency of the *propagative modes* of the waveguide. Therefore in this case, for each wavenumber ξ_i as an input, n_{dof} circular frequencies ω_i will be obtained. The added information that we obtain, beside all the propagative modes, by solving the problem $\xi(\omega)$ are the so called *nonpropagative modes* (evanescent and end modes). In this case, for each frequency ω_i real positive quantity in input, $2n_{dof}$ wavenumbers ξ_i will be obtained. It should be noted that the wavenumbers obtained by solving the eigenvalue problem as $\xi(\omega)$ can be real, complex or purely imaginary. The real wavenumbers, that are the same solutions of the $\omega(\xi)$ case, carry the information of the propagative modes, while the complex and imaginary solutions are related to nonpropagative modes. A nonpropagating mode can be thought of as a local vibration of the structure which does not propagate away from the point where it is generated. In summary, if there are n_{dof} degrees of freedom in the model, there are n_{dof} solutions for the $\omega(\xi)$ problem and $2n_{dof}$ solutions for $\xi(\omega)$, (including infinite valued ξ):

$$\omega(\xi) = n_{dof} \text{ solutions} \rightarrow \text{propagative modes}$$

$$\xi(\omega) = 2n_{dof} \text{ solutions} \rightarrow \begin{cases} \text{propagative modes} \\ \text{non-propagative modes} \end{cases}$$

3.1 Propagative modes only

When the interest is only on the propagative modes, the $\omega(\xi)$ approach is preferred than the $\xi(\omega)$ since the former calculation is made at a numerically lower cost. Equation (2.42) can be solved for a given real wavenumber, ξ, as a linear generalized eigenvalue problem in ω. The system (**??**) can be rewritten as:

$$\left[\mathbf{K}(\xi)-\omega^2\mathbf{M}\right]\hat{\mathbf{U}}=\mathbf{0} \tag{3.1}$$

where:

$$\mathbf{K}(\xi) = \mathbf{K}_1 + \xi\hat{\mathbf{K}}_2 + \xi^2\mathbf{K}_3 \tag{3.2}$$

Observe that for real ξ, the left hand side of eq.(3.1) is real and symmetric, since $\mathbf{K}_1$, $\hat{\mathbf{K}}_2$, $\mathbf{K}_3$ and $\mathbf{M}$ are real and symmetric matrices. Consequently, all the eigenvalues and eigenvectors are real. The eigennvalues represent the temporal circular frequencies for all the propagating modes, while the corresponding eigenvectors represent the wavestructures or cross-sectional mode shapes:

$$\omega_i = a, \qquad \hat{\mathbf{U}}i = \Phi$$

where a is a positive scalar while Φ indicates a real vector.

3.2 Propagative and nonpropagative modes

In order to characterized both propagative and nonpropagative modes the first order eigenvalue problem (2.44) must be solved. The numerical problem here consists of finding the set of generally complex valued scalars $\xi_i = a_i + ib_i$ and the set of corresponding complex eigenvectors $\mathbf{Q}_i = \Phi_i + i\Psi_i$ for a given excitation frequency ω_i (a and b are real arbitrary values; Φ and Ψ are real vectors). The number of eigenvalues and corresponding eigenvectors is equal to twice the number of degrees of freedom of the system (n_{dof}). The problem $\xi(\omega)$ has a numerically higher cost than $\omega(\xi)$. The $2n_{dof}$ eigenvalues and eigenvectors can be real, imaginary or complex.
Real eigenvalues correspond to the wavenumbers of propagative elastic waves ξ_i, and the upper part of the corresponding eigenvector $\mathbf{Q}_i = [\ \hat{\mathbf{U}}_i^{\mathrm{T}} \quad \xi_i\hat{\mathbf{U}}_i^{\mathrm{T}}\ ]^{\mathrm{T}}$, i.e. $\hat{\mathbf{U}}_i$, describe the wavestructure or cross-sectional mode shapes of the waveguide of the corresponding propagative mode. The real eigenvalues appear in pairs of opposite sign, which indicate two waves propagating in opposite directions:

$$\xi_i = \pm a, \qquad \hat{\mathbf{U}}_i = \pm\Phi \tag{3.3}$$

The purely imaginary solutions for the wavenumber are representative of the exponentially decaying near fields, which generally do not transport any appreciable mechanical energy, unless the length of the waveguide is small. This type of evanescent modes are also known as *end modes,* referring to the fact that their presence is necessary to satisfy the condition of traction free in a boundary problem or in the study of wave reflection. They also appear in pairs of opposite sign along with the corresponding eigenvectors:

$$\xi_i = \pm ib, \qquad \hat{\mathbf{U}}_i = \pm i\Psi \tag{3.4}$$

The complex wavenumbers and corresponding complex modes appear in groups of four and are the *evanescent modes.* Each of the waves has two components. The real part of the solution represents the propagative part of the field while the imaginary part is an exponential envelope of the wave:

$$\xi_i = \pm(a \mp ib), \qquad \hat{\mathbf{U}}_i = \pm(\Phi \mp i\Psi) \tag{3.5}$$

3.3 Considerations on the solutions

Now, since the displacement vector is in general expressed as:

$$\mathbf{u} = \mathbf{Nq}e^{i(\xi z - \omega t)} = \mathbf{Nq}e^{i((\xi_{\mathrm{Re}} + i\xi_{\mathrm{Im}})z - \omega t)} = \mathbf{Nq}e^{i(\xi_{\mathrm{Re}} z - \omega t)}e^{-\xi_{\mathrm{Im}} z} \tag{3.6}$$

it is possible to observe that the real part of the solution represents the harmonic propagative part of the field while the imaginary part is an exponential envelope of the wave. The propagative harmonic part of the field is described by a constant phase argument $e^{i\phi} = e^{i(\xi_{\mathrm{Re}} z - \omega t)}$. Thus for a given real wavenumber and frequency of propagation, while time is increasing the value of z has to change in order to accommodate the property of constant phase ϕ. For example for a wave with positive real part of the wavevector $\xi_{\mathrm{Re}} > 0$, if the time increases ($t \uparrow$) the value of z has to increase ($z \nearrow$) in order to maintain a constant phase:

$$\underset{t \uparrow \text{ time is increasing}}{\phi = \text{constant}} \longrightarrow \begin{cases} if \;\; \xi_{\mathrm{Re}} > 0 \longrightarrow z \nearrow \\ if \;\; \xi_{\mathrm{Re}} < 0 \longrightarrow z \searrow \end{cases}$$

It the light of this, we can assure that the sign of ξ_{Re} indicates the wave propagation direction, thus a positive ξ_{Re} indicates a wave traveling in the positive direction of the z-axis, vice-versa for a negative value of ξ_{Re}. The solution is symmetric, in the sense that for each wave propagating in the positive direction ($\xi_{\mathrm{Re}} > 0$) a corresponding wave is propagating in the negative direction ($\xi_{\mathrm{Re}} < 0$). This confirm what stated for the propagative modes $\xi_i = \pm a$, i.e. that both solutions are acceptable and carry the same information. Regarding the end modes, it can be seen that:

$$\mathbf{u} = \mathbf{Nq}e^{i(\xi z - \omega t)} = \mathbf{Nq}e^{i((i\xi_{\mathrm{Im}})z - \omega t)} = \mathbf{Nq}e^{-i\omega t}e^{-\xi_{\mathrm{Im}} z} \tag{3.7}$$

Analyzing the solutions for the evanescent modes, turns out that only the eigenvalues for which the exponential term $e^{-\xi_{\mathrm{Im}} z} < 1$ are solutions that deserve interest. In fact, since we are considering that no energy is added to the system during the propagation phenomena, all the solutions with $e^{-\xi_{\mathrm{Im}} z} > 1$ corresponding to waves increasing in magnitude while propagating need to be discarded because physically meaningless. Thus, for a wave propagating in the positive z-axis direction, an acceptable real solution needs ξ_{Im} to be bigger or equal than zero. In the opposite case, for wave that propagate in the negative z-axis direction a possible physical solution needs ξ_{Im} to be smaller or equal than zero.

$$\underset{\text{time is increasing} \longrightarrow z > 0}{\overset{\xi_{\mathrm{Re}}>0}{\text{propagation in the positive direction}}} \begin{cases} \xi_{\mathrm{Im}} > 0 & \text{decaying waves} \searrow \\ \xi_{\mathrm{Im}} = 0 & \text{standing waves} \rightarrow \\ \xi_{\mathrm{Im}} < 0 & \text{MEANINGLESS} \nearrow \end{cases}$$

$$\underset{\text{time is increasing} \longleftarrow z<0}{\overset{\xi_{\mathrm{Re}}<0}{\text{propagation in the negative direction}}} \begin{cases} \xi_{\mathrm{Im}} > 0 & \text{MEANINGLESS} \nearrow \\ \xi_{\mathrm{Im}} = 0 & \text{standing waves} \rightarrow \\ \xi_{\mathrm{Im}} < 0 & \text{decaying waves} \searrow \end{cases}$$

where here the arrows indicate the trend of the level of energy while the wave is propagating, e.g. the arrow that point down $\searrow$ represents a decaying wave. Thus, among all the possible solutions for ξ_i for complex wavenumbers:

$$\xi_i = +a + ib \rightarrow \mathbf{u} = \mathbf{N}\mathbf{q}e^{i((a+ib)z-\omega t)} = \mathbf{N}\mathbf{q}e^{i(az-\omega t)}e^{-bz} \quad \searrow \tag{3.8}$$
$$\xi_i = -a - ib \rightarrow \mathbf{u} = \mathbf{N}\mathbf{q}e^{i((-a-ib)z-\omega t)} = \mathbf{N}\mathbf{q}e^{i(-az-\omega t)}e^{bz} \quad \searrow \tag{3.9}$$
$$\xi_i = +a - ib \rightarrow \mathbf{u} = \mathbf{N}\mathbf{q}e^{i((a-ib)z-\omega t)} = \mathbf{N}\mathbf{q}e^{i(az-\omega t)}e^{bz} \quad \nearrow \tag{3.10}$$
$$\xi_i = -a + ib \rightarrow \mathbf{u} = \mathbf{N}\mathbf{q}e^{i((-a+ib)z-\omega t)} = \mathbf{N}\mathbf{q}e^{i(-az-\omega t)}e^{-bz} \quad \nearrow \tag{3.11}$$

the solutions $\xi_i = \pm(a + ib)$ are physically meaningful while the type $\xi = \pm(a - ib)$ are physically meaningless and for this reason will be discarded.

3.4 Frequency cut-on

The waves change their nature with frequency. Some of them are real, i.e. propagative, for all excitation frequencies. They have zero cut-on frequency (e.g. the low order symmetric s_0 and anti-symmetric a_0 modes in plates). Generally, as the frequency increases the number of propagative waves also increases. The waves which are evanescent at low frequencies become propagative when the excitation frequency reaches their respective *cut-on frequencies*. At a cut-on frequency (wavenumber zero) the wavelength is infinite and all the points along the waveguide vibrate in phase. If the wavenumber is taken to have zero value, equation (2.42) becomes:

$$\left[\mathbf{K}_1 - \omega_c^2\mathbf{M}\right]\hat{\mathbf{U}} = \mathbf{0} \tag{3.12}$$

where ω_c is the cut-on frequency. To compute cut-on frequencies this simple eigenvalue problem has to be solved. Some of the eigenvalues will be zero. They correspond to the rigid body modes of the cross-section. The non-zero solutions give the cut-on frequencies ω_c and corresponding cross-section modes Φ_c of the higher order waves which generally deform the cross-section while propagating. It should be noted that the matrices $\mathbf{K}_1$ and $\mathbf{M}$ in equation (3.12) are not equal to the stiffness and mass matrices of corresponding plane-stress or plane-strain dynamic problems.

Partial wave theory gives extra physical insight on how to relate the phenomenon of nonpropagating modes to the frequency of the wave. In fact, if we consider a propagating mode and we decrease the frequency, the angle of incidence of the partial wave decreases, becoming zero at the cut-on frequency, that in this case is named as cut-off frequency. At the cut-on (cut-off) frequency, the partial waves simply reflect back and forth across the thickness of the waveguide and there is no variation of the stress and displacement field along the direction of propagation.

3.5 Group velocity

It is well known that the wave group velocity c_g for each propagative mode is the derivative of the frequency-wavenumber dispersion relation $\partial\omega/\partial k$. This imply that once the solution of the eigenvalue problem is obtained and categorized for all the propagative modes, so that the different branches belonging to the same mode are collected in the same vector, the group velocity can be calculated by a simple derivative. In the light of the complexity of dispersion curves, this is not straightforward when one branch approaches another. One technique to track the same mode in the whole frequency range, consists in looking at the cross-sectional mode shape of the propagating mode in proximity of the overlay between two different branches (Pavlakovic et al., 1997). It is obvious that correspondent branches before and after the cross will show similarities in the wavestructure allowing thus to identify the propagating modes behavior. This tedious procedure is not necessary here since the SAFE method allows calculating the group velocity directly from eq.(2.42) by using some algebra manipulation. Let's start from eq.(2.42):

$$\left[\mathbf{K}(\xi) - \omega^2\mathbf{M}\right]\Phi_R = \mathbf{0}, \qquad \mathbf{K}(\xi) = \mathbf{K}_1 + \xi\hat{\mathbf{K}}_2 + \xi^2\mathbf{K}_3 \tag{3.13}$$

where Φ_R indicates the right eigenvector of the system. Let's derive the first equation in (3.13) with respect to the wavenumber:

$$\frac{\partial}{\partial\xi}\left(\left[\mathbf{K}(\xi) - \omega^2\mathbf{M}\right]\Phi_R\right) = \mathbf{0} \tag{3.14}$$

then:

$$\left[\frac{\partial}{\partial\xi}\mathbf{K}(\xi) - 2\omega\frac{\partial\omega}{\partial\xi}\mathbf{M}\right]\Phi_R + \left[\mathbf{K}(\xi) - \omega^2\mathbf{M}\right]\frac{\partial}{\partial\xi}\Phi_R = \mathbf{0} \tag{3.15}$$

Multiplying eq.(3.15) by the transpose of the left-eigenvector, Φ_L^{T}, and in force of the definition:

$$\Phi_L^{\mathrm{T}}\left[\mathbf{K}(\xi) - \omega^2\mathbf{M}\right] = \mathbf{0} \tag{3.16}$$

eq.(3.15) yields:

$$\Phi_L^{\mathrm{T}}\left[\frac{\partial}{\partial\xi}\mathbf{K}(\xi) - 2\omega\frac{\partial\omega}{\partial\xi}\mathbf{M}\right]\Phi_R = 0 \tag{3.17}$$

Since $\partial\omega/\partial\xi$ is a scalar and represents the group velocity, eq.(3.17) can be placed in the form:

$$c_{gr} = \frac{\partial\omega}{\partial\xi} = \frac{\Phi_L^{\mathrm{T}}\mathbf{K}'\Phi_R}{2\omega\Phi_L^{\mathrm{T}}\mathbf{M}\Phi_R} \tag{3.18}$$

where:

$$\mathbf{K}' = \frac{\partial}{\partial\xi}\mathbf{K}(\xi) = \hat{\mathbf{K}}_2 + 2\xi\mathbf{K}_3 \tag{3.19}$$

Equation (3.18) can be evaluated considering one solution of the dispersion relations at the time. It is on a general form and may be applied to any structure built up by a semi-analytical finite element model. It is possible to simplify eq.(3.18) by showing the identity of the left and right eigenvector. Let's start from eq. (3.16):

$$\begin{aligned}
\Phi_L^{\mathrm{T}}\mathbf{K}(\xi) - \omega^2\Phi_L^{\mathrm{T}}\mathbf{M} &= \mathbf{0} \\
\Phi_L^{\mathrm{T}}\mathbf{K}(\xi) &= \omega^2\Phi_L^{\mathrm{T}}\mathbf{M} \\
\mathbf{K}^{\mathrm{T}}(\xi)\Phi_L &= \omega^2\mathbf{M}^{\mathrm{T}}\Phi_L \\
\left(\mathbf{K}^{\mathrm{T}}(\xi) - \omega^2\mathbf{M}^{\mathrm{T}}\right)\Phi_L &= \mathbf{0}
\end{aligned} \tag{3.20}$$

The matrices $\mathbf{K}(\xi)$ and $\mathbf{M}$ are real and symmetric. Therefore $\mathbf{K}^{\mathrm{T}}(\xi) = \mathbf{K}(\xi)$ and $\mathbf{M}^{\mathrm{T}} = \mathbf{M}$ and the above eq.(3.20) yields:

$$\left[\mathbf{K}(\xi) - \omega^2 \mathbf{M}\right] \Phi_L = \mathbf{0} \tag{3.21}$$

Now comparing the first equation in (3.13) and (3.21) it can be seen that $\Phi_R = \Phi_L$.

Thus, showed the coincidence between the right and left eigenvector, the group velocity formula can be rewritten as:

$$c_{gr} = \frac{\partial \omega}{\partial \xi} = \frac{\Phi_R^{\mathrm{T}} \mathbf{K}' \Phi_R}{2\omega \Phi_R^{\mathrm{T}} \mathbf{M} \Phi_R} \tag{3.22}$$

Based on the value of the wavenumber, that can be real, complex or purely imaginary, the formula (3.22) returns the following type of result:

$$\begin{array}{rcl} \text{propagative modes} & \rightarrow & \text{real}(\xi) \rightarrow \text{real(group velocity)} \\ \text{non-propagative modes} & \rightarrow & \left\{ \begin{array}{c} \text{imag}(\xi) \rightarrow \text{imag(group velocity)} \\ \text{complex}(\xi) \rightarrow \text{imag(group velocity)} \end{array} \right. \end{array}$$

As previously showed, until the cut-on frequency the mode is evanescent and presents an imaginary group velocity. When the cut-on frequency is reached, the imaginary part of the wavenumber becomes equal to zero and the mode converts from evanescent to propagative. This can also be observed from the group velocity, in fact the imaginary group velocity drops to zero and the group velocity for that specific mode becomes real, meaning that the mode starts propagating.

Acknowledgement

This topic is one of the subjects of the Centre of Study and Research for the Identification of Materials and Structures (CIMEST) "M. Capurso". The numerical tests presented were carried out using the facilities of the Laboratory of Computational Mechanics (LAMC) "A. A. Cannarozzi" of the University of Bologna.

Bibliography

B. Aalami. Waves in prismatic guides of arbitrary cross section. *Journal of Applied Mechanics*, 40:1067–1072, 1973.

P.K. Banerjee and S. Kobayashi. *Advanced Dynamic Analysis by Boundary Element Methods.* Elsevier Applied Science, 1990.

A. Bernard, M.J.S. Lowe, and M. Deschamps. Guided waves energy velocity in absorbing and non-absorbing plates. *Journal of the Acoustical Society of America*, 110(1):186–196, 2001.

C.A. Brebbia and S. Walker. *Boundary element techniques in engineering.* Butterworths and Co, 1980.

S.B. Dong and K.H. Huang. Edge vibrations in laminated composite plates. *Journal of Applied Mechanics*, 52:433–438, 1985.

S. Finnveden. Evaluation of modal density and group velocity by a finite element method. *Journal of the Acoustical Society of America*, 273:51–75, 2004.

L. Gavrić. Finite element computation of dispersion properties of thin-walled waveguides. *Journal of Sound and Vibration*, 173(1):113–124, 1994.

L. Gavrić. Computation of propagative waves in free rail using a finite element technique. *Journal of Sound and Vibration*, 185(3):531–543, 1995.

X. Han, G.R. Liu, Z.C. Xi, and K.Y. Lam. Characteristics of waves in a functionally graded cylinder. *Int. J. for Numerical Methods in Eng.*, 53:653–676, 2002.

T. Hayashi, W.J. Song, and J.L. Rose. Guided wave dispersion curves for a bar with an arbitrary cross-section, a rod and rail example. *Ultrasonics*, 41(3):175–183, 2003.

A.C. Hladky-Hennion. Finite element analysis of the propagation of acoustic waves in waveguides. *Journal of Sound and Vibration*, 194(2):119–136, 1996.

A.C. Hladky-Hennion, P. Langlet, and M. de Billy. Finite element analysis of the propagation of acoustic waves along waveguides immersed in waters. *Journal of Sound and Vibration*, 200(4):519–530, 1997.

K.H. Huang and S.B. Dong. Propagating waves and edge vibrations in anisotropic composite cylinders. *Journal of Sound and Vibration*, 96(3):363–379, 1984.

P.E. Lagasse. Higher-order finite element analysis of topographic guides supporting elastic surface waves. *Journal of the Acoustical Society of America*, 53(4):1116–1122, 1973.

M.J.S. Lowe. Matrix techniques for modeling ultrasonic waves in multilayered media. *IEEE Trans. UFFC*, 42(4):525–542, 1995.

O.M. Mukdadi and S.K. Datta. Transient ultrasonic guided waves in layered plates with rectangular cross section. *Journal of Applied Physics*, 93(11):9360–9370, 2003.

O.M. Mukdadi, Y.M. Desai, S.K. Datta, A.H. Shah, and A.J. Niklasson. Elastic guided waves in a layered plate with rectangular cross section. *Journal of the Acoustical Society of America*, 112(5):1766–1779, 2002.

O.Jr. Onipede and S.B. Dong. Propagating waves and end modes in pretwisted beams. *Journal of Sound and Vibration*, 195(2):313–330, 1996.

B.N. Pavlakovic. Leaky guided ultrasonic waves in NDT. *PhD thesis, Imperial College of Science, Technology and Medicine, University of London*, 1998.

B.N. Pavlakovic, M.J.S. Lowe, D.N. Alleyne, and P. Cawley. Disperse: A general purpose program for creating dispersion curves. *Review of Progress in Quantitative NDE*, 16: 185–192, 1997.

H. Taweel, S.B. Dong, and M. Kazic. Wave reflection from the free end of a cylinder with an arbitrary cross-section. *International Journal of Solids and Structures*, 37: 1701–1726, 2000.

V.V. Volovoi, D.H. Hodges, V.L. Berdichevsky, and V.G. Sutyrin. Dynamic dispersion curves for non-homogeneous, anisotropic beams with cross-section of arbitrary geometry. *Journal of Sound and Vibration*, 215(5):1101–1120, 1998.

O.C. Zienkiewicz and R.L. Taylor. *The finite element method, Volume 2, Solid Mechanics.* John Wiley and Sons, 5th edition, 2000.

Numerical Evaluation of Semi-analytical Finite Element (SAFE) Method for Plates, Rods and Hollow Cylinders

Erasmo Viola, Alessandro Marzani and Ivan Bartoli

Dipartimento di Ingegneria delle Strutture, dei Trasporti, delle Acque, del Rilevamento, del Territorio - University of Bologna, Viale Risorgimento 2, 40136 Bologna, Italy

Abstract Here, some applications of the Semi-analytical (SAFE) formulation presented in a previous paper of this book (Viola et al., 2005) are shown. In order to emphasize the potentiality of the SAFE method presented, guided wave features are calculated for several isotropic waveguides immersed in vacuum. In particular a plate, a rod and a hollow cylinder are considered. The dispersion results are presented in terms of phase velocity and group velocity along with the wavestructures or cross-sectional mode shapes.

1 Modeling assumptions

In the applications of the SAFE method presented in this lecture, the material considered, steel, is supposed to be linearly elastic, isotropic and homogeneous at the wavelength scale. The waveguides that will be studied, such as plates, rods and hollow cylinders, are considered of infinite length in the propagation direction and surrounded by vacuum. The coordinate reference system will always be placed in such a way that the z-axis coincides with the wave propagation direction. The acoustic properties of the material are the density $\mu = 7800 \quad Kg/m^3$, and its longitudinal and shear bulk wave speeds, $c_L = 5960 \quad m/s$ and $c_T = 3260 \quad m/s$, respectively.

The SAFE formulation computes the material properties via the constitutive elastic stiffness tensor $\mathbf{C}$ expressed in Voight notation, as shown in section 2 of Ref. (Viola et al., 2005). For an isotropic material, the elastic stiffness tensor depends only by two constants, the Young's modulus E and the Poisson's ratio ν, or, equivalently, by the two Lame's constants λ and G.

The bulk velocities, c_L and c_T, are defined in terms Young's modulus E, Poisson's ratio ν and density μ, as:

$$c_L = \left[\frac{E(1-\nu)}{\mu(1+\nu)(1-2\nu)}\right]^{1/2}, \qquad c_T = \left[\frac{E}{2\mu(1+\nu)}\right]^{1/2} \tag{1.1}$$

Inverting the above relations, we can express the mechanical properties of the material, Young's modulus E and Poisson's ratio ν, as a function of its acoustic properties:

$$E = \mu c_T^2 \left[\frac{3c_L^2 - 4c_T^2}{c_L^2 - c_T^2} \right], \qquad \nu = \frac{1}{2} \left[\frac{c_L^2 - 2c_T^2}{c_L^2 - c_T^2} \right] \tag{1.2}$$

The elastic stiffness tensor for the isotropic material can now be written as:

$$\mathbf{C} = \begin{bmatrix} \lambda{+}2G & \lambda & \lambda & & & \\ \lambda & \lambda{+}2G & \lambda & & & \\ \lambda & \lambda & \lambda{+}2G & & & \\ & & & G & & \\ & & & & G & \\ & & & & & G \end{bmatrix} \tag{1.3}$$

where the Lame's constants, λ and G:

$$\lambda = \frac{E\nu}{(1+\nu)(1-2\nu)}, \qquad G = \frac{E}{2(1+\nu)} \tag{1.4}$$

have been introduced.

2 Guided wave in isotropic elastic plate

In this first example, wave propagation in a 1-mm thick steel plate is investigated. A clockwise (x, y, z) reference Cartesian coordinate system is located at the top of the plate with the y-axis normal to the plate surfaces, that are parallel to the x-z reference plane (see Figure 1). The wave propagation direction is along the z-axis. The plate is considered infinite in the x-direction. This assumption is equivalent to stating that no edge effects along the x-direction will be involved. In practice, the y-z plane of wave propagation can be chosen arbitrarily among the infinite parallel y-z planes, since the wave behavior is exactly alike in each of them. In accordance with the SAFE model (Viola et al., 2005), a section of the plate in the x-y plane should be discretized by planar finite elements. However, the discretization over a one-dimensional line through the thickness of the plate (see Figure 1) is sufficient to represent the displacement field completely for the whole plate. Here, quadratic mono-dimensional elements with 3 degree of freedom (d.o.f.) per node, namely the displacements in the x, y and z-direction, are used. It is worth remembering that if the plate has a finite dimension in the x-direction, a two-dimensional mesh has to be used to discretize the plate section in the x-y plane.

2.1 Lamb and Shear Horizontal guided waves

For a plate-like system, there are particular cases of wave propagation that can be represented by considering only some of the d.o.f. used to discretize the plate thickness. This in turn allows us to reduce the dimension of the discrete system making the study of the solution easier. In isotropic plates two different types of wave, the wavevectors of which lie in the propagation plane y-z, can be identified. Plane waves that are polarized along the x-direction will have no component of particle motion normal to the

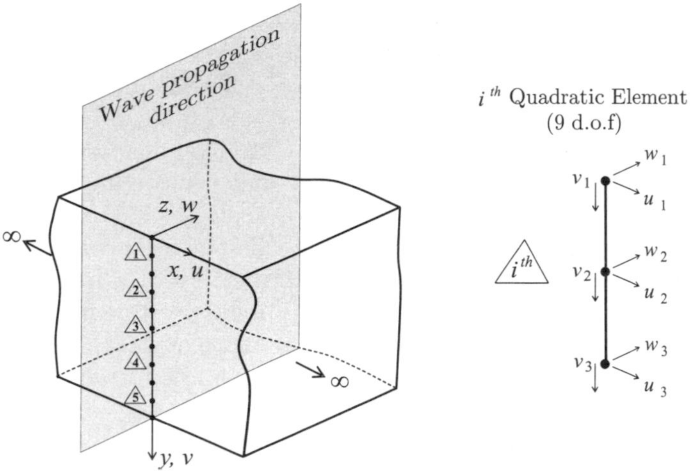

Figure 1. Through thickness discretization and particular of the 9 degrees of freedom quadratic mono-dimensional element

plate surfaces. In isotropic media such horizontally polarized shear, *ShearHorizontal* or *SH*, waves do not mode convert in reflection from the planar x-z surfaces. On the other hand, longitudinal or shear waves polarized in the vertical y-z plane, *Lamb* waves, will, in general, be coupled to each other at the plate surfaces, but will result decoupled to the *SH* waves. In short, the solution for these two types of waves, *SH* and *Lamb*, can be obtained separately. Among the *Lamb* waves, two different types of modes, symmetric S and anti-symmetric A, can be highlighted. Finally, in isotropic plates three different kinds of modes exist: *ShearHorizontal*, *Symmetric* and *Antisymmetric Lamb* modes.

The only non zero displacement component for the *SH* modes is u, that is out-of phase of $\pi/2$ respect the plane of wave propagation y-z. Therefore, even though the energy is traveling in the propagation direction z, these modes induce a particle displacement only in the x direction.

The symmetric modes S_i activate the displacement in the y and z directions, with zero displacement in the x-direction. They are generally characterized by an in-plane displacement w predominant with respect to the out-of-plane displacement v.
In the Antisymmetric modes A_i, the out-of-plane displacement v is predominant with respect to the in-plane displacement w.

Using the SAFE approach, the solution for the *SH* modes can be obtained just considering the degrees of freedom (d.o.f.) in the x-direction, while the *Lamb* modes will result considering only the d.o.f. in the y and z-directions. The dimension of the discrete system is equal to the number of d.o.f. used to discretize the plate thickness (Viola et al., 2005). For example, considering five quadratic beam elements to discretize the thickness of the plate:

$n_{el} = 5$ number of finite elements

$n_{nod} = 3$ number of nodes per element
$n = 3$ number of degrees of freedom per node
$n_{dof} = [n_{el} \times (n_{nod} - 1) + 1] \times n = 33$ degrees of freedom of the system

the dimension of the discrete system is 33. In the light of what was stated above for isotropic plates, the SH modes can be obtained by only considering the 11 d.o.f. in the x-direction while the remaining 22 d.o.f. in the $y - z$ plane will characterize the *Lamb* modes. The choice of using 5 finite elements depends on the accuracy needed for the solution. It is worth remembering that the quality of the solution is frequency dependent, thus 5 finite elements can provide a very good solution for a certain frequency range, but the accuracy can turn out to be poor for a higher frequency range. Some comments about the accuracy of the solutions will be given later on in a further section.

2.2 Propagative modes: Solution for $\omega(\xi)$

Let us briefly remember that the dispersion equation comes from the following two-parameters $(\omega - \xi)$ eigenvalue problem:

$$det\left[\mathbf{K}_1 + \xi\hat{\mathbf{K}}_2 + \xi^2\mathbf{K}_3 - \omega^2\mathbf{M}\right]_{n_{dof}} = 0 \tag{2.1}$$

where the detailed calculation of the matrices in eq.(2.1) is given, for example, in Viola et al. (2005). In this case, $n_{dof} = 33$, for each wavenumber ξ_i assigned to the system, the corresponding frequency values ω_i $(i = 1, 2, .., 33)$ are obtained by solving the eigenvalue problem (2.1). Among these 33 solutions, 22 values represent the *Lamb* modes and the remaining 11 values characterize the SH modes. This shows the potentiality of the SAFE method compared to the Superposition of Partial Bulk Wave (SPBW) methods, in the sense that for the same wavenumber the SPBW has to perform a zero-finding algorithm 33 times, while the SAFE approach has to solve the eigenvalue problem once to get the same type of solution. The eigenpairs wavenumber-frequency (ξ_i-f_i), where $f_i = \omega_i/2\pi$, solution of eq.(2.1) are represented in Figure 2 (a) for the SH modes and in Figure 2 (b) for the *Lamb* modes. The real wavenumber projection displays the relationship between the temporally and spatially varying wave characteristics of the guided mode along the direction of propagation. The wavenumber is inversely related to the wavelength of the guided wave by the equation $\lambda = 2\pi/\xi$, where λ is the wavelength and ξ is the real wavenumber in radians per meter. Real wavenumbers represent propagative modes. From these two plots it can be seen that for zero wavenumber $\xi_i = 0$ as input 4 SH and 6 *Lamb* modes are present in the displayed frequency range. It is also true that for each wavenumber as an input the solution of the eigenvalue problem (2.1) provides all the 33 ω_i values. The remaining $33 - 4 - 6 = 23$ solutions for zero wavenumber not represented in these two plots have a frequency bigger than the upper limit displayed of 5 MHz. The SH modes are indicated in Figure 2 (a) by the label S_{Hi}. In Figure 2 (b) the symmetric and antisymmetric *Lamb* modes are indicated by S_i and A_i, respectively. The subscript i refers to the location of the mode with respect to the others of the same kind in a frequency scale. This means that scanning the frequency range from zero and going up, for example the S_{Hi} mode will be found before S_{Hi+1}. The modes with index

$i = 0$ are named *Fundamental modes* and they are the only modes that exist at zero frequency. A more common way to represent frequency-wavenumber relation is in terms of phase velocity $c_p = \omega/\xi$ versus frequency f. The phase velocity of a guided wave describes the rate at which individual crests of the wave move. Because of this relation, the phase velocity shows the same information as the real wavenumber display and can be used interchangeably. However, the phase velocity view is more convenient to use for realistic ultrasonic testing. It is easy to compare the phase velocity to the bulk velocities of the materials. In addition, the phase velocity plot emphasizes the velocity changes due to the guided nature of the modes. These curves are named *Dispersion curves*. The dispersion curves in terms of phase velocity are represented for the *SH* and *Lamb* modes in Figure 2 (c) and (d), respectively. In the bottom part of Figure 2 the group velocity $c_g = \partial\omega/\partial\xi$ information for both the modes, *SH* and *Lamb*, is presented. The group velocity is the feature that plays a fundamental role for NDE aims. In fact, for lossless waveguide, the group velocity coincides with the velocity at which the energy travels along the structure. If the mode is not dispersive, then the group velocity is equal to the phase velocity. From the plots of Figure 2 some interesting remarks, regarding the behavior of the guided modes in plates, can be made:

- in isotropic plates three fundamental modes exist, S_{H0}, S_0 and A_0.
- It can be seen that among all the modes, the *Lamb* and the *SH* ones, the only nondispersive mode in the whole frequency range is the low order SH mode S_{H0}.
- Below 1.5 MHz, among the SH modes only the S_{H0} appears. This can be also useful from an NDE application since working below that threshold allows one to deal with only one guided mode.
- Cut-off frequencies for the Lamb and SH modes, that are evaluated by solving the eigenvalue problem (2.1) for a zero wavenumber (Viola et al., 2005) can easily be seen in Figure 2 (a) and (b).
- The plate velocity $c_{plate} = E^{1/2}\left[\mu\left(1-\nu^2\right)\right]^{-1/2}$ (Rose, 1999) can be seen in the phase and group velocity plot and coincide with the velocity of the fundamental S_0 mode at zero frequency (E and ν can be defined from the acoustic properties of the material by the relations given in Appendix A).
- For increasing frequency the A_0 and S_0 modes converge to c_R, the Rayleigh surface wave velocity, while the higher *Lamb* modes converge to c_T the bulk shear speed of the material (steel) (Lowe, 1992). The expression for the Rayleigh velocity is available in many wave propagation textbooks (Achenbach (1973), Auld (1990)).
- It is interesting to note in Figure 2 (f) that mode S_1 has negative group velocity at its lowest frequencies. This implies that the energy of the S_1 mode travels in the direction opposite to the phase velocity (Lowe, 1992).
- Furthermore, close examination of the data of Figure 2 (f) reveals that mode A_2 also has negative group velocity over a very small range of frequencies. The slight extension under consideration of the end of the curve below the zero group velocity has the same shape as the one of the mode S_1 and similar considerations hold.

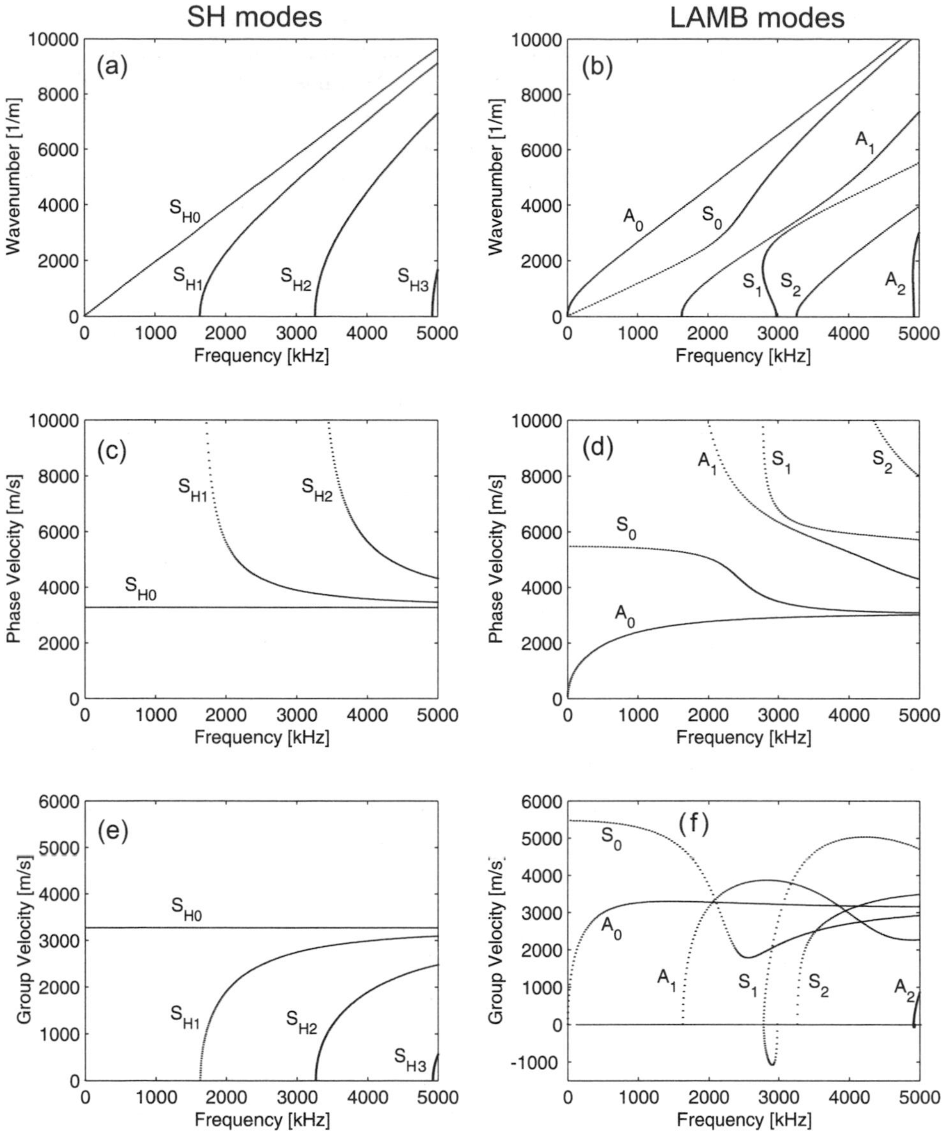

Figure 2. SH and Lamb modes up to 5 MHz for a 1-*mm* steel plate

2.3 Case of the evanescent SH modes: solution for $\xi(\omega)$

In this section, the solution for propagative and evanescent SH modes in plates will be obtained and some results given. Since the evanescent modes are beyond of the scope of this lecture, this will be the only case of evanescent modes considered.
Propagative and evanescent modes can be obtained by solving for a given real frequency ω_i the following $2n_{dof}$ eigenproblem (Viola et al., 2005):

$$det\,[\mathbf{A}-\xi\mathbf{B}]_{2n_{dof}} = \mathbf{0} \tag{2.2}$$

where **A** and **B** are real symmetric matrices:

$$\mathbf{A}=\begin{bmatrix} \mathbf{0} & \mathbf{K}_1-\omega^2\mathbf{M} \\ \mathbf{K}_1-\omega^2\mathbf{M} & \hat{\mathbf{K}}_2 \end{bmatrix}, \qquad \mathbf{B}=\begin{bmatrix} \mathbf{K}_1-\omega^2\mathbf{M} & \mathbf{0} \\ \mathbf{0} & -\mathbf{K}_3 \end{bmatrix} \tag{2.3}$$

In this case the solution is in terms of wavenumber ξ_l that can be real, complex or purely imaginary. Solving the problem as $\xi(\omega)$ yields a symmetric solution, in the sense that for a positive wavenumber there is also its negative correspondent. The real wavenumbers correspond to the propagative modes while complex wavenumbers pertain to evanescent modes. The positive real wavenumbers obtained by solving the system as $\xi(\omega)$ are coincident with the one obtained by solving the eigenvalue problem as $\omega(\xi)$ (see Figure 2 (a) and Figure 3 (a)). For complex wavenumbers, it can be seen how the imaginary part of the wavenumber represents the attenuation of the waves. The displacement field in the SAFE method has been defined as (Viola et al., 2005):

$$\mathbf{u} = \mathbf{N}\mathbf{q}e^{i(\xi z-\omega t)} = \mathbf{N}\mathbf{q}e^{i((\xi_{\mathrm{Re}}+i\xi_{\mathrm{Im}})z-\omega t)} = \mathbf{N}\mathbf{q}e^{i(\xi_{\mathrm{Re}}z-\omega t)}e^{-\xi_{\mathrm{Im}}z} \tag{2.4}$$

where it can be seen that the imaginary part $\xi_{\mathrm{Im}} = \alpha$ of the wavenumber describes the decay of the waves. The coefficient α is also termed as absorption coefficient in Nepers/meter. $1/\alpha$ is the distance, expressed in meters, over which the amplitude of the wave drops by $1/e$ (-8.6859 dB). The attenuation may also be expressed in other units such as decibels per meter that may be more familiar to some readers (1 Np = 8.6859 dB).
Using 5 finite elements, the dimension of the system is $2n_{dof} = 66$. Among these 66 d.o.f., $22 \times 2 = 44$ d.o.f. will describe the propagative and nonpropagative *Lamb* modes, while $11 \times 2 = 22$ d.o.f. will be used to trace the propagative and nonpropagative SH modes. Overall, for each ω_i as an input, 44 solutions for the *Lamb* modes and 22 solutions for the SH modes will be obtained. Up to 1.5 MHz it has been showed that among the SH modes only the zero order propagative mode S_{H0} exists (see Figure 2 (a)). Now, since the solution is symmetric, two roots $\pm S_{H0}$, exist. Thus, in the frequency range $0.0 - 1.5$ MHz, among the 22 solutions for the SH modes, 2 propagative and 20 nonpropagative modes exist: this can be seen in Figure 3 (a) and (b), where the $1 + 10 = 11$ positive roots at 1 MHz are highlighted. In this particular case all the nonpropagative modes are end modes (zero real wavenumber) (Viola et al., 2005). It can be noted that while the frequency is increasing, the end modes gradually lose in attenuation. A value of zero attenuation corresponds to a cut-on frequency point, where the end non propagative mode starts to propagate assuming real wavenumber. In Figure 3 (a) and (b) the cut-on frequency for the S_{H1} mode around 1.6 MHz is indicated by the arrow. The phase

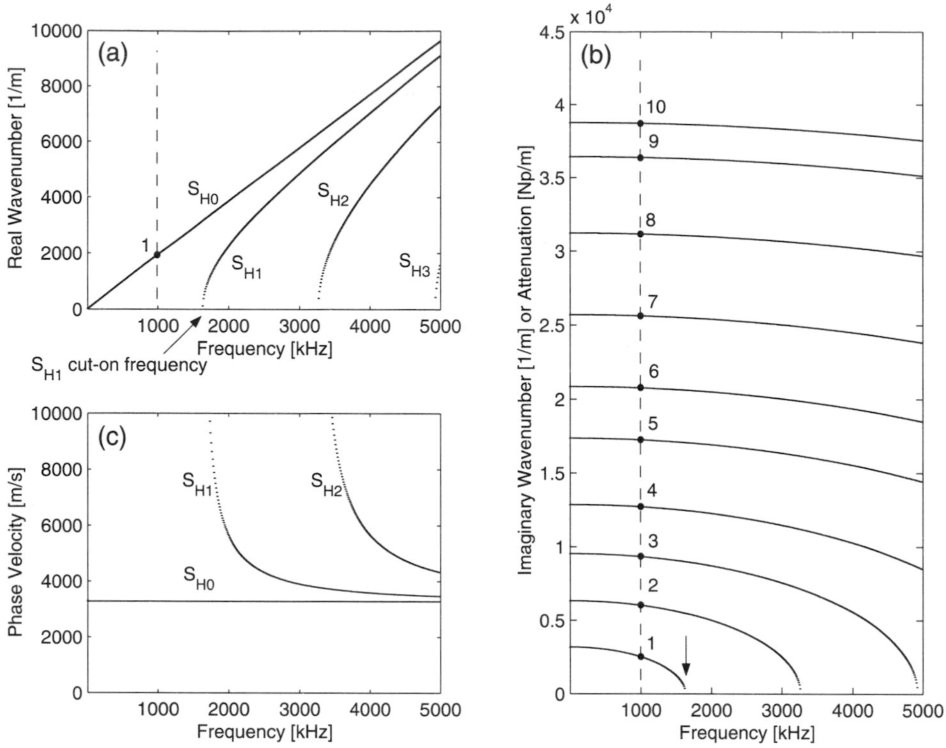

Figure 3. Propagative and evanescent SH modes up to 5 MHz for a 1-mm steel plate

and group velocities of the propagative modes are represented in Figure 3 (a) and (c), respectively. These results are coincident with the one displayed in Figure 2 (a) and (c). The group velocity for the evanescent modes is imaginary and so means less and for this reason is not represented.

2.4 Mode shapes or wavestructures of plates

Figure 4 shows all possible cross-sectional mode shapes at 1 MHz: the zero order *Lamb* modes, A_0 and S_0, and the zero order *SH* mode S_{H0}. These three modes that exist at zero frequency are named Fundamental modes. For the A_0 and S_0 modes, the only non zero displacements are the *in-plane* displacement $w(y)$ (i.e. the displacement along the direction of propagation z) and the *out-of-plane* displacement $v(y)$ (i.e. the displacement normal to the plane of the plate in y-direction). For these two modes the displacement in x-direction $u(y)$, named $90-deg$ displacement, is zero. The only non zero displacement for the S_{H0} mode is the $90-deg$ displacement $u(y)$. In particular the fundamental non dispersive S_{H0} mode shows a constant $90-deg$ displacement ($u(y)$). Also from these plots it can be noted that the displacement field of the *Lamb* modes does

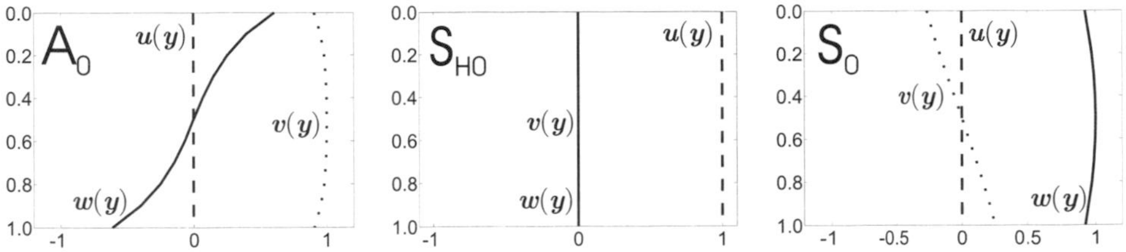

Figure 4. Wavestructures at 1 MHz for a 1-*mm* steel plate: A_0, S_{H0} and S_0 modes

not couple the displacement field of the SH modes. As for the *Lamb* modes, considering the $v(y)$ and $w(y)$ displacements through the thickness of the plate, also the S_{Hi} waves can be either symmetric or antisymmetric. However, no differentiation is made here and a simple counter variable is used to distinguish the different modes. Nevertheless, if the plate is geometrically and mechanically symmetric with respect to its mid plane, it can be assured that:

- the displacement field $u(y)$ for the S_{Hi} modes is symmetric for the even order modes $i = 0, 2, 4...$ and anti-symmetric for the odd order modes $i = 1, 3, 5...$ (see Figure 10).
- The in-plane displacement $w(y)$ for the S_i is symmetric across the thickness of the plate while the out-of-plane displacement $v(y)$ is anti-symmetric.
- The A_i modes show an anti-symmetric in-plane displacement $w(y)$ across the thickness and a symmetric out-of-plane displacement $v(y)$.

3 Guided wave in isotropic elastic rod

Here the dispersion curves will be given for a 1-*mm* radius steel rod. According to reference (Viola et al., 2005), an orthogonal coordinate system will be used. However, a cylindrical coordinate system (r, θ, z) could also be used as in reference Viola and Marzani (2005). The cross-section of the rod is discretized via triangular constant strain CST elements. Each element has three nodes with three d.o.f. per node, namely the displacements in the x, y and z-direction, and the shape functions are linear over the element domain. The mesh is automatically generated by using the *Pdetool* of Matlab. In cylindrical structures three different families of propagative guided modes exist: Longitudinal, Torsional and Flexural (Viola and Marzani, 2005) indicated by L, T and F, respectively. In addition to the type of mode, a dual index system helps track the modes. The first index, n, refers to the circumferential order of the mode, which describes the integer number of wavelengths around the circumference of the cylinder. For example, the displacement amplitudes for zero order modes ($n = 0$) are constant with angle and the displacements on the top and bottom of the cylinder are 180 degrees out of phase for first order modes ($n = 1$). In other words, when $n = 0$ the projection of the displacement vector on the $x - y$ plane is always direct along a radial line, for all the points belonging to the cross-section. Instead, for $n = 1$, the projections of the

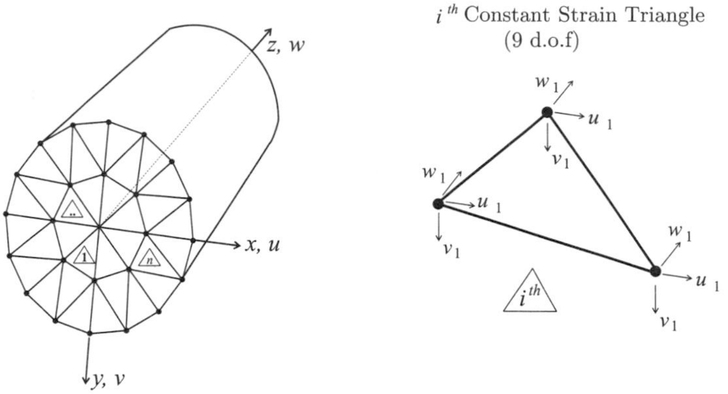

Figure 5. Cross section discretization and particular of the 9 degrees of freedom Linear Strain triangle LST

displacement vector on the $x-y$ plane are in the same direction, for all the points of the cross-section. Therefore, all longitudinal and torsional modes are of circumferential order $n=0$, and all of the flexural modes have a circumferential order greater than or equal to $n=1$. The second index, i, is a counter variable. The fundamental modes (those modes that can propagate at zero frequency) are given the value $i=1$ and the higher order modes are numbered consecutively. This index roughly reflects the modes of vibration within the wall of the cylinder. For example, using this convention, the third flexural mode of circumferential order $n=1$ would be named F(1,3). Due to the complexity and to the secondary experimental importance of the nonpropagative modes, only the $\omega(\xi)$ solution will be performed. In this example a mesh with 81 nodes for 128 CST triangles is assumed. From Figure 6 some interesting remarks can be made:

- in isotropic rods three fundamental modes exist, $T(0,1)$, $L(0,1)$ and $F(1,1)$.
- $L(0,1)$ behaves as a pure extensional mode at near zero frequency. The particle motion consists of a uniform axial traction or compression in the z-direction.
- The fundamental flexural mode, $F(1,1)$, begins as a pure bending, or transverse, mode at zero frequency.
- It can be seen that among all the modes, the only nondispersive mode in the whole frequency range is the fundamental torsional mode $T(0,1)$. Its velocity coincides with the shear bulk velocity c_T of the material.
- The bar velocity $c_{bar} = [E/\mu]^{-1/2}$ (Pavlakovic, 1998) can be seen in the phase and group velocity plot indicated with c_{bar}, and coincides with the velocity of the fundamental $L(0,1)$ mode for zero frequency.
- For increasing frequency the $L(0,1)$ and $F(1,1)$ modes converge to c_R, the Rayleigh surface wave velocity.

It is worth noting here that the group velocity in Figure 6 is displayed only for positive values. However some modes show a negative group velocity at their lowest frequencies, following the behavior observed for the S_1 mode in isotropic plates.

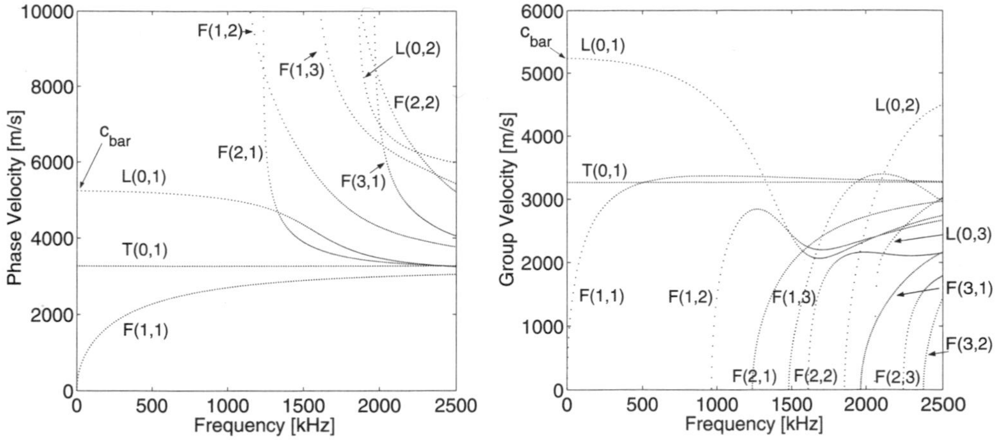

Figure 6. Guided modes phase and group velocity for 1-mm radius steel bar

3.1 Mode shapes or wavestructures of rods

The dispersion curves provide a lot of information about the velocities and wavenumbers of the guided modes. However, this information does not provide a full description of how the mode behaves. In order to better understand the nature of the various guided waves and the names that are assigned to them, Figure 7 shows some of the mode shapes or wavestructures for the 1-mm radius rod under consideration. In Figure 7 (a), the exact points in terms of phase velocity and frequency where the modes have been calculated is indicated. Two different representations are given for each mode: a plane view of the cross section of the waveguide and a three dimensional view. In each representation the dashed grid represents the undeformed finite element mesh over the cross-section. It can be easily seen from Figure 7 that $T(0,1)$ and $L(0,1)$ are axisymmetric modes. The fundamental torsional mode corresponds to a uniform twisting of the entire bar around the z-axis. The fundamental longitudinal mode mainly displaces the cross section in the longitudinal z-direction and the consequent lateral contraction can be viewed from the plane view. The comparison among the three flexural modes $F(1,1)$, $F(2,1)$, and $F(3,1)$ is useful to better understand the meaning of the circumferential order n. Bar modes that have an even circumferential order resemble the symmetric modes of a plate, for which the displacements at the top and the bottom are in phase. On the other hand, modes that have an odd circumferential order resemble anti-symmetric plate modes, the top and bottom displacements of which are out of phase. In isotropic rods these general trends exist:

- Longitudinal modes ($L(0,i)$) are longitudinal axially symmetric modes. The fundamental longitudinal mode $L(0,1)$ is purely extensional (displacements in the axial direction) at zero frequency;
- Torsional modes ($T(0,i)$) are rotational modes the displacement of which is primarily in the θ direction. The fundamental torsional mode $T(0,1)$ is nondispersive in

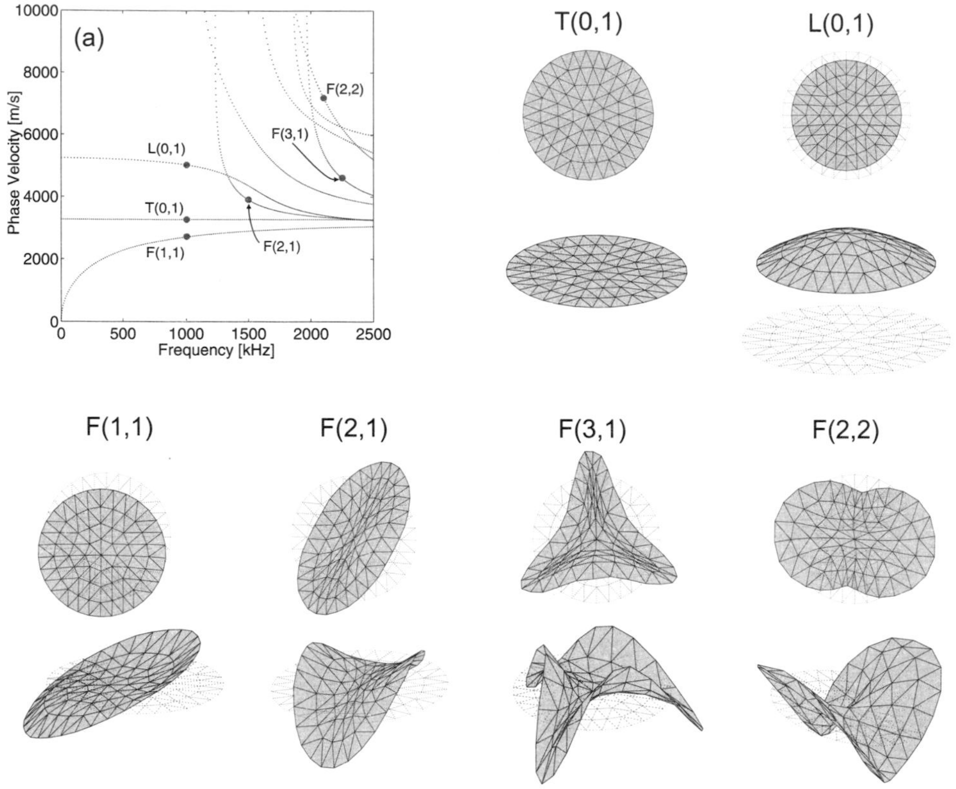

Figure 7. Plane and 3D Wavestructure representation for the guided modes indicated in the Phase Velocity plot (a)

the whole frequency range and it is the only nondispersive mode in rods. Torsional modes correspond to the SH modes in plates;

- Flexural modes ($F(n,i)$) are non-axially symmetric bending modes, and for this reason the circumferential order n has to be ≥ 1. The fundamental flexural mode $F(1,1)$ is a pure bending mode at zero frequency.

4 Guided wave in isotropic elastic hollow cylinder

Compared to wave propagation in rods, the behavior of the modes changes when the inside of the cylinder is hollow. A 1-mm thick steel pipe that has an inner radius of 3-mm and is in vacuum is considered. The cross section is discretized by using the *pdetool* toolbox of Matlab in 140 CST using 105 nodes. In Figure 8 the phase and group velocity curves up to 500 kHz are represented. Several modes appear in the studied frequency range. As for rods, some important remarks can be highlighted:

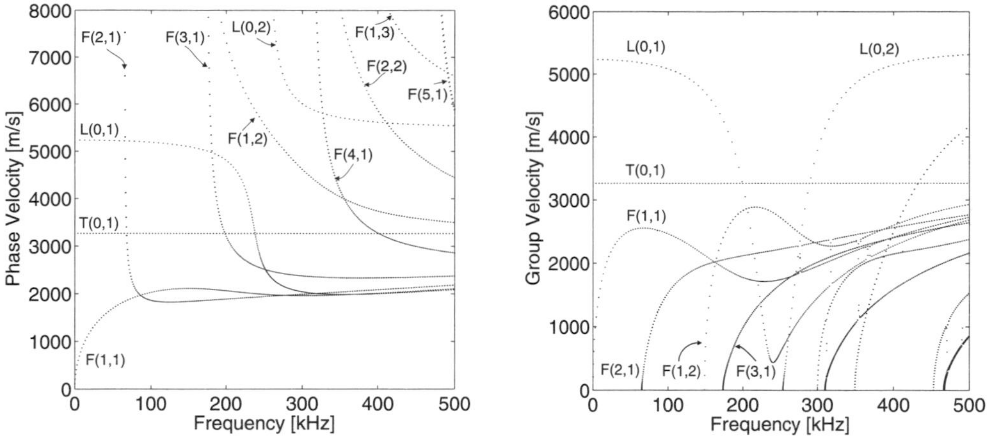

Figure 8. Guided modes phase and group velocity for a steel Hollow Cylinder with 3-mm inner radius and 4-mm outer radius

- in isotropic hollow cylinders three fundamental modes exist, $T(0,1)$, $L(0,1)$ and $F(1,1)$.
- The fundamental longitudinal mode, $L(0,1)$, of a hollow cylinder begins at zero frequency, with a phase velocity equal to the bar velocity, c_{bar}.
- The fundamental flexural mode, $F(1,1)$, begins at zero frequency, phase velocity, and wavenumber. At very low frequency it represents a bending mode of the entire pipe that is similar to the $F(1,1)$ mode of a solid bar that has the same outer radius.
- The fundamental torsional mode of circumferential order zero propagates at a constant phase velocity that is equal to the bulk velocity of the shear wave in the material c_T.
- Like in rods, $L(0,1)$ and $F(1,1)$ tend to the Rayleigh velocity at high frequency.

4.1 Mode shapes or wavestructures of hollow cylinder

In Figure 9 the cross-section wavestructures or mode shapes for the hollow cylinder are represented in the same manner as previously done for the rod. For each mode a top view along with a 3-dimensional representation is given. The 3-dimensional representation is given scaling the z-axis in order to emphasize the deformation of the mode in the propagation direction. The fundamental longitudinal mode $L(0,1)$ is an extensional mode, as can in fact be seen from Figure 9, the whole section is displaced in the propagation z-direction. The torsional mode shape $T(0,1)$ consists primarily in a rotation of the cross section around the center of the hollow cylinder, i.e. the displacement is primarily in θ-direction. This can be seen from Figure 9, where the plane view shows no appreciable displacement of the cross-section except for a rotation of the wavestructure with respect to the dashed grid. From Figure 9, it can be seen the axisymmetric behav-

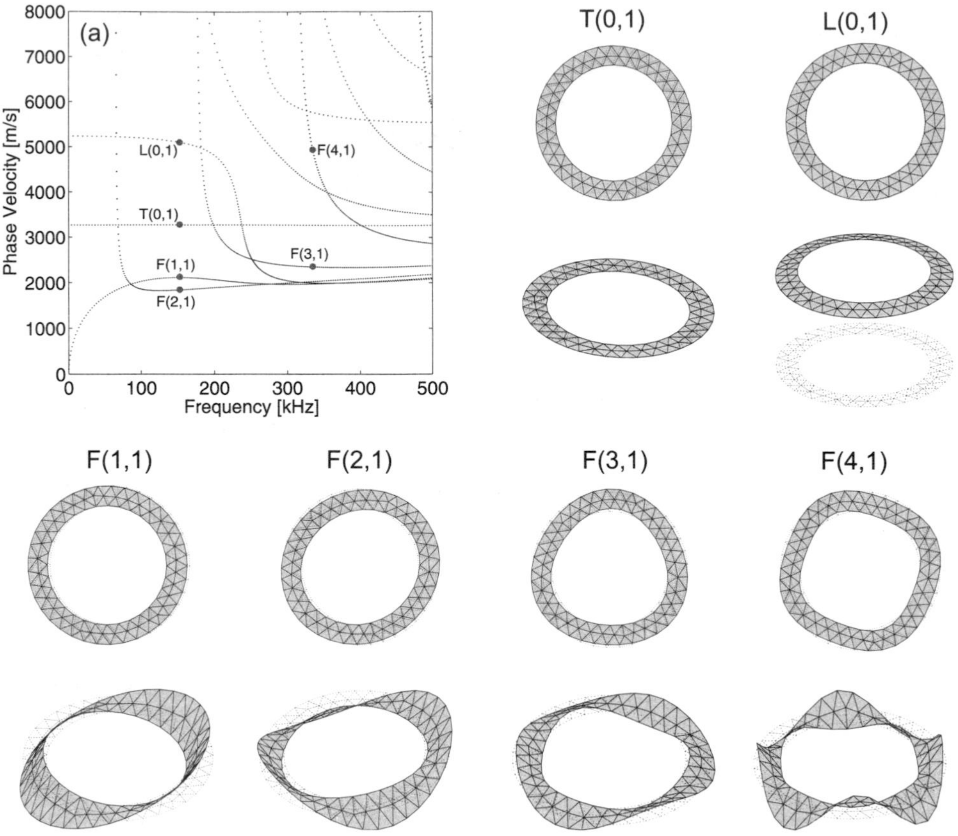

Figure 9. Plane and 3D Wavestructure representation for the guided modes indicated in the Phase Velocity plot (a)

ior of the zero order modes $n = 0$ ($T(0,1)$ and $L(0,1)$) too. The displacements at the top and bottom of the cylinder are 180 degrees out of phase for the $F(1,1)$ mode, but they are in phase for the $L(0,1)$ mode. The sequence of flexural mode shapes $F(1,1)$, $F(2,1)$, $F(3,1)$ and $F(4,1)$ reported in Figure 9 can also be used to clarify what was stated about the guided modes label. The circumferential order of the mode n, which describes the integer number of wavelength around the circumference can easily be seen from these wavestructures. For example looking at the $F(4,1)$, four wavelengths $n = 4$ can be counted around the circumference.

5 Convergence criteria

As already shown, the approximation of the semi-analytical finite element method is only related to the discretization of the waveguide in the plane perpendicular to the direction of propagation. Obviously the accuracy of the solution increases by increasing the number of finite element employed in the procedure. Nevertheless, an increasing number of elements produces an increasing number of unknowns in the problem. In order to limit the size of the problem and to reduce the computational cost, it is important to consider the frequency range of interest and the number of modes that are present in such an interval. Unfortunately it is difficult to establish a general rule for the estimation of the minimum number of elements necessary in the SAFE analysis and only some general comments can be made. Increasing the frequency value, each mode presents an increasing displacement gradient. Consequently, if for small value of the frequency a small number of elements is sufficient to approximate the mode deformation, at higher frequencies, the same number of elements can produce noticeable errors in the dispersion curves given by the poor representation of the mode shapes. For the same reason, if several higher order modes are expected in the frequency interval considered, an high number of elements has to be considered to predict their propagation properties. Let us clarify what has just been stated with this simple example. In Figure 10 the first

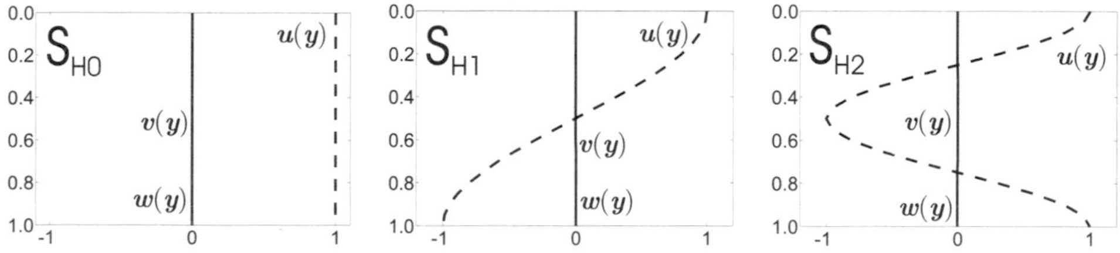

Figure 10. First three SH mode shapes at 4 MHz for a 1-*mm* steel plate

three Shear Horizontal mode shapes for the 1-*mm* steel plate at 4 MHz are represented. As can be seen, the low order mode shape S_{H0} presents a constant $u(y)$ displacement through the thickness. The S_{H1} mode presents one zero-displacement cross while the S_{H2} shows two zero-displacement crosses. In general the S_{Hm} mode will have m zero-displacement crosses. Generally, as the frequency increases, the number of propagative modes also increases. In general, to better interpolate the cross section displacement, the number of finite elements has to increase when the frequency is increasing because more propagative modes will show up. Thus a certain mesh, for a given frequency, can give a good accuracy for some modes but it could be too coarse to interpolate adequately the displacement field of the higher modes at the same frequency. A further refinement of the mesh is then required when an accurate evaluation of the stress and strain fields associated to each mode has to be obtained. In fact, in the considered finite element formulation (Viola et al., 2005) the unknowns are represented by the displacement components. The strain and stress components are obtained by deriving the displacement field and consequently their accuracy is much smaller. As has been shown (Viola et al., 2005), the group velocity is a function of the eigenvectors. For this reason phase and group velocity curves present the same order of precision because they depend on the same eigenvectors. The only way to guarantee that an appreciable accuracy is obtained, is to perform two analyses with a different number of elements, if the differences in the results are negligible the convergence can be considered achieved, or via the comparison with the exact solution when exists. Convergence analyses have been done by using different meshes and different finite elements by Marzani (2005). In particular, triangular and quadrilateral elements, with linear and quadratic interpolation functions, have been used. The meshes for quadratic elements have been generated by using GID, a free software available on line (http://gid.cimne.upc.es/). Comparison was also made with the exact solution obtained by the exact analysis (Marzani, 2005). It can be assured that the 5 finite elements used to represent the solutions for wave propagation in plates gives an accurate result for all the modes considered in the frequency range $0 - 5$ MHz. In rods and hollow cylinders, a good agreement between the exact solution and the numerical solution provided by the meshes used in this note exist for the low order modes (the first non-dispersive torsional modes $T(0,1)$, two longitudinal modes $L(0,1)$ and $L(0,2)$ and three flexural modes $F(1,1)$, $F(1,2)$ and $F(1,3)$). However, a noticeable error is made for the higher order modes even if they appear at low frequency, as for example the $F(2,1)$ mode. The error becomes larger when the order of the modes increases. The maximum error appears for the fifth order flexural mode $F(5,1)$.

Acknowledgement

This topic is one of the subjects of the Centre of Study and Research for the Identification of Materials and Structures (CIMEST) "M. Capurso". The numerical tests presented were carried out using the facilities of the Laboratory of Computational Mechanics (LAMC) "A. A. Cannarozzi" of the University of Bologna.

Bibliography

J.D. Achenbach. *Wave Propagation in Elastic Solids.* North Holland, 1973.

B.A. Auld. *Acoustic Fields and Waves in Solids.* Krieger, 1990.

M.J.S. Lowe. Plate waves for the NDT of diffusion bonded titanium. *PhD thesis, Imperial College of Science, Technology and Medicine, University of London*, 1992.

A. Marzani. Numerical modeling in engineering: Flutter of beams and plates, and guided wave in structural components. *PhD thesis, Department of Structures, University of Calabria*, 2005.

B.N. Pavlakovic. Leaky guided ultrasonic waves in NDT. *PhD thesis, Imperial College of Science, Technology and Medicine, University of London*, 1998.

J.L. Rose. *Ultrasonic waves in solid media.* Cambridge Press, 1999.

E. Viola and A. Marzani. Exact analysis of wave motions in rods and hollows cylinders. In I. Elishakoff, editor, *Mechanical Vibrations: Where Do We Stand?* Springer, 2005.

E. Viola, A. Marzani, and I. Bartoli. Semi-analytical formulation for guided wave propagation. In I. Elishakoff, editor, *Mechanical Vibrations: Where Do We Stand?* Springer, 2005.

Smart Materials and Structures

Daniel J. Inman

Center for Intelligent Material Systems and Structures, Department of Mechanical Engineering,
Virginia Tech, Blacksburg, VA. USA

Abstract. This chapter introduces smart materials suitable for use in vibration related problems. The remaining chapters in this section integrate smart structures into the context of vibration analysis, vibration prevention and structural health monitoring (diagnostics).

1 Introduction

Smart materials is the name given to a class of materials that exhibit the ability to change mechanical force and motion into some other form of energy and vice versa. It also includes other types of interesting materials such as fiber optics, which have unique sensing properties. Several names have been given to the study of such materials in applications: adaptronics, intelligent materials, active materials and adaptive materials. Unlike a simple transducer however, many of these materials are able to transform energy into motion and mechanical motion into energy. Examples of such materials are those that exhibit the piezoelectric effect, shape memory alloys, magnetorheological fluids, and ionic polymers. The most common examples of such materials are those based on the familiar piezoelectric effect. Piezoelectric materials produce a voltage when strained and a strain when an electric field is applied across them. This means, in particular, that a piezoelectric material can be used both as a sensor and as an actuator. Most of the examples used in this article employ piezoceramic materials known as PZT. However, piezoelectric polymers are also very useful. More detailed descriptions of smart materials can be found in the books by: Banks, et al (1996), Curlshaw (1996), Leo (2006), Clark et al. (1998), Srinivasan and McFarland (2001), Gandhi and Thompson (1992) and Smith (2005).

A smart structure is a structure that hosts smart materials in some embedded or layered way and performs some sort of control function of sensing and actuating. A precise definition is not appropriate but many examples are given by Banks, et al (1996), Smith (2005) and in Leo (2006). The word smart is an over statement but the field remains meaningful because of the ability to create structures with highly integrated sensing and actuation functions. Numerous other names have been attached to such systems including active structures, adaptive structures, intelligent structures, multi-functional structures and structronics. The algorithms for control and sensing, the associated hardware and power considerations are all included within the discipline. An important characteristic of smart materials and structures is that they provide an unobtrusive, integrated and distributed way to add actuation and sensing to a structure. This characteristic is a perfect match for use in health monitoring and in structural control applications where space and mass provide heavy penalties. In the following, several examples are given of using smart structures, mostly piezoelectric based, to solve vibration, vibration control and structural health monitoring problems.

Table 1. Materials Common to the Discipline of Smart Structures, All Capable of Converting Mechanical Force or Motion into Some Other form of Energy and Vice Versa.

Material	Energy Field
Piezoelect ric	Electric
Magnetost rictive	Magnetic
Elcctrorhc ological	Electric
Magnetorh eological	Magnetic
Shape Memory Alloy	Thermal

Table 1 lists some common smart materials and associated fields. In addition, the field of fiber optics is often considered to fall in the domain of smart materials. Of course there are many others and occasionally a new material is invented that falls into the domain of smart materials. Here we will focus mainly on the piezoelectric effect, and give brief examples using the shape memory effect and the magnetorheological effect.

A key element in recognizing the utility of integrating smart materials into vibration problems is that the unobtrusive and integrated nature of smart materials substantially changes the way one thinks about measurement and actuation. Smart materials:

- Allow increased numbers of sensors and excitation sources
- Allow almost distributed sensing and excitation
- Remove much of the mass loading and local stiffness changes associated with conventional devices
- Allow the sensing and actuation system to be fully integrated into the structure

A main goal here is to encourage the use of smart materials in future vibration problems, especially those involving vibration and control.

A key element in using smart materials as the excitation and sensing system for vibration is to match the choice of material with the dynamic response of the structure or machine under study. In particular, in order to excite the desired dynamics of the system containing the desired parameters or information, the stroke length (displacement or strain), force and time constant of the smart material to be used as the excitation source must be such that the structural response is large enough and of the correct frequency range to be excited. While this may seem like common sense, there are numerous papers in the literature suggesting impossible schemes because the excitation material is not physically capable of disturbing the structure. The use of smart materials as exciters must also incorporate the energy source for the excitation. Again this may seem obvious, but the literature contains numerous papers with unrealistic energy requirements. For example, the piezoelectric effect is relatively fast yet provides relatively low strain (a fraction of a percent), while the shape memory effect provides relatively high strain (about 8%) but is very slow in responding. The same matching must be carefully analyzed for the sensor material as well. However, most those working on the experimental side of identification are well aware of the process from years of experience in sizing strain gauges and accelerometers. Location is also a key issue in using smart materials as the excitation and sensing system. In fact, the ability to integrate a smart material fully into a structure is one of the key advantages. Smart materials can often be used in locations not accessible by conventional sensing and excitation devices.

In summary, smart materials can solve several of the modern problems facing vibration, vibration control and vibration measurement systems:

- Need to provide sensor and actuator redundancy
- Need for lots of sensors in observation (observability)
- Need to control the input in known ways in online systems
- Need to place actuators in key places (controllability)

2 Piezoelectric Based Systems

There are numerous materials that exhibit the piezoelectric effect including common salt crystals. Piezoelectric materials are materials that when given a mechanical strain produce an electric field and when an electric field is applied, they produce a strain. This effect has been successfully implemented into lightweight flexible structures for vibration suppression, noise suppression shape control and position control.

First it is important to realize that piezoelectric actuation is limited in force and stroke length and devices made of such materials are not replacements for hydraulic actuators. On the other hand they respond very quickly and hence are ideal for controlling transient vibrations. They are also relatively unobtrusive and hence can be easily integrated into a structure or machine. In use piezoelectric materials in the direct effect, the magnitude of the voltage generated by an applied pressure ranges from a fraction of a volt to several thousand volts. The factors influencing the voltage generated include the macroscopic size and the particular microstructure of the material. In the converse effect application of piezoelectric materials, the magnitude of the strain is likewise a function of the material geometry and microstructure. Strain values are directly related to the magnitude of the applied electric field.

There are several types of piezoelectric materials used in applications. The most notable of these materials are lead metaniobate (PMN) and lead zirconate titanate (PZT) and to date are among the most used of piezoelectric materials in structural and acoustic applications. These developments were followed by the development of piezoelectric semiconductor film transducers and piezoelectric polymers. The piezoelectric polymer in longest use is polyvinylidene fluoride film (PVDF), and other copolymer films with piezoelectric properties are being produced. The polymer-based materials (PVDF) have the advantage of being flexible, the disadvantage of having lower electromechanical coupling. On the other hand, the ceramic-based materials (PZT) have a higher electromechanical coupling, but are brittle.

As a convention, the orthogonal rectangular Cartesian axes 1, 2 and 3 (analogous to X, Y and Z) are assigned to the material as shown in Figure 1. The 3 axis, known as the polar axis, is always defined to be parallel to the direction of the polarization within the ceramic. This direction is set during the manufacturing process as a permanent polarization is induced into the material by the application of a high static DC voltage. Often, a polarization vector is shown on manufacturer data sheets. This vector points from the positive to the negative poling electrode. Standard notation sets the positive 3 axes in the direction opposite the poling vector. Care should be taken to observe the exact notation employed in manufacturer data sheets as some manufacturers set the positive direction of 3 axes in the same direction as the poling vector.

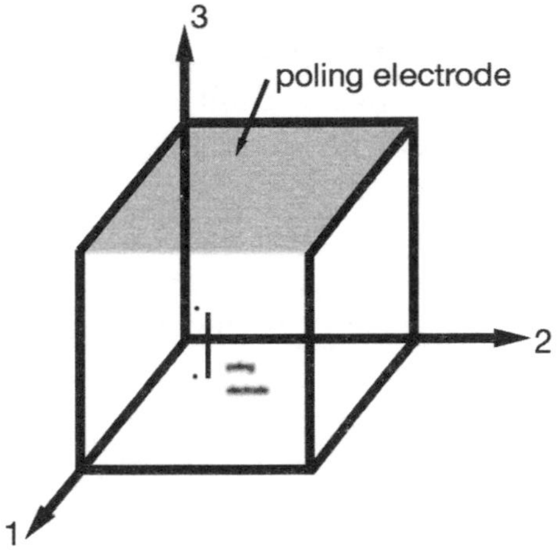

Figure 1. Designation of the material axes.

All piezoelectric properties depend on the direction within the material. Many properties also depend on both the mechanical and electrical boundary conditions. To completely identify the descriptive coefficients, super and subscripts are attached to each coefficient. The directionality of a coefficient is indicated by the subscripts 1, 2, 3, 4, 5 and 6 attached to the coefficient. Subscripts 1, 2 and 3 indicate the principal axes; 4, 5 and 6 denote shear around the 1, 2 and 3 axes respectively. Because polar symmetry exists, the 1 and 2 axes are identical (differing only by a right angle). For simplicity, reference is made only to the 3 and 1 directions and it is understood that 1 also implies 2. Occasionally the subscripts "*p*" and "*h*" are used to denote specific conditions. "*p*" signifies that the stress (or strain) is equal in all directions perpendicular to the 3 axis and that the electrodes are on the faces perpendicular to the 3 axes. "*h*" indicates the electrodes are again on the faces perpendicular to the 3 axes (for ceramics) and that the stress is hydrostatic in nature (equal magnitude in the 1, 2 and 3 directions). The boundary conditions, when applicable, are indicated by a superscript attached to the coefficient. The boundary condition superscripts used are T, S, D and E signifying:

T = constant stress (mechanically free)
S = constant strain (mechanically constrained)
D = constant electrical displacement (open circuit)
E = constant field (short circuit).

In general, the superscripts describe external factors that affect the property while subscripts describe the relationship of the property to the principal axes of the material. Because piezoelectric materials are electro-mechanical coupled materials, the descriptive symbols differ slightly from those traditionally used in linear elastic mechanics. Typical symbols are shown in Table 2.

Most piezoelectric constants have double subscripts. These subscripts denote coupled electrical and mechanical interactions. The first subscript gives the direction of the electrical field associated with the applied voltage or charge produced. The second subscript gives the direction of the mechanical stress or strain. Examples of constants used to describe a piezoceramic material are given in Table 2.

Table 2. Notation Used for Piezoelectric Materials

S_{33}^{E} Indicates that the property is measured with the electrode circuit closed Indicates that the <u>stress</u> or strain is in the 3 direction Indicates that the stress or <u>strain</u> is in the 3 direction Compliance = strain / stress Note: all stresses, other than the stress indicated in one subscript, are held constant.	S_{36}^{D} Indicates that the property is measured with the electrode circuit open Indicates that the <u>stress</u> or strain is shear around the 3 direction Indicates that the stress or <u>strain</u> is in the 3 direction Compliance = strain / stress Note: all stresses, other than the stress indicated in one subscript, are held constant.
K_{3}^{T} Indicates that the stresses on the material are constant (example: no applied forces) Indicates that the electrodes are perpendicular to the 3 axes Relative Dielectric Constant	K_{1}^{S} Indicates that all strains in the material are constant (example: completely blocked boundaries) Indicates that the electrodes are perpendicular to the 1 direction Relative Dielectric Constant
k_{31} Indicates that the electrodes are perpendicular to the 3 axes Indicates that the stress or strain is in the 1 direction Electromechanical Coupling	k_{p} Indicates that the electrodes are perpendicular to the 3 axes and the stress or strain is equal in all directions perpendicular to the 3 axes Electromechanical Coupling
g_{31} Indicates that the electrodes are perpendicular to the 3 axes Indicates that the stress or strain is in the 1 direction $\frac{\text{field}}{\text{applied stress}} = \frac{\text{strain}}{\text{applied charge/electrode area}}$	g_{h} Indicates that the electrodes are perpendicular to the 3 axes (ceramics only) and the stress is equal in all directions (hydrostatic stress) $\frac{\text{field}}{\text{applied stress}} = \frac{\text{strain}}{\text{applied charge/electrode area}}$
d_{31} Indicates that the electrodes are perpendicular to the 3 axes Indicates that the stress or strain is in the 1 direction $\frac{\text{strain}}{\text{applied field}} = \frac{\text{short circuit charge/electrode A}}{\text{applied stress}}$	d_{33} Indicates that the electrodes are perpendicular to the 3 axes Indicates that the stress or strain is in the 3 direction $\frac{\text{strain}}{\text{applied field}} = \frac{\text{short circuit charge/electrode A}}{\text{applied stress}}$

Table 3. Piezoceramic Coupling Coefficients

k_{31} — 3: Indicates that the electrodes are perpendicular to the 3 axes; 1: Indicates that the stress or strain is in the 1 direction; k: Electromechanical Coupling	k_p — p: Indicates that the electrodes are perp the 3 axes and the stress or strain is e directions perpendicular to the 3 axes; k: Electromechanical Coupling
k_r — r: Indicates that ceramic is disk shaped, the electrodes are perpendicular to the 3 axes and the stress or strain is radial; k: Electromechanical Coupling	$k^2 = \frac{\text{electrical energy converted to mecha}}{\text{input electrical energy}}$ $k^2 = \frac{\text{mechanical energy converted to elec}}{\text{input mechanical energ}}$

The electromechanical coupling coefficient is arguably the most important of all of the properties. It is the best single measurement of the strength of a piezoelectric effect. It describes the conversion of energy by the ceramic element from mechanical to electrical form (or from electrical to mechanical form). In essence, it measures the fraction of the electrical energy from an applied field that is converted to mechanical energy (or vice versa when the piezoelectric is stressed). Table 3 illustrates various coupling coefficients.

Because the conversion from one energy form to another is always incomplete, k^2 is always less than unity. Typical values of the electro-mechanical coupling coefficient k^2 range from 0.40 to 0.70 for most piezoceramics. Values as high as 0.90 have been measured for Rochelle salt at its Curie temperature. The magnitude of the coupling factors depends upon the degree of poling that exists in the material. The most common coupling factors are derived from the constitutive equations and have the form as shown below:

$$k_{33} = \frac{d_{33}}{\sqrt{s_{33}^{E}\,\varepsilon_{3}^{T}}} \qquad k_{31} = \frac{d_{31}}{\sqrt{s_{11}^{E}\,\varepsilon_{3}^{T}}}$$
$$k_{15} = \frac{d_{15}}{\sqrt{s_{44}^{E}\,\varepsilon_{1}^{T}}} \qquad k_{p} = \frac{d_{31}}{\sqrt{2/\,(s_{11}^{E} + s_{11}^{E})\,\varepsilon_{3}^{T}}} \tag{1}$$

While the coupling factor does quantify the fraction of energy converted from one form to another, it is not to be confused with conversion efficiency. In the quasi-static energy conversion form, if there are no dissipation in the system, the net efficiency is 100%. This is a result of recoverable energy being stored in the material and returned to the source upon completion of the cycle. Therefore, the coupling coefficient k^2 can be thought of as a measure of the ineffective energy or unavailable energy.

Further understanding of this term can be gained by considering a single piezoelectric element under applied pressure. Part of the energy from the applied pressure is converted into an electrical charge that appears on the electrodes (analogous to a charged capacitor), and the remaining energy is stored in the element as mechanical energy (analogous to a compressed spring). When the pressure is removed, the element will return to its original dimensions and the charge will disappear from the electrode surfaces.
The coupling coefficient is physically related to the piezoceramic's resonance and anti-resonance frequencies. Although analytical methods for predicting the electro-mechanical coupling exist, standard test methods are easily applied for measuring the coupling coefficient and are therefore often the determination

d_{31} — 3: Indicates that the electrodes are perpendicular to the 3 axes; 1: Indicates that the piezoelectrically induced strain or the applied stress is in the 1 direction	d_{33} — 3: Indicates that the electrodes are perpend the 3 axes; 3: Indicates that the piezoelectrically induce the applied stress is in the 3 direction
d_{15} — 1: Indicates that the electrodes are perpendicular to the 1 axes; 5: Indicates that the piezoelectrically induced strain or the applied stress is shear around the 2 axes	$d = \dfrac{\text{strain}}{\text{applied field}} = \dfrac{\text{short circuit charge/e}}{\text{applied stress}}$ Typical units: $\dfrac{m/m}{V/m}$ - OR - $\dfrac{\text{coulombs} / m^2}{\text{Newton} / m^2}$

method chosen. Because of the directionality of the material, subscripts denote the directionality of the coupling coefficient and the type of motion involved. Examples of the common electro-mechanical coefficients are given above for clarity.

Table 4. Notation for Piezoelectric Field-Strain Coefficients.

...ance of a transducer. It indicates the utility of the material in the transducer in the sense of quantifying how much energy remains in the material rather than being converted to another form. A high k^2 value indicates a measure of the magnitude of the transducer bandwidth and is a performance measure of the transducer.
The electro-mechanical coupling coefficient is related to the elastic constants: stiffness c and compliance s. The relationships between the stiffness constants are

$$\begin{aligned} s_{11}^{D} &= s_{11}^{E}(1 - k_{31}^{2}) \qquad & s_{33}^{D} &= s_{33}^{E}(1 - k_{33}^{2}) \\ s_{11}^{D} &= s_{12}^{E} - k_{31}^{2} s_{11}^{E} \qquad & s_{44}^{D} &= s_{44}^{E}(1 - k_{15}^{2}) \end{aligned} \tag{2}$$

Insight into the physical meaning of the electro-mechanical coefficients is realized through the following simple example. Consider a piezoceramic that has a k_{31} value of 0.35. If 1 erg of electrical energy is applied to the 3 axis of the ceramic, the mechanical energy stored in the resulting deformation along the 1 axis is: 1 erg x $(0.35)^2 = 0.1225$ erg.

The piezoelectric "*d*" constants or "strain" constants relate the mechanical strain produced by an applied electric field. When used in this manner, the action is often referred to as a "motor", and the effect is called the "motor effect." The direct relationship may also be defined for the constant relating the short circuit charge density to the applied mechanical stress. This is the well-known piezoelectric effect and is occasionally called the "generator" action. The *d* constants directly relate properties of the material and are not affected by the boundary conditions. The *d* constants directly relate the magnitude of the strain, and thus the displacement, to the applied electric field. Materials with large *d* constants are typically sought after for use as actuators.

The *d* constants can be used directly to calculate the change in dimensions of a piezoelectric element. Consider an element that has dimensions of length *L*, width *W*, and thickness *t*. The free change in the dimensions are calculated as:

$$\Delta W = \frac{d_{31}\, W\, V}{t}, \quad \Delta L = \frac{d_{31}\, L\, V}{t}, \quad \text{and } \Delta t = d_{33}\, V. \tag{3}$$

Table 5. Notation for Piezoelectric Electric Field-Stress Coefficients.

g_{31} — Indicates that the electrodes are perpendicular to the 3 axes; Indicates that the applied stress or piezoelectrically induced strain is in the **1** direction	g_{33} — Indicates that the electrodes are perpendicular to the 3 axes; Indicates that the applied stress or piezoelectrically induced strain is in the 3 direction
g_{15} — Indicates that the electrodes are perpendicular to the 1 axes; Indicates that the applied stress or piezoelectrically induced strain is shear around the 2 axes	$g = \dfrac{\text{strain developed}}{\text{applied charge density}} = \dfrac{\text{open circuit electric field}}{\text{applied stress}}$ Typical units: $\dfrac{\text{m/m}}{\text{coulombs / m}^2}$ - OR - $\dfrac{\text{V/m}}{\text{Newton / m}^2}$

The "*g*" constants or "voltage" constants are related to the electric field produced by a mechanical stress. These constants define the sensing action of a piezoelectric material. The converse relationship may also be defined for the constant relating the strain developed to the applied charge density. Like the *d* constants, these constants are not affected by the boundary conditions. The most common forms of the *g* constants and the appropriate units are shown in Table 5.

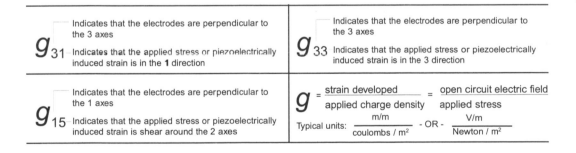

The "*d*" and "*g*" constants are related through the dielectric constant. If any two of the "*d*", "*g*" and dielectric constants are known for a particular mode, the third is uniquely determined. The relations between these constants are

$$d_{33} = K_3\,\varepsilon_o\,g_{33} \qquad d_{31} = K_3\,\varepsilon_o\,g_{31} \qquad d_{15} = K_3\,\varepsilon_o\,g_{15}. \tag{4}$$

An extended version of the above can be found in Inman and Cudney (2000).

3 Packaging of Piezoelectric Materials

The most commonly used mode for piezoelectric materials in vibration related applications is in the d_{31} bending mode in the form of thin, beam like wafer. Piezoceramics, especially PZTs are preferred over piezofilms because their *d* coefficients are larger providing better actuation. However, the ceramics are brittle and break easily when being handled. Thus several manufacturers offer piezoceramics encased in Kapton or mounted on substrates. PZT mounted this way is still stiff and cannot conform to irregular surfaces. Hence films, PVDF being the most common, are used as actuation devices in such circumstances.

Recently active fiber composites have become popular. A commercially available version is called a Macro Fiber Composite. Macro fiber composites, or MFC, claim to have superior electro mechanical coupling properties when compared with monolithic piezoceramics that are more commonly available. In addition the MFC is very flexible. This flexibility will allow applications in areas where curved surfaces are prominent, such as wrapped around a compressor or other source of vibration. By using interdigitated electrode patterns and uniform fibers the MFC is able to capitalize on the larger d_{33} coefficient, but yet provide actuation and sensing in the d_{31} mode. This is the reason that the MFC is able to provide additional power compared to monolithic PZT actuation materials. Figure 2 is a comparison of various piezoceramic based actuation materials and is provided by NASA. The plot in Figure 2 is stress versus induced strain for a variety of voltages and different piezoceramic materials. The higher the plot is above the axis the "stronger" the actuator is. These plots indicate the nature of the electro mechanical coupling provided by each material. Again the higher the plot, the better this coupling should be. Higher coupling means that a greater percentage of the mechanical motion that can be converted in electrical energy and the greater percentage of applied voltage that is converted into mechanical motion. The plots of Figure 2 show clearly that MFC should provide double the coupling that monolithic PZT material.

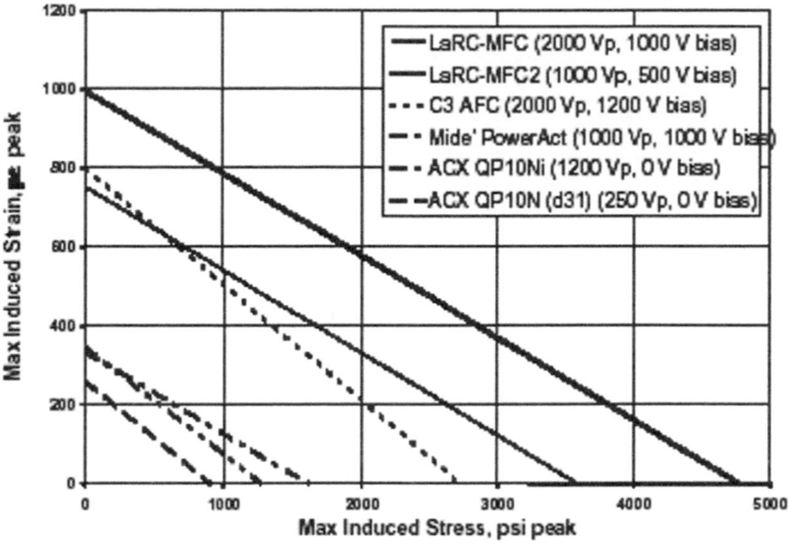

Figure 2. Comparisons of the stress strain curves of various piezoceramic based materials.

More details on the MFC can be found in Sodano et al (2004).

The performance of both monolithic PZT and MFC wafers can be improved by using a bimorph configuration. This amounts to placing a matched but out of phase pair on either side of the neutral axis of a structure to produce twice the moment. This approach is common in structural control applications.

References

Banks, H. T., Smith, R. and Wang, Y. (1996), *Smart Materials and Structures: Modeling, Estimation and Control*, Wiley

Clark, R., Saunders, W and Gibbs, B. (1998) *Adaptive Structures: Dynamics and Control*, Wiley,

Culshaw, B. (1996), *Smart structures and Materials*, Artech House.

Gandhi and Thompson, B.S. (1992) *Smart Materials and Structures*, Chapman and Hall.

Inman, D. J. and Cudney, H. H. (2000) *Structural and Machine Design Using Piezoceramic Materials: a Guide for Structural Design Engineers*, Final Report NASA Langley Grant NAG-1-1998

Leo,D.J. (2006) *Smart Material Systems: Analysis, Design, and Control*, John Wiley & Sons., in press.

Smith, R. C. (2005) *Smart Material Systems: Model Development*, Society for Industrial and Applied Mathematics, Philadelphia.

Sodano, H. A., G. Park, and D. J. Inman, 2004. "An Investigation into the Performance of Macro-Fiber Composites for Sensing and Structural Vibration Applications," *Mechanical Systems and Signal Processing*; Vol. 18, pp. 683-697.

Srinivasan, S. and McFarland, M. (2001) *Smart Structures: Analysis and Design*, Cambridge University Press.

Applications of Smart Materials for Vibration Suppression

Daniel J. Inman

Center for Intelligent Material Systems and Structures, Department of Mechanical Engineering,
Virginia Tech, Blacksburg, VA. USA

Abstract. This chapter examines several examples of using smart materials for suppressing vibrations in structures. In particular, it is shown that smart materials can greatly increase controllability and observability and can be used for both passive and active control.

1 Introduction

From the point of view of control theory, the use of smart structures is extremely beneficial because it allows the control designer to approach full state feedback. Full state feedback allows the designer to shape the response of the structure to almost any specification without having to resort to building a state estimator. In addition, the use of smart structures allows the placement of sensors and actuators at almost any location allowing a maximum use of the concepts of controllability and observability. From a mechanical design point of view, use of smart structures offers an order of magnitude increase in settling times for a relatively small expenditure of power.

The ideas from control theory of controllability and observability are simple and attempt to quantify where the best place to drive a system is and where the best place to measure the system is. A primitive example of controllability comes from thinking about a door moving on its hinges and where to locate the door knob in order to perform the most efficient control. The answer of course is to place the knob as far away from the hinge as possible. In terms of mechanics we do this because the applied moment is position crossed into the force. In this simple analogy the moment arm (distance from the knob to the hinge) is a measure of controllability. For a given force, the larger the moment arm is, the larger the moment is. In more complex structures, such as the frame illustrated below, the indicator of controllability, and hence where to place the actuator, becomes a little more complex, but proceeds along the same idea. That is in control, we would like to place actuators where the controllability is the highest.

2 A Slewing Frame

To solidify these concepts consider trying to suppress the vibrations of a rotating frame (meant to simulate a solar panel in slewing). A frame rotated by a dc motor is presented in Figure 1. Due to its configuration, the action of rotating the frame about an axis (slewing) causes both bending and torsional vibrations. The frame consists of individual elements of thin-walled circular aluminum tubing. Each member is 0.635 cm in diameter and has wall thickness of 0.124 cm. The elements are joined at octagonal nodes that are also made of aluminum. Each member is pinned and bolted into the node to eliminate looseness in the joints. The frame is mounted onto the larger steel shaft by bolting two of the nodes into aluminum clamps.

The frame represents a complex structure as it has bending, plate and torsional modes of vibration. Here the smart structure solution is superior to a conventional solution because piezoceramics greatly improve the controllability of the closed loop system yet add very little additional mass and require very little additional energy. The primary action of slewing induces both bending and torsional vibrations in the structure. Two active members that can be used as collocated sensor/actuators in feedback control loops are inserted into the frame along diagonal elements. Theoretical and experimental control laws are applied that simultaneously slew the frame and suppress the residual vibrations. Controllability results indicate that the dc motor is effective in slewing the frame and suppressing the bending motion, but not very effective in suppressing the torsional motion. Hence, the torsional vibrations are suppressed using the active members in collocated feedback loops.

Slewing a flexible structure involves vibration suppression as well as accurate pointing and tracking. Vibration suppression and accurate pointing is accomplished using Multiple-Input-Multiple-Output (MINO) control. This is compared to the single input single output (SISO) results of conventional structural control. The ability to implement MIMO control is provided by replacing two passive members of the frame with active elements. The active members can be used in conjunction with the slewing actuator as independent collocated sensor/actuators.

The slewing actuator is an Electro-Craft 670 dc motor. The shaft of the motor is coupled to a steel shaft with a diameter of 0.635 cm, which in turn is connected to another steel shaft of diameter 1.270 cm. The smaller shaft can easily be removed so that gears can be placed between the motor and the structure. A tachometer housed inside the motor measures angular rate, and a potentiometer attached to the bottom of the larger steel shaft produces a signal proportional to angular position. The whole slewing rig is attached to a large concrete block that serves as ground. Figure 1 is a photo of the slewing frame testbed.

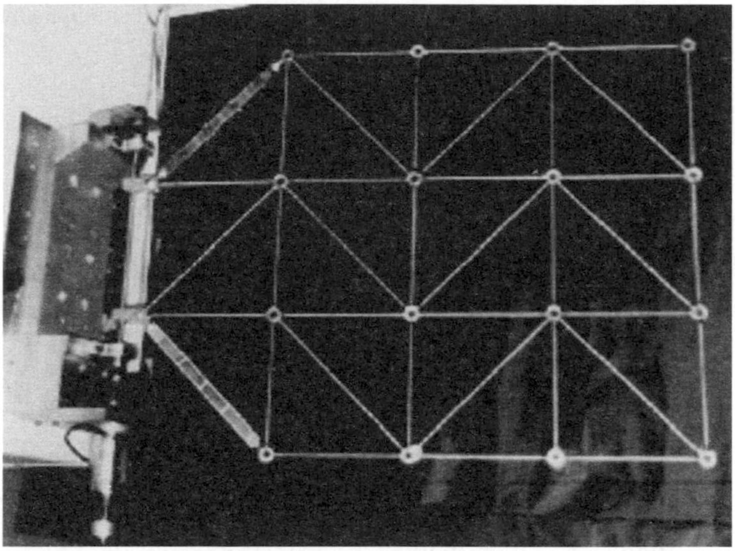

Figure 1 Slewing frame testbed showing the active members, angular rate and position sensors, and the dc motor.

Several controllability measures are available. These measures can provide an indication of where to place actuators on a structure to achieve the "best" closed loop behavior. In particular a low controllability for a specific actuator location indicates the need for additional actuators. An experimental model of the closed loop response of a simple proportional derivative (PD) control is listed in Table 1. The closed loop modal damping ratio (ζ) is used here as a performance index. Note that the closed loop damping ratio for the two torsional modes and the 1st plate mode are much lower than the damping ratios for the two bending modes. This indicates that the motor control alone is not able to damp the torsional modes of vibration and suggests a controllability problem.

Because of the inability of the motor actuator by itself to control the torsional modes, the secondary piezoceramic actuators are added to the frame as illustrated in Figure 1. Adding this secondary actuator, improves the controllability as can be seen by examining any number of controllability measures. Consider for instance the controllability measure given by

$$C_i = \sum_{j=1}^{m} [C_{ij}^2]^{1/2} \quad i = 1,2,\ldots m$$

where C_{ij} is a modal controllability measure defined in the previous chapter. Applying this controllability measure applied to the two different systems -

motor control of the frame and motor control plus piezoceramic member of the frame (active member) - is illustrated in Table 2. Note that the active member produces a large controllability index in the troubled modes.

Table 1 Inability of the motor to add damping to the torsional and plate modes.

	Experimental	
Mode	ω_d (Hz)	ζ% Critical
1st Torsional	4.32	0.82
1st Bending	7.68	9.18
2nd Torsional	14.11	1.26
1st Plate	20.76	0.94
2nd Bending	26.25	1.32

Table 2 A per mode look at controllability.

	Gross Controllability	
Mode	Slewing Actuator	+ Active Member
1st torsional	8.62	109.55
1st bending	155.16	156.42
2nd torsional	1.11	39.81
1st plate	17.28	54.31
2nd bending	80.29	91.61

These two tables (1 & 2) basically illustrate the difficulty in controlling such a system and how embedded piezoceramic actuators provide a straightforward, logical solution. The rigid body actuation induces motion in the torsional and plate modes of the structure, yet this actuator as a control device offers very little control authority over these modes. The addition of the embedded piezoceramics on the "other side" of the hinge placed at an angle renders these troublesome modes controllable. The result is a faster response. In terms of the material from the previous chapter, the effect of the active member is to increase the rank of the B and C matrices.

It is also important to note that the cost of this tremendous gain in performance is very little. In fact for a total power increase from 121.53 watts to 121.84 watts the damping ratio increases by a factor of 15 in the 1st torsional mode 2.3 times in the second torsional mode and 6 times in the 1st plate mode. Thus a very inexpensive addition in control complexity and effort results in a tremendous increase in performance providing an excellent example of the utility of the smart structures concept. The key point illustrated by this example is that the usefulness of "smart materials" is the ability to place sensors and actuators at preferred locations (those that increase controllability and observability). The result is a much more efficient vibration suppression scheme than can be achieved with conventional actuators and sensors.

3 Passive Control Using PZT

PZT may also be used for vibration suppression in a passive way, by connecting the PZT to an external shunt circuit to dissipate energy. Shunt designs are reviewed in Lesieutre (1998). Piezoceramic patches connected to electronic shunt circuits have formed successful vibration reduction devices. The basic idea is that a PZT patch is mounted on the structure and then connected to a resistive circuit. As the PZT strains, it produces an electric field (sensor effect), which appears across the resistor and dissipates energy through heat. If an inductor is added to the circuit, an electronic vibration absorber results, which may be tuned to remove energy at a specific frequency.

Passive electronic damping is becoming a viable alternative to passive viscoelastic damping treatments. Recently, snow ski designers used a resistive/capacitive (RC) shunt circuit to dissipate the vibration energy absorbed by the piezoceramic and showed reasonable vibration suppression performance at the first vibration mode. Several researchers have shown that the use of tuned electronic damping is a potential way to suppress vibrations. A passive shunt circuit is tuned to suppress a target mode much like a classic mechanical vibration absorber. Following the principle of a mechanical absorber, the electric resonant frequency should be tuned to be very near or equal to the mechanical frequency of interest.

The conventional shunt circuit consists of three electrical components: a capacitor C (PZT), an inductor L and resistor R. The two external terminals of the PZT, modeled as a capacitor (since, electrically, it behaves similarly), are connected to the series inductor and resistor shunt circuit. The piezoceramic element is used to convert mechanical energy of a vibrating structure to electrical energy by direct piezoelectric effect. This electric energy is dissipated as heat through the shunt resistor efficiently when the electrical resonant frequency matches the mechanical frequency. At resonance, the reactive components between the inductor and capacitor cancel each other and the phase between the current and voltage is zero. As a result, the power factor at resonance becomes one.

By adjusting a variable resistor in the electronic inductor circuit, a spike (electrical resonant frequency) moves to be tuned to the mechanical resonant frequency. When it is tuned correctly, there are no peaks around the second mode (58.031 Hz) and the third mode (149.75). This indicates that the shunt RLC circuit is tuned at the first mechanical resonant frequency, where the imaginary components between the inductor and capacitor cancel each other i.e. the voltage and current are in phase. The experimental results cover two ranges of frequencies. The first range (9-12 Hz) includes the first bending mode and the second range (52-66 Hz) covers the second bending mode. Comparisons are shown between the amplitudes of vibration when the shunt circuit is not activated (open circuit) and when it is activated using different resistor values at a fixed room temperature.

After measuring the frequency response function between output and input voltage, the open and short circuit frequencies measured are 10.625 Hz and 10.5 Hz for the first mode and 58.625 and 58.247 for the second mode. The generalized electro-mechanical coupling coefficients, calculated for a PZT pair, are 0.15 and 0.11 for the first and second mode frequency, respectively.

The shunt electrical passive absorber is tested experimentally to reduce the vibration amplitudes of the first and the second modes of a cantilever beam with the half and full inductances. It is evident that decreasing resistance results in improving the vibration attenuation characteristics of the first bending mode and the second bending mode. When the resistance is decreased below the optimal resistance, the two peaks rise up, which is a similar phenomenon of mechanical absorber damping. The shunted piezoelectric pair is found to produce about 25 dB drop from the peak vibration amplitude of the open circuit case at first mode frequency. For the second mode after shunting, a peak amplitude reduction of about 20dB is obtained.

A technique that is capable of reducing the structural vibration amplitude using the electrical passive absorber has been presented. This technique was experimentally tested with a simple beam. The experiments demonstrated successful results with a full and a half inductance at the first mode and the second mode frequency. The vibration amplitudes of the beam were decreased about 25dB and 20dB at the first and second mode frequencies respectively. When the shunt circuit with a half inductance was used, the electric damping was doubled. Hence, piezoceramic based shunting presents a suitable passive tool for suppressing structural vibrations that occur in many engineering applications.

4 Vibration Suppression Across a Temperature Range

Panel vibration remains a serious source of fatigue in large aircraft. The usual approach is to add constrained layer damping material to the panels to reduce vibration levels. This section presents the possibility of using active control implemented though a small piezoceramic actuator and tiny fiber optic sensor to perform vibration suppression in a representative plate. To model an actual aircraft panel, a plate is constructed with similar frequencies, with the control targeting modes in the low frequency range.

Because modeling and boundary conditions in a real aircraft have significant variance in aircraft panels, a control method is chosen that is based only on knowing frequency data for the structure. This approach also provides a good starting point for a follow on effort to use adaptive control to provide damping in the presence of large temperature changes. Positive position feedback control based on a measured frequency response to the structure is chosen as the control law. The tests were performed in a standard test fixture commonly used in the auto industry for evaluation damping materials. The method, implemented through a piezoceramic actuator is referred to as smart damping.

The tests show that smart damping materials have substantial performance benefits in terms of providing effective noise and vibration reduction at a frequency range that is often outside of the effective range of passive damping materials. Further, judging by vibration reduction per added weight, the test results indicate that the smart damping materials can provide substantial vibration reduction at selected frequencies, without adding any appreciable amount of weight to the substrate structure. For example, smart damping can decrease the vibration peak of a steel panel at 47 Hz by up to 20 dB with an additional weight of only 0.11 lb. This feature of smart damping materials is particularly useful for applications that involve vehicles, where the constraint requires a particular noise or vibration cancellation at a specified frequency, without adding any weight to the vehicle or requiring any change to the vehicle structure.

Overall, the test results show that the application of smart damping materials and fiber optic sensors can be used for active control of aircraft style panels. The smart damping materials combined with fiber optics can be viewed as a new technology that, once developed and optimized, can extend future life by providing more effective noise and vibration solutions for new or existing aircraft structures.

A test stand was designed and fabricated for testing and evaluating the effectiveness of piezoelectric damping materials for reducing vibration and structure-borne noise. The test stand enables vibration and acoustic measurements and analysis on a plate with clamped-clamped boundary conditions. The plate, simulating an aircraft panel, is clamped rigidly around its edges and excited over a frequency range of 0-400 Hz. Various passive damping materials (shunts, viscoelastic layers), and smart damping materials, can be added to the panel in order to evaluate their effect on reducing the panel vibration and the noise that it causes.

The test stand (the mid section is shown in Figure 2), includes an electromechanical shaker, a panel excitation frame, a sound-insulating enclosure, and data acquisition equipment. Measurements are taken with two accelerometers, located on the plate and excitation frame, and a microphone positioned in the upper reception chamber. The reception chamber and bottom enclosure are designed to eliminate background noise and isolate the noise generated by the vibrating panel.

To begin validation testing, a standard test plate was clamped into place with 14 bolts tightened to a torque of 25 N-m. The standard plate was a 500mm X 600mm, 20-guage, galvanized steel plate. The plate was bolted as in Figure 2 such that the outside 10 cm along the edges were clamped and the remaining test plate area was 400mm X 500mm. The bolts were always tightened in a pattern similar to that for lug nuts on a car, to improve the repeatability of the boundary conditions for the plate.

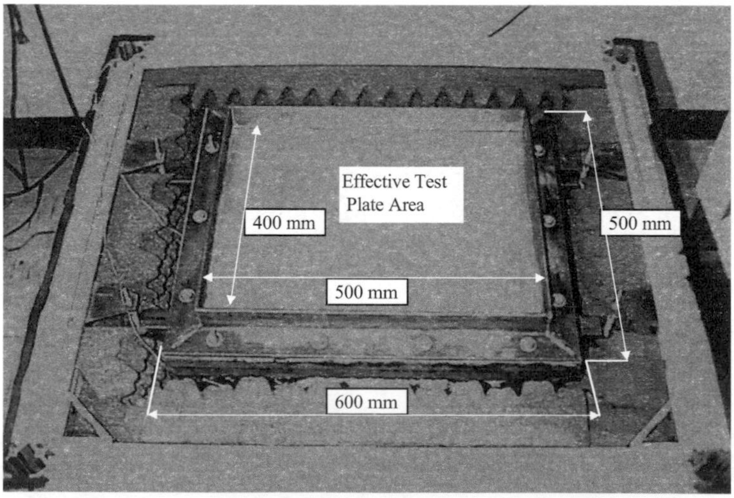

Figure 2 Standard Test Plate in Testing Position

Accelerometers were attached in the center underneath the top beam of the excitation frame and on the test plate underneath its center. A microphone was hung in the reception chamber such that it was 0.5m above the center of the test plate when the test stand was fully assembled.

The data acquisition system consisted of a Hewlett Packard dynamic signal analyzer, a fast Fourier transform, a band pass filter, and a signal generator for controlling the shaker. Initial tests and experiments were performed with a number of different excitation functions and sampling techniques. Although the generated input signal is an ideal signal for testing the frequency response for a plate, the direct input excitation for the plate is from the frame not from the HP Analyzer. The desired excitation range for the plate is between 0 and 400 Hz, and poor data will result if there are any resonant frequencies of the frame within this range. The frequency response of the frame was then analyzed to ensure that this was not the case. Data was first collected for the excitation frame and clamping frame without the plate in place. This test clearly shows that the major frame structural resonant frequencies occur above 500 Hz. The frequency spectrum of the frame within the 0-400 Hz range is relatively constant as well. Fiber optic sensors are widely used as physical parameter gauges in various structural applications, such as strain and vibration sensing and damage detection. After examining the various types of sensors, we concluded that the Fabry-Perot fiber sensors demonstrated the best response for use in an active feedback control system.

The Fabry-Perot (FP) interferometric strain sensors can be classified into two main types: intrinsic FP interferometers (IFPI) and Extrinsic FP interferometers (EFPI). The EFPI sensor is constructed by fusion splicing a glass capillary tube with two optical fibers. Compared to the IFPI sensor, the EFPI-based sensor is relatively simple to construct and the FP cavity length can be accurately controlled. The EFPI sensor can also be easily configured to suit different applications with desired strain range and sensitivity by altering the type of fibers, the capillary tube, air-gap distance and the length of the sensor. In addition, a major advantage of the EFPI sensor is its low temperature sensitivity, which makes it possible to interrogate the EFPI sensor with simple signal processing techniques. An EFPI sensor can be constructed using either single-mode or multi-mode fibers. The single-mode design offers higher accuracy and low insensitivity to unwanted disturbance while the multi-mode design offers higher power coupling efficiency. In our experiment, single-mode fibers are used to deliver and collect the light, and are used as an internal reflector as well. Two EFPI sensors were fabricated and measured under different circumstances, both on a bench and the vibration test stand.

The goal of the active control tests was to establish a technique for vibration suppression in a representative aircraft component. This technique combined the technology of optical fiber sensors with piezoelectric actuators to minimize the vibration levels in the test article. The specific control law used for the active control tests is called Positive Position Feedback (PPF). This control law uses a generalized displacement measurement from the test article to accomplish the vibration suppression.

Two 2.85in X 2.85in piezoelectric actuators bonded to its surface of the plate. These two locations were chosen for the actuators because the plate has a large amount of strain energy in these regions for the modes that needed to be controlled. In general, the control authority of an actuator is increased when it is placed in a region of high strain energy. Although the test plate has two piezoelectric actuators attached to its surface, it was determined that the center PZT was more effective at minimizing the levels of vibration.

The sensor used for control was a modal domain optical fiber sensor for vibration monitoring. This sensor is based upon a laser that focuses coherent light through a lens into one end of a multimode optical fiber. One end of the fiber was attached to the plate and the other end of the optical fiber passed through a spatial filter and into a photodetector. The output of the photodetector is a variable voltage that is fed into a monitoring unit such as an oscilloscope.

Positive position feedback (PPF) is a technique for vibration suppression that uses a displacement measurement to accomplish control. PPF control is a stable and relatively simple control method for vibration suppression (Friswell and Inman, 1999). The control law for a PPF controller consists of two equations, one describing the structure and one describing the compensator:

$$\text{Structure: } \ddot{\xi} + 2\zeta\omega\dot{\xi} + \omega^2\xi = g\omega^2\eta \tag{1}$$

$$\text{Compensator: } \ddot{\eta} + 2\zeta_f\omega_f\dot{\eta} + \omega_f^2\eta = \omega_f^2\xi \tag{2}$$

where g is the scalar gain > 0, ξ is the modal coordinate (structural), η is the filter coordinate (electronic), ω and ω_f are the structural and filter frequencies, respectively, and ζ and ζ_f are the structural and filter damping ratios, respectively. The positive position terminology in the name PPF comes from the fact that the position coordinate of the structure equation is positively fed to the filter, and the position coordinate of the compensator equation is positively fed back to the structure. In effect, a PPF controller behaves much like an electronic vibration absorber.

The control system was set up such that the optical fiber sensor sends a signal through a signal conditioner and amplifier into the dSPACE board. Once the signal was processed through the PPF control law it was sent out to an amplifier to drive the PZT actuator. The system's parameters and the overall control system's performance were measured using a two channel HP Dynamic Signal Analyzer. This signal analyzer was used to get the frequency response function between the plate and the clamping frame using two accelerometers, one on the bottom center of the plate and one on the clamping frame.

A series of tests were run to determine the optimum parameters for the active vibration suppression of a representative aircraft panel. These parameters, ranging from the gains assigned to the various amplifiers to the damping ratios of the PPF filters, were based upon the tuned frequency and the number of modes assigned to each PPF filter. The final Simulink Model used for the active control tests uses three PPF filters to provide active damping to most of the modes from 0-400 Hz. However, you can achieve significant active damping of a mode or modes with one PPF filter tuned properly.

There were two different means of supplying disturbance energy to the plate for the active control tests. The first method used a 100 lb shaker to excite the plate mechanically, and the second method used a 10" sub-woofer to excite the plate acoustically. Both the shaker and the speaker were driven with a 0-400 Hz periodic chirp input signal. The initial design procedure for the active control system was to create a Simulink model with one PPF filter to control as many modes as possible. This design method involved determining the frequency range at which the system had control authority over multiple modes. The tests results show that the control system with one PPF filter was most effective when tuned to a frequency between 60 and 70 Hz. Figures 4 show some of the best results obtained with the one PPF filter controller. In each of these tests the 100 lb shaker excited the plate mechanically. The active control test was run with a PPF filter damping ratio of 0.02, a Simulink gain of –80,000, a sampling time of 0.0001 seconds, and tuned to a frequency of 60 Hz.

Additional tests were run using pure tone inputs at the resonant frequencies of the plate. These tests were performed to determine the effectiveness of the controller in the suppression of the structure born noise resulting from the vibration of the test plate. Most of these tests were run at the first resonant frequency of the plate (47 Hz) because this mode produced the most audible structure born noise. The test results proved that the controller reduced the noise from the plate significantly. Figure 4 shows the best reduction achieved by the controller with a 47 Hz pure tone input signal.

The goal here was to show experimentally that active vibration control is possible under varied temperatures. Emitting infrared radiation a 250 W heat lamp, placed about 30 cm (12 in) above the plate, was able to heat the center of the plate up to 70°C (160°F). To measure the temperature a 10kΩ-thermistor was attached to the plate. Due to the location of the heat lamp over the center of the plate the distribution of the heat was not very uniform. While the highest temperature was always reached in the center, the temperature decreased significantly towards the boundaries of the plate.

The purpose of the baseline tests was to examine the influence of the increased temperature on the plate and on the sensor/actuator pair. Using a dimmer in the power cord of the heat lamp, the temperature in the center of the plate could be increased in 2.5°C steps from room temperature up to 65°C (150°F). The frequency response function of the plate was taken every time a steady state of the desired temperature was reached. Figure 3 shows the transfer function of the uncontrolled plate at two different temperatures. Note the significant change in the frequency response of the test specimen at increased temperature. The first measured vibration mode shifted from 53.5 Hz at room temperature to 108 Hz at 65°C (150°F). The extreme change in the properties of the clamped plate is thought to be the result of stiffening from the mechanical strain and tension in the plate induced by a temperature gradient over the plate area. The infrared lamp, located over the center of the test specimen, hardly increased the temperature on the edges of the plate or on the clamping frame itself. Therefore the thermally induced material expansion remained mainly in the center part and even caused the plate to warp out at temperatures as high as 65°C (150°F). For later experiments, the temperature was kept below 45°C to avoid the extensive bending stress for the structure and the actuator bond. Figure 3 illustrates the drastic effect that temperature has one the frequencies.

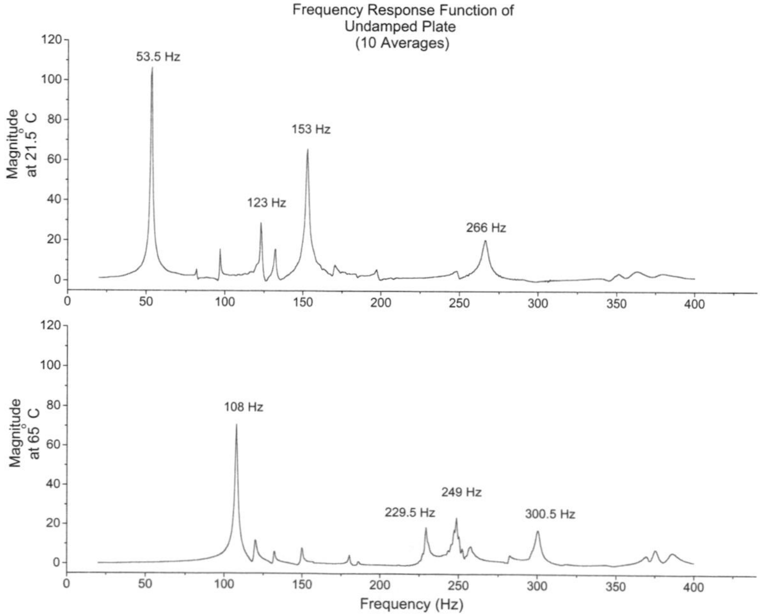

Figure 3 Frequency Response of the Uncontrolled Plate at 21.5°C (70°F) and 65°C (150°F).

The optimal design parameters for the positive position feedback filters depend on the zero/pole spacing of the structural transfer function. The PZT actuator and the accelerometer form a collocated sensor/actuator pair. Although the amplitude of the first pole is very low, the collocated transfer function agrees sufficiently with the frequency response of the plate at room temperature. The accelerometer provided a displacement to acceleration collocated transfer function together with the PZT actuator. A double integrator was implemented on the accelerometer signal using a DSP board to get the required displacement signal for the positive position feedback controller. The integrator was designed as a pseudo-integrator having one zero at 0 Hz and an extra pole at 5 Hz to counter the effects of the zero. The use of a standard integrator would have caused the integration signal to increase up to the saturation limits of the DSP at the slightest DC offset in the sensor signal.

The frequency responses of the plate in Figure 4 show that three PPF filters with open-loop parameter adaptation reduced the magnitude of the resonance peaks sufficiently, especially for temperatures up to 30°C (86°F). For higher temperatures, the PPF filter could not reduce the magnitude of the first mode significantly. The reason is that the zero, following the first mode pole, moves towards this pole and therefore decreases the pole-zero spacing $\omega_{zp} = \omega_z / \omega_p$. As mentioned above, the pole-zero spacing determines the maximum amount of damping, which can be added by PPF filters.

These series of design and experiments clearly indicate the feasibility of a smart structure's concept to control vibration in the presence of substantial temperature effects. The method does not require an analytical model of the structure and hence has the potential to be applied to real aircraft and automobile components. When combined with transient methods the method suggested here has the potential for active vibration suppression across a wide range of temperature changes, not possible with passive constrained layer treatments.

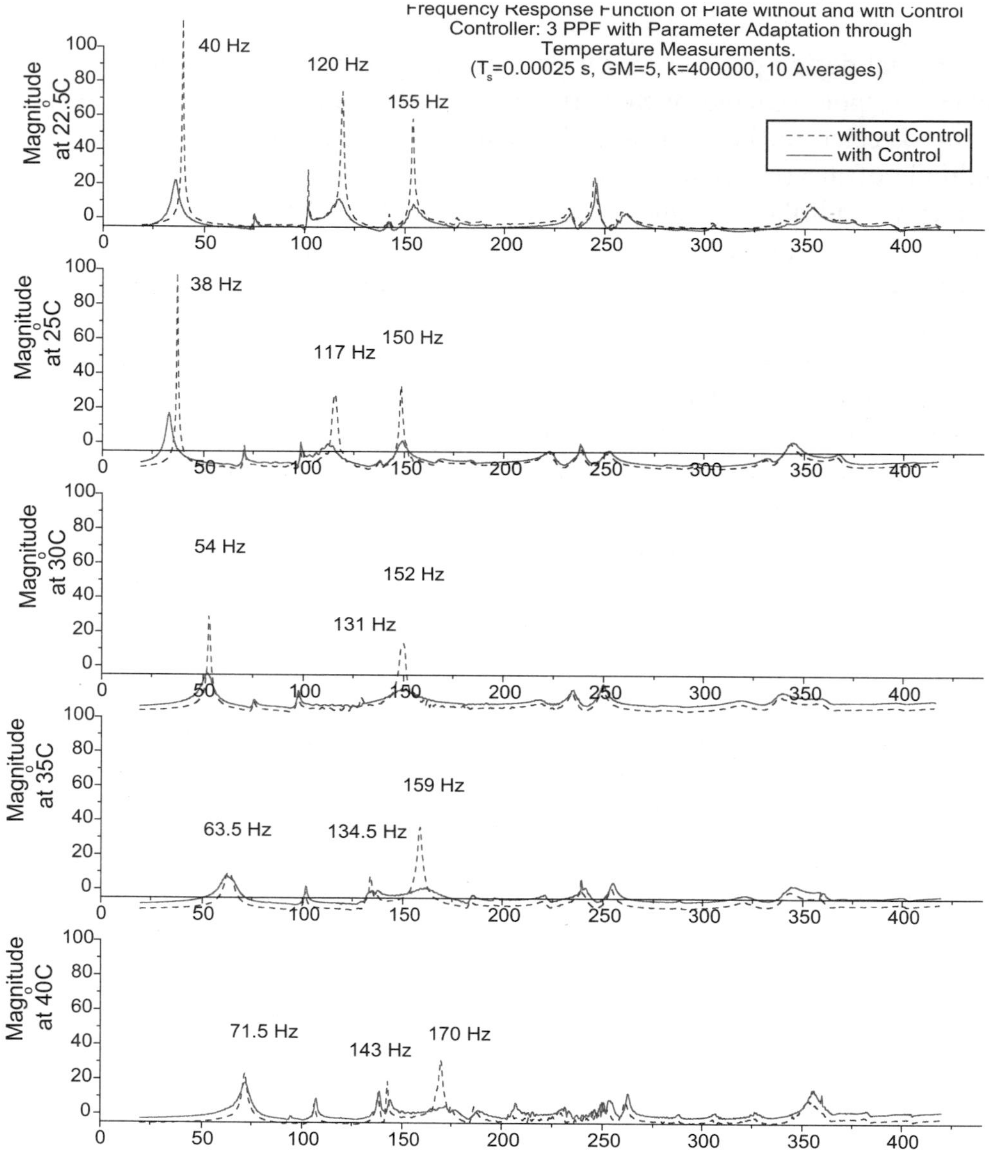

Figure 4 Active control results at various different temperatures.

5 Summary

Several examples have been presented to illustrate the utility of using smart materials in vibration suppression schemes. Both fiber optics and piezoceramics have been integrated into a variety of simple structures. Both passive and active control methods have been presented. The advantage of smart materials over conventional senor and actuator systems are their unobtrusive nature and resulting ability to be placed almost anywhere increasing controllability and observability.

References

Friswell, M. I. and Inman, D. J. (1999) "The Relationship Between Positive Position Feedback and Output Feedback Controllers," *Smart Materials and Structures*, Vol. 8, pp. 285-291.

Hegewald, T., and Inman, D. J. (2001) "Vibration Suppression via Smart Structures Across a Temperature Range", *Journal of Intelligent Material Systems and Structures*, Vol. 12, No. 3, pp. 191-203.

Leo, D. J. and Inman, D. J. (1994)"Pointing Control and Vibration Suppression of a Slewing Flexible Frame", *AIAA Journal of Guidance Control and Dynamics*, Vol. 7, No. 3, pp. 529-536.

Lesieutre, G. A. (1998) "Vibration Damping and Control Using Shunted Piezoelectric Materials", *The Shock and Vibration Digest*, Vol. 30, No 3, pp. 181-190.

Worden, K., Bullough, W.A. and Haywood, J. (2003) *Smart Technologies*, World Scientific, London.

Basics of Control for Vibration Suppression

Daniel J. Inman

Center for Intelligent Material Systems and Structures, Department of Mechanical Engineering,
Virginia Tech, Blacksburg, VA. USA

Abstract. This chapter introduces basic concepts from control for suppressing vibrations in structures. These concepts set the background for integrating smart materials into vibration problems for measurement, suppression and monitoring.

1 Introduction to Control

The choice of the physical parameters of mass, damping and stiffness (*m, c,* and *k*) determines the shape of the response of a vibrating system. In this sense, the choice of these parameters can be considered as the design of the structure. Passive control can also be considered as a redesign process of changing these parameters on an already existing structure to produce a more desirable response. For instance, some mass could be added to a given structure to lower its natural frequency. Although passive control or redesign is generally the most efficient way to control or shape the response of a structure, the constraints on *m, c,* and *k* are often such that the desired response cannot be obtained. Then the only alternative, short of starting over, is to try active control.

There are many different types of active control methods, and only a few will be considered to give the reader a feel for the connection between the vibration and control disciplines as they pertain to smart structures. First, a clarification of the difference between active and passive control is in order. Basically, an active control system uses some external adjustable or active (for example, electronic) device, called an actuator, to provide a means of shaping or controlling the response. Passive control, on the other hand, depends only on a fixed (passive) change in the physical parameters of the structure. Active control depends on current measurements of the response of the system and passive control does not. Active control requires an external energy source and passive control typically does not.

Feedback control consists of measuring the output, or response, of the structure and using that measurement to determine the force to apply to the structure to obtain a desired response. The device used to measure the response (sensor) and the device used to apply the force (actuator) together make up the control hardware. Smart materials will be used in later chapters to form the control hardware. This is illustrated in Figure 1 by using a block diagram. Systems with feedback are referred to as closed-loop systems, while control systems without feedback are called open-loop systems, as illustrated in Figures 1 and 2. The difference between open loop and closed loop control is simply that closed-loop control depends on information about the system's response and open-loop control does not. In the figures, all quantities are described in the Laplace domain where s indicates the Laplace transform variable. The function $F(s)$ denotes the Laplace Transform of the applied input force to the system and $X(s)$ denotes the Laplace Transform of the response of the system.

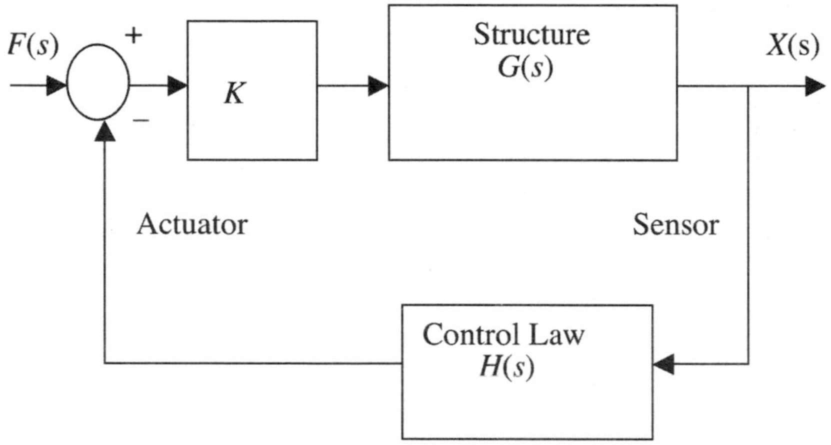

FIGURE 1 The block diagram of a closed-loop system.

The rule that defines how the measurement from the sensor is used to command the actuator to effect the system is called the *control law*, denoted $H(s)$ in Figure 1. Much of control theory focuses on clever ways to choose the control law to achieve a desired response. The dynamics of the structure are contained in the transfer function of the structure denoted by $G(s)$.

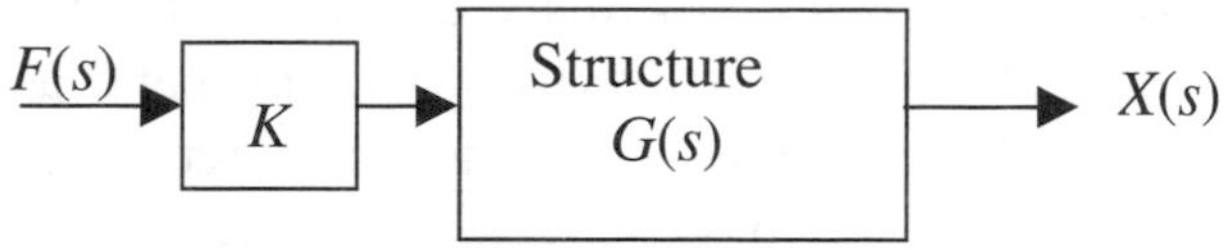

FIGURE 2 The block diagram of an open-loop system.

A simple open-loop control law is to multiply (or amplify) the response of the system by a constant. This is referred to as constant gain control. The magnitude of the frequency response function for the system in Figure 2 is multiplied by the constant *K*, called the gain. The frequency domain equivalent of Figure 2 is

$$\frac{X(s)}{F(s)} = KG(s) = \frac{K}{(ms^2 + cs + k)} \tag{1}$$

where the plant, *G*(*s*), is taken to be a single degree of freedom model of structure. In the time domain, this becomes

$$m\ddot{x}(t) + c\dot{x}(t) + kx(t) = Kf(t) \tag{2}$$

The effect of this open-loop control is simply to multiply the steady-state response by *K* and to increase the value of the peak response, M_p.

On the other hand, the closed-loop control, illustrated in Figure 1, has the equivalent frequency domain representation given by

$$\frac{X(s)}{F(s)} = \frac{KG(s)}{(1 + KG(s)H(s))} \tag{3}$$

If the feedback control law is taken to be one that measures both the velocity and position, multiplies them by some constant gains g_1 and g_2, respectively, and adds the result, the control law *H*(*s*) is given by

$$H(s) = g_1 s + g_2 \tag{4}$$

As the velocity and position are the state variables for this system, this control law is called *full state feedback*. As these feedback terms are also the position and the velocity, this is also called PD control for position and derivative feedback. In this case equation (3) becomes

$$\frac{X(s)}{F(s)} = \frac{K}{ms^2 + (Kg_1 + c)s + (Kg_2 + k)} \tag{5}$$

The time domain equivalent of this equation is (obtained by using the inverse Laplace Transform)

$$m\ddot{x}(t)+(c+Kg_1)\dot{x}(t)+(k+Kg_2)x(t)=Kf(t) \tag{6}$$

By examining the coefficients the versatility of closed-loop control versus open loop, or passive, control is evident. In many cases the choice of values of K, Kg_1 and Kg_2 can be made electronically using smart materials. By using a closed-loop control, the designer has the choice of three more parameters to adjust than are available in the passive case in order to meet the desired specifications.

On the negative side, closed-loop control can cause some difficulties. If not carefully designed, a feedback control system can cause an otherwise stable structure to have an unstable response. For instance, suppose the goal of the control law is to reduce the stiffness of the structure so that the frequencies are lower. Stability requires that g_2 be a negative number. Then suppose that the value of k was over estimated and g_2 calculated accordingly. This could result in the possibility that the coefficient of $x(t)$ becomes negative, causing instability. That is, the response of (6) would be unstable if $(k + Kg_2) < 0$. This would amount to positive feedback and is not likely to arise by design on purpose but can happen if the original parameters are not well known. On physical grounds, instability is possible because the control system is adding energy to the structure. One of the major concerns in designing high-performance control systems is to maintain stability. This introduces another constraint on the choice of the control gains. Of course, closed-loop control is also expensive because of the sensor, actuator (smart materials) and electronics required to make a closed-loop system. On the other hand, closed-loop control will always provide better performance provided the appropriate hardware is available (Inman, 1989, Preumont, 2002).

Feedback control uses the measured response of the system to modify and add back into the input to provide an improved response. Another approach to improving the response consists of producing a second input to the system that effectively cancels the disturbance to the system. This approach, called feedforward control, uses knowledge of the response of a system at a point to design a control force that when subtracted from the uncontrolled response yields a new response with desired properties, usually a response of zero. Feedforward control is most commonly used for high frequency applications and in acoustics (for noise cancellation) and is not considered here. An excellent treatment of feedforward controllers can be found in Fuller, et. al. (1996).

2 Control of multiple Degree of Freedom Systems

Feedback control of a vibrating structure or machine requires measurements of the response (by using sensing transducers) and the application of a force to the system (by using force transducers) based on these measurements. The mathematical representation of the method of computing how the force is applied based on the measurements is called the control law. As is often the case in modeling physical systems, there are a variety of mathematical representations of feedback control systems. The use of control forces proportional to position (denoted by $K_p\mathbf{q}$) and velocity (denoted by $K_v\dot{\mathbf{q}}$) are used to shape the response of the structure. The d constants defined in the previous section would be contained in the matrices K_p and K_v if piezoelectric materials were used as the actuation material. The closed-loop system is modeled by

$$M\ddot{\mathbf{q}} + (D+G)\dot{\mathbf{q}} + (K+H)\mathbf{q} = -K_p\mathbf{q} - K_v\dot{\mathbf{q}} + \mathbf{f}$$

where $\mathbf{q}$ represents the position vector, the over dots are time derivatives, and M, D, and K are the mass, damping and stiffness matrices respectively. This form represents state variable feedback (or position and velocity feedback).
Another form of feedback, called *output feedback*, which as the name implies is a function of only the measured output (or measured states) and results if the closed-loop system (with $\mathbf{f} = \mathbf{0}$) is written as

$$M\ddot{\mathbf{q}} + A_2\dot{\mathbf{q}} + K\mathbf{q} = -B_f\mathbf{u}$$

In this case, the control vector $\mathbf{u}$ is a function of the response coordinates of interest, denoted by the vector $\mathbf{y}$, i.e., $\mathbf{u}(t) = G_f\mathbf{y}$. This form of control is called output feedback. The vector $\mathbf{y}$ can be any combination of state variables (i.e., position and velocities), as denoted by the output equation, which is

$$\mathbf{y} = C_p\mathbf{q} + C_v\dot{\mathbf{q}}$$

The matrices C_p and C_v denote the locations of and the electronic gains associated with the transducer used to measure the various state variables (the g constants defined in the previous section would be contained in these matrices if the sensors were piezoelectric materials).
Each of these two mathematical formulations can also be expressed in state space or first order form. The important point is that each of these various mathematical models is used to determine how to design a system with improved vibration performance using active feedback control, which provides an alternative to passive design methods.

Active control is most often formulated in the state space by

$$\dot{\mathbf{x}} = A\mathbf{x} + B\mathbf{u}$$

where

$$\dot{\mathbf{x}}(t) = \begin{bmatrix} 0 & I \\ -M^{-1}(K+H) & -M^{-1}(D+G) \end{bmatrix} \mathbf{x}(t) + \begin{bmatrix} 0 \\ M^{-1} \end{bmatrix} \mathbf{f}(t),$$
$$\mathbf{x} = [\mathbf{x}_1 \quad \mathbf{x}_2]^T = [\mathbf{q} \quad \dot{\mathbf{q}}]^T$$

with output equation

$$\mathbf{y} = C\mathbf{x}$$

The relationships between the physical coordinates (M, D, K, and B_f) and the state space representation (A, B, and C) are given above. Most control results are described in the state space coordinate system. The symbols for **y** and **u** in both the physical coordinate system and the state space coordinate system are the same because they represent different mathematical models of the same physical control devices.

3 Controllability and Observability

It is not always possible to find a control law of a given form that causes the system to behave as desired. This inability to find a suitable control law raises the concept of *controllability*. A closed-loop system, meaning a structure and the applied control system, is said to be completely *controllable*, or state controllable, if every state variable (i.e., all positions and velocities) can be effected in such a way as to cause it to reach a particular value within a finite amount of time by some unconstrained (unbounded) control, $\mathbf{u}(t)$. If one state variable cannot be affected in this way, the system is said to be *uncontrollable*. The formal definition of controllability for linear time invariant systems is given in state space rather than in physical coordinates. In particular, consider the first-order system defined as before by

$$\dot{\mathbf{x}}(t) = A\mathbf{x}(t) + B\mathbf{u}(t) \tag{7}$$
$$\mathbf{y}(t) = C\mathbf{x}(t) \tag{8}$$

Recall that $\mathbf{x}(t)$ is the $2n \times 1$ state vector, $\mathbf{u}(t)$ is an $r \times 1$ input vector, $\mathbf{y}(t)$ is $p \times 1$ output vector, A is the $2n \times 2n$ state matrix, B is a $2n \times r$ input coefficient matrix, and C is a $p \times 2n$ output coefficient matrix. The control influence matrix B is determined by the position of control devices on the structure (i.e., the location of the PZT patches in the case of using piezoceramic control actuators). The number of outputs is p, which is the same as $2s$, where s is defined in physical coordinates as the number of sensors (again a function of PZT patch location in the case of using piezoceramic sensors). In state space the state vector includes velocity and position coordinates and hence has twice the size ($p = 2s$) since the velocities can only be measured at the same locations as the position coordinates. The state, $\mathbf{x}(t)$, is said to be controllable at $t = t_0$ if there exists a piecewise continuous input $\mathbf{u}(t)$ that causes the state vector to move to any final value $\mathbf{x}(t_f)$ in a finite time $t_f > t_0$. If each state $\mathbf{x}(t_0)$ is controllable, the system is said to be completely state controllable, which is generally what is implied when a system is said to be *controllable*.

The standard check for the controllability of a system is a rank test of a certain matrix. That is, the system of equation (7) is completely state controllable if and only if the $2n \times 2nr$ matrix R, defined by

$$R = [B \;\; AB \;\; A^2B \cdots A^{2n-1}B] \tag{9}$$

has rank $2n$. In this case the pair of matrices $[A, B]$ is said to be controllable. The matrix R is called the controllability matrix for the matrix pair $[A, B]$. The power of smart materials in control application can be realized here. For example, additional small unobtrusive piezoceramic wafers can be included and this has the effect of increasing the rank of the matrix B and hence of R and the controllability of the system.

A similar concept to controllability is the idea that every state variable in the system has some effect on the output of the system (response) and is called *observability*. A system is observable if by examination of the response (system output) information about each of the state variables can be determined. The linear time-invariant system of equations (7) and (8) is completely observable if for each initial state $\mathbf{x}(t_0)$, there exists a finite time $t_f > t_0$ such that knowledge of $\mathbf{u}(t)$, A, C, and the output $\mathbf{y}(t)$ is sufficient to determine $\mathbf{x}(t_0)$ for any (unbounded) input $\mathbf{u}(t)$. The test for observability is very similar to that for controllability. The system described by equations (7) and (8) is completely observable if and only if the $2np \times 2n$ matrix O defined by

$$O = \begin{bmatrix} C \\ CA \\ \vdots \\ CA^{2n-1} \end{bmatrix} \tag{10}$$

has rank $2n$. The matrix O is called the *observability matrix*, and the pair of matrices $[A, C]$ are said to be observable if the rank of the matrix O is $2n$. Again, the use of smart materials as sensors provides simple way to improve the observability of the system but increasing the rank of O.

The amount of effort required to check controllability and observability can be substantially reduced by taking advantage of the physical configuration rather than using the state space formulation of equations (7) and (8). Consider for example the controllability and observability of conservative systems of the form:

$$M\ddot{\mathbf{q}}(t) + K\mathbf{q}(t) = \mathbf{f}_f(t) \tag{11}$$

with observations defined by

$$\mathbf{y} = C_p\mathbf{q} + C_v\dot{\mathbf{q}}. \tag{12}$$

Here $\mathbf{f}_f(t) = B_f\mathbf{u}(t)$ and $\mathbf{u}(t) = -G_f\mathbf{y}(t)$ defines the input. In this case it is convenient to assign $G_f = I$. Then, the system of equation is controllable if and only if the $n \times 2n$ matrix R_n defined by

$$R_n = [\tilde{B}_f \;\; \Lambda_K\tilde{B}_f \ldots \Lambda_K^{n-1}\tilde{B}_f] \tag{13}$$

has rank n, where $\tilde{B}_f = P^T B_f$ and $\Lambda_K = P^T KP$, P is the modal matrix and Λ_K is the diagonal matrix of eigenvalues of K. Thus, controllability for a conservative system can be reduced to checking the rank of a smaller order matrix than the $2n \times 2nr$ matrix R of equation (9).

This condition for controllability can be further reduced to the simple statement that system is controllable if and only if the rank of each matrix B_q is n_q, where B_q are the partitions of the matrix $\tilde{B}_f$ according to the multiplicities of the eigenvalues of K. Here, $d \leq n$ denotes the number of distinct eigenvalues of the stiffness matrix and $q = 1, 2, \ldots, d$. The integer n_q refers to the order of a given multiple eigenvalue. The matrix $\tilde{B}_f$ is partitioned into n_q rows. For example, if the first eigenvalue is repeated ($\lambda_1 = \lambda_2$), then $n_1 = 2$ and B_1 consists of the first two rows of the matrix $\tilde{B}_f$. If the stiffness matrix has distinct eigenvalues, the partitions B_q are just the rows of $\tilde{B}_f$. Thus, in particular if the eigenvalues of K are distinct, then the system is controllable if and only if each row of $\tilde{B}_f$ has at least one nonzero entry.

For systems with repeated roots, this last result can be used to determine the minimum number of actuators required to control the response. Let d denote the number of distinct eigenvalues of K, and let n_q denote the multiplicities of the repeated roots so that $n_1 + n_2 + \ldots + n_q = n$, the number of degrees of freedom of the system, which corresponds to the partitions B_q of $\tilde{B}_f$. Then the minimum number of actuators for the system to be controllable must be greater than or equal to the maximum of the set $\{n_1, n_2, \ldots, n_d\}$. Note that in the case of distinct roots, this test indicates that the system could be controllable with one actuator. Similar results for general asymmetric systems can also be stated.

If the controllability or observability matrix is square, then the rank check consists of determining if the determinant is nonzero. The usual numerical question then arises concerning how to interpret the determinant having a very small value of say 10^{-6}. This situation raises the concept of "degree of controllability" and "degree of observability". One approach to measuring the degree of controllability is to define a *controllability norm*, denoted by C_q, of

$$C_q = [\det(B_q B_q^T)]^{1/(2n_q)} \tag{14}$$

where q again denotes the partitioning of the control matrix $\tilde{B}_f$ according to the repeated eigenvalues of K. According to equation (7.8), the system is controllable if and only if $C_q > 0$ for all q. In particular, the larger the value of C_q is, the more controllable the modes associated with the qth natural frequency are. Unfortunately, the definition in (14) is dependent on the choice of coordinate systems. Non-the-less, such norms are used to successfully in the applications section to follow as a means to place smart material actuators and sensors.

4 Optimal Control

One of the most commonly used mcthods of modern control theory is called optimal control. Like optimal design methods, optimal control involves choosing a cost function or performance index to minimize. Although this method again raises the issue of how to choose the cost function, optimal control is a powerful method of obtaining a desirable vibration response. Optimal control formulations also allow a more natural consideration of constraints on the state variables as well as consideration for reducing the amount of time, or final time, required for the control to bring the response to a desired level.

Consider again the control system and structural model of a damped system and its state space representation given in equations (7) and (8). The *optimal control problem* is to calculate the control $\mathbf{u}(t)$ that minimizes some specified performance index, denoted by $J = J(\mathbf{q},\ \dot{\mathbf{q}},\ t,\ \mathbf{u})$, subject to the constraint that

$$M\ddot{\mathbf{q}} + A_2\dot{\mathbf{q}} + K\mathbf{q} = B_f\mathbf{u}$$

is satisfied, subject to the given initial conditions $\mathbf{q}(t_0)$ and $\dot{\mathbf{q}}(t_0)$. This last expression is called a differential constraint and is usually written in state space form. The cost function is usually stated in terms of an integral. The design process in optimal control consists of the judicious choice of the function J. The function J must be stated in such a way as to reflect a desired performance, yet have a nicely behaved optimum. The best, or optimal, $\mathbf{u}$, denoted by $\mathbf{u}^*$, has the property that

$$J(\mathbf{u}^*) < J(\mathbf{u}) \tag{15}$$

for any other choice of $\mathbf{u}$. Solving optimal control problems extends the concepts of maximum and minimum from calculus to functionals J. Designing a vibration control system using optimal control involves deciding on the performance index J. Once J is chosen, the procedure is systematic.

Before proceeding with the details of calculating an optimal control, $\mathbf{u}^*$, several examples of common choices of the cost function J are given corresponding to various design goals. The *minimum time* problem consists of defining the cost function by

$$J = t_f - t_0 = \int_{t_0}^{t_f} dt$$

which indicates that the state equations take the system from the initial state at time t_0 – i.e., $\mathbf{x}(t_0)$ – to some final state $\mathbf{x}(t_f)$ at time t_f, in a minimum amount of time.

Another common optimal control problem is called the *linear regulator problem*. This problem has particular application in vibration suppression. In particular, the design objective is to return the response (actually, all the states) from the initial state value $\mathbf{x}(t_0)$ to the systems equilibrium position (which is usually $\mathbf{x}_e = \mathbf{0}$ in the case of structural vibrations) as fast as possible. The performance index for the linear regulator problem is defined to be

$$J = \frac{1}{2}\int_{t_0}^{t_f} (\mathbf{x}^T Q\mathbf{x} + \mathbf{u}^T R\mathbf{u})\,dt \tag{16}$$

where Q and R are symmetric positive definite weighting matrices. The larger Q is, the more emphasis optimal control places on returning the system to zero, since the value of $\mathbf{x}$ corresponding to the minimum of the quadratic form $\mathbf{x}^T Q\mathbf{x}$ is $\mathbf{x} = \mathbf{0}$. On the other hand, increasing R has the effect of reducing the amount, or magnitude of the control effort allowed. Note that positive quadratic forms are chosen so that the functional being minimized has a "nice" minimum. Using both nonzero Q and R represents a compromise in the sense that, based on a physical argument, making $\mathbf{x}(t_f)$ zero requires $\mathbf{u}(t)$ to be large. By also minimizing R there is a penalty placed on using too much control effort. For example, in the case of using piezoceramics as the control actuator, R would limit the voltage applied to the piezoceramic and hence avoid destroying the piezoelectric effect.

The linear regulator problem is an appropriate cost function for control systems that seek to eliminate, or minimize, transient vibrations in a structure. The need to weight the control effort (R) results from the mathematical fact that no solution exists to the variational problem when constraining the control effort. That is the problem of minimizing J with the inequality constraint $\|u(t)\| \le c$, where c is a constant, is not solved. Instead, R is adjusted in the cost function until the control is limited enough to satisfy $\|u(t)\| \le c$.

On the other hand, if the goal of the vibration design is to achieve a certain value of the state response, denoted by the state vector $\mathbf{x}_d(t)$, then an appropriate cost function would be

$$J = \int_{t_0}^{t_f} (\mathbf{x} - \mathbf{x}_d)^T Q(\mathbf{x} - \mathbf{x}_d)dt \qquad (17)$$

where Q is again symmetric and positive definite. This problem is referred to as the *tracking problem*, since it forces the state vector to follow, or track, the vector $\mathbf{x}_d(t)$.

In general, the optimal control problem is difficult to solve and lends itself very few closed-form solutions. With the availability of high-speed computing, the resulting numerical solutions do not present much of a drawback. The following illustrates the problem by analyzing the linear regulator problem.

Consider the linear regulator problem for the state space description of a structure given by equations (7) and (8). That is, consider calculating $\mathbf{u}$ such that $J(\mathbf{u})$ given by (16) is a minimum subject to the constraint that (7) is satisfied. A rigorous derivation of the solution is available in most optimal control texts. Proceeding less formally, assume that the form of the desired optimal control law will be

$$\mathbf{u}^*(t) = -R^{-1}B^T S(t)\mathbf{x}^*(t) \qquad (18)$$

where $\mathbf{x}^*(t)$ is the solution of the state equation with optimal control $\mathbf{u}^*$ as input and $S(t)$ is a symmetric time-varying $2n \times 2n$ matrix to be determined. Equation (18) can be viewed as a statement that the desired optimal control be in the form of state feedback. With some manipulation (see Kirk (1970), for instance) $S(t)$ can be shown to satisfy what is called the *Matrix Riccati Equation* (MRE) given by

$$Q - S(t)BR^{-1}B^T S(t) + A^T S(t) + S(t)A + \frac{dS(t)}{dt} = 0 \qquad (19)$$

subject to the final condition $S(t_f) = 0$. This calculation is a backward-in-time matrix differential equation for the unknown time-varying matrix $S(t)$. The solution for $S(t)$ in turn gives the optimal linear regulator control law (16), causing $J(\mathbf{u})$ to be a minimum. Unfortunately, this calculation requires the solution of $2n(2n + 1)/2$ nonlinear ordinary differential equations simultaneously, backward in time (which forms a difficult numerical problem). In most practical problems – indeed, even for very simple examples – the MRE must be solved numerically for $S(t)$, which then yields the optimal control law via equation (18).

The Riccati equation, and hence the optimal control problem, becomes simplified if one is interested only in controlling the steady-state vibrational response and controlling the structure over a long time interval. In this case, the final time in the cost function $J(\mathbf{u})$ is set to infinity. In this case the Riccati matrix $S(t)$ is constant for completely controllable, time-invariant systems. Then, $dS(t)/dt$ is zero and the Riccati equation simplifies to

$$Q - SBR^{-1}B^T S + A^T S + SA = 0 \tag{20}$$

which is now a nonlinear algebraic equation in the constant matrix S which can be solved numerically.

4 Summary

The basic notation and concepts of feedback control commonly used in vibration suppression have been introduced. Controllability and observability matrices (B and C) are key to having being able to control a system to a desired response. Smart materials can be used to fill out the rank of these matrices and to increase there corresponding values, greatly enhancing the closed loop performance. In addition, the choice of the weighting matrix, R, can be used to limit the amount of voltage applied to actuators reducing the possibility of staining a smart material beyond its physical limits. In the following chapter, these concepts are integrated with smart materials to create vibration suppression systems for various different structures.

References

Fuller, C. R., Elliot, S. J. and Nelson, P. A. (1996) *Active Control of Vibration*, London, Academic Press.
Inman, D. J. (1989) *Vibrations: Control, Measurement and Stability*, New Jersey, Prentice Hall.
Kirk, D. E. 1970. *Optimal Control Theory – An Introduction*. Englewood Cliffs, N.J.: Prentice Hall.
Preumont, A.. (2002) *Vibration Control of Active Structures: An Introduction*, 2nd ed, Kluwer Academic Publishers, London.

Smart Structures in Structural Health Monitoring

Daniel J. Inman

Center for Intelligent Material Systems and Structures, Department of Mechanical Engineering,
Virginia Tech, Blacksburg, VA. USA

Abstract. This chapter introduces smart materials for problems related to structural health monitoring and diagnostics. Specifically PZT based materials form a natural actuator and sensor for impedance based to structural health monitoring.

1 Introduction

Smart materials are extremely useful in structural health monitoring applications. This discussion examines the hardware systems relating to damage detection, or structural health monitoring (SHM) and prognosis, particularly in terms of solutions offered by smart structures. First a quick review of smart materials is provided. This is followed by a review of the goals of damage prognosis and the requirements these goals make on sensors and actuators. The remainder of the chapter examines several examples of the integration of smart materials in damage related problems.

2 Monitoring Using Smart Materials

A key element in recognizing the utility of integrating smart materials into structural health monitoring problems is that the unobtrusive and integrated nature of smart materials substantially changes the algorithms for damage identification and monitoring because they:

- Allow increased numbers of sensors and excitations sources
- Allow almost distributed sensing and excitation
- Remove much of the mass loading and local stiffness changes associated with conventional devices
- Allows the sensing and actuation system to be fully integrated into the structure

A key element in using smart materials as the excitation and sensing system for identification and monitoring is to match the choice of material with the dynamic response of the structure or machine under study. In particular, in order to excite the desired dynamics of the system containing the desired parameters or information, the stroke length (displacement or strain), force and time constant of the smart material used as the excitation source must be such that the structural response is large enough and of the correct frequency range to be excited. While this may seem like common sense, there are numerous papers in the literature suggesting impossible schemes because the excitation material is not physically capable of disturbing the structure. The use of smart materials as exciters must also incorporate the energy source for the excitation. Again this may seem obvious, but the literature contains numerous papers with unrealistic energy requirements.

The same matching must be carefully analyzed for the sensor material as well. However, most those working on the experimental side of identification are well aware of the process from years of experience in sizing strain gauges and accelerometers. Location is also a key issue in using smart materials as the excitation and sensing system. In fact, the ability to integrate a smart material fully into a structure is one of the key advantages. Smart materials can often be used in locations not accessible by conventional sensing and excitation devices.

In summary, smart materials can solve several of the problems facing structural health monitoring and damage prognosis tasks:

- Need to provide sensor redundancy
- Need for lots of sensors in observation
- Need to control the input in known ways in online systems
- Need for wireless transmission, on board computing and self power

Damage prognosis (DP) is the prediction in near-real-time of the remaining useful life of an engineered system given the measurement and assessment of its current damaged (or aged) state and accompanying predicted performance in anticipated future loading environments. A key element in damage prognosis is obviously that of structural health monitoring. The added effort in damage prognosis is that of organizing the ability to make a decision based on the current assessment of damage by assuming future loads and predicting how the damaged system will behave. This prediction is then used to make a decision about how to use the damaged structure (or if to use it) going forward. A clear example of a prognosis system is given by a military aircraft hit by enemy fire. The ideal prognosis system would detect the damage then inform the pilot if the should bail out, ignore the damage or perhaps continue to fly by under reduced flight conditions. The battery indicator on a lap top performs a similar prediction in the since that it measures your current usage and estimates the remaining time left before required shut down.

Damage prognosis and damage mitigation are a natural extension to structural health monitoring and can be viewed as the next steps. In order of increasing difficulty then, damage problems can be categorized as the following problems:

1) Determining the existence of damage
2) Determining the existence and location of damage
3) Determining the existence, location and characterization (quantification) of damage
4) All of the above and predicting the future behavior under various loads (Damage Prognosis)
5) All of the above and mitigating the effects of damage (Self Healing Structures)
6) Combine problems 1, 2, 3 or 4 with smart materials to form Self-Diagnosing Structures
7) Simultaneous structural control and damage detection

Smart materials and structures integrate very nicely into all seven of these problems. In the following several examples are given to illustrate the effect that integration of these two disciplines has on solving problems arising in damage prognosis and mitigation with the goal of eventually producing and entirely "chip" based, stand alone system not much larger then a sensor.

3 Impedance Based Methods

The basic principle behind this technique is to apply high frequency structural excitations (typically higher than 30 kHz) through surface-bonded piezoelectric transducers, and measure the impedance of structures by monitoring the current and voltage applied to the piezoelectric materials. Changes in impedance indicate changes in the structure, which in turn can indicate that damage has occurred. Three examples, including a bolted joint, gas pipeline and composite structure, are presented to illustrate the effectiveness of this health monitoring technique to the wide variety of practical field applications.

Although many proof-of-concept experiments have been performed using impedance methods, the impedance-measuring device (HP4194A) is still bulky and expensive rendering it useless for our goal of a self-contained sensor system. Therefore, we have developed an operational amplifier-based device that can measure and record the electric impedance of a PZT. The performance of this miniaturized and portable device has been compared to our previous results and its effectiveness has been demonstrated. The impedance measurements, when combined with statistical methods can easily provide for the detection and quantification of damage.

The impedance-based health monitoring method is made possible through the use of piezoelectric patches bonded to the structure that act as both sensors and actuators on the system. When a piezoelectric is stressed it produces an electric charge. Conversely when an electric field is applied the piezoelectric produces a mechanical strain. The patch is driven by a sinusoidal voltage sweep. Since the patch is bonded to the structure, the structure is deformed along with it and produces a local dynamic response to the vibration. The area that one patch can excite depends on the structure and material. The response of the system is transferred back from the piezoelectric patch as an electrical response. The electrical response is then analyzed where, since the presence of damage causes the response of the system to change, damage is shown as a phase shift or magnitude change in the impedance. A more detailed explanation of the technique can be found in Park et al. (2000). Inman et al. (2005) reviews other structural health monitoring methods.

The solution to the wave equation gives the following equation for electrical admittance

$$Y(\omega) = i\omega\, a(\varepsilon_{33}^{T}(1 - i\delta) - \frac{Z_s(\omega)}{Z_s(\omega) + Z_a(\omega)} d_{3x}^{2} Y_{xx}^{E}) \tag{1}$$

In equation (1), Y is the electrical admittance (inverse of impedance), Z_a and Z_s are the PZT material's and the structure's mechanical impedances, respectively, Y_{xx}^{E} is the complex Young's modulus of the PZT with zero electric field, d_{3x} is the piezoelectric coupling constant in the arbitrary x direction at zero stress, ε_{33}^{T} is the dielectric constant at zero stress, δ is the dielectric loss tangent of the PZT, and a is a geometric constant of the PZT. This equation indicates that the electrical impedance of the PZT bonded onto the structure is directly related to the mechanical impedance of a host structure.

The impedance method has many advantages compared to global vibration based and other damage detection methods. Low excitation forces, combined with high frequencies (typically greater than 30 kHz), produce power requirements in the range of micro Watts. The small wavelengths at high frequencies also allow the impedance method to detect minor changes in structural integrity and in some cases detect imminent damage. The impedance measurement used for the health monitoring technique contains values of impedance for a range of frequencies. To allow the use of impedance methods in a chip based environment, a new method of generating impedance measurements utilizing an FFT analyzer and small current measuring circuit has been developed. FFT analyzers, such as those used in modal analysis, currently available on a computer chip the size of a postage stamp. This it may be possible to produce an impedance-based method that would be suitable for combining directly into a compact sensor system.

The electrical impedance of the bonded PZT is equal to the voltage applied to the PZT divided by the current through the PZT. An approximation of the impedance is generated by taking the ratio with the analyzer of the voltage supplied to the circuit, V_i, to the voltage, V_o, across a sensing resistor, R_s, in series with the PZT as seen in Figure 1.

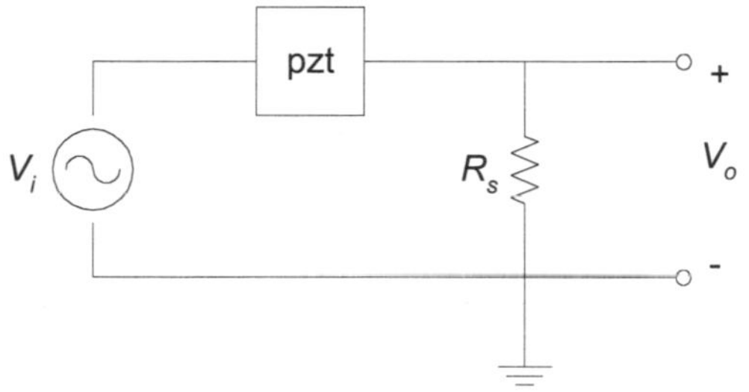

Figure 1 Circuit for approximating PZT impedance.

This circuit is a simple voltage divider. The output voltage is proportional to the current through the sensing resistor, which, if the sensing resistor is small, is approximately the current through the PZT if the sensing resistor was not included (as when measuring with a normal impedance analyzer). The circuit is described by the following equations:

$$I = \frac{V_o}{R_s} \tag{2}$$

where I is the current through the sensing resistor. The approximated impedance (Z) is:

$$Z = \frac{V_i}{I} = \frac{V_o}{V_i / R_s} \tag{3}$$

Since PZT's are a capacitive element the current through them increases with frequency. Conversely, at low frequencies the circuit has very high impedance. In this case an inverting amplification circuit can be used to provide a larger output voltage. The size of the sensing resistor could be increased, however, this reduces the voltage applied to the PZT (a larger voltage drop occurs on the sensing resistor). The circuit for approximating the impedance along with the amplification circuit is shown in Figure 2.

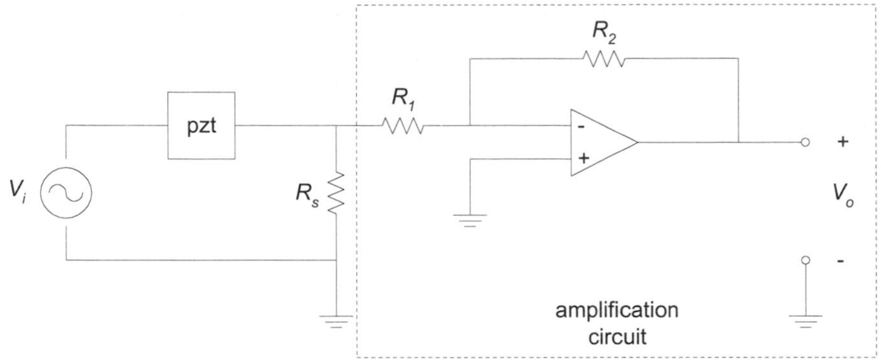

Figure 2 Impedance approximating circuit with amplification.

The gain, *G*, provided by the amplification circuit is shown in the following equation. Since an inverting amplifier is shown, the gain is negative.

$$G = -\frac{R_2}{R_1} \tag{4}$$

At high frequencies the op-amp is ineffective due to roll-off of the output signal.

The loosening mode of failure of bolted joints is a common form of failure in many structures. Real-time condition monitoring and active control of critical joints will improve the reliability and safety of many structures where bolted joints are used. A test specimen was constructed using two 1/4 x 2 x 9 inches sections with an overlap of 3 inches. The joint was fastened using a 3/4-inch bolt with washers. Six 7/8 x 3/4-inch PZT sensors, cut from 0.01 inch thick sheets were bonded to the beam. Three were located 0.5, 2.75 and 5 inches from one free end of the beam and three were similarly placed from the other free end of the beam. In the actual implementation of a self-monitoring system only one PZT would be needed, however, more were bonded to the beam for use in additional experiments. The beam was suspended vertically from one loop of fishing line to simulate a free-free boundary condition. A schematic of the beam is shown in Figure 3.

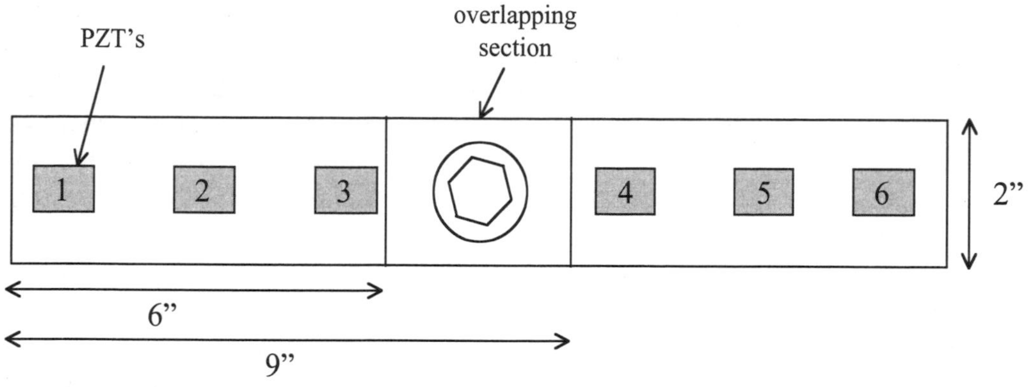

Figure 3 Bolted beam schematic.

In order to verify that the new, low cost measuring method and traditional method using an HP 4194A impedance analyzer would give the same results, the joint was tested using both methods. For the low cost method a 1 V peak chirp signal was applied using a DSPT SigLab™ 20-42 dynamic signal analyzer and an approximately 20 ohm sensing resistor used without the amplification circuit. Ten runs were averaged. The resulting impedance measurements are shown in Figure 4.

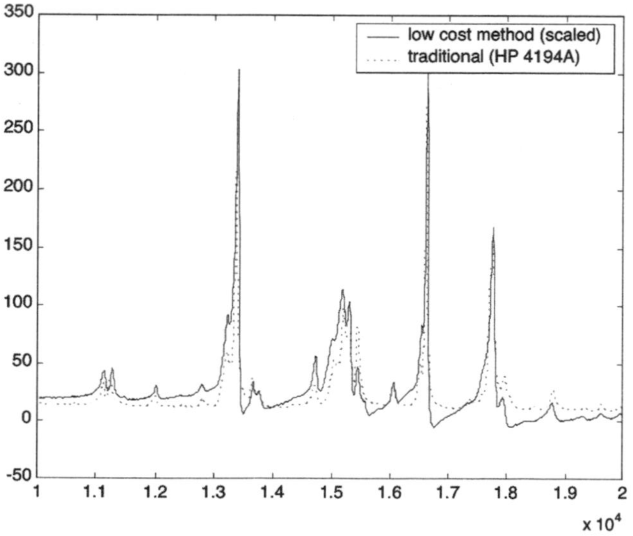

Figure 4 Impedance Method Measurement Comparison.

The peaks in the response are exactly the same, indicating that the measurements are the same even though small differences are seen. These differences could be as a result of differences in the windowing, A/D conversion, sampling frequency and excitation. Even though the shapes of the two measurements are nearly identical it is important to remember that the impedance method works by comparing a change in impedance from one measurement to another. However, as long as the same method is used to make the two measurements being compared it is irrelevant if the measurements using different methods are different. The bolt was also loosened from 20 ft-lbs and the change in impedance compared between the two methods. As can be seen in Figure 5 and Figure 6, the change in impedance was also approximately the same for each method.

The shift in the three largest resonant peaks for the new method were 1.16%, 1.40%, and 2.85% compared to 1.30%, 1.50% and 2.92% for the traditional method giving a maximum difference of 0.14 % shift. This again confirms that the new method gives the same results as the traditional method. A more practical method of making measurements for the impedance method has been developed and tested to allow greater accessibility to the technology. The experimental investigations of low cost impedance-based health monitoring techniques on various components were presented showing that the small device is as effective as a traditional impedance analyzer in making impedance measurements for health monitoring. Several issues currently under investigation include the effect of varying the driving voltage due to different sensing resistances, the development of miniaturized impedance measuring device without the use of a separate FFT analyzer and the combination of the impedance method with wireless communication technology. Additional examples and comparisons are given in Park et al. (2000).

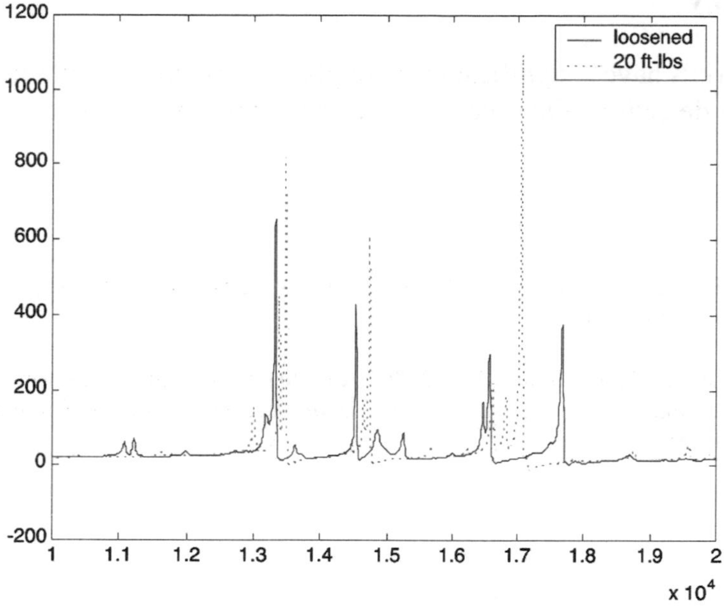

Figure 5 Low Cost Response to Simulated Damage.

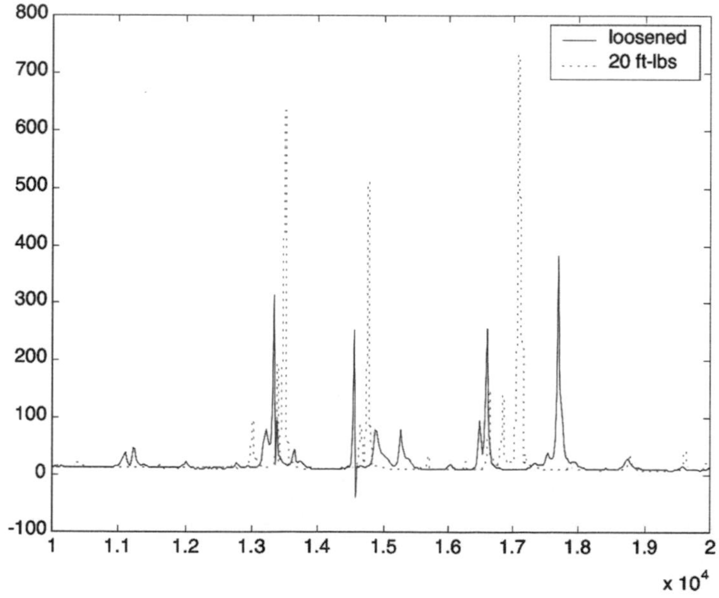

Figure 6 Traditional Response to Simulated Damage.

3 Summary

Smart materials have a significant role to play in structural health monitoring and damage detection. One such method, the impedance method is shown here.

References

Inman, D. J., Farrar, C. R., Steffen, Jr., V., and Lopes, V., Eds. (2005) *Damage Prognosis*, John Wiley and Sons, New York, NY.

Park, G., Cudney, H.H., and Inman, D.J.(2000)"An Integrated Health Monitoring Technique using Structural Impedance Sensors," *J. of Intelligent Material Systems and Structures* Vol. 11, No. 6, pp. 448-455.

Vibrational Mechanics - a General Approach to Solving Nonlinear Problems

Iliya I. Blekhman [*‡]

[*] Laboratory of Vibrational Mechanics, Inst. of Problems of Mechanical Engineering, Russian Academy of Sciences
&
"Mekhanobr-Tekhnika" Corp.

1 The Most Important Phenomena Accompanying the Action of Vibration on Nonlinear Mechanical Systems and their Application in Technology

1.1 Introduction

This study has two aims. The first aim is to describe a number of wonderful phenomena caused by the action of vibration[1] on nonlinear mechanical systems along with important applications of these effects in technology. The second aim is to present a general mechanical-mathematical approach to the description and investigation of this class of phenomena. We call the approach to be described herein "Vibrational Mechanics". While it is a new approach, it is based on the classical idea of averaging in the theory of nonlinear oscillations and in the theory of the stability of motion.

This study is a brief introduction into vibrational mechanics. A detailed presentation of the fundamentals of vibrational mechanics and of its numerous applications can be found in the monographs (Blekhman, 2000; ed. Blekhman, 2004) . The later also contains a detailed list of publications in which this approach is successfully used for the solution of new applied problems. These works were performed by scientists from Germany, Denmark, Russia and Ukraine.

1.2 Some Paradoxical Effects, Caused by the Action of Vibrations in Nonlinear Systems

The specific effects which appear under the action of vibration in nonlinear mechanical systems can, on the one hand, induce unwanted consequences and even catastrophes, and, on the other hand, can be used quite efficiently in modern technique.

Let us dwell on three groups of such effects.

[1] By *vibration* we mean the mechanical oscillations whose period is much shorter than the characteristic interval in which the motion of the system is considered and whose swing is far smaller than the characteristic size of the system.

1) The change in the behavior of oscillatory systems and mechanisms under the action of vibrations.

Some examples are as follows:

a) *The classical example*: under the action of vibrations of the support the inverted position of the pendulum can become stable (Fig. 1.1).

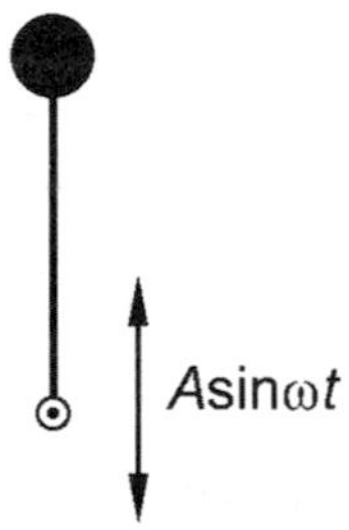

Figure 1.1. Pendulum with a vibrating axis of suspension.

The lower position can also remain stable, but the frequency of free oscillations increases:the pendulum clocks on a vibrating support are always fast.

b) *The second example*:under the action of vibration the same pendulum (now represented by the unbalanced rotor) can rotate steadily around its axis (Fig. 1.2).

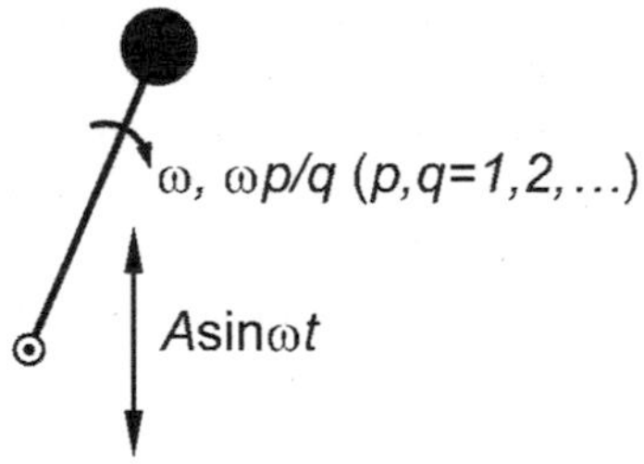

Figure 1.2. Vibrational maintenance of rotation of unbalanced rotor.

A considerable power can be transferred to the rotor. Due to the transfer of energy from the vibrating support to the rotor, it may seem that the efficiency of the motor with the unbalanced rotor is more than unity (Fig. 1.3).

This circumstance has more than once caused various misunderstandings, as will be discussed below.

The above phenomena are called *vibrational maintenance of rotation of unbalanced rotors.*

An example of this phenomenon is a well-known game hula hoop (Fig. 1.4). In this case the hoop plays the role of the rotor and the person's waist plays the role of a vibrating

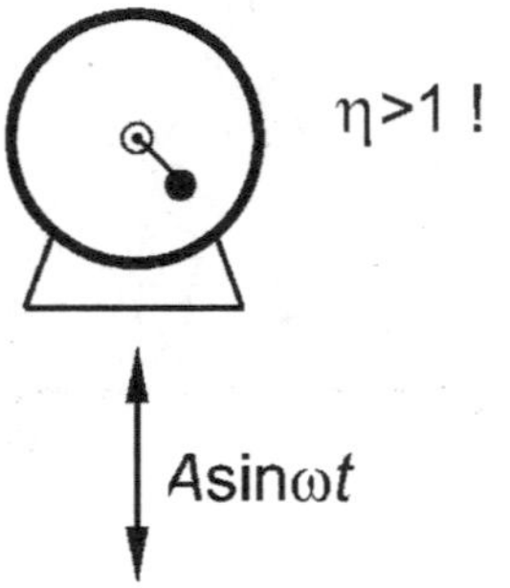

Figure 1.3. The action of vibration on the rotation of an unbalanced rotor. Due to vibration the measured efficiency may prove to be either smaller or greater than its actual value and even more than unity.

axis. It is remarkable that the power which vibration can transmit to the rotating rotor or ring can be quite large. That is why the phenomenon under consideration has found a wide utility in technology. Radically new machines for crushing the ore, namely, conical inertial crushers (Fig. 1.5), have been constructed on that basis at the "Mekhanobr" Institute in Russia. The role of the hoop in those crushers is played by the massive outer body 1 , while the role of the person's waist is immitated by the inner cone 2, whose vibrations are excited by the unbalanced rotor 3. The crushing of the ore takes place in the space between the cone and the body.

Figure 1.4. The game-exercise hula hoop – a simple example of vibrational maintenance of rotation.

c) *The third example*, which is more complicate, is the *self-synchronization of mechanical vibro-exciters*. It consists in the following: let a pair of unbalanced rotors be rotating as a result of being driven by the independent induction motors (Fig. 1.6).

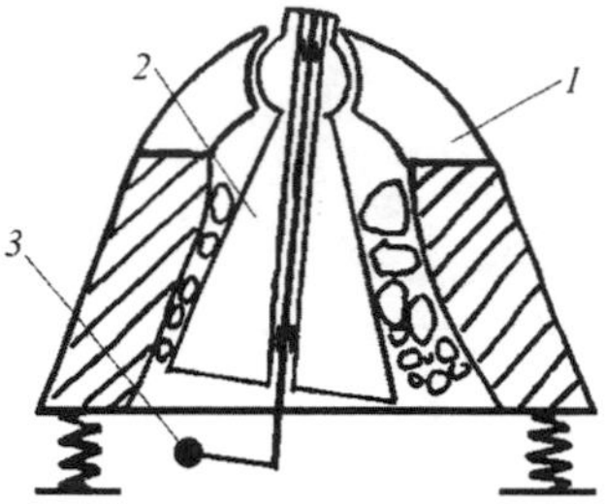

Figure 1.5. Inertial crusher of the "Mekhanobr" Institute is an example of using the "hula hoop effect".

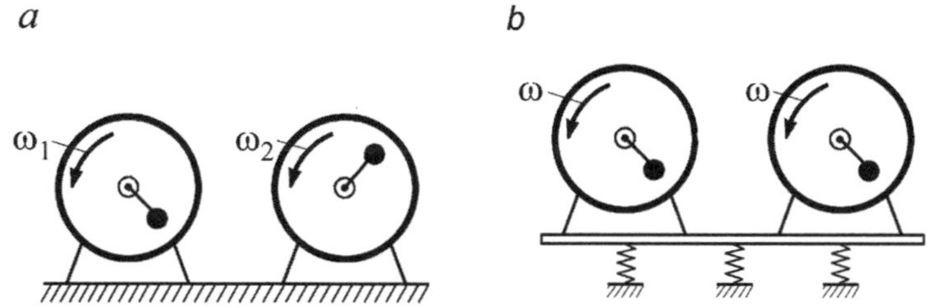

Figure 1.6. Self-synchronization of the mechanical vibro-exciters: a – two or more unbalanced rotors on a fixed base, driven by the asynchronous engines, have different frequencies of rotation ω_1 and ω_2; b – when placed on a common mobile base they rotate with the same average frequency ω.

These devices are often used for exciting mechanical vibration and so they are called *vibro-exciters.* If the rotors are installed on a fixed base (see (Fig. 1.6,a), they rotate in the general case with different frequencies ω_1 and ω_2 in spite of the fact that they are nominally equal. This can be explained by the inaccuracy of the manufacture of motors and rotors. If the rotors are installed on a common movable base, then , under certain conditions, they rotate with the same frequency , i.e. their *self-synchronization* takes place (Fig. 1.6,b). The tendency to synchronization is so strong that even when one of the motors is switched off, the synchronization remains on place, and the rotation goes on. The energy comes from the motor which is switched on. The transfer of the energy takes place as a result of an almost invisible vibration of the base.

Special attention in this study is paid to the effect of self-synchronization and its numerous manifestations.

2) The effects of vibrational displacement and drift.

Here are examples of the above effects.

a) *Effects of vibrational transportation of solid bodies and granular material* (Fig. 1.7).

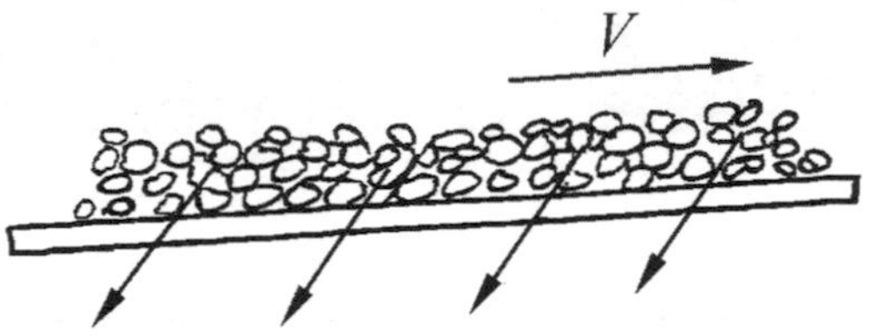

Figure 1.7. Vibrotransportation of a granular material.

A layer of granular material moves along the vibrating plain as a result of a hardly appreciable vibration. Transportation is possible upwards the plain. The velocity of transportation in the industrial mountings reaches 0.5 m/s, the productivity is about 200-300 t/h.

b) *The separation of solid particles on the vibrating surfaces in accordance with their physical properties.* (Fig. 1.8). Particles having different properties get into different cells.

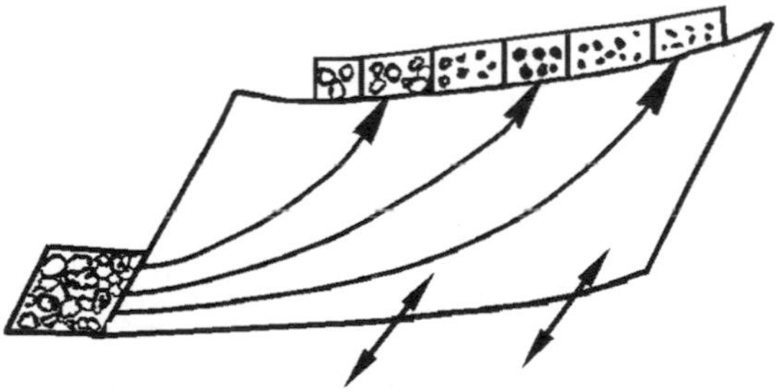

Figure 1.8. Separation of solid particles on a vibrating surface.

c) *Vibro-bunkering of granular materials.*(Fig. 1.9)

When the direction of vibration is as shown in Fig. 1.9, a, then the ordinary vibro-transportation takes place. Should, however, the direction of vibration be changed, as is shown in Fig. 1.9, b, the material will fill the bunker moving upwards. This effect is called *vibro-bunkering.*

d) *The scales, installed on a vibrating support, can show that a pound of steel is lighter than a pound of cotton.* (Fig. 1.10)

The action of vibration can change the readings of the pointer-indicators and axes of gyroscopes in a similar way. These effects can be called the *vibrational drift.*

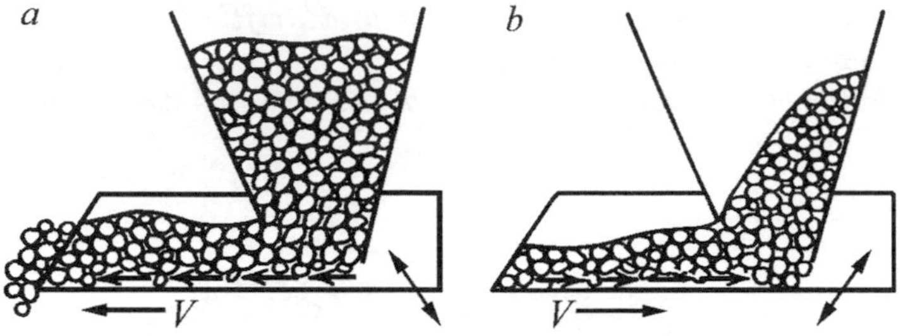

Figure 1.9. Two cases of vibrotransportation of a granulated material: a – vibrodischarge ; b – vibrobunkering.

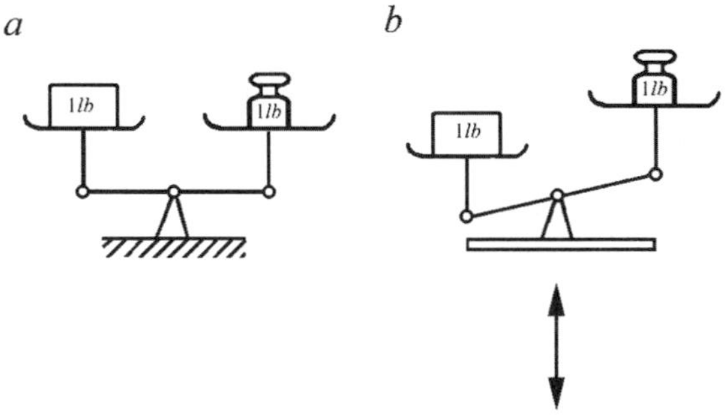

Figure 1.10. The action of vibration on the date of scale: a – the scale on a fixed base; b – on a vibrating base.

e) *A steel ball, contained in a vessel with sand, can float up under the action of vibration*(Fig. 1.11, a).
At the same time, bubbles of the air in the vibrating vessel with a liquid can sink to the bottom (see Fig. 1.11, b).
f) *Effects of vibrational displacement in liquid. The outflow of fluid from the vibrating vessels.* A body, oscillating in a fluid or gas, excites in them slow streams. A vibrational pump, shown in Fig. 1.12, is an example of it.Such a pump was suggested by Lishansky (Blekhman, 2000). Under the oscillations of plate 1, a heightened pressure appears in space 3 in vessel 2, as a result of which fluid flows out of the vessel more intensively than it would have been in the absence of the oscillating plate.

Peculiar effects are observed when the fluid flows out of the vibrating vessels. One of them is the so called *vibro-jet effect* (ed. Blekhman, 2004). It consists in the fact that when a plate with conic holes vibrates in liquid, slow flows of the fluid appear in the direction of the narrowing of the holes (Fig. 1.13, a)

The vibro-jet effect is successfully used in a number of technical devices (Blekhman,

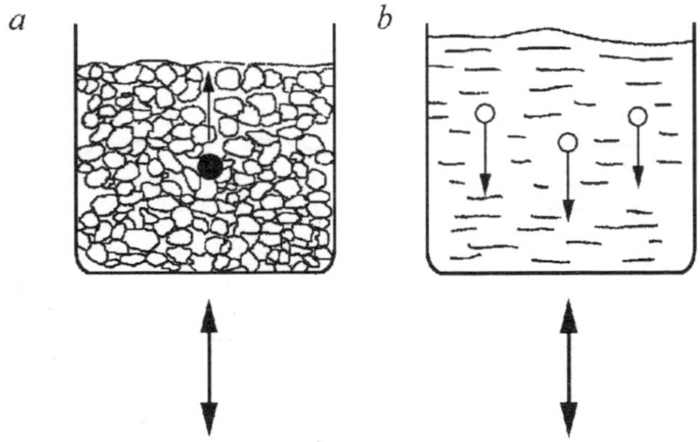

Figure 1.11. A steel ball in a vibrating vessel with sand can float up (1) and the bubbles of air in a such vessel with fluid can sink to the bottom (b).

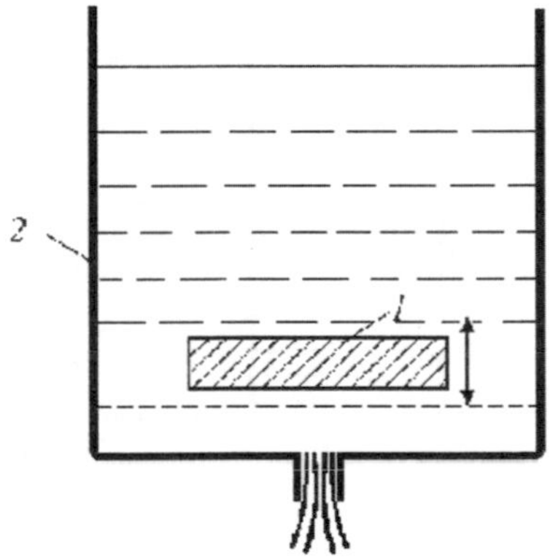

Figure 1.12. Lishansky's vibrational pump.

2000; ed. Blekhman, 2004). Along with this, there are data indicating that this phenomenon was the cause of some aviation catastrophes: due to vibration the fuel stopped coming from tanks (Fig. 1.13, b), i. e. there was an effect of *vibrational closure of the holes.* In this case the pressure, facilitating the discharge of the fuel, is balanced by the counter-pressure, appearing due to vibration.

The second phenomenon – the *vibrational injection of gas into fluid* has been uncovered recently (ed. Blekhman, 2004). It consists in the sucking of the air or another gas into a vessel with the fluid, vibrating in this gas, through the holes in its lower part. It is possible also the injection of fluid into another fluid (Fig. 1.14).

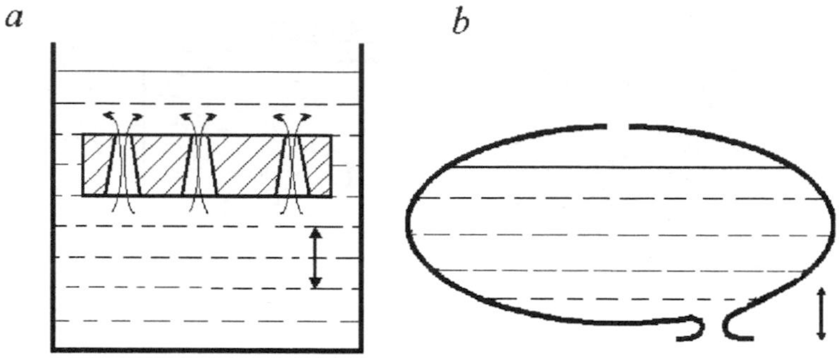

Figure 1.13. Vibro-jet effect: a - the generation of slow flows of the fluid through the conic holes in a vibrating plate, b - the effect of closing the holes in the vibrating tank with the fluid.

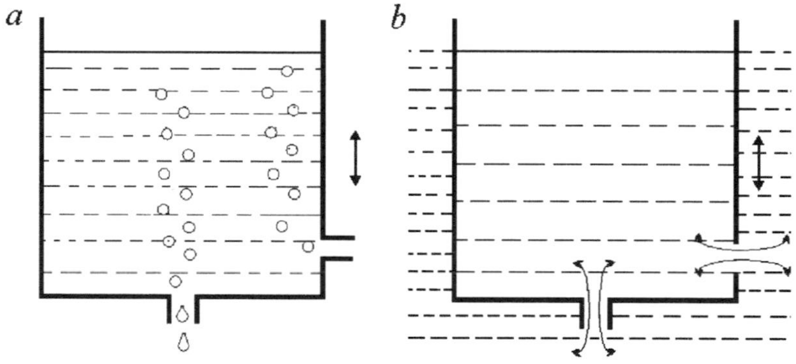

Figure 1.14. Phenomenon of vibrational injection: a - the injection of gas into a fluid, b - the injection of fluid into a fluid.

g) *Vibrational coaches.* If a body which is oscillating in a fluid excites there streams, then in accordance with the laws of hydro-dynamics the body is acted upon by a certain force. This circumstance underlies the *Vibrational coaches* – the bodies, moving in a certain medium or in a force field due to the periodic motions of bodies, connected with the moving one.

Figure 1.15 shows a scheme of such a coach, investigated by Nagaev and Tamm and named by them *vibro-flier* (Blekhman, 2000).The main element of such a "coach" is a body, whose form is such that the resistance of the medium to the motion of the body upward is considerably less than that to its motion downward (in Fig. 1.15 the body is shown in form of an umbrella). As result of the axial oscillations of the body, excited in one way or another, there appears a lifting force, which may under certain conditions exceed the weight of the system mg, so that the apparatus will be lifted upwards. The same principle serves to transport objects not only vertically, but in any arbitrary direction as well. This principle underlies the motion of birds and insects which are, in fact, vibrational coaches in the above sense.

Figure 1.15. Vibro-flier by Nagaev and Tamm.

h) *Exotic examples of vibrational displacement.* In Fig. 1.16 a body lies on a horizontal plane, performing longitudinal harmonic oscillations.

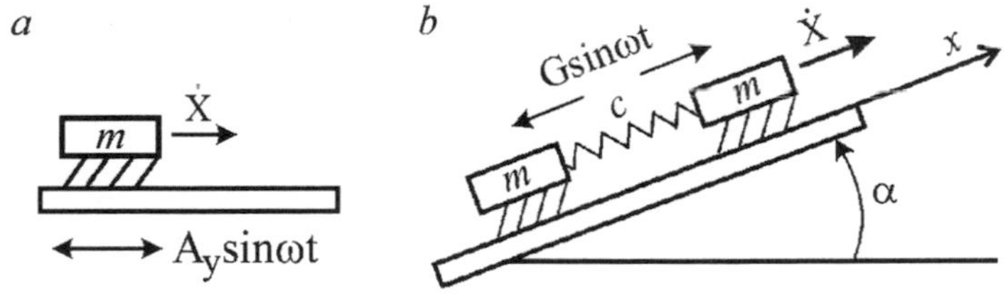

Figure 1.16. "Shaggy" body on a rough vibrating surface moves in the direction of a lesser resistance (a); the model of a "warm-like" motion (b).

We assume that a body or a plane has inclined "hairs", so the resistance to the motion of the body will be greater than that when the body moves in the direction of the inclined hairs. As a result, when the hairs are directed as shown in Fig. 1.16, a and the vibration is intensive enough, the body will move in the direction of the axis x i.e. in the direction where the resistance will be the weakest. This diagram simulates the work of an effective device for processing small fish (fins instead of hairs!). The fish must move head forward in this device.

The velocity of vibro-transportation can be increased if we use the diagram, given in Fig. 1.16, b. Two similar bodies are linked here by a spring whose rigidity is c. A harmonic exciting mass, exerted by an electro-magnet, acts between the masses. The frequency of that force can be selected from the condition of the resonance. This diagram simulates the "worm-like" motion (Zimmerman et al., 2004). An adequate device can be used to move a "shuttle" along the blood vessels when the latter are being investigated or in surgeries.

Fig. 1.17 and Fig. 1.18 show the diagrams of two vibrational coaches which are rather of a basic than of applied interest.

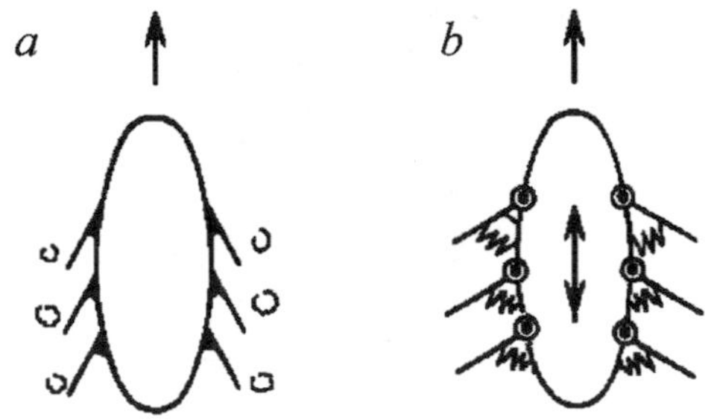

Figure 1.17. Vibroship: a –passive; b – active, resonant.

According to the schemes in Fig. 1.17 the coaches (the so called vibro-ships) also have a weaker resistance to the motion forward than that to the motion backward due to the presence of fins. It is remarkable that those ships can move without any consumption of the internal energy and without any internal source of excitation. It will move forward if it is regularly acted upon by accidental turbulent pushes on the part of the fluid, equally probable in every direction.

The velocity of motion of vibro-ships may be increased in case a vibro-exciter is placed on it, and also if the swivel-mounted fins are linked with the body by elastic elements, whose stiffness is chosen from the resonance condition with the frequency of excitation (Fig. 1.17,b).

Figure. 1.18 presents an idea of a cosmic ship, proposed by Beletsky and Ghiverts, and named by them *graviflier*. The ship was revolving as a satellite of the Earth in the elliptic orbit.

The authors have shown that if the masses $m/2$, which are inside the ship are periodically in a certain way brought close together and drifted apart, then it is possible to accelerate the satellite and even make it leave the area of the Earth's attraction. That will take place despite the fact that the ship has not got any traditional reactive engine!

One of the engineers responds to it that according to the theorem of the motion of an isolated system that was absolutely impossible, and that the authors' ignorance was quite outrageous. But that engineer was quite wrong. The ship as well as the loads were affected by the external force of gravity. The non-linear dependence of that force on the

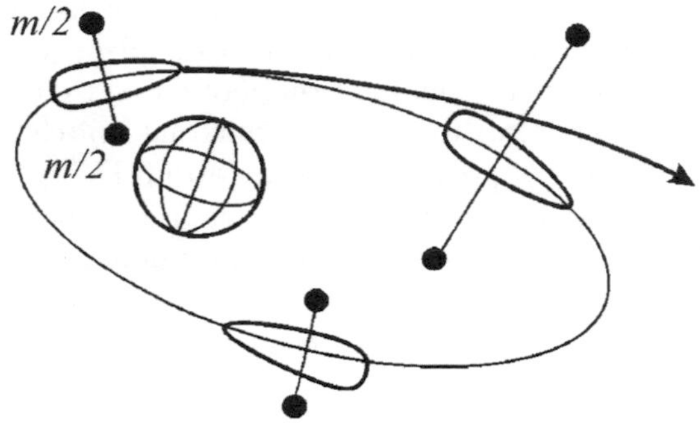

Figure 1.18. Graviflier by Beletsky and Giverts.

distance from the Earth's center causes the above effect. It is not difficult to show that the manipulation with the weights requires a certain expenditure of the energy and that it results in the appearance of forces similar to those of thrust.

Unfortunately, it is not easy to use this idea in actual practice. According to the calculations of the authors it will take the gravi-flier about 10 thousand years to leave the Earth's gravitational field and the distance between the masses must be of an order of one kilometer. But that does not diminish the basic interest which the gravi-flier presents from the point of view of mechanics.

3) Vibrorheological effects.

Speaking of such effects we mean the seeming changes of rheological properties of physical bodies caused by vibration. Here we mean the change of properties with regard only to slow forces.

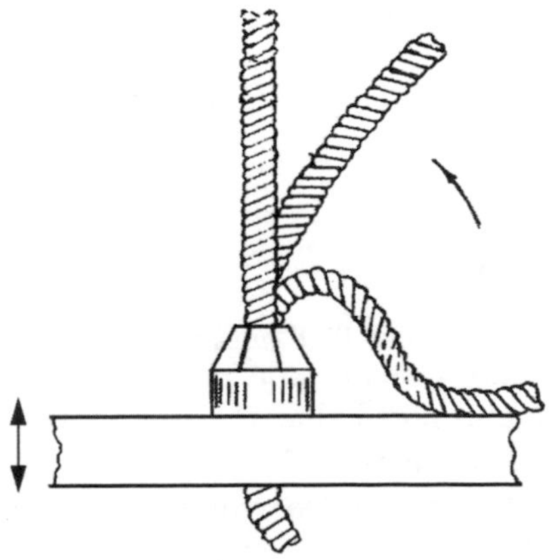

Figure 1.19. Effect of the Indian magic rope.

The seeming transformation of dry friction forces into viscous forces under vibration is a classical example of such transformations. Another example which is usually interpreted in a different way is the turbulent regime of the motion of water in tubes. Under the action of vibration the water seems to be transformed into pitch. Its effective viscosity with regard to slow differences of pressure increases abruptly.

On the other hand the pulsating difference of pressure in the tube may result in the increase of average flow rate, i.e. to the seeming reduce of efficient viscosity.

To the rheological effects one can relate also the sufficient decrease of strength of materials under vibrational loading (phenomena of fatigue), and also both effects of vibrocreeping and vibrorelaxation.

An example of vibrational effect is the behavior of the so-called Indian magic rope: under the action of vertical vibration a "soft" rope seems to acquire some additional rigidity and takes a stable vertical position (Fig. 1.19). The effect shows itself rather vividly. In some popular papers one can even see an affirmation (a little bit questionable) that a small monkey can climb such a stabilized vertical rope.

This effect may be regarded as referring to the extreme case of the problem of the so-called dynamic materials, the idea to create them was advanced by K.Lurie and the author (Blekhman, 2000; ed. Blekhman, 2004). With the change of the amplitude A and the frequency Ω of vibration we obtain the seemingly different materials (with different rigidity EJ). (Fig. 1.20)

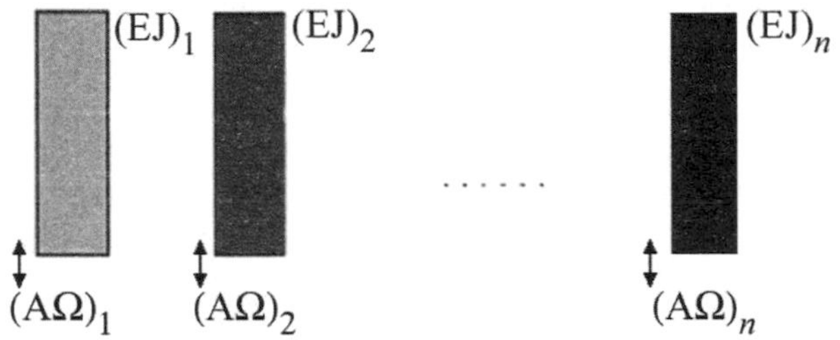

Figure 1.20. To the problem of vibrational dynamic material.

1.3 Conclusion to Section 1

Above we have enumerated some surprising effects, connected with the action of vibration on nonlinear mechanical systems. The number of such examples may be increased considerably. It should be also noted that the list of such examples is far from being exhausted, it keeps on being supplemented.

The approach, presented below, called vibrational mechanics, makes it possible to explain physically in a rather simple form and describe mathematically all the effects, considered above. Some of them, though, are still awaiting their investigators and the inventors who will put them to practical use. The frames of this course allow me to touch upon the investigation of some most important effects and their applications. As was already mentioned, numerous other investigations can be found in the books (Blekhman, 2000; ed. Blekhman, 2004).

2 Vibrational Mechanics – a New Approach to the Explanation and Description of the Nonlinear Vibrational Effects

2.1 The main Idea of Vibrational Mechanics. The Observer "O" and the Observer "V"

Let us now consider briefly and schematically the main idea of vibrational mechanics. Let the following differential equation describe the motion of the system:

$$\mathbf{m}\ddot{\mathbf{x}} = \mathbf{F}(\dot{\mathbf{x}}, \mathbf{x}t) + \mathbf{\Phi}(\dot{\mathbf{x}}, \mathbf{x}t, \omega t) \tag{2.1}$$

Here $\mathbf{x}$ is the n-dimensional vector of the generalized coordinates, $\mathbf{m}$ is a certain positive matrix whose dimension is $n \times n$, $\mathbf{F}$ is a "slow" force and $\mathbf{\Phi}$ is a "fast" force, t is the "slow" time, and $\omega t = \tau$ is the "fast" time, $\omega > 0$ is a large positive parameter (the frequency of vibration).

The functions $\mathbf{F}$ and $\mathbf{\Phi}$ are the n-dimensional vectors, with $\mathbf{\Phi}$ being an almost periodical function of the argument $\omega t = \tau$ (in particular—the periodical function τ with a period of 2π). Though, to make it more understandable, we can assume that equation (2.1) is scalar. The exact meaning of the adjectives "fast" and "slow" is explained below. And here we will perceive them by intuition.

All the above mentioned examples are characterized by the following facts:

1) By the essential nonlinearity of the corresponding equation (2.1)

2) The motion , caused by vibration, is in fact fast oscillations imposed on a certain slow motion.

As a result, the vector of the generalized coordinates in the case we have considered can be presented in the following way:

$$\mathbf{x} = \mathbf{X}(t) + \mathbf{\Psi}(t, \omega t) \tag{2.2}$$

where $\mathbf{X}$ is a slow component and $\mathbf{\Psi}(t, \omega t)$ is a fast periodically or almost periodically component with respect to ωt.

Therewith we can assume for the sake of definiteness that

$$\langle \mathbf{\Psi}(t, \tau) \rangle = \frac{1}{T} \int_0^T \mathbf{\Psi}(t, \tau) d\tau = 0, \quad \langle \mathbf{x} \rangle = \dot{\mathbf{X}}(t) \tag{2.3}$$

Here and below the angular brackets indicate averaging with respect to the fast time τ for the period $T = 2\pi/\omega$ in the case of periodical function and for the infinitely large period $T \to \infty$ in the case of almost periodical motions. Schematically relation (2.2) is presented in Fig. 2.1. It should be noted that $\mathbf{\Psi}$ is not necessarily to be small as compared to $\mathbf{X}$ as is shown in the Figure.

The main idea of vibrational mechanics consists in the transition from the differential equation (2.1) *for the initial variable* $\mathbf{x}$ *to the differential equation for the slow component*

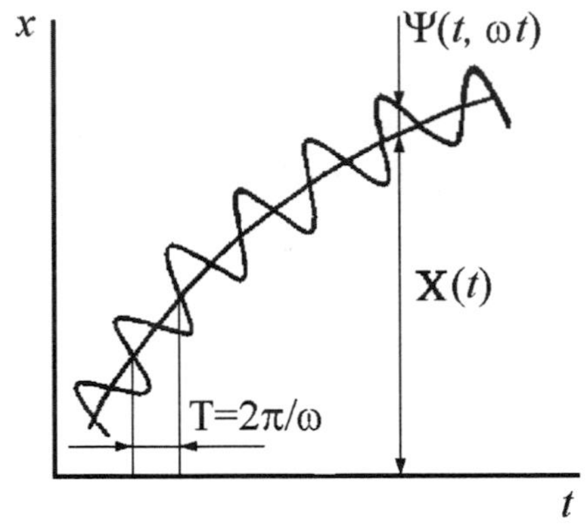

Figure 2.1. The real motion (**x**) and its slow (**X**) and fast (**Ψ**) components.

X *which interests us.* As we will see, this transition is quite possible under certain conditions. As this takes place, we obtain a differential equation for **X**

$$\mathbf{m}\ddot{\mathbf{X}} = \mathbf{F}(\dot{\mathbf{X}}, \mathbf{X}, t) + \mathbf{V}(\dot{\mathbf{X}}, \mathbf{X}, t) \tag{2.4}$$

We call this equation (2.4) the *main equation of vibrational mechanics.* Besides the "usual" slow force **F**, it also contains a certain additional force **V**. We call it a *vibrational force.* By doing that we show our respect for P.L.Kapitza, who introduced this term when solving a problem of the pendulum with a vibrating axis. We will consider this problem below. This situation is shown schematically in Fig. 2.2.

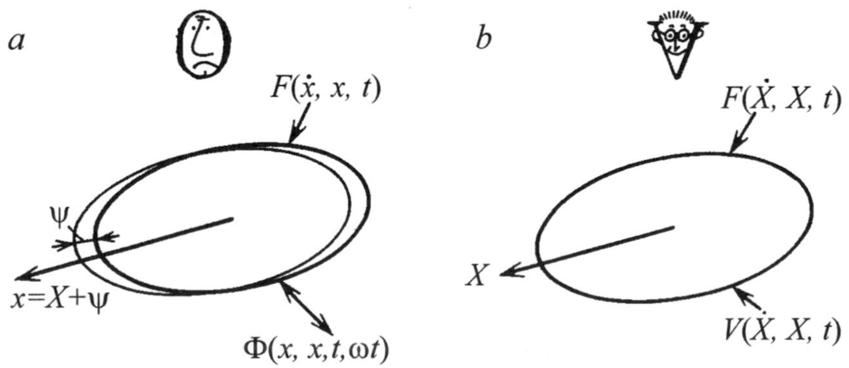

Figure 2.2. To the main idea of vibrational mechanics: a — the force and motion perceived by the ordinary observer **O**; b — the same, perceived by the observer **V**.

Vibrational mechanics can be defined as the mechanics for the observer who doesn't notice any "fast" forces or " fast " motions. We name him the observer **V** *unlike the ordinary observer, named by us the observer* **O**: this observer sees everything. The observer **V** behaves as if he were wearing special glasses which prevent him from seeing any fast motions. We can say that this observer sees the motions of the systems illuminated by a stroboscope whose frequency of flashes are equal to that of vibrations. Vibrational mechanics is analogous to the mechanics of the relative motion. Indeed, in the mechanics of relative motion the addition of the forces of inertia to all the usual forces seems to be a kind of a fine for the use of the non-inertial system of coordinates, i.e. for ignoring the accelerated character of the transportational motion. And in the vibrational mechanics the addition of vibrational forces seems to be a kind of a fine for ignoring the fast motions.

All the effects under consideration can be explained in a simple way by the appearance of vibrational forces. System of equations (2.4) is much more simple than the initial equation (2.1). It does not carry any "excessive" information which the initial equation contains. In particular, the system of equations (2.4) can be of a lower order than (2.1) . The system of equations (2.4) can be smooth even if the initial equation is discontinuous or not smooth. The system of equations (2.4) can be autonomous, where the equation (2.1) is non-autonomous. Finally, equations (2.4) can be potential in spite of the fact that the initial equation is non-potential. So the world seen by the observer **V** is much simpler that that seen by the observer **O**. Therefore the observer **V** looks merry and cheerful while the observer **O** is very sad. All these circumstances will be illustrated below both in the generalized form and by concrete examples.

Vibrational mechanics can be considered within the frames of a much more general conception — the conception of systems with the ignored (hidden) motions (Blekhman, 2000). Figure 2.3 shows the relation between the enumerated above sections of mechanics.

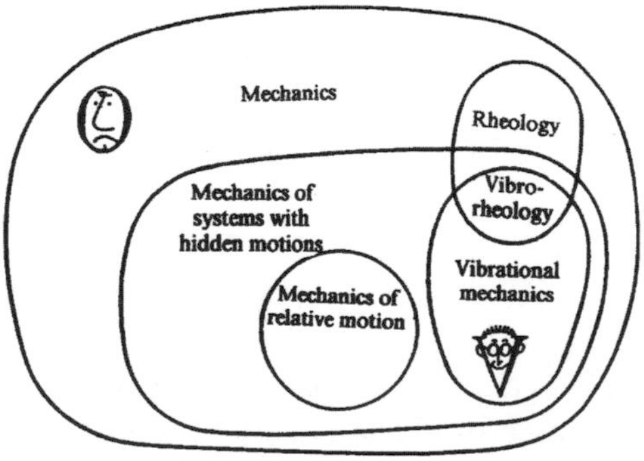

Figure 2.3. The relation between mechanics, mechanics of systems with hidden motion, mechanics of relative motion, rheology, vibrational mechanics and vibrorheology.

2.2 The Initial Equation and its Reduction to a System of Integro-differential Equations

So we will assume that the motion of the system is described by differential equation (2.1). As for the smoothness of the functions $\mathbf{F}$ and $\boldsymbol{\Phi}$, the usual conditions are supposed to be satisfied, which provides the existence of all the solutions of differential equations considered below. The conditions of the existence of the average values, introduced below, are also believed to be satisfied. All the variable and constant values, involved in the equations, might be considered, whenever necessary, dimensionless. So in the inequality $\omega >> 1$ the frequency ω is believed to be dimensionless: $\omega = \omega_*/\omega_0$ where ω_* is the dimensional frequency and ω_0 is a certain frequency characteristic of the system.

Now it is necessary to transfer from differential equation (2.1) to the system of two integro-differential equations for the functions $\mathbf{X}$ and $\boldsymbol{\Psi}$. To do this we will substitute expression (2.2) into equation (2.1):

$$\mathbf{m}(\ddot{\mathbf{X}} + \ddot{\boldsymbol{\Psi}}) = \mathbf{F}(\dot{\mathbf{X}} + \dot{\boldsymbol{\Psi}}, \mathbf{X} + \boldsymbol{\Psi}, t) + \boldsymbol{\Phi}(\dot{\mathbf{X}} + \dot{\boldsymbol{\Psi}}, \mathbf{X} + \boldsymbol{\Psi}, t, \tau) \tag{2.5}$$

Introducing two functions $\mathbf{X}$ and $\boldsymbol{\Psi}$ instead of the one unknown function $\mathbf{x}$, we can to a certain extent, split equation (2.5) into two equations, i.e. in fact write down another additional relation. Thus, we seem to separate the slow motions from the fast ones. For this purpose we will demand that equation (2.5) should be satisfied on the average, i.e. equation

$$\mathbf{m}\ddot{\mathbf{X}} = \left\langle \mathbf{F}(\dot{\mathbf{X}} + \dot{\boldsymbol{\Psi}}, \mathbf{X} + \boldsymbol{\Psi}, t) \right\rangle + \left\langle \Phi(\dot{\mathbf{X}} + \dot{\boldsymbol{\Psi}}, \mathbf{X} + \boldsymbol{\Psi}, t, \tau) \right\rangle \tag{2.6}$$

should be satisfied. Then, however, for the initial equation (2.5) to be satisfied, the equation

$$\begin{aligned} \mathbf{m}\ddot{\boldsymbol{\Psi}} = {} & \mathbf{F}(\dot{\mathbf{X}} + \dot{\boldsymbol{\Psi}}, \mathbf{X} + \boldsymbol{\Psi}, t) - \left\langle \mathbf{F}(\dot{\mathbf{X}} + \dot{\boldsymbol{\Psi}}, \mathbf{X} + \boldsymbol{\Psi}, t) \right\rangle + \\ & \boldsymbol{\Phi}(\dot{\mathbf{X}} + \dot{\boldsymbol{\Psi}}, \mathbf{X} + \boldsymbol{\Psi}, t, \tau) - \left\langle \boldsymbol{\Phi}(\dot{\mathbf{X}} + \dot{\boldsymbol{\Psi}}, \mathbf{X} + \boldsymbol{\Psi}, t, \tau) \right\rangle \end{aligned} \tag{2.7}$$

must also be satisfied. When averaging equations (2.5) we have used the obvious relations

$$\left\langle \ddot{\mathbf{X}} \right\rangle = \ddot{\mathbf{X}}, \quad \left\langle \ddot{\boldsymbol{\Psi}} \right\rangle = 0, \tag{2.8}$$

which are valid as well as equalities

$$\left\langle \dot{\mathbf{X}} \right\rangle = \dot{\mathbf{X}}, \quad \left\langle \dot{\boldsymbol{\Psi}} \right\rangle = 0 \tag{2.9}$$

Combining equations (2.6) and (2.7) we, as it must, obtain equation (2.5).

Therefore the system of integro-differential equations (2.6), (2.7) that has been obtained, is equivalent to the initial equation (2.1), at least in the sense that if there is a certain solution $\mathbf{X}$, $\boldsymbol{\Psi}$ of this system, the function $\mathbf{x} = \mathbf{X} + \boldsymbol{\Psi}$ will be the solution of equation (2.1). In other words, for the existence of the solution of equation (2.1) of type (2.2) it is sufficient that there should be the corresponding solutions $\mathbf{X}$, $\boldsymbol{\Psi}$ of the system (2.6), (2.7).

Introducing for briefness sake the designations

$$\mathbf{F}_{\mathrm{X}} = \mathbf{F}(\dot{\mathbf{X}}, \mathbf{X}, t), \quad \mathbf{F}_{\mathrm{X}+\boldsymbol{\Psi}} = \mathbf{F}(\dot{\mathbf{X}}+\dot{\boldsymbol{\Psi}}, \mathbf{X}+\boldsymbol{\Psi}, t), \quad \Phi_{\mathrm{X}+\boldsymbol{\Psi}} = \Phi(\dot{\mathbf{X}}+\dot{\boldsymbol{\Psi}}, \mathbf{X}+\boldsymbol{\Psi}, t, \tau) \quad (2.10)$$

we will write down equations (2.6) and (2.7) in the form

$$\mathbf{m}\ddot{\mathbf{X}} = \mathbf{F}_{\mathrm{X}} + \langle \mathbf{F}_{\mathrm{X}+\boldsymbol{\Psi}} - \mathbf{F}_{\mathrm{X}} \rangle + \langle \Phi_{\mathrm{X}+\boldsymbol{\Psi}} \rangle \quad (2.11)$$

$$\mathbf{m}\ddot{\boldsymbol{\Psi}} = \mathbf{F}_{X+\boldsymbol{\Psi}} - \langle \mathbf{F}_{\mathrm{X}+\boldsymbol{\Psi}} \rangle + \Phi_{\mathrm{X}+\Psi} - \langle \Phi_{\mathrm{X}+\boldsymbol{\Psi}} \rangle \quad (2.12)$$

It has been taken into consideration here that

$$\mathbf{F}_{\mathrm{X}} = \langle \mathbf{F}_{\mathrm{X}} \rangle \quad (2.13)$$

since $\mathbf{F}_{\mathrm{X}} = \mathbf{F}(\dot{\mathbf{X}}, \mathbf{X}, t)$ does not depend on τ.

Equation (2.11) can be named (for the time being conditionally, since we have not yet used anywhere the suppositions about the value ω and about the rate of change of the components $\mathbf{X}$ and $\boldsymbol{\Psi}$) the *equation of slow motions*, and equation (2.12) the *equation of fast motions*.

2.3 The Case when a Separate Equation for the Slow Component is Obtained

As was mentioned, solutions of the system (2.11), (2.12), in which $\boldsymbol{\Psi}$ are periodical or almost periodical with respect to $\tau = \omega t$, are of a considerable interest. Let us assume that we have managed to find such solutions

$$\boldsymbol{\Psi} = \boldsymbol{\Psi}^*(\dot{\mathbf{X}}, \mathbf{X}, t, \tau) \quad (2.14)$$

which are isolated at the given $\mathbf{X}$ and $\dot{\mathbf{X}}$. Let them also be asymptotically stable with respect to the generalized coordinates $\bar{\boldsymbol{\Psi}}$ (i.e. to the coordinates which do not contain the components of $\mathbf{X}$), and also with respect to the corresponding generalized velocities $\dot{\bar{\boldsymbol{\Psi}}}$ all over the region of the change of other generalized coordinates and velocities which is of interest to us. Then for every such solution $\boldsymbol{\Psi} = \boldsymbol{\Psi}^*$ a certain additional force can be obtained:

$$\mathbf{V}(\dot{\mathbf{X}}, \mathbf{X}, t) = \langle \mathbf{F}_{\mathrm{x}+\boldsymbol{\Psi}} - \mathbf{F}_{\mathrm{x}} \rangle + \langle \Phi_{\mathrm{x}+\boldsymbol{\Psi}} \rangle, \quad (2.15)$$

and equation (2.12) can be written as

$$\mathbf{m}\ddot{\mathbf{X}} = \mathbf{F}(\dot{\mathbf{X}}, \mathbf{X}, t) + \mathbf{V}(\dot{\mathbf{X}}, \mathbf{X}, t) \quad (2.16)$$

The fulfillment of this condition of asymptotic stability provides, under rather general assumptions, the definiteness of the expression for the additional force within a certain range of changing the initial conditions for the coordinates $\bar{\boldsymbol{\Psi}}$ at sufficiently large values of t. It also provides a mutual conformity of properties of the stability of motions of the initial system (2.1) with respect to $\dot{\mathbf{x}}, \mathbf{x}$ and of system (2.16) with respect to $\dot{\mathbf{X}}, \mathbf{X}$. And namely, if the transformation of the variables $\dot{\mathbf{x}}, \mathbf{x}$ to the variables $\dot{\mathbf{X}}, \mathbf{X}, \dot{\bar{\mathbf{\Psi}}}, \bar{\mathbf{\Psi}}$, defined by the relations

$$\begin{aligned} x_s &= X_s(t) + \psi_s(\dot{\mathbf{X}}, \mathbf{X}, t, \omega t) \quad (s = 1, ..., k); \\ x_s &= \bar{\psi}_s(\dot{\mathbf{X}}, \mathbf{X}, t, \omega t) \quad (s = k+1, ..., k+r = n), \end{aligned} \tag{2.17}$$

and also if the inverse transformation possess the properties of uniqueness and continuity in the vicinity of the non-perturbed motions under consideration, then the stable (asymptotically stable) with respect to $\dot{\mathbf{X}}$ and $\mathbf{X}$ solutions of (2.16) correspond to the stable (asymptotically stable) solutions $\mathbf{x} = \mathbf{X} + \mathbf{\Psi}$ of the initial system (2.1). And vice versa, the stable(asymptotically stable) solutions $\mathbf{x}$ of the initial system (2.1) correspond to the stable (asymptotically stable) solutions $\mathbf{X}$ of the system (2.16).

This statement can also be formulated in the following way: if we "do not want to notice either the component of the motion $\mathbf{\Psi}$ or the force Φ, which depend on the argument $\tau = \omega t$, then at the assumed version of "splitting the initial equation (2.1), we must first find from equation (2.6) the component $\mathbf{\Psi}$, possessing the above-mentioned properties, and, secondly, we must determine the component $\mathbf{X}$ from (2.16) in which the force Φ is not given in its direct form and a certain additional force $\mathbf{V}$ is added to the force $\mathbf{F}$.

It should be noted that the present section is directly connected with a more general conception—that of the mechanics of the systems with the hidden motions (Blekhman, 2000).

So far, we have not yet made use of the assumption about the value of the parameter ω and, accordingly, about the rate of change of the forces $\mathbf{F}$ and Φ as well as about the components $\mathbf{X}$ and $\mathbf{\Psi}$. At the same time without that assumption the relations (2.6)—(2.16) have a formal rather than a constructive meaning, since the solution of system (2.11), (2.12) is, in the general case, not simpler than that of the initial equation (2.1).

2.4 The Main Assumption of Vibrational Mechanics and Conditions of its Fulfillment

The main assumption of vibrational mechanics implies that the initial equation (2.1) may have solutions of type (2.2) or that system (2.11), (2.12) may have adequate solutions $\mathbf{X}$, $\mathbf{\Psi}$ with the component $\mathbf{X}$ in those solutions being "actually slow" in comparison with $\mathbf{\Psi}$.

The formal mathematical treatment of the question of under what conditions this supposition can be fulfilled can be found in the books (Blekhman, 2000; ed. Blekhman, 2004). These conditions in particular place certain constraints on the relative orders of the right sides of equations (2.11) and (2.12) depending on the relative scales of the values $\mathbf{X}$ and $\mathbf{\Psi}$. This investigation is mainly of a theoretical interest and therefore is not stated in the present brief course. When the method of direct separation of motions is used in actual practice, the main supposition is usually accepted a priori and the result of solving problems is analyzed a posteriori particularly by comparing it to the results of the numerical solution or of the experiment.

It is natural that the main condition of the validity of the main supposition is the requirement that the dimentionless frequency ω should be large enough, practically more than 3 (it is assumed that the dimentional frequency of disturbation ω_* is referred to the characteristic frequency of the slow motions of the system). If the system (either discrete or continuous), describing the slow motions, is close to the linear system with the eigenfrequencies $\lambda_1, \lambda_2, ...$, then it is natural to expect equation (2.16) to describe sufficiently well the motion with the frequencies $\lambda_i << \omega_*$ and to be not valid for the motions with higher frequencies. The consideration of a number of concrete problems shows that it is usually sufficient that the condition $3\lambda_i < \omega_*$ should be valid.

2.5 The Main Equation of Vibrational Mechanics. Vibrational Forces, the Observer O and the Observer V

Let the main assumption of vibrational mechanics and the conditions of i. 2.3 be satisfied. Then we will call equation (2.16), containing only the slow component of motion, the *main equation of vibrational mechanics*, or the *equation of slow motions*, and the expression $\mathbf{V}(\dot{\mathbf{X}}, \mathbf{X}, t)$ for the additional force will be called by us the *vibrational force.*

Equation (2.12) will then be called *equation of fast motions.* As has been mentioned in i. 2.1, vibrational mechanics can be considered to be a mechanics in which the observer does not notice any fast forces or fast motions, while he perceives only the slow forces and slow motions. Such an observer (the *observer* **V**) can be contrasted to the "ordinary *observer* **O** who perceives both the slow and fast forces and motions. It is convenient to use these images, taking, as may be required, the positions of either the first or the second observer. In accordance with what has been said, the observer **V**, so as not to come in conflict with the laws of mechanics, must take into consideration not only the "ordinary" slow forces **F**, but also the additional slow forces i.e. the vibrational forces **V**.

Note that according to equations (2.15), the vibrational force is obtained by means of averaging the "eigenfast" force $\mathbf{\Phi}$ and the fast contribution $\tilde{\mathbf{F}} = \mathbf{F}_{X+\Psi} - \langle \mathbf{F}_{X+\Psi} \rangle$ from the slow force. In accordance with it, we will distinguish the *eigenvibrational force*

$$\mathbf{V}^s = \langle \mathbf{\Phi} \rangle \tag{2.18}$$

and the *induced vibrational force*

$$\mathbf{V}^i = \left\langle \tilde{\mathbf{F}} \right\rangle \tag{2.19}$$

Mark that the induced vibrational force may be different from zero even in the absence of a fast external perturbation due to the fast self-excited oscillations which may appear in systems with slow forces.

One can see (and it will be demonstrated by numerous examples) that equation (2.16), at least in its adequate approximate version, is much simpler than the initial equation (2.1); the slow component **X**, whose change this equation describes, is of utmost interest for the researchers.

2.6 Method of an Approximate Derivation of the Expression of Vibrational Forces and of Composing the Main Equation of Vibrational Mechanics

As before, we will consider valid both the main assumption of vibrational mechanics and the conditions of i. 2.3. Then it seems natural to use the following method of an approximate finding of the vibrational force $\mathbf{V}$ and of composing the main equation (2.16). We will describe it here at an heuristic level. We begin by solving the equation of fast motions (2.12). Since the change of the values $\dot{\mathbf{X}}$, $\mathbf{X}$ and t for the typical period of the fast motion $2\pi/\omega$ is relatively small (see Fig. 2.1), in solving Eq. (2.12) those values are regarded as "frozen", i.e. as fixed parameters.

Let Eq. (2.12), with $\dot{\mathbf{X}}$, $\mathbf{X}$ and t being frozen, admit either one or several almost-periodical (particularly 2π-periodical) with respect to $\tau = \omega t$ solutions $\mathbf{\Psi} = \mathbf{\Psi}^*(\dot{\mathbf{X}}, \mathbf{X}, t, \tau)$, satisfying condition (2.3) and asymptotically stable with respect to all the fast generalized coordinates and velocities, while all the other fast variables are changing and $\dot{\mathbf{X}}$, $\mathbf{X}$ and t are frozen all over the region under consideration. This assumption is usually checked up very easily and it is really valid under the conditions of i. 2.3. Besides, Eq. (2.12) is such that the necessary condition (2.3) of the existence of almost-periodical (particularly 2π-periodical with respect to τ) solutions is automatically fulfilled for it. To make sure of it, it is enough to average Eq.(2.12) with respect to $\tau = \omega t$ (with $\dot{\mathbf{X}}$, $\mathbf{X}$ and t being frozen). What has been said holds the key to the way adopted by us of "splitting" the initial equation (2.1) into two equations (2.11) and (2.12).

Substituting the definite solution $\mathbf{\Psi} = \mathbf{\Psi}^*(\dot{\mathbf{X}}, \mathbf{X}, t, \tau)$ into expression (2.15), we will find an approximate expression for the vibrational force $\mathbf{V}(\dot{\mathbf{X}}, \mathbf{X}, t)$ after which the main equation (2.16) can be composed. This equation will, of course, also be approximate and must be valid on the one hand with $t > t_0$ where t_0 is the time of achieving a steady state of the fast motions, and on the other hand — with $t < T_0$, where T_0 is the boundary of the interval of validity of the asymptotic approximation (it is naturally assumed that $T_0 >> t_0$).

The mathematical foundation of this approximate method for equations of different types is given in the books (Blekhman, 2000; ed. Blekhman, 2004).

2.7 On the Other Simplifications of Solving Equations for the Fast Component of Motion. Purely Inertial Approximation

Apart from the above-mentioned method, there are other expedient methods of an approximate solution of the equation for the fast component $\mathbf{\Psi}$. From general positions, the rightfulness of such approximate methods follows from the fact that the information which is contained in the initial equation (2.1) and also in the system (2.11), (2.12) is excessive if we are interested only in the change of the slow component $\mathbf{X}$. It is natural that then we can restrict ourselves to an approximate finding of the fast component $\mathbf{\Psi}$. To be concrete, the possibility of an approximate finding of $\mathbf{\Psi}$ follows from the fact that this function appears in equation (2.15) for the vibrational force only under the integration sign.

In particular, when solving equations of fast motions, it is possible to neglect some other, not very significant terms. One of such approximations is of a special practical significance. It may be called a *purely inertial approximation* and it is based on the assumption that in this equation the "oscillating component" of the fast force is much greater than the that of the slow force, i.e.

$$|\,\Phi_{X+\Psi} - \langle\Phi_{X+\Psi}\rangle| >> |\,\mathbf{F}_{X+\Psi} - \langle\mathbf{F}_{X+\Psi}\rangle| \tag{2.20}$$

and equation (2.12) can be roughly written as

$$\mathbf{m}\ddot{\boldsymbol{\Psi}} = \Phi_{X+\Psi} - \langle\Phi_{X+\Psi}\rangle \tag{2.21}$$

From the examples, given below, we will see that this approximate solution of the equation of fast motions leads to quite satisfactory results when calculating the vibrational force and deriving equations of slow motions.

We must emphasize that here we mean an approximate calculation of $\boldsymbol{\Psi}$ when $\dot{\mathbf{X}}$, $\mathbf{X}$ and t are "frozen", i.e. within the frames of the method given in i. 2.6.

2.8 Summary: on the Procedure of the Practical Use of the Method

The use of any method, including even the strictly mathematical method or result, when solving an applied problem, comprises, as a rule, a number of stages which do not belong to those mathematically strict, but can be referred to the methods realized at the so-called rational level of rigour. This circumstance and some other specific features of applied mathematical research are discussed in detail in the book (Blekhman et al., 1990). This refers, of course, to the method of direct separation of motions as well. The procedure of using that method, at least in the form stated above, comprises a number of rational elements, including the heuristic ones. We will describe this procedure schematically. We will emphasize that it allows various modifications and improvements, in future, however, we will try to keep to it when solving concrete problems.

1. A preliminary conclusion based on the experimental data and/or on the heuristic grounds, deals with the question whether the motion of the system under consideration presents the imposing of fast oscillations upon the slow motion. In other words, a prognosis is made about the fulfillment of the main assumption of vibrational mechanics.

2. A system of equations (2.11), (2.12) is composed, based on the initial equations of motion. It should be noted that the right side of the equation of slow motions (2.11) presents an averaging on the fast time $\tau = \omega t$ of the right side of the initial equation (2.1), while the right-hand side of the equations of fast motions (2.12) is the right side of the initial equation from which the result of averaging the right side has been extracted.

3. Almost-periodic (in particular 2π-periodic) with respect to $\tau = \omega t$ solutions $\boldsymbol{\Psi} = \boldsymbol{\Psi}^*(\dot{\mathbf{X}}, \mathbf{X}, t, \omega t)$ of the equation of fast motions at the frozen (fixed) $\dot{\mathbf{X}}, \mathbf{X}$ and t are sought as a rule approximately. Among them such solutions are selected which are asymptotically stable with respect to all the fast generalized coordinates and velocities all over the region of values of the variables which interest us. These solutions are used while finding approximate expressions for the vibrational force by expression (2.15), after which the corresponding main equations of vibrational mechanics (2.16) are composed.

4. Solutions of equations (2.16) which are of interest are being studied.

5. In questionable cases the initial suppositions and the results of the investigation are checked a posteriori.

2.9 Brief Historical Reference

The impetus to the development of vibrational mechanics was given by the work of P.L.Kapitsa "Dynamic stability of the pendulum when the point of suspension is oscillating [Journ. of Exp. and Theor. Phys., **21**, 5, 1951 (in Russian)]. I learned about that work when I was just beginning my scientific researches. I was delighted by the fact how simple and physically transparent — by means of direct separation of slow and fast motions — a bright result was achieved: was established the stability of the upper position of the pendulum and was found the corresponding condition of the stability.

In those years many outstanding scientists, mostly mathematicians, were of the opinion that this result was obtained in a non-strict way and does not have a sufficiently general significance. Therefore in my theses, devoted to the theory of self-synchronization of rotating bodies and to the theory of vibrational displacement, I preferred to use the Poincare-Lyapunov methods and, wherever possible, the exact methods.

How surprised I was when it turned out that the results, found by means of a rather complicated investigation can be easily obtained by means of the method (though not grounded mathematically!), generalizing the method of P.L.Kapitsa for more complicated systems. It became obvious that the success of that method was not accidental. And I made it my task to extend the method of Kapitsa to the sufficiently general nonlinear systems and to prove it mathematically, as far as possible, in the lines of asymptotic methods of nonlinear mechanics. This brought me, first, to the solution of a number of problems, important for applications, about the action of vibration on nonlinear systems – problems in the range from the theory of mechanisms and machines to celestial mechanics. Secondly, the apparatus of the method was formalized mathematically. And, eventually, thirdly, conceptions were formulated which were called "vibrational mechanics" and "vibrorheology". They were a special case of a more general conception—of the mechanics of the systems with the hidden (ignored) motions.

It is the enumerated results that made the subject of the book (Blekhman, 2000). The method of direct separation of motions as well as the conceptions of vibrational mechanics and the mechanics of systems with hidden motions are comparatively universal (within the frames, of course, of certain natural suppositions). That, however, cannot be said about the methods of an approximate solution of equations of fast motions, used by P.L.Kapitsa and his followers. The establishing of the region of the applicability of those methods (when in the equations of fast motions the slow variables are considered to be "frozen", and also when they are limited by the so-called purely inertial approximation) remains the task which has not yet been completely solved. Elaborating efficient means of solving problems beyond the region of the above mentioned approximate methods is also an important problem.

In recent years considerable progress has been made in solving the above problems. Besides, a number of interesting applied problems have been solved. The main of those investigations, which belong to the scientists of Germany, Denmark, Russia, and Ukraine are stated in the book (ed. Blekhman, 2004). The books (Blekhman, 2000; ed. Blekhman, 2004; Thomsen, 2003) reflect also the contribution of other authors into the development of vibrational mechanics.

2.10 Conclusion to Section 2

Section 2 contains the approach of vibrational mechanics and the method of solving problems on the effect of vibration on nonlinear mechanical systems. We have specially omitted a number of elements of mathematical character which the reader may find in the cited literature.

The main result of the above approach consists in the transfer from a rather complicated system of differential equations to a much simpler system, describing the slow motions alone. It is these motions that are of the utmost interest.

The advantages of this approach will be seen below when we consider some concrete technical problems. They will be discussed in detail in the final lecture of the course.

3 The Motion of a Particle in a Field of a Fast Oscillatory Standing Wave. On the Potential of the Average Dynamic Systems

3.1 The Case of a Solid Particle

The problem on the behavior of a particle in the field of a fast oscillating standing wave is of both applied and theoretical interest. The equation of motion of such particle is

$$m\ddot{x} = -h\dot{x} + \Psi(x)\sin\omega t \tag{3.1}$$

Here x is the coordinate of the particle, m is its mass, h is the viscous resistance coefficient, $|\Psi(x)|$ is the amplitude of the force with the frequency ω, dependent on the coordinate x. Thus, the fast oscillating field is of the character of a standing wave (Fig. 3.1).

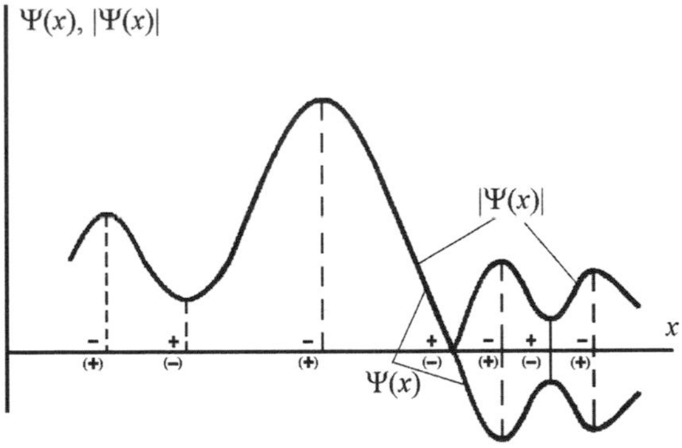

Figure 3.1. The particle in a field of a fast-oscillating standing wave.

The points x where $\Psi(x) = 0$, will, as usual, be called *nodes*, and the points x where $\Psi(x)$ has maximums or minimums will be called the *anti-nodes* of the wave; at such points we have $\Psi'(x) = 0$.

Assuming that the solution of equation (3.1) looks as

$$x = X(t) + \psi(t, \omega t) \tag{3.2}$$

where X is a slow component, and ψ is a fast component 2π-periodical over ωt with a zero average over $\tau = \omega t$, we will derive equations (2.11), (2.12):

$$m\ddot{X} = -h\dot{X} + \langle \Psi(X+\psi) \sin \omega t \rangle \tag{3.3}$$

$$m\ddot{\psi} = -h\dot{\psi} + \Psi(X+\psi) \sin \omega t - \langle \Psi(X+\psi) \sin \omega t \rangle \tag{3.4}$$

We will assume that ψ is small as compared to X and will limit ourselves to the linear components alone in the decomposition $\Psi(x)$ according to the degrees of Ψ:

$$\Psi(X+\psi) \approx \Psi(X) + \psi \Psi'(X) \tag{3.5}$$

Then in view of the equalities

$$\begin{aligned} \langle \Psi(X) \sin \omega t \rangle &= \Psi(X) \langle \sin \omega t \rangle = 0, \\ \langle \Psi'(X) \psi \sin \omega t \rangle &= \Psi'(X) \langle \psi \sin \omega t \rangle \end{aligned} \tag{3.6}$$

equations (3.3), (3.4) will take the form

$$m\ddot{X} = -h\dot{X} + \Psi'(X) \langle \psi \sin \omega t \rangle \tag{3.7}$$

$$m\ddot{\psi} = \Psi(X) \sin \omega t - h\dot{\psi} + \Psi'(X) \psi \sin \omega t - \langle \Psi'(X) \psi \sin \omega t \rangle \tag{3.8}$$

Now taking into consideration that we assume that Ψ is small, and that the equation of fast motions can be solved approximately, we will neglect in the right-hand side of equation (3.8) all the terms except the first. Then this equation will take the following simple form

$$m\ddot{\psi} = \Psi(X) \sin \omega t \tag{3.9}$$

which corresponds to the so called purely inertial approximation (see i. 2.6). Besides, when solving this equation, the slow variable X can be regarded as constant (frozen). Then the periodic solution of the equation will be

$$\psi = -\frac{\Psi(X)}{m\omega^2} \sin \omega t \tag{3.10}$$

Substituting this expression into equation (3.7), we will perform averaging in view of the equality

$$\langle \sin^2 \omega t \rangle = \frac{1}{2} \tag{3.11}$$

Then equation (3.7) (the main equation of vibrational mechanics for the problem under consideration) will have the form

$$m\ddot{X} + h\dot{X} = V(X) \tag{3.12}$$

where $V(X)$ is the vibrational force, defined by the expression

$$V(X) = -\frac{1}{2m\omega^2}\Psi(X)\Psi'(X) = -\frac{1}{4m\omega^2}\left[\Psi^2(X)\right]' = -\frac{d\Pi_V}{dX} \tag{3.13}$$

Here Π_V designates the function

$$\Pi_V = \frac{1}{4m\omega^2}\Psi^2(X), \tag{3.14}$$

which may be named the *potential energy of the vibrational forces.* As a result, equation (3.12) can be also presented as

$$m\ddot{X} + h\dot{X} = -\frac{d\Pi_V}{dX} \tag{3.15}$$

The relations obtained are quite remarkable and deserve a discussion.

First of all, it should be noted that in the absence of the wave $[\Psi(X) = 0, V(X) = 0]$ every point of the axis x is a position of a stable equilibrium of the particle. This follows both from the initial equation (3.1) and from equation (3.12). The situation will change abruptly in the presence of a standing wave. According to the classical theorem of Thomsen-Tait-Chetayev, the points $X = X_0$ of the strict minimum of the function Π_V corresponds to the asymptotically stable equilibrium position of the particle, while the points of maximum correspond the positions of the unstable equilibrium. That means that the points of the minimum of the amplitude of the wave $|\Psi(x)|$ (including the nodes of the function $\Psi(x)$) correspond to the positions of the stable quasi-equilibrium, while the points of the maximum correspond to the positions of the unstable quasi-equilibrium of the particle. In other words, in the force field of type of the standing wave particles are "attracted" to the points of the minimum of the amplitude of the field $|\Psi(x)|$ and are repulsed from the points of the maximum $|\Psi(x)|$.

Summing up, we may also say that for the slow motions of the particle, the presence of the standing wave leads to the appearance of the "potential pits", corresponding to its stable states. These pits are located at the points of the minimum of the amplitude of the wave $|\Psi(x)|$ including the nodes of the wave where $\Psi(x) = 0$.

In Fig. 3.1 the signs "+" and "-" mark the stable and unstable positions of the particle respectively.

The result obtained is remarkable also by the fact that equation (3.12) answers the potential system in case there is dissipation energy, while the initial system, described by equation (3.1), is essentially non-conservative. Such systems were called by us *potential on the average dynamic systems.* The property of potentiality on the average was discovered for several important classes of systems, in particular, for the systems of objects which are being synchronized (see Section 4), and also for the systems with kinematic and dynamic excitation of oscillations (Blekhman, 2000).

The presence of potentiality on the average makes it much easier to solve respective problems and is therefore one of the greatest advantages of the approach of vibrational mechanics.

It is of interest to consider the case when apart from the field of a standing wave, the particle is in the stationary force field $F(x)$, say, in the gravitational field. In that case the motion of the particle will be described not by equation (3.1) but by

$$m\ddot{x} = -h\dot{x} + F(x) + \Psi(x)\sin\omega t \tag{3.16}$$

With the same assumptions as before, for the slow component X instead of (3.15) we will obtain the equation

$$m\ddot{X} + h\dot{X} = -\frac{d(\Pi_F + \Pi_V)}{dX} \tag{3.17}$$

where Π_F denotes the potential energy, corresponding to the force F:

$$F(X) = -\frac{d\Pi_F}{dX} \tag{3.18}$$

If the force F is much smaller than the vibrational force V, then the presence of the force F will cause but slight shifts of the particle from the positions of a stable equilibrium, found before. However, another circumstance is important: the equilibrium positions, unstable without the force $\Psi(x)\sin\omega t$, may become stable in its presence and vice versa. This is clear if we consider the right side of equation (3.17). Indeed, the points of the maximums of the function Π_F may become the points of the minimums of the function $D = \Pi_F + \Pi_V$ which plays the role of the general potential energy of the system in the presence of a fast oscillating field (Fig. 3.2). It is this effect that takes place in the classical problem when under the effect of vibration of the axis of the pendulum its upper position becomes stable (see i. 4.1; a review of this problem is given, in particular, in Blekhman, 2000).

This effect is also connected with the effect of levitation and stability of the charged particle in the oscillating field. In 1989 the discovery of this effect (the so called Paul trap) was awarded the Nobel prize.

3.2 The Case of the Deformable Particle

It is of interest that in case of a deformable particle in a certain range of frequencies the stable positions of the particle in Fig.3.1 can change their places: the stable positions correspond to the points of the maximum of the amplitude $|\Psi(x)|$, while the unstable positions correspond to the points of its minimum (Blekhman, 2004; Kremer, 2004). The corresponding points are marked by the signs "+" and "-".

Finally it should be noted that if the wave is not a standing one, but a slowly running, say, in the direction of the axis x, then the particle which is at some stable point of the wave is driven by it in the direction of its motion (in this case the function Ψ in equation (3.1) depends both on x, and on the slow time t.

3.3 Conclusion

In lecture 3 we have considered an example of using the approach of vibrational mechanics for the solution of an important applied problem on the behavior of the particle in a fast oscillatory field of a standing wave. In doing that we paid attention to a remarkable fact: the slow motion of an essentially non-conservative initial system is described by the equation, corresponding to a certain conservative system.

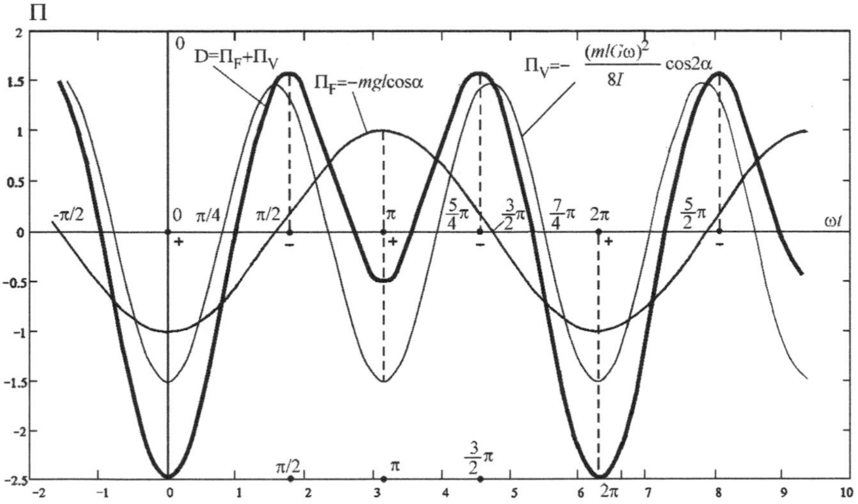

Figure 3.2. Potential energy $\Pi_{\mathrm{F}}, \Pi_{\mathrm{V}}$ and $\Pi = \mathrm{D} = \Pi_{\mathrm{F}} + \Pi_{\mathrm{V}}$.

4 The Action of Vibration on Mechanisms. Self-synchronization of the Rotating Bodies, Applications of Self-synchronization to Technology

4.1 Pendulum with a Vibrating Axis of Suspension (the Stephenson-Kapitsa Pendulum)

This problem is classical in the theory of nonlinear oscillations.

The main conclusion from its consideration is that the upper ("overturned") position of the pendulum which is unstable without vibration, becomes stable in certain conditions under vibration. This effect must have been first observed by Stephenson as far back as 1908, but his publications were practically forgotten and different authors considered it in different ways. Best known among them are the publications by P.L.Kapitsa, mentioned in i. 2.9 which gave impetus to the development of vibrational mechanics (see i. 2.9 and Blekhman and Sperling, 2004).

Here we will show that this problem can be regarded as a special case of the problem on the behavior of the particle in a field of a standing wave, discussed in Section 3.

Indeed, the differential equation of the motion of a pendulum with a vertically vibrating axis of suspension (Fig. 4.1) has the form

$$I\ddot{\varphi} + h\dot{\varphi} + ml(g + A\omega^2 \sin \omega t) \sin \varphi = 0 \tag{4.1}$$

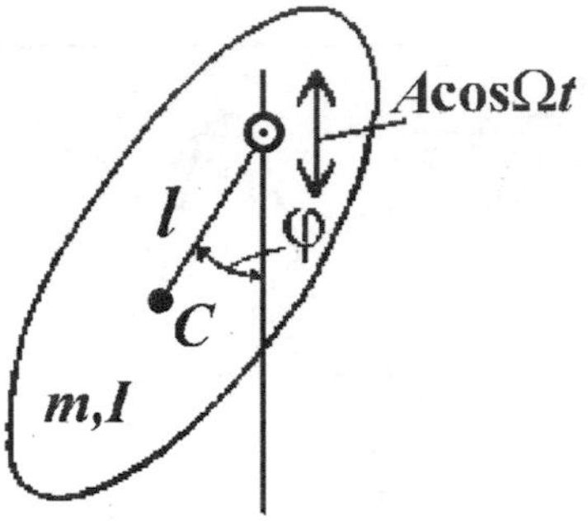

Figure 4.1. Pendulum with a vibrating axis of suspension.

Comparing this equation with (3.16), we come to the conclusion that in this case

$$F(\varphi) = -mgl\sin\varphi, \quad \Psi(\varphi) = -mA\omega^2 l\sin\varphi \tag{4.2}$$

and the role of the unknown function is played by the angle φ.

The expressions for the potential energies Π_F and Π_V will be in this case

$$\Pi_F = -mgl\cos\alpha, \quad \Pi_V = -\frac{(mlA\omega)^2}{8I}\cos 2\alpha \tag{4.3}$$

Here the angle α plays the role of the slow component X. The total potential energy

$$\Pi = D = \Pi_F + \Pi_V = -\left[mgl\cos\alpha + \frac{(mlA\omega)^2}{8I}\cos 2\alpha\right] \tag{4.4}$$

has a minimum at the point $\alpha = \pi$ (the upper position of the pendulum) if

$$\left.\frac{d^2D}{d\alpha^2}\right|_{\alpha=\pi} = -mgl + \frac{(mlA\omega)^2}{2I} > 0 \tag{4.5}$$

or

$$A\omega > \sqrt{2gl_0} \tag{4.6}$$

where $l_0 = I/ml$ denotes the given length of the physical pendulum. Inequality (4.6) is the classic condition of the stability of the upper position of the pendulum in case there is vibration. It should be noted, as it must, that condition (4.6) can be also obtained on the basis of investigating the equilibrium position for the differential equation of slow motions

$$I\ddot{\alpha} + h\dot{\alpha} + mgl\sin\alpha + \frac{(mlA\omega)^2}{4I}\sin 2\alpha = 0 \tag{4.7}$$

It is obtained from (3.17), using expression (4.4) where X is changed for α.

The result obtained is graphically explained in Fig. 3.2.

More general cases, when the axis of the pendulum vibrates in two mutually perpendicular directions, are discussed in books (Blekhman, 2000; ed. Blekhman, 2004), book (ed. Blekhman, 2004) studying the case of arbitrary periodic oscillations. Work (Blekhman and Sperling, 2004) discusses the behavior of the pendulum with a vibrating axis of suspension when it has an "inner degree of freedom". Similarly to the problem on a deformable particle (i. 3.2), in this case within a certain frequency range ω the stable positions of quasi-equilibrium become unstable and vice versa. The condition of stability (4.6) is also considerably softened: the stability of the upper position of the pendulum can be achieved at much smaller values of the amplitude of vibration velocity $A\omega$.

4.2 Vibrational Maintenance of Rotation

The vibration of the axis of an unbalanced rotor with a certain frequency ω can steadily maintain its stationary rotation with a mean angular velocity $<\dot{\varphi}>= \pm(p/q)\omega$ where p and q are positive integers. This wonderful phenomenon has already been discussed in the first section (see also Figs. 1.2, 1.4 and 1.5).

It's natural that the equation of motion of the rotor coincides with that of the corresponding pendulum. So, in the case of vertical harmonic oscillations of the axis of the rotor this equation will look as (4.1).

The solution of that equation will in this case be sought as

$$\varphi = \omega t + \alpha(t) + \psi(t, \omega t) \tag{4.8}$$

where ω is the mean angular velocity of rotation, coinciding with the oscillation frequency of the axis, $\alpha(t)$ is the slow component and ψ is the fast one with the zero mean value. The rotation of the rotor ωt is also fast.

We will transfer from equation (4.1) to the equations of slow and fast motions (those of the type (2.11) and (2.12)):

$$\begin{aligned} &I\ddot{\alpha} + h\dot{\alpha} + h\omega + mgl\langle\sin(\omega t + \alpha + \psi)\rangle + \\ &mlA\omega^2\langle\sin\omega t \sin(\omega t + \alpha + \psi)\rangle = 0 \end{aligned} \tag{4.9}$$

$$I\ddot{\psi} + h\dot{\psi} = \mu\Psi(\alpha, \psi, \omega t) \tag{4.10}$$

Here $\mu\Psi$ designates the right-hand side of the equation of fast motions which will be considered to be small (as was mentioned more than once, this equation may be solved approximately). Then as the first approximation we obtain

$$\psi = 0 \tag{4.11}$$

In fact, according to (4.8), we believe that in the first approximation the rotor rotates uniformly with a constant phase shift α.

Substituting the value $\psi = 0$ into equation (4.9), we arrive at the following final equation of slow motions (the main equation of vibrational mechanics)

$$I\ddot{\alpha} + h\dot{\alpha} + h\omega + \frac{1}{2}mlA\omega^2\cos\alpha = 0 \tag{4.12}$$

When deriving that equation we take into consideration the equalities

$$\langle \sin(\omega t + \alpha) \rangle = 0, \qquad \langle \sin \omega t \sin(\omega t + \alpha) \rangle = \frac{1}{2} \cos \alpha \tag{4.13}$$

The stationary solutions $\alpha = \alpha^* = const$ of the equation (4.12) are determined from the equation

$$\frac{1}{2} ml A \omega^2 \cos \alpha = -h\omega \tag{4.14}$$

When the following conditions are satisfied

$$h\omega < \frac{1}{2} ml A \omega^2 \tag{4.15}$$

there will be two essentially different solutions

$$\begin{aligned} \alpha_1^* &= \arccos(-2h/mlA\omega), \\ \alpha_2^* &= 2\pi - \alpha_1^* = 2\pi - \arccos(-2h/mlA\omega) \end{aligned} \tag{4.16}$$

one of which is stable, and the other – unstable. This can be easily seen from investigating equation (4.12).

Designating the power, spent to overcome the moment of resistance to the rotor's rotation, by $N = h\omega^2$, we can present the condition of maintaining the rotation of the rotor with the oscillation frequency of its axis ω as follows

$$N < N_{\max} \tag{4.17}$$

where

$$N_{\max} = \frac{1}{2} ml A \omega^3 \tag{4.18}$$

denotes the maximum value of the power to be transmitted to the rotating unbalanced rotor by means of the vibration of its axis.

As has already been mentioned, this power, the real values of the parameters being quite real, can be large enough, which provides the applied significance of this effect. So, say, at $ml = 10$ kg·m, $A = 0.5 \cdot 10^{-2}$ m, $\omega = 147$ s^{-3} we obtain

$$N_{\max} = \frac{1}{2} \cdot 10 \cdot 0.5 \cdot 10^{-2} \cdot 147^3 \cong 0.8 \cdot 10^5\,\mathrm{N \cdot m/s} \approx 80\,\mathrm{kw!}$$

It should be noted that this value is doubled in the case of circular oscillations of the point of suspension, characteristic of applications, presented in Fig. 1.5 (see also Blekhman (2000); ed. Blekhman (2004)).

Apparently, the first solution of this problem was performed by N.N. Bogolubov in 1950 by means of an asymptotic method.

The review of that investigation as well as the references to that and to other investigations are given in the book (Blekhman, 2000). The more general cases of a two component vibration as well as of an arbitrary periodic vibration by means of the method of direct separation of motions are discussed in the books (Blekhman, 2000; ed. Blekhman, 2004).

4.3 Self-synchronization of Unbalanced Rotors (Mechanical Vibro-exciters)

Introduction.Short historical remark. The discovery of complex chaotic motion in low-dimensional systems as well as the detection of simple regular motion in high-dimensional systems turned out to be one of the most important recent achievements of non-linear mechanics.

Particularly, it was established that many interacting dynamical objects, even those that are loosely linked, may demonstrate a tendency towards mutual synchronization, i.e. to some mutual regulation of their behavior.Synchronization is seen in such varied examples as weakly coupled pendulums, organ tubes, celestial bodies, electrical, electromagnetic and quantum generators, mechanical vibroexciters, turbine blades, populations of cells, swarms of fireflies, flocks of birds, schools of fish and in the in the applause or marches of people.

Synchronization of periodic self-oscillations in coupled systems has been known for a long time. Huygens (1673) was probably the first scientist who observed the synchronization phenomenon as early as the 17th century. He discovered that a couple of pendulum clocks which ran with different rates became synchronized when attached to a light beam instead of a wall. The pendulums oscillated with equal frequencies and opposite phases (4.2, a). Huygens rightly understood that the synchronization had been caused by imperceptible oscillation of the beam. However Huygens did not notice that the synphased motion can be stable (4.2, b)

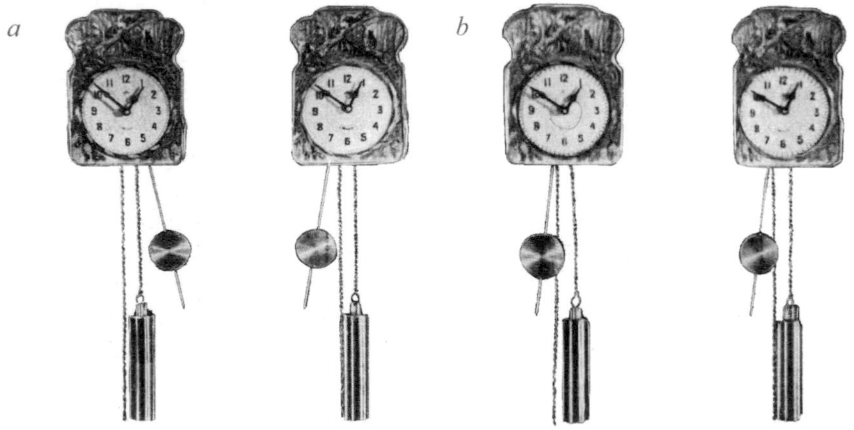

Figure 4.2. Selfsynchronization of pendulum clocks: a – antiphase motion; b – synphase motion.

In the middle of the 19th century, in his famous treatise "The Theory of Sound" Rayleigh (1845) described several interesting phenomena of synchronization in acoustical systems. He discovered that two organ tubes with closely disposed outlets begin to sound in unison, i.e. their mutual synchronization takes place. A similar effect was observed by Rayleigh for two electrically or mechanically connected tuning forks.

At the end of the 19th and the beginning of the 20th centuries synchronization phenomena were discovered in some electrical and electrical-mechanical systems, and in lamp generators in particular (van der Pol, 1920).

Rotating bodies are rated in a special class of synchronizing objects. From time immemorial it has been known that only one side of the Moon is turned towards the Earth. This is the evidence of the equality of mean frequencies of rotation of the Moon around its axis and around the Earth, i.e. of synchronization. Later many similar relations between the frequencies of revolution of celestial bodies became known. However, these relations have been considered as some separate phenomena of no general importance. The situation changed after the discovery of the synchronization of unbalanced rotors in 1948. The theory of this phenomenon, created by the author, his co-workers and other investigators has made it possible to understand that the tendency to synchronization is a general feature of a joint operation of material objects, no matter whatever the nature of those objects might be. It was established that the synchronization phenomenon is a manifestation of the tendency toward self-organization in complex systems. As a result, the general theory of synchronization has had an essential progress. The main achievements of this theory are presented in the books (Blekhman, 2000; ed. Blekhman, 2004; Blekhman, 1998). Recently there have appeared many detailed investigations concerning the synchronization in systems with both regular and chaotic behavior. They refer to a great variety of scientific areas including physics, biology, medicine and so on.

Selfsynchronization of unbalanced rotors (mechanical vibro-exciters) phenomenon The phenomenon of self-synchronization of unbalanced rotors was briefly described in Section 1. It is based on the fact that two or more rotors, placed on a common movable base and excited by independent asynchronous motors, rotate with equal multiple frequencies and with certain phase differences. It can take place even if the rotors are not linked kinematically or electrically (see Fig. 1.6). It is essential that this conformity of rotor motions occurs in spite of the difference between the partial angular velocities (by partial we meant here the angular velocities of the rotors placed on a fixed base). The quest for synchronous rotation is so strong that even if one or several motors are switched off, the synchronization is not broken, i.e. the rotors which are not excited can continue to move for an indefinitely long time. The energy supporting their motion is transmitted from the motors, still switched on, through the vibrating base.

The occasional observation of this effect at the "Mekhanobr" Institute (St.Petersburg, 1948) gave impetus to the investigation of this phenomenon. During a prolonged testing of vibro-machines with two vibro-exciters, i.e. with two unbalanced rotors, driven by two asynchronous electric motors, a wire, feeding one of motors, was broken. This open-circuit was detected only several hours later, because the machine continued operating normally. The physical explanation and the mathematical description of this fact was given by author. Later these investigations were continued and extended mainly by Russian and German scientists. The review and the presentation of the main results on the subject are given in (Blekhman, 2000; ed. Blekhman, 2004; Blekhman, 1998).

The discovery of the phenomenon of self-synchronization of the unbalanced rotors and the subsequent elaboration of the theory and methods of designing devices with self-synchronized vibro-exciters led to the creation of a new class of vibrational machines —

conveyers, feeders, screens, crushers, mills, concentration tables, special stands, etc., By now about three hundred inventions have been registered, (mostly in the former USSR and in Russia), based on the use of the effect of self-synchronization. Many of those inventions could not have been made without the theoretical investigations or methods of calculation.

The simplest case: self-synchronization of unbalanced vibro-exciters on a platform with one degree of freedom. The basic peculiarities of both the formulation and solution of the synchronization problem for mechanical vibro-exciters, as well as many features of the phenomenon, may be discovered by using the simplest model, related to the self-synchronization of unbalanced vibro-exciters, located on the absolutely rigid platform with one degree of freedom (Fig. 4.3). Rigid platform *1* (*the supporting body*) is movable in relation to the stationary foundation *2*in the fixed direction Ox. The platform is connected with the foundation by elastic elements *3* (stiffness c_x) and a linear damping element *4* (resistance coefficient k_x). There are a certain number k of unbalanced vibro-exciters *5* mounted on the platform whose axes are perpendicular to the direction of the platform oscillations and which are driven by the induction motors.

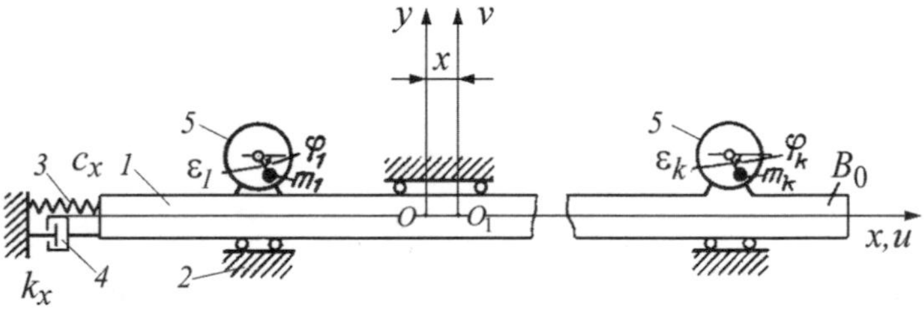

Figure 4.3. Self-synchronization of vibro-exciters on a platform with one degree of freedom.

The displacement x of the platform from a position, corresponding to the nonstrained elastic elements, and the rotor rotation angles $\varphi_s (s = 1, \dots, k)$ counted from O_x-axis direction in a clockwise direction, are assigned as the generalized coordinates of the system. Then the expressions for the kinetic and potential energy of the system will be as follows:

$$T = \frac{1}{2}\mathcal{M}\dot{x}^2 + \frac{1}{2}\sum_{s=1}^{k} J_{C_s}\dot{\varphi}_s^2 + \frac{1}{2}\sum_{s=1}^{k} m_s(\dot{x}_{C_s}^2 + \dot{y}_{C_s}^2), \tag{4.19}$$

$$\Pi = \frac{1}{2}c_x x^2 + \sum_{s=1}^{k} m_s g \varepsilon_s (1 - \sin\varphi_s). \tag{4.20}$$

Here $\mathcal{M}$ is the mass of the platform, m_s and J_{C_s} are respectively the mass and the moment of inertia of the rotor of the s-th vibro-exciter with respect to the axis, passing through its center of gravity C_s, g is the free fall acceleration, ε_s is the eccentricity, and

$$x_{C_s} = u_s + x + \varepsilon_s \cos\varphi_s, \quad y_{C_s} = v_0 - \varepsilon_s \sin\varphi_s \tag{4.21}$$

are the coordinates of the center of gravity of the s-th rotor in the system of the stationary axes xOy (uO_1v being the axes, rigidly connected with the platform and coinciding with the axes xOy at $x = 0$; u_s and v_0 being constants, representing the coordinates of the axes O_s of the rotor rotation in the axes uO_1v).

Considering formulas(4.21), expression (4.19)for the kinetic energy can be reduced to the form

$$T = \frac{1}{2}M\dot{x}^2 + \frac{1}{2}\sum_{s=1}^{k} I_s\dot{\varphi}_s^2 - \dot{x}\sum_{s=1}^{k} m_s\varepsilon_s\dot{\varphi}_s \sin\varphi_s, \tag{4.22}$$

The notations here are

$$M = \mathcal{M} + \sum_{s=1}^{k} m_s, \quad I_s = J_{C_s} + m_s\varepsilon_s^2 . \tag{4.23}$$

Composing Lagrange equations

$$\frac{d}{dt}\frac{\partial T}{\partial \dot{q}_j} - \frac{\partial T}{\partial q_j} = -\frac{\partial \Pi}{\partial q_j} + Q_j .$$

where q_j are the generalized coordinates of the system and Q_j are the generalized non-conservative forces, the following differential equations of motion for the system can be obtained:

$$I_s\ddot{\varphi}_s = L_s(\dot{\varphi}_s) - R(\dot{\varphi}_s) + m_s\varepsilon_s(\ddot{x}\sin\varphi_s + g\cos\varphi_s) \\ (s = 1, \ldots, k), \tag{4.24}$$

$$M\ddot{x} + k_x\dot{x} + c_x x = \sum_{j=1}^{k} m_j\varepsilon_j(\ddot{\varphi}_j \sin\varphi_j + \dot{\varphi}_j^2 \cos\varphi_j) . \tag{4.25}$$

In these equations the nonconservative forces $Q_x = -k_x\dot{x}$ and $Q_s = L_s(\dot{\varphi}_s) - R_s(\dot{\varphi}_s)$ are considered. They represent, respectively, the force of viscous resistance to the platform oscillation and the torque, acting on the rotor of the s-the vibro-exciter. In this case $L_s(\dot{\varphi}_s)$ is the driving torque of the induction motor (it is so-called static characteristic)and $R_s(\dot{\varphi}_s)$ is the resistance torque which is usually determined by the bearing friction. In this case

$$R_s(\dot{\varphi}_s) = R_s^{\circ}(\,|\,\dot{\varphi}_s\,|\,)\, sgn\, \dot{\varphi}_s \ , \tag{4.26}$$

where R_s° is the magnitude of the resistance torque.

The equations (4.24) describe the motion of the exciter rotors, and equation (4.25) is the equation for the platform oscillations. This system is essentially nonlinear.

The principal problem of the self-synchronization of vibro-exciters consists in the obtaining conditions, compliance with which would make the rotors of all exciters rotate with equal, by magnitude, average angular velocities, in spite of the absence of any direct connections between them and in spite of the differences in the parameters, characterizing the exciters and the torques acting on them. In other words, it is determination of the conditions of the existance and stability of solutions for the systems (4.24) and (4.25) which have the form:

$$\varphi_s = \sigma_s[\omega t + \alpha_s + \psi_s(\omega t)] \qquad (s = 1, \ldots, k), \qquad x = x(\omega t), \tag{4.27}$$

where ω is the magnitude of the average rotor rotational velocity, α_s are the constants (initial rotation phases), ψ_s and x_s are the periodic functions of the time t with the period $T = 2\pi/\omega$ and each of the values σ_s is equal to either 1 or -1. The first case corresponds to the positive and the second to the negative direction of the rotation of the rotor in the s-th exciter. The value of the synchronous angular velocity ω is not known beforehand, and it has to be determined in the process of solution.

The motions of (4.27) are *simple synchronous*. Sometimes the *multiple-synchronous* motions are also of interest [8](Blekhman, 1998). Since the latter case is more complicated, it requires, as a rule, a special consideration. As was shown by practical experience and analytical evaluation in the considered synchronous motions of the system, the rotation of the rotors differs only slightly from the uniform rotation. Therefore, the functions $L_s(\dot\varphi_s)$ and $R_s(\dot\varphi_s)$ may be linearized in the vicinity of $\dot\varphi_s = \sigma_s\omega$:

$$\begin{aligned} L_s(\dot\varphi_s) &= L_s(\sigma_s\omega) - k_s^*(\dot\varphi_s - \sigma_s\omega), \\ R_s(\dot\varphi_s) &= \sigma_s R_s^\circ(\omega) + k_s(\dot\varphi_s - \sigma_s\omega) \end{aligned} \tag{4.28}$$

where

$$k_s^* = -\left(\frac{dL_s}{d\dot\varphi_s}\right)_{\dot\varphi_s = \sigma_s\omega}, \quad k_s^\circ = \left(\frac{dR_s^\circ}{d\,|\,\dot\varphi_s\,|}\right)_{\dot\varphi_s = \sigma_s\omega} = \frac{dR_s^\circ(\omega)}{d\omega} \tag{4.29}$$

are, respectively, *coefficients of electrical and mechanical damping*. Both of these coefficients are usually positive.

It can be noted, however, that the final result would be the same if relations (4.28) were not used but it would be achieved in a more complicated way.

Considering the solution of the problem, we will introduce the small parameter μ into system (4.24) and (4.25), thus presenting it in the form:

$$\begin{aligned} &I_s\ddot\varphi_s + k_s(\dot\varphi_s - \sigma_s\omega) = \mu\Phi_s(\varphi_s, \ddot x) \quad (s = 1, \ldots, k). \\ &M\ddot x + c_x x = \sum_{j=1}^{k} m_j\varepsilon_j(\ddot\varphi_j \sin\varphi_j + \dot\varphi_j^2 \cos\varphi_j) - \mu k_x' \dot x. \end{aligned} \tag{4.30}$$

where the parameter $k_s = k_s^* + k_s^0 > 0$ will be called the *total damping coefficient* and

$$k_x = \mu k'_x \quad \mu\Phi_s = L_s(\sigma_s\omega) - \sigma_s R_s^\circ(\omega) + m_s\varepsilon_s(\ddot{x}\, sin\varphi_s + g\cos\varphi_s)\,. \tag{4.31}$$

Such a way for introducing the small parameter corresponds to an assumption of an almost uniform rotation of rotors in the analyzed synchronous motion, and also to an assumption that the *motion is analyzed far from the resonance when the friction force $k_x\dot{x}$ in the supporting system may be considered as small compared to the elastic or inertia forces, as well as the exciting force.*

We will solve the equations (4.30) of the type (4.27) using the method of direct separation of motions. In this case $\alpha_s = \alpha_s(t)$ is the main slow component, ψ_s and x are the fast 2π-periodical components dependent on ωt and they satisfy the conditions

$$< \psi_s >= 0\,, \quad < x >= 0 \tag{4.32}$$

The torques L_s and R_s^0 can be considered to be slow; the torques $m_s\varepsilon_s(\ddot{x}\sin\psi_s + g\cos\psi_s)$ and forces $m_s\varepsilon_s(\ddot{\psi}\sin\psi_s + \dot{\psi}_s^2\cos\psi_s)$ can be considered to be fast.

In the first approximation the solution of the equations of motions is $\psi_s = \psi_s^0 = 0$. Then the expressions (4.27) take the form

$$\varphi_s = \varphi_s^0 = \sigma_s(\omega t + \alpha_s)\,, \quad (s = 1, \ldots, k); \quad x = x^0(\omega t) \tag{4.33}$$

where

$$x^0(\omega t) = -\frac{\omega^2}{\omega^2 - p^2}\sum_{j=1}^{k}\frac{m_j\varepsilon_j}{M}\cos(\omega t + \alpha_j)\,, \quad p^2 = c_x/M \tag{4.34}$$

is the $2\pi/\omega$-periodical solution of the equation

$$M\ddot{x}^0 + c_x x^0 = \sum_{j=1}^{k} m_j\varepsilon_j\omega^2\cos(\omega t + \alpha)\,, \tag{4.35}$$

Equation (4.35) and expressions (4.33), (4.34) correspond to the assumption that the rotors are uniformly rotating with the unknown initial phases α_s and the platform is oscillating steadily under the action of the corresponding unbalanced forces.

As a result we obtain the main equation of the vibrational mechanics in the form

$$I_s\ddot{\alpha}_s + k_s\dot{\alpha}_s = k_s(\omega_s - \omega) + V_s(\alpha_1 - \alpha_k, \ldots, \alpha_{k-1} - \alpha_k, \omega) \tag{4.36}$$

where ω_s are the partial angular velocities, i.e. the velocities of the exciters if they are mounted on a fixed foundation, and

$$V_s = \frac{1}{2}\frac{\omega^4}{\omega^2 - p^2}\frac{m_s\varepsilon_s}{M}\sum_{s=1}^{k} m_j\varepsilon_j\sin(\alpha_s - \alpha_j) \tag{4.37}$$

are vibrational torques. Expressions (4.37) can be easily obtained by means of averaging the first k equations (4.30), taking into account both expressions (4.33),(4.35) and obvious equalities

$$\sigma_s = 1/\sigma_s, \quad \sin\sigma_s\alpha_s = \sigma_s \sin\alpha_s, \quad cos\sigma_s\alpha_s = \cos\alpha_s,$$
$$< \sin(\omega t + \alpha_s)\cos(\omega t + \alpha_j) >= \frac{1}{2}\sin(\alpha_s - \alpha_j),$$
$$< \sin(\omega t + \alpha_s)\sin(\omega t + \alpha_j) >=< \cos(\omega t + \alpha_s)\cos(\omega t + \alpha_j) >$$
$$= \frac{1}{2}\cos(\alpha_s - \alpha_j) \tag{4.38}$$

Integral sign of stability (extremal property) of synchronous motion. Expression (4.37) can be presented in the form

$$V_s = -\frac{\partial\Lambda^{(I)}}{\partial\alpha_s} \tag{4.39}$$

where

$$\Lambda^{(I)} =< L^{(I)} >=< T^{(I)} - \Pi^{(I)} >= \frac{1}{4M}\frac{\omega^4}{\omega^2 - p^2}\sum_{s=1}^{k}\sum_{j=1}^{k} m_s\varepsilon_s m_j\varepsilon_j \cos(\alpha_s - \alpha_j) \tag{4.40}$$

is the averaged value of the Lagrange function of the supporting body, calculated in the assumption that $\varphi_s = \varphi_s^0$, $x = x^0$ in accordance with (4.33) and (4.34), i. e.

$$T^{(I)} = \frac{1}{2}M(\dot{x}^0)^2, \quad \Pi^{(I)} = \frac{1}{2}c(x^0)^2. \tag{4.41}$$

And then if we introduce a function

$$D = \Lambda^{(I)}(\alpha_1 - \alpha_k, \ldots, \alpha_{k-1} - \alpha_k, \omega) - \sum_{s=1}^{k} k_s(\omega_s - \omega)(\alpha_s - \alpha_k) \tag{4.42}$$

the main equations (4.36) can be written in the form

$$I_s\ddot{\alpha}_s + k_s\dot{\alpha}_s = -\frac{\partial D}{\partial\alpha_s} \tag{4.43}$$

Thus, according to the classical theorem of the stability of equilibrium positions, the stable synchronous motions correspond to the minimums of the function D; we call it, as before, *potential function.* In the case when the partial velocities are the same $\omega_1 = \ldots = \omega_k = \omega$ it is

$$D = \Lambda^{(I)}(\alpha_1 - \alpha_k, \ldots, \alpha_{k-1} - \alpha_k) \tag{4.44}$$

It is remarkable that this feature proves to be applicable not only to much more general problems on the synchronization of rotors, but also to the orbital motions of the celestial bodies (Blekhman, 1998). Recall that in Section 3 we came across the case when an essentially non-conservative system corresponded to the equation of slow motions with potential forces. Such systems were named by us *potential on the average systems.* The stable synchronous motions in this system correspond to the minimums of the Lagrange function $\Lambda^{(I)}$. This is one of the main results of the theory of synchronization of unbalanced rotors. We call it *extreme sign of the stability of synchronous motions* (Blekhman, 2000, 1998). From (4.37) it follows

$$\sum_{s=1}^{k} V_s \cong 0 \tag{4.45}$$

and for the stationary motion ($\alpha_s = const$) we obtain (in the general case)

$$\omega = \sum_{s=1}^{k} k_s \omega_s \,/ \sum_{s=1}^{k} k_s \tag{4.46}$$

i.e. the synchronous angular velocity ω is equal to the weighted mean of the partial velocities ω_s .

The case of the two vibroexciters. In the simplest case of the two identical vibroexciters ($k = 2,\quad \omega_1 = \omega_2 = \omega,\quad m_1\varepsilon_1 = m_2\varepsilon_2,\quad k_1 = k_2$) we have in accordance with (4.40)

$$\Lambda^{(I)} = \frac{1}{2M}\frac{\omega^4}{\omega^2 - p^2} m^2 \varepsilon^2 \cos(\alpha_1 - \alpha_2) + const \tag{4.47}$$

and before resonance ($\omega < p = \sqrt{c_x/M}$) the synphase motion ($\alpha_1 - \alpha_2 = 0$) is stable and the antiphase motion ($\alpha_1 - \alpha_2 = \pi$) is unstable; after the resonance ($\omega > p$), on the contrary, synphase motion is unstable and the antiphase motion is stable.

Example of the use of the integral sign of stability:self-synchronizing of two vibro-exciters on a soft-suspended solid body performing planar vibrations. The system under consideration is presented in Fig. 4.4,a.

The supporting body B_0 can move parallel to a plane which is perpendicular to rotational axes of the exciter rotors which are assumed identical. The center of gravity of the body lies in the plane passing through these axes at a distance r from both of them. The elastic supports are assumed to be very soft; thus the frequencies of the free vibrations of the body B_0 on the supports are much lower that the synchronous angular velocity ω of the rotors. In such case, the potential energy $\Pi^{(I)}$ and the corresponding members in the equations of motion for B_0 can be neglected. Thus, if the rotors rotate as described by (4.33), the vibrations of B_0 would be described by the equations

$$\begin{aligned} M\ddot{x}^0 &= F[cos(\omega t + \alpha_1) + \cos(\omega t + \alpha_2)]\,, \\ M\ddot{y}^0 &= -F[\sigma_1 sin(\omega t + \alpha_1) + \sigma_2 \sin(\omega t + \alpha_2)]\,, \\ I\ddot{\varphi}^0 &= Fr[\sigma_1 sin(\omega t + \alpha_1) - \sigma_2 \sin(\omega t + \alpha_2)]\,, \end{aligned} \tag{4.48}$$

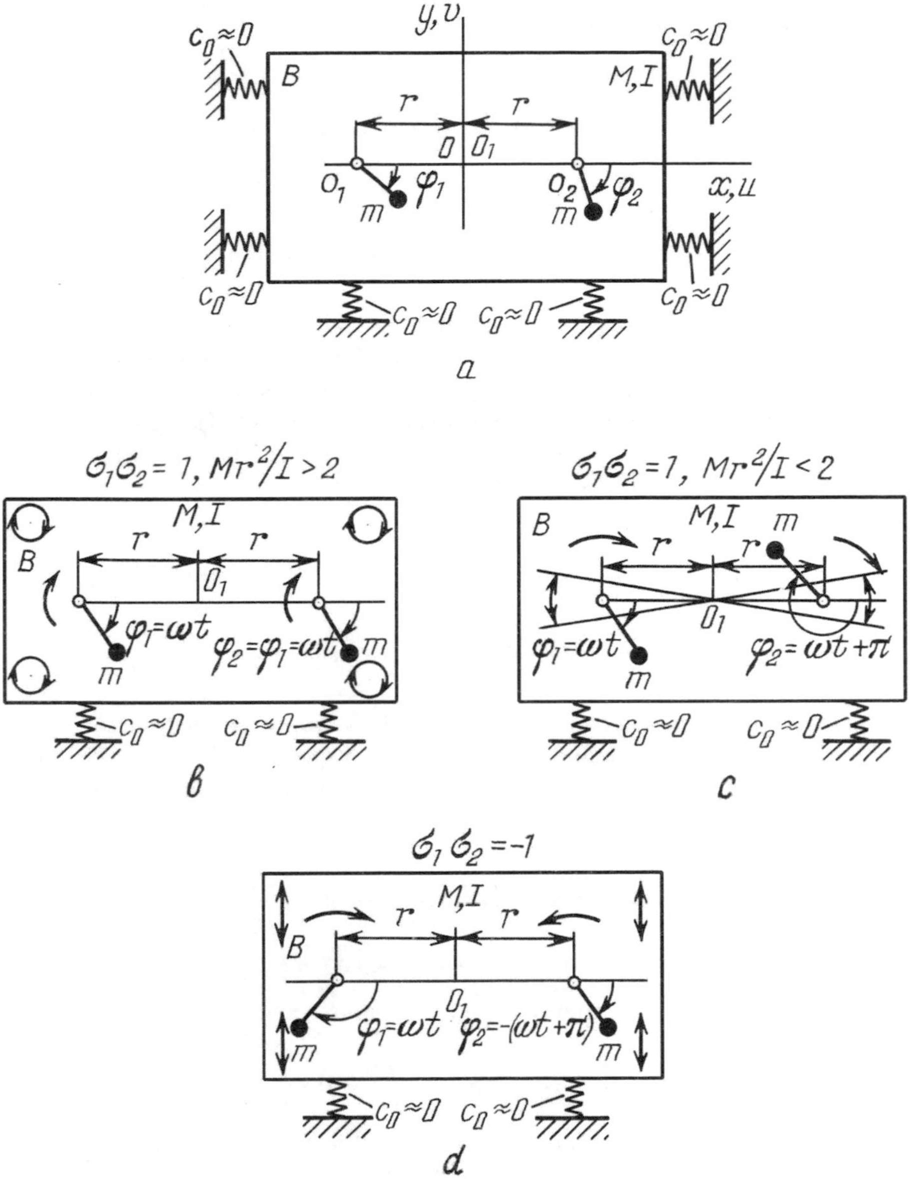

Figure 4.4. Self-synchronization of two identical unbalanced vibro-exciters, located of a softly-vibro-isolated flatly oscillating solid body.

where M and I are respectively the mass and moments of inertia of the body B_o together with exciters and $F = m\varepsilon\omega^2$ is the amplitude of the force developed by each exciter. Steady periodic solution of (4.48) is

$$
\begin{aligned}
x^0 &= -\frac{F}{M\omega^2}[cos(\omega t + \alpha_1) + \cos(\omega t + \alpha_2)]\,, \\
y^0 &= \frac{F}{M\omega^2}[\sigma_1 sin(\omega t + \alpha_1) + \sigma_2 \sin(\omega t + \alpha_2)]\,, \\
\varphi^0 &= -\frac{Fr}{I\omega^2}[\sigma_1 sin(\omega t + \alpha_1) - \sigma_2 \sin(\omega t + \alpha_2)]\,. \qquad (4.49)
\end{aligned}
$$

To derive the basic equations, the integral stability criterion will be used. Since the partial velocities of the exciters are assumed to be identical and positive, the role of the potential function is played in this case by an average value of the Lagrange function of the supporting body. Due to the assumed softness of the elastic supports, the latter is simply equal to the average value of the kinetic energy of the body. Performing the averaging in view of (4.49) and equalities (4.38) we obtain

$$
\begin{aligned}
D = \Lambda^{(I)} = < T^{(I)} > = \frac{1}{2} < M\left[(\dot{x}^0)^2 + (\dot{y}^0)^2\right] + I\,(\dot{\varphi}^0)^2 > = \\
\frac{F^2}{2M\omega^2}\left(1 + \sigma_1\sigma_2 - \sigma_1\sigma_2\frac{Mr^2}{I}\right)\cos(\alpha_2 - \alpha_1) + C_1 \qquad (4.50)
\end{aligned}
$$

Here C_1 is a value not dependent on α_1, α_2. As could be expected the function D depends in this case only on the phase difference $\alpha = \alpha_2 - \alpha_1$. Equating $\partial D/\partial\alpha$ to zero, we arrive at the following equation for the phase difference α in possible synchronous motions

$$
\sin\alpha = 0 \qquad (4.51)
$$

This equation has two essentially different solutions:

$$
(\alpha)_1 = (\alpha_2 - \alpha_1)_1 = 0\,, \quad (\alpha)_2 = (\alpha_2 - \alpha_1)_2 = \pi\,, \qquad (4.52)
$$

i.e. synphase and antiphase solutions. If the conditions stands

$$
1 + \sigma_1\sigma_2 - \sigma_1\sigma_2\frac{Mr^2}{I} < 0\,, \qquad (4.53)
$$

then the function D has a minimum at $\alpha = (\alpha)_1 = 0$ and a maximum at $\alpha = (\alpha)_2 = \pi$. Thus, according to the extreme sign of stability, synphaseous motion is stable and antiphaseous is unstable.

Let exciter rotors first rotate in the same directions, $\sigma_1\sigma_2 = 1$. Then, if

$$
Mr^2/I > 2\,, \qquad (4.54)
$$

synphase rotation will be stable and, as can be seen from (4.49), the body B_0 would perform circular translational vibrations with the amplitude $r_0 = 2m\varepsilon/M$ (Fig. 4.4, b). If the opposite inequality $Mr^2/I < 2$ occurs, the antiphaseous rotation would be stable, and the body B_0 would perform angular vibrations with the amplitude $\varphi_0 = 2m\varepsilon r/I$ (Fig. 4.4, c).

When the rotors are rotating in opposite directions, $\sigma_1\sigma_2 = -1$, condition (4.53) is never complied with. Thus, antiphaseous rotation is always stable. This leads, as can be seen from (4.49) to rectlinear translational harmonic vibrations of B_0 perpendicular to the plane containing the rotor axes (Fig. 4.4, d), with the amplitude $A_0 = 2m\varepsilon/M$.

The features of self-synchronization of the two vibro-exciters in the simple system established above are widely used in vibratory machines and devices (Blekhman, 1998).

4.4 Conclusion to Section 4

In Section 4 we considered the applications of the approach of vibrational mechanics and of the method of direct separation of motions to the problems on the action of vibration in mechanisms. Particularly, we discussed the problems which refer to the mechanisms of the type of pendulums and rotors. Special attention was given to the problem of the synchronization of vibro-exciters which has many technical applications.

5 On the Vibrational Displacement and Vibrational Rheology

5.1 On the Effect of Vibrational Displacement

Introduction. Definition of vibrational displacement. The effects of vibrational displacement and their applications were discussed in the first section. *By vibrational displacement we mean the appearance of the "directed on the average", as a rule slow, change of the state of the system (in particular of motion) under the action of the "non-directed on the average", oscillatory, as a rule fast, action.* Examples of vibrational displacement are the transportation of single bodies and granular materials in the vibrating vessels or trays; the work of the devices, called vibrational transformers of motion and vibro-engines; vibrational driving of piles, of sheet piles, of shells; vibrational separation of granular materials; the motion of vibrational coaches; flight, swimming and crawling of living organisms. By now there have been several thousand investigations of this effect, both purely theoretical and devoted to numerous technical applications. A brief review of those investigations can be found in the book (Blekhman, 2000).

The majority of problems on the theory of vibrational displacement are reduced to the investigation of the solutions of nonlinear differential equations with the periodic over the fast time $\tau = \omega t$ right sides, for which the velocities of the change of the generalized coordinates have the form

$$\dot{\mathbf{x}} = \dot{\mathbf{X}}(t) + \dot{\mathbf{\Psi}}(t, \omega t), \tag{5.1}$$

where $\dot{\mathbf{X}}(t)$ is a slowly changing component, and $\dot{\mathbf{\Psi}}$ is a fast changing component, with

$$< \dot{\mathbf{\Psi}}(t, \omega t) >= 0. \tag{5.2}$$

The component $\dot{\mathbf{X}}(t)$ is called the *velocity of vibro-displacement*; in most cases its determination in the stable stationary motions (that is when $\dot{\mathbf{X}} = \text{const}$) is of utmost interest for applications. In a wide class of systems an essential non-linearity of differential equations is caused by the presence in the system of dry friction forces and of unilateral constraints, though motions of such systems in certain regions of phase space can be linear as well.

The form of the solutions being sought (5.1) makes its quite natural and expedient to use concepts and methods of vibrational mechanics when solving problems of the theory of vibro-displacement.

Simplest model of the vibrational displacement. Some important regularities of vibrational displacement can be made clear by solving a simplest problem — about the motion of a flat solid body (particle) whose mass is m over a rough horizontal plane under the action of a longitudinal harmonic exciting force $\Phi_0 \sin \omega t$ (Fig. 5.1, a) or a similar problem about the motion of a body over such plane, performing longitudinal harmonic oscillations. To be specific, we will first consider the first of these problems.

The equation of motion of the body is

$$m\ddot{x} = F(\dot{x}) + \Phi_0 \sin \omega t, \tag{5.3}$$

where m is the mass of the body and

$$F(\dot{x}) = \begin{cases} -m\,g\,f_+ & \text{at} \quad \dot{x} > 0 \\ m\,g\,f_- & \text{at} \quad \dot{x} < 0 \end{cases}$$
$$-m\,g\,f_+ < F(\dot{x}) < m\,g\,f_- \quad \text{at} \quad \dot{x} = 0 \tag{5.4}$$

is the dry friction force, corresponding to the characteristic shown in Fig. 5.1, a. The case $f_+ \neq f_-$ may seem artificial, however, this is not so: it corresponds e.g. to the vibro-transportation of fish for which the friction coefficients are quite different when the sliding takes place in the direction of the scalcs and when it is opposite to them. The same refers to the sliding of a body over a fleecy surface or to the simplest model of the resistance of the ground in the problem of the vibrational sinking of piles and sheet pile (see Blekhman, 2000).

Equation (5.3) can be solved exactly by using either the method of fitting or the method of point mapping. However, a much more simple and clear approximate solution can be obtained by the method of direct separation of motions. The dry friction force $F(\dot{x})$ which can essentially change during the period, as well as the exciting force $\Phi_0 \sin \omega t$, will be referred to fast forces. Searching the solution of equation (5.3) of type (5.1), we will write down equations (2.11) and (2.12) as

$$m\ddot{X} = V \tag{5.5}$$

$$m\ddot{\psi} = F(\dot{X} + \dot{\psi}) - < F(\dot{X} + \dot{\psi}) > + \Phi_0 \sin \omega t \tag{5.6}$$

where

$$V = < F(\dot{X} + \dot{\psi}) > \tag{5.7}$$

is the vibrational force which, as can be seen, presents in this case the average for the period dry friction force acting upon the body.

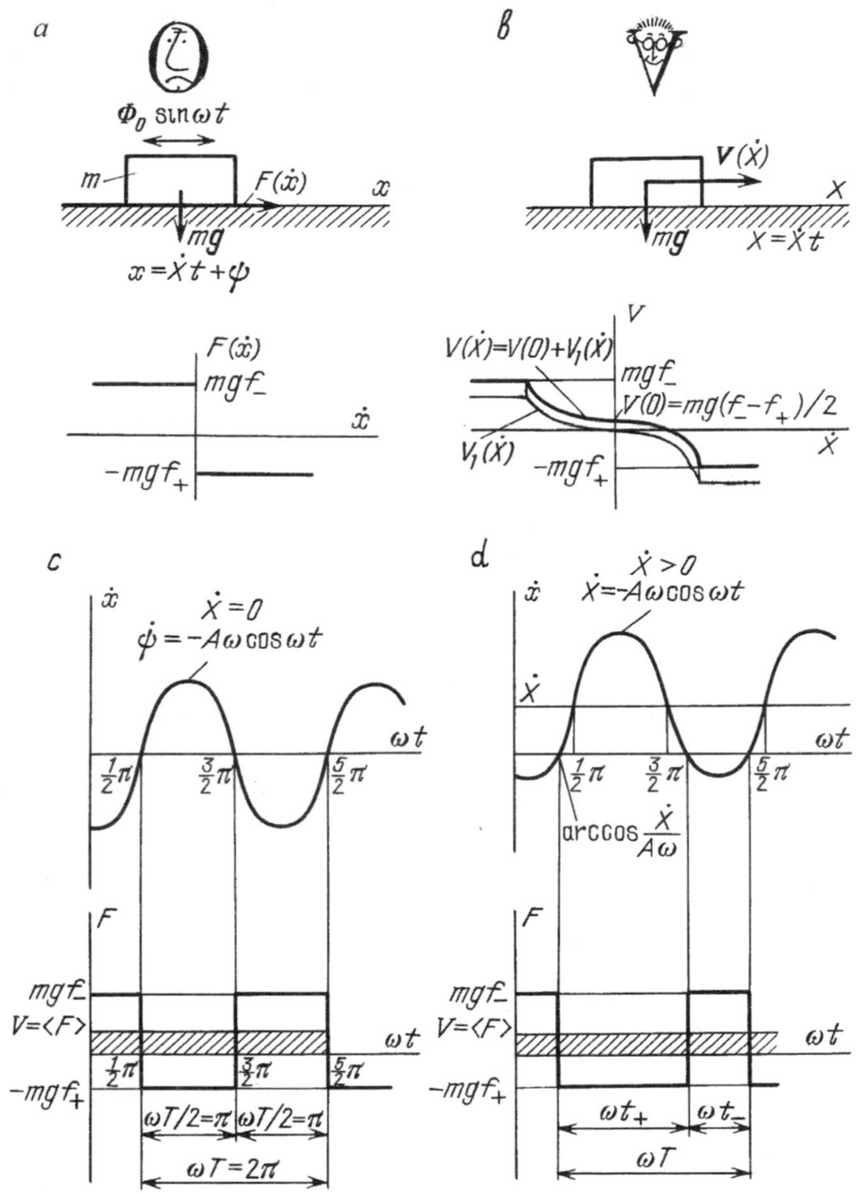

Figure 5.1. A flat body on a rough horizontal plane at the asymmetrical law of dry friction under the action of a harmonic longitudinal vibration.

We will solve this problem in a crude way, considering that the amplitude of the exciting force Φ_0 is much greater than the limiting values of the dry friction forces $F_+ = m\,g\,f_+$ and $F_- = m\,g\,f_-$, so that the latter can be neglected when solving the equations of fast motions (5.6). Then the periodic solution of this equation, satisfying condition (5.2), will be

$$\psi = -A\sin\omega t \tag{5.8}$$

where

$$A = \Phi_0/m\omega^2. \tag{5.9}$$

Now let us perform the averaging according to formula (5.7) in view of expressions (5.4) and (5.8). The process of this averaging is presented graphically in Fig. 5.1, c at $\dot{X} = 0$, and in Fig. 5.1, d — in the general case $\dot{X} \neq 0$. These figures illustrate very clearly the mechanism of appearance of the vibrational force. As a result we obtain

$$V = V(\dot{X}) = \begin{cases} -m\,g\,f_+ & \text{at} \quad \dot{X} \geq A\omega, \\ \dfrac{\omega}{2\pi} m\,g\,(f_- t_- - f_+ t_+) & \text{at} \quad |\dot{X}| \leq A\omega, \\ m\,g\,f_- & \text{at} \quad \dot{X} \leq -A\omega \end{cases} \tag{5.10}$$

where t_+ and t_- denote the intervals of time during which the body moves over the plane to the right ($\dot{x} = \dot{X} + \dot{\psi} > 0$) and to the left ($\dot{x} = \dot{X} + \dot{\psi} < 0$) respectively, with

$$\omega t_+ = 2\left(\pi - \arccos\frac{\dot{X}}{A\omega}\right), \qquad \omega t_- = 2\arccos\frac{\dot{X}}{A\omega}. \tag{5.11}$$

In view of those expressions formula (5.10) for the vibrational force acquires the following form:

$$V(\dot{X}) = \begin{cases} -m\,g\,f_+ & \text{at} \quad \dot{X} \geq A\omega, \\ \dfrac{m\,g}{\pi}\left[(f_+ + f_-)\arccos\dfrac{\dot{X}}{A\omega} - f_+\pi\right] & \text{at} \quad |\dot{X}| \leq A\omega, \\ m\,g\,f_- & \text{at} \quad \dot{X} \leq -A\omega, \end{cases} \tag{5.12}$$

and the main equation of vibrational mechanics will be written as

$$m\ddot{X} = V(\dot{X}). \tag{5.13}$$

The velocity of the stationary motion of the particle over the plane $\dot{X} = \dot{X}_*$ is the *velocity of vibro-displacement* which will be found from the equation $V(\dot{X}_*) = 0$, from which we obtain

$$\dot{X}_* = A\omega\cos\frac{\pi f_+}{f_+ + f_-} = \frac{\Phi_0}{m\omega}\cos\frac{\pi f_+}{f_+ + f_-}, \tag{5.14}$$

and since $V'(\dot{X}_*) < 0$, the motion with this velocity is stable. At $f_+ = f_-$ according to formula (5.14) we obtain $\dot{X}_* = 0$ that is, as it must, the body "on the average " remains motionless (is in the state of quasi-equilibrium). It should be also noted that the value of the velocity of vibro-displacement $\dot{X} = 0$ is answered by the vibrational force

$$V(0) = \frac{1}{2} m g (f_- - f_+). \tag{5.15}$$

All the obtained formulas remain also valid in the case when the body is acted upon by the exciting force $\Phi_0 \sin \omega t$ and the plane performs the preassigned harmonic oscillations with the amplitude A and the frequency ω.

Let us turn to the analysis of the dependences we have obtained. First of all it should be noted that according to (5.10) the dependence $V(\dot{X})$ whose graph is presented in Fig. 5.1, b is continuous — there has been a *vibrational smoothening of the discontinuous characteristic*of dry friction)(Fig. 5.1, a) — the effect well known in the theory of automatic control. It should be noted further that the vibrational force $V(\dot{X})$ can be presented as a sum of two terms

$$V(\dot{X}) = V(0) + V_1(\dot{X}), \tag{5.16}$$

where $V(0)$ is defined by equation (5.15). The force $V(0)$ can be called the *driving vibrational force* and the force $V_1(\dot{X})$ — the *vibro-transformed resistance force*, and since according to (5.16), $V_1(0) = 0$, the latter displays the nature of viscous friction if in order to make a body move it is necessary to have a certain finite force. In case the motion can be caused by a force as small as small can be, we will speak about a *resistance force of the type of viscous friction.* Thus as a result of the action of vibration on the system that is being considered, there is not only a seeming (that is seen only by the observer **V**) transformation of dry friction into viscous, but also the generation of the driving vibrational force. The latter circumstance is often forgotten, and only the effect of fluidization of the system with dry friction under vibration is spoken about. We will again emphasize that this fluidization occurs only with regard to slow motions; "in reality" (for the observer **O**), i.e., in fact the system remains that with dry friction (Fig. 5.1, a, b).

Thus, the consideration of this simplest system leads to the explanation and description not only of the effect of vibrational displacement, but also of other regularities which may be referred to vibrorheological effects.

In conclusion we will mark that in case $\dot{X}/A\omega \ll 1$ when in expression (5.12) we can restrict ourselves to the first two members of the expansion of $\arccos \dot{X}/A\omega$ into a power series by $\dot{X}/A\omega$ (i.e. $\arccos(\dot{X}/A\omega) \approx \pi/2 - \dot{X}/A\omega$), the expression for the vibrotransformed vibrational force takes the form

$$V_1(\dot{X}) = -\frac{m g}{\pi}(f_+ + f_-)\frac{\dot{X}}{A\omega}, \tag{5.17}$$

while the expression (5.15) for the driving vibrational force $V(0)$ remains the same.

It should be noted that the results of the above investigations can be used for the investigation of more complicated models, in particular for the model of worm-like motion, presented in Fig. 1.14, b (see also Zimmerman et al., 2004).

Types of asymmetry of the system causing vibrational displacement. The simplest model considered above shows the main condition of appearance of the effect of vibrational displacement — the asymmetry of the system: when $f_+ = f_-$, this effect is absent. However, the case considered here presents only one possible type of asymmetry, resulting in vibrational displacement. Other types of asymmetry and the physical mechanisms corresponding to them can be also characterized by the model of a flat body (material particle), moving relative to the vibrating rough plane; as was already marked, such a model is of basic importance for investigating the processes of vibrational displacement.

Figure 5.2, a shows schematically six types of asymmetry and accordingly six ways of realizing the process of vibrational displacement: I — force (three versions), II — kinematic, III — structural or constructive, IV — gradient, V — wave and VI — initial asymmetry.
In this figure F_+ and F_- denote the forces of resistance when the body moves forward and backward respectively. The arrows on the axis x which have the same length in both directions show arbitrarily that the law of the vibration of the points of the surface is absolutely symmetrical, i.e. it presents e.g. purely harmonic oscillations.

It should be also noted that the cinematic asymmetry (type II) can also be created when the trajectories are non-rectilinear, say, elliptical or circular. A detailed description of every type of asymmetry can be found in the book (Blekhman, 2000). It is easy to see what type of asymmetry each of the examples of vibrational displacement, given in the first section, belongs to.

5.2 On Vibrational Rheology

Introduction. Definition of vibro-rheology. *Vibro-rheology is a section of nonlinear mechanics which studies the changes, caused by vibration,in the rheological properties of bodies with respect to slow forces and also the relevant slow motions of bodies* (Blekhman, 2000).

The vibro-rheological effects were briefly discussed in the first section. Figure 2.3 shows schematically the relation of vibro-rheology, rheology and vibrational mechanics.

Two essential moments should be emphasized once more: 1) in most cases it is expedient to speak of vibrorheological effects as of the seeming ones, which take place only for the observer **V** (but very often this point of view is actual for the applications); 2) as a result of the action of vibration on the nonlinear mechanical systems, in the general case there is not only a change in the rheological characteristics or properties of the body with respect to slow actions, but there also appear either driving or shifting forces or torques. Both of these circumstances were illustrated by the example, discussed in i. 5.1.

The effect of vibration on the dissipative and elastic characteristics of bodies with respect to the slow effects is of a considerate applied interest. These vibro-rheological effects will be discussed below by two examples.

Effective coefficients of dry friction under the action of vibration. The effect of the seeming reduction of the friction coefficient of rest is the simplest manifestation of vibrorheological regularities, allowing the elementary consideration (Blekhman, 2000).

Let an absolutely solid body be pressed by a force **N** to a rough surface and let it be acted upon by a longitudinal harmonic force **S** directed along the plane (Fig. 5.3, a). Let the body be also acted upon by a force $\Phi = \Phi_0 \sin \omega t$; then for the body to begin moving along the plane it is necessary that there should be not the force $S = S_0 = f_1 N$ like it is in the absence of the force Φ, but only the force $S = f_1 N - \Phi_0$ (f_1 being the

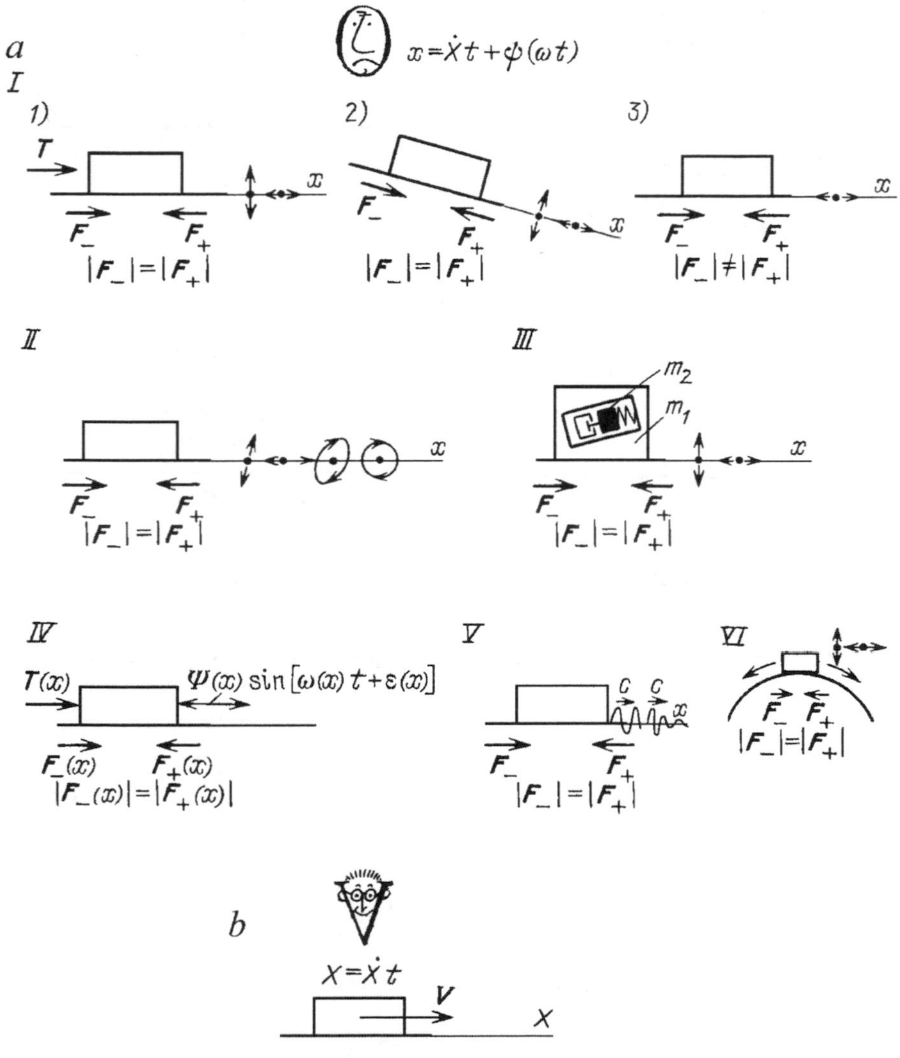

Figure 5.2. The main types of asymmetry, causing vibrational displacement.

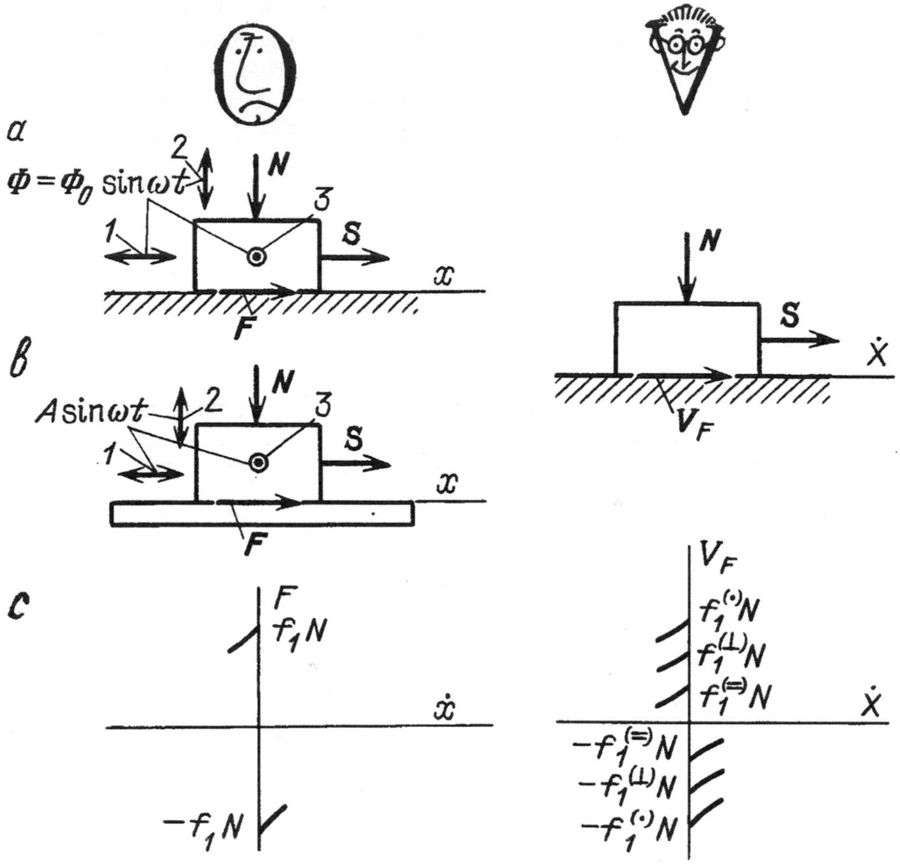

Figure 5.3. Effective coefficients of dry friction under the action of vibration.

friction coefficient of rest). Therefore it will seem to the observer **V** "who does not see" the fast force Φ (Fig. 5.3, c) that the friction coefficient of rest has reduced with respect to the slow force **S**, and has become

$$f_1^{(=)} = \frac{S^{(=)}}{N} = f_1\left(1 - \frac{\Phi_0}{N}\right). \tag{5.18}$$

Similarly when the force Φ acts perpendicularly to the plane,

$$f_1^{(\perp)} = f_1\left(1 - \frac{\Phi_0}{N}\right). \tag{5.19}$$

If the force Φ is parallel to the plane and is perpendicular to the force **S**, i.e. to the plane of the Figure, then

$$f_1^{(\cdot)} = f_1\sqrt{1-(\Phi_0/f_1N)^2}\,. \tag{5.20}$$

Let us introduce the parameter

$$w = \Phi_0/N\,, \tag{5.21}$$

characterizing the relative intensity of vibration and called the *parameter of overload*, or, simply, *overload*. Then formulas (5.18)–(5.20) will be written as

$$f_1^{(=)} = f_1\left(1-\frac{w}{f_1}\right), \quad f_1^{(\perp)} = f_1(1-w)\,, \quad f_1^{(\cdot)} = f_1\sqrt{1-(w/f_1)^2}\,. \tag{5.22}$$

Formulas (5.22) remain valid in case the force Φ is absent, but the plane performs harmonic oscillations in the adequate directions (Fig. 5.3, b); then it is only necessary to calculate the parameter of the overload according to the formula

$$w = mA\omega^2/N\,, \tag{5.23}$$

i.e. to assume that in (5.21) $\Phi_0 = mA\omega^2$ where m is the mass of the body, A is the amplitude and ω is, as before, the frequency of vibration. Finally, if the normal force N is the weight of the body $m\,g$, then

$$w = A\omega^2/g, \tag{5.24}$$

Formulas (5.18)–(5.20) and (5.22) make sense only while the values $f_1^{(=)}$, $f_1^{(\perp)}$ and $f_1^{(\cdot)}$ called the *effective coefficients of friction under vibration* are positive; at the larger values of the overload parameter w there is a seeming change in the character of friction (see above). In that case the effective coefficients of friction must be considered to be equal to zero.

Vibrational transformation of the characteristic of positionally-viscous resistance. We will consider an oscillatory system with one degree of freedom, described by the equation of the type

$$\ddot{x} + \mu f(x)\dot{x} + p^2x = A\omega^2\cos\omega t \tag{5.25}$$

where

$$f(x) = a_0 + a_1x + a_2x^2 + a_3x^3 + a_4x^4, \tag{5.26}$$

μ, p^2, A and ω are the positive constants, with μ being a small parameter; we also assume that the vibration frequency ω is much greater than the frequency of free oscillations of the system p in the absence of the force of resistance; the constants $a_0,\ldots,a_4$ can have either positive or negative values; some of them may even be zeros. The force of resistance which in equation (5.25) corresponds to the term $\mu f(x)\dot{x}$ will be arbitrarily called by us *positionally-viscous resistance*; the polynomial $f(x)$ is in this case "cut off" by us at the fourth degree of x since that is enough to reveal the main regularities, essential for applications.

We are interested in the solutions of equation (5.25) of the type

$$x = X(t) + \psi(t, \omega t) \tag{5.27}$$

where X is the slowly changing and $\psi(t, \omega t)$ —- the fast changing functions of time, with ψ periodic with respect to ωt with a period of 2π and it satisfies the condition

$$< \psi(t, \omega t) >= 0\,. \tag{5.28}$$

Using the method of direct separation of motions, we will write equations (2.11) and (2.12) as

$$\ddot{X} + p^2 X + \mu < (\dot{X} + \dot{\psi}) f(X + \psi) >= 0, \tag{5.29}$$

$$\ddot{\psi} + p^2\psi + \mu[(\dot{X} + \dot{\psi}) f(X + \psi) - < (\dot{X} + \dot{\psi}) f(X + \psi) >] = A\omega^2 \cos\omega t\,. \tag{5.30}$$

With X and $\dot{X}$ frozen, and μ small enough, the equation of fast motions (5.30) has a single 2π-periodic with respect to ωt solution

$$\psi = \frac{A\omega^2}{p^2 - \omega^2} \cos\omega t + O(\mu), \tag{5.31}$$

satisfying condition (5.28). Therefore with an accuracy to the terms of the order of μ and in view of the obvious equality

$$< \psi^n \dot{\psi} >= 0 \tag{5.32}$$

we will obtain

$$\begin{aligned} &< (\dot{X} + \dot{\psi}) f(X + \psi) >= \dot{X} f(X) \\ &+ \left[\frac{1}{2} B^2 a_2 + \frac{3}{8} B^4 a_4 + \frac{3}{2} X B^2 a_3 + 3X^2 B^2 a_4\right] \dot{X} \end{aligned} \tag{5.33}$$

where

$$B = A\omega^2/(p^2 - \omega^2) = A/(\lambda^2 - 1), \quad \lambda = p/\omega\,. \tag{5.34}$$

As a result, the vibrorheological equation (5.29) (the main equation of vibrational mechanics) will be presented in the form

$$\ddot{X} + \mu f_1(X)\dot{X} + p^2 X = 0. \tag{5.35}$$

Here $f_1(X)\dot{X}$ is the function which characterizes the positional- viscous resistance to the slow motions of the system, i.e. it is the *effective characteristic of friction*. As a matter of convenience for comparison with the initial characteristic $f(x)\dot{x}$ we will write out these characteristics one under the other

$$\begin{aligned} f(x)\dot{x} &= (a_0 + a_1 x + a_2 x^2 + a_3 x^3 + a_4 x^4)\dot{x}, \\ f_1(X)\dot{X} &= (A_0 + A_1 X + A_2 X^2 + A_3 X^3 + A_4 X^4)\dot{X}. \end{aligned} \tag{5.36}$$

Hence we see the character of transformation, suffered by the initial characteristic of resistance: the coefficient of the linear part of the resistance a_0 has been changed for the expression $A_0 = [(a_0 + \frac{1}{2}B^2 a_2 + \frac{3}{8}B^4 a_4)$, the coefficient a_1 at $x\dot{x}$ has been changed for the term $A_1 = a_1 + \frac{3}{2}B^2 a_3$, $A_2 = a_2 + 3B^2 a_4$, $A_3 = a_3$ $A_4 = a_4$ etc. These changes explain the important regularities of slow motions of the system which are considered below.

Equation (5.35) is answered by the only position of equilibrium $X = 0$; this position will be either stable or unstable depending on the type of the function $f_1(X)$. If all coefficients $a_0, \ldots, a_4$ are positive, then from (5.36) it follows that the effective damping in the system may become much greater than the initial, and if they are negative, then, accordingly, the "negative resistance" grows and it leads to the instability of the equilibrium position $X = 0$. Most interesting, however, are cases when the indicated coefficients have different signs. In such cases the positive damping near the point $x = 0$ in the initial system (5.25) ($a_0 > 0$) may be answered by the negative damping near $X = 0$ in the system, described by equation (5.34) ($a_0 + \frac{1}{2}B^2 a_2 + \frac{3}{8}B^4 a_4 < 0$), and vice versa.

Two such cases — the asynchronous suppression and excitation of self-excited oscillations are considered in (Blekhman, 2000).

Vibrational transformation of the elastic properties of nonlinear systems (on the theory of the Indian Magic Rope). V.N. Chelomey was probable the first to pay attention to the possibility to increase the rigidity of elastic systems by means of a high frequency vibration (ed. Blekhman, 2004).

An interesting effect of this type is the behavior of the so-called Indian magic rope: under the action of the vertical vibration a "soft" rope seems to acquire some additional rigidity and takes a stable vertical position (Fig. 5.4). The effect shows itself rather vividly. In some popular papers one can even see an affirmation (a little bit questionable) that a small monkey can climb such a stabilized vertical rope.

The effect under consideration attracted the attention of a member of investigators. The review of the relevant publications is given in the book (Blekhman, 2000). The book also renders a simple solution of the problem by method of direct separation of motions, performed by H.Dresig, E.V. Shishkina and the author. This solution is given below.

We will start from the differential equation of the bending of the rod in view of the action of the longitudinal forces:

$$EI\frac{\partial^4 u}{\partial s^4} = q + \frac{\partial}{\partial s}(N\frac{\partial u}{\partial s}) \tag{5.37}$$

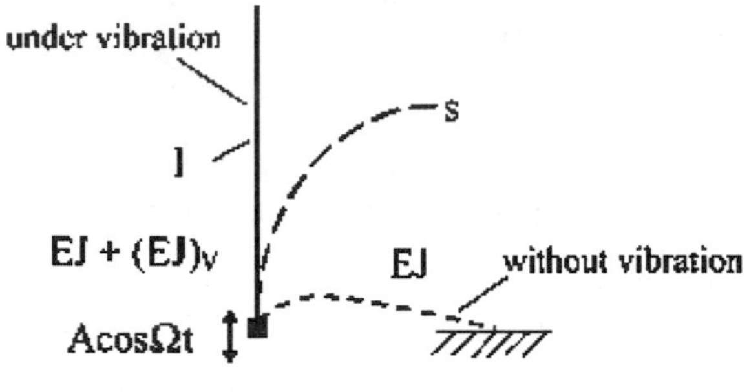

Figure 5.4. The scheme of the system.

where u is the flexure (transverse shift) of the rod (rope),s is the longitudinal coordinate, EI is the bending rigidity, q is the transverse distributed load, N is the longitudinal force. Counting off u from the vertical position and the coordinate s upward from the lower end of the rod, oscillating according to the law

$$s_0 = A\cos\omega t \tag{5.38}$$

where A is the amplitude and Ω is the oscillation frequency, we will have

$$q = -\rho F\frac{\partial^2 u}{\partial t^2}, \quad N = -\rho F(g - A\Omega^2\cos\omega t)(l - s), \tag{5.39}$$

Here ρ is the density of the material of the rod, F is its cross section area, g is the free fall acceleration, l is the length of the rod.

In view of equalities (5.39) the equation of motion (5.37) will have the form

$$EIds\frac{\partial^4 u}{\partial s^4} = -\rho F\frac{\partial^2 u}{\partial t^2} - \rho F(g - A\Omega^2\cos\omega t)\left[-\frac{\partial u}{\partial s} + (l - s)\frac{\partial^2 u}{\partial s^2}\right] \tag{5.40}$$

or, after the transfer to the dimensionless values

$$\frac{\partial^2 u}{\partial t_1^2} = (1 - \eta^2\gamma\cos\tau)\left[\frac{\partial u}{\partial\sigma} + (\sigma - 1)\frac{\partial^2 u}{\partial\sigma^2}\right] - \kappa\frac{\partial^4 u}{\partial\sigma^4} \tag{5.41}$$

where

$$\begin{aligned}&\sigma = s/l\,, \quad \eta = \Omega/\omega\,, \gamma = A/l\, quad \omega = \sqrt{g/l}\,, \quad t_1 = \omega t\,,\\ &\tau = \eta t_1 = \omega t\,, \quad k = EI/\rho F g l^3\end{aligned} \tag{5.42}$$

Solving the problem by method of direct separation of motions. Let us assume that

$$\eta >> 1\,, \quad \gamma << 1\,, \tag{5.43}$$

i.e. the vibration frequency Ω is much greater than the frequency ω, and the oscillation amplitude A is much less than the length of the rod l . To solve this problem we will use the approach of vibrational mechanics and method of direct separation of motions. Considering t_1 to be the "slow" and τ - the "fast" time, we will assume that

$$u = U(t_1) + \psi(t_1, \tau)\,, \quad \langle u \rangle = U\,, \quad \langle \psi \rangle = 0 \tag{5.44}$$

where U is the main slow component of motion- and ψ is the fast one (small as compared to U); the pointed brackets denote the averaging for the period $T = 2\pi/\Omega$ over the fast time τ.

To find the functions U and ψ, the following equations of the type (2.11, 2.12) are obtained:

$$\frac{\partial^2 U}{\partial t_1^2} = \frac{\partial U}{\partial \sigma} + (\sigma - 1)\frac{\partial^2 u}{\partial \sigma^2} - \kappa \frac{\partial^4 U}{\partial \sigma^4} - \eta^2 \gamma \left\langle \cos\tau \left[\frac{\partial \psi}{\partial \sigma} + (\sigma - 1)\frac{\partial^2 \psi}{\partial \sigma^2}\right]\right\rangle \tag{5.45}$$

$$\begin{aligned}\frac{\partial^2 \psi}{\partial t_1^2} &= -\eta^2 \gamma \cos\tau \left[\frac{\partial U}{\partial \sigma} + (\sigma - 1)\frac{\partial^2 U}{\partial \sigma^2}\right] + \frac{\partial \psi}{\partial \sigma} + (\sigma - 1)\frac{\partial^2 \psi}{\partial \sigma^2} - \kappa \frac{\partial^4 \psi}{\partial \sigma^4} \\ &-\eta^2 \gamma \cos\tau \left[\frac{\partial \psi}{\partial \sigma} + (\sigma - 1)\frac{\partial^2 \psi}{\partial \sigma^2}\right] + \eta^2 \gamma \left\langle \cos\tau \left[\frac{\partial \psi}{\partial \sigma} + (\sigma - 1)\frac{\partial^2 \psi}{\partial \sigma^2}\right]\right\rangle\end{aligned} \tag{5.46}$$

Equation (5.45) is the equation of slow motions, while equation (5.46) is the equation of fast motions. In accordance with the method used, equation (5.46) can be solved approximately. When solving this equation we will consider the function U and its derivatives to be independent of t_1 (to be "frozen") and will neglect the relatively small terms in that equation (mind that we have assumed that $\psi << U$). Then to determine ψ we will have the approximate equation

$$\frac{\partial^2 \psi}{\partial t_1^2} = -\eta^2 \gamma \cos\tau \left[\frac{\partial U}{\partial \sigma} + (\sigma - 1)\frac{\partial^2 U}{\partial \sigma^2}\right] \tag{5.47}$$

whose periodic solution will have the form

$$\psi = \gamma \left[\frac{\partial U}{\partial \sigma} + (\sigma - 1)\frac{\partial^2 U}{\partial \sigma^2}\right] \cos\tau \tag{5.48}$$

About the extent of sufficiency of such quite approximate solution see below.

Substituting expression (5.48) into equation (5.45), and having done the averaging in view of the equality $\langle \cos^2\tau \rangle = 1/2$, we come to the following equation of slow motions (the main equation of vibrational mechanics):

$$\begin{aligned}\frac{\partial^2 U}{\partial t_1^2} &= \frac{\partial U}{\partial \sigma} + \left[(\sigma - 1) - \underline{\gamma^2 \eta^2}\right] \frac{\partial^2 U}{\partial \sigma^2} - \underline{2\gamma^2 \eta^2 (\sigma - 1)\frac{\partial^3 U}{\partial \sigma^3}} - \\ &\left[\kappa + \underline{\frac{1}{2}\gamma^2 \eta^2 (\sigma - 1)^2}\right] \frac{\partial^4 U}{\partial \sigma^4}\end{aligned} \tag{5.49}$$

In equation (5.49) we have underlined the additional (as compared to equation (5.41)) coefficients, reflecting the action of vibration. We see, in particular, that instead of the usual dimensionless rigidity κ the equation contains a greater rigidity

$$k_* = k + k_v \tag{5.50}$$

where the additional ("vibrational") rigidity

$$k_v = \frac{1}{2}\left(\frac{A}{l}\right)^2\left(\frac{\Omega}{\omega}\right)^2\left(\frac{s}{l} - 1\right)^2 \tag{5.51}$$

increases quadratically from the free (upper) end of the rope. Equality (5.50) can be written for the dimensional rigidity in the following way

$$(EI)_* = EI + (EI)_v \tag{5.52}$$

where

$$(EI)_v = \frac{1}{2}\rho F[A\Omega(l - s)]^2 \tag{5.53}$$

is the additional rigidity.

So we can say that the physical explanation of the effect under consideration lies in the fact that the effective rigidity of the rope (rod) increases due to vibration.

It should be noted that there is a qualitative correspondence of expression (5.53) to the expression for the additional angular rigidity, which seems to be appearing due to the harmonic vibration of the axis of suspension of the mathematical pendulum. In accordance with (Blekhman, 2000) (see also i. 4.1) we obtain for this rigidity the expression

$$(c_v)_p = \frac{3}{2}m(A\Omega)^2 \tag{5.54}$$

while in our case the value is

$$(c_v)_r = \left.\frac{(EI)_v}{l}\right|_{s=0} = \frac{1}{2}m(A\Omega)^2 \tag{5.55}$$

Here $m = \rho Fl$ is the mass of the rope.

It should be emphasized that according to equation (5.49) the stability of the rod is determined by two dimensionless parameters $k = EI/\rho Fgl^3$ and $q^2 = \eta^2\gamma^2 = (\Omega A)^2/gl$, while in the static case there is but one parameter k.

Let us consider the stability of the rope. In this case it is sufficient to consider the static stability, i.e. to consider the equation in full derivatives

$$-\frac{d^2}{d\sigma^2}\left(k_*\frac{d^2U}{d\sigma^2}\right) + \frac{d}{d\sigma}\left((\sigma - 1)\frac{dU}{d\sigma}\right) = 0 \tag{5.56}$$

which is obtained from (5.49) at $\partial^2U/\partial t_1^2 = 0$.

Equation (5.56) coincides with an equation describing the behaviour of a column with rigidity equal to k and loaded with a gravity force. At the same time one end of the column should be clamped, another one - free. We shell use Bubnov-Galerkin method to obtain the boundary of an area of stability for this column.

The main equation of Bubnov-Galerkin method is the following:

$$\int_0^1 \left(\frac{d}{d\sigma} \left[k_* \frac{d^2U}{d\sigma^2} \right] + (1-\sigma)\frac{dU}{d\sigma} \right) \frac{d\eta}{d\sigma} d\sigma = 0 \tag{5.57}$$

According to the method the function η should satisfy both geometric and static boundary conditions. Assuming $\eta = 6\sigma^2 - 4\sigma^3 + \sigma^4$, $U = c(6\sigma^2 - 4\sigma^3 + \sigma^4)$ and substituting U and η to (5.57) we obtain the relationship between k and q which gives the boundary of an area of stability

$$k = -0.357q^2 + 0.125 \tag{5.58}$$

The diagram of stability of the rope in the plane of the parameters k and q is shown in Figure 5.5, a, where the value $k = 0.125$ at $q = 0$, corresponds to the absence of vibration (to the static stability of the column). Figure 5.5, b for the sake of comparison shows a diagram, obtained by means of more exact but much more complicated investigation (Fraser and Champneys, 2002). As we can see, the qualitative character of both diagrams is the same, but there is a considerable quantitative distinction: the boundary of stability area is intersected by the axis q in the first case at the point $q = 0.592$and in the second case at the point $q \approx 1.13$. This differences is explained by a rather approximate character of the solution (5.48) of equation of fast motions (5.46) and, mainly, apparently, by the fact that the boundary conditions are satisfied but approximately - only for the slow component U, i. e. only on the average for the period $2\pi/\Omega$. Along with that, the rough solution leads, as we have seen above, to a distinct physical interpretation and to the analytical description of this effect by means of a very simple investigation. At the same time the more exact solution does not possess such clear physical illustration.

It is interesting to compare the results given in Fig. 5.5 with the condition of stability of the upper position of a simple pendulum of the length l which axis vibrates according

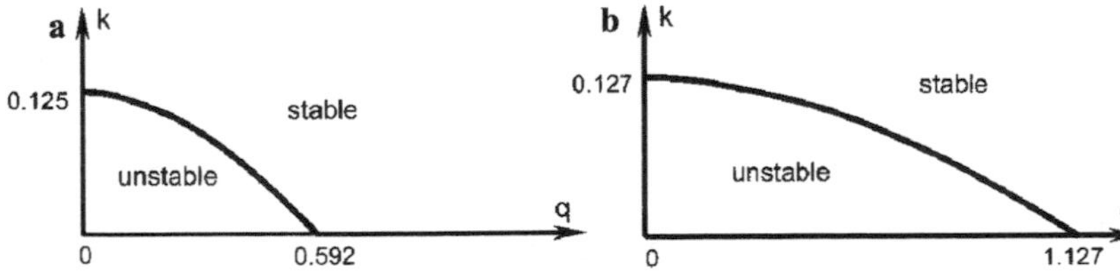

Figure 5.5. The diagram of stability of the rope upper vertical position: a - according to the approximate solution, b- according to the more exact solution (Fraser and Champneys, 2002).

to the same law (5.38). In accordance with Fig. 5.5, b in the case of negligibly small rigidity k we have the following condition of stability

$$(A\Omega) > 1.13\sqrt{gl} \tag{5.59}$$

At the same time for a pendulum

$$(A\Omega) > 1.41\sqrt{gl} \tag{5.60}$$

The condition (5.60) is stronger than (5.59). It should be noted that the difference between these conditions is quite natural: in our case we consider a rather flexible rod with a clamped lower end while the pendulum is an absolutely rigid rod with the hinged end.

Let us consider as an example the rope of the length $l = 100$ mm, which rigidity k is negligibly small. Let the oscillation frequency Ω be equal to 200 rad/s. Then from (5.59) follows that to provide the stability the amplitude A should be greater than 5.6 mm.

Thus, one can see that even for a rope without flexural rigidity its vertical position can be stabilized by means of an action of vibration. And the magnitude of vibration parameters is easy realizable in practice.

The results obtained may be regarded as referring to the extreme case of the problem of the so-called dynamic materials (Blekhman, 2000; ed. Blekhman, 2004 and see also Section 1).

5.3 Conclusion to Section 5

Section is devoted to two related groups of effects: — those of vibrational displacement and the vibro-rheological ones. The use of the approaches of vibrational mechanics when solving relative problems seem to have essential advantages.

6 Comments on Peculiarities and Advantages of the Approaches of Vibrational Mechanics.

6.1 Peculiarities and Limitations

1. The approach set forth in this course called vibrational mechanics and the method of direct separation of motions have been adapted to investigate the processes presentable in form of the sum of the main "slow" motion $X(t)$ and the fast (not necessarily small) oscillation $\psi(t, \omega t$, i.e. in the form

$$x = X(t) + \psi(t, \omega t) \tag{6.1}$$

Here lies its distinction from, e.g., the classical version of Van der Pol's method when the solution sought has the form

$$x = A(t)\sin[\omega(t)t + \beta(t)] \tag{6.2}$$

where $A(t)$, $\omega(t)$ and $\beta(t)$ are the slowly changing functions of time.

As can be seen from the content of this course and books (Blekhman, 2000; ed. Blekhman, 2004), the cases when solutions of the type of (6.1), that are of interest, are quite numerous and highly diversified.

2. When using in practice the method of direct separation of motions, the way of obtaining an approximate solution of the equation of fast motions is of special importance. The main limitation in using this method lies in the fact that the component X changes slowly enough as compared to the component ψ, i.e. that the frequency ω is sufficiently large. Usually it is sufficient that the typical period of changing the component X should be at least three times larger than that of the component ψ. Another limitation lies in the requirement that the equations of fast motions should have solutions asymptotically stable by the corresponding variables.

6.2 Advantages

Many of the advantages of the method and of the approach listed below are connected with their peculiarity that only the slow component X of the motion is of interest. These advantages are as follows:

1. The main advantage lies in the fact that the main equations (the equations of slow motions) are obtained in the form of equations of dynamics. The result of the action of vibration on the nonlinear system is expressed in these equations by means of adding to the "ordinary" slow forces the so- called vibrational forces V. The "appearance" of these forces, as well as the process of obtaining them, have in this case a distinct physical meaning.
   The above superiority of the proposed methodology is essential in the following circumstances.

   a) when revealing the so-called potential on the average dynamic systems — systems for which vibrational forces have a potential. Slow motions are described in this case by the potential system, while the initial system is essentially non-potential.
   b) when solving complicated problems in case it is difficult even to compose the initial differential equations of motion of the system. In some cases the expressions for vibrational forces, found as a result of solving simple "model" problems, can be used with certain modifications to solve much more complicated problems. (Problems on slow flows of the granular material, caused by vibration, are considered by (Blekhman, 2000)).
   c) when solving problems on the action of vibration on systems with dry friction and impacts. In these cases, instead of the initial non-smooth or discontinuous system, for the description of slow motion one gets a system with "smooth" vibrational forces. Such systems are considered in i. 5.1.
   d) when solving problems of controlling the properties of non-linear mechanical systems, including vibro-rheological properties of non-linear media by means of vibration.
   e) when solving problems of the inventors. For the creators of new machines and technologies, it is very convenient and expedient to take the position of

the observer **V** and add to all the ordinary slow forces the additional force V. It gives an opportunity "to neglect" the laws of mechanics which prevent achieving the aim. Thus the inventor can obtain quite unexpected positive effects. This approach — the temporary ignoring in the process of inventing the physical laws — is known under the name of "*fantastic analogy*" suggested by Gordon and proved rather effective (see, e.g. Blekhman, 2000).

2. Unlike the asymptotic methods the method of direct separation of motions does not require a transformation of the initial equations to the so-called canonic form. Such a transformation is not always easy and/or possible.
3. The order of the system of slow motion can be much less that the order of the initial system.
4. The method of direct separation of motions is "not very sensitive" to the accuracy of finding the fast component of motion ψ when obtaining the equation for the slow component X (the main equation of vibrational mechanics). This is connected with the fact that the fast component enters into the expression for the vibrational force only under the sign of integration. Due to that, to obtain an acceptable result it is often sufficient to find a crudely approximate solution of the equation of fast motions. In particular, we can confine ourselves to the so-called purely inertional approach.
5. As a result, instead of difficulties of solving the initial differential equations, we face the difficulties, connected with solving separately the equations of fast motions and those of slow motions. As follows from what has been said, these latter difficulties are far lesser that the first.
6. Vibrational mechanics can be considered within the frames of a much more general conception — the conception of systems with the ignored (hidden) motions (Blekhman, 2000; see Fig. 2.3). Hence vibrational mechanics can be considered to be quite well grounded. The method of direct separation of motions is also well justified, at least with regard to the majority of problems, met in applications.

6.3 Final Remarks

The contents of this study show that the enumerated peculiarities combined with limitations of the method do not prevent one from successful solving numerous applied problems on the action of vibration on nonlinear systems. Their merits make it possible to get considerable advantages when solving such problems. A natural question arises: is it possible to get the same results by other methods? For most of the problems the answer will be definitely positive. In particular, some results, stated in the book (Blekhman, 2000), were obtained by the author by means of other methods. For instance the solution of the problems on self-synchronization of mechanical vibro-exciters was obtained by Poincare-Lyapunov's method of small parameter, and the solution of the problems of the theory of vibrational displacement was got by exact methods. The reader, just like the author, can also make sure that by method of direct separation of motions the solution can be obtained in a much simpler and often in a much more general form.

At present there are a number of complicated problems of the theory of vibrational displacement and of the theory of synchronization of vibro-exciters whose solution has been obtained by the method of direct separation of motions and which has not yet been obtained (and apparently cannot be obtained) by any other method.

Bibliography

I. I. Blekhman. *Vibrational Mechanics.* World Scientific, Singapore, 2000, 509 pp.

I. I. Blekhman, editor. *Selected Topics in Vibrational Mechanics.* World Scientific, Singapore, 2004, 410 pp.

I. I. Blekhman, A. D. Myshkis, and Ya. G. Panovko. *Mechanics and applied mathematics: Logic and peculiarities of applications of mathematics.* Nauka, Moscow, 1990, 2nd ed.(in Russian), 356 pp; Germany translation: Angewandte Mathematik. Gegenstand, Logik, Besonderheiten. Veb Deutscher Verlag der Wissenschaften, Berlin, 1989, 350 S.

I. I. Blekhman. On two resonant effects under the action of a high frequency vibration on nonlinear systems. *Khimicheskaya promyshlennost*, 81(7):329–331, 2004 (in Russian).

I. I. Blekhman and L. Sperling. Behavior of the pendulum od Stephenson-Kapitsa with inner degrees of freedom. In *Proc. of the XXII Summer School "Advanced Problems in Mechanics"(APM'2004)*, June 24–July 1, St.Petersburg (Repino), Russia, p. 59-67, 2004.

I. I. Blekhman. *Synchronization in Science and Technology.* New York: ASME Press, 1998, 255 pp.

W. B. Fraser and A. R. Champneys. The "Indian rope trick" for a parametrically excited flexible rod: nonlinear and subharmonic analysis. *Proc. Roy. Soc.*, London, A458, p. 1353–1373, 2002.

E. Kremer. Vibrational liquid-bubble interaction and acoustic induced flow in gas suspension. In *Proc. of the IUTAM Symposium "Liquid-Particle in Suspension Flows", Grenoble, France, April, 1994.*

J. J. Thomsen. *Vibration and Stability: Advanced Theory, Analysis and Tools.* Springer-Verlag, Berlin-Heidelberg, 2003, 403 pp.

K. Zimmerman, I. Zeidis, J. Steigenberger and M. Pivovarov. An approach to worm-like motion. In *Proc. of the XXI ICTAM*, Warsaw, Aug., 2004.

Linearization Techniques in Stochastic Dynamic Systems

Leslaw Socha*‡

* Faculty of Mathematics and Natural Sciences, Collage of Sciences, Cardinal Stefan Wyszyński University, ul. Dewajtis 5, 01-815 Warsaw, Poland

‡ Institute of Physics, University of Silesia, Katowice, Poland

Abstract The purpose of this part of the book is to provide a review of main linearization methods in analysis of stochastic dynamic systems. Two basic groups of linearization methods namely statistical and equivalent linearization are presented. In particular moment criteria, energy criteria, linearization criteria in the space of power spectral density functions and probability density functions are discussed. Applications of linearization methods to the response analysis and control design for mechanical and structural systems subjected to earthquake excitation or road irregularities are also included.

1 Introduction

Linearization methods are the most versatile methods for analysis of nonlinear systems and structures under stochastic excitation. The concept of linearization methods is old and well described in the literature. Historically the earliest work in the theory of statistical linearization was carried out virtually simultaneously by Booton (1954) and Kazakov (1954). The objective of this method is to replace the nonlinear elements in a model by corresponding linear forms where the coefficients of linearization coefficients can be found based on a specified criterion of linearization. These methods were extended and developed mainly by Russian authors in the 1950s and 1960s in connection with the modeling of automatic control systems and next developed in mechanical and structural engineering.

A different philosophy of the replacement of a nonlinear oscillator under Gaussian excitation by a linear one under the same excitation for which the coefficients of linearization can be found from a mean–square criterion was proposed by Caughey (1959, 1960). Similarly to Krylov and Bogolubov who studied deterministic vibration systems by asymptotic methods in Krilov and Bogoliuboff (1943). Caughey called his approach by *Equivalent linearization.* Since Caughey has used the same name of method for a few version of linearization techniques with deferent criteria some misunderstandings regarding the derivation of formulas for linearization coefficients appeared in the literature. These misunderstandings caused that statistical linearization proposed by Kazakov and equivalent linearization are mainly treated in the literature as the same methods. However, some authors in their papers or books introduce different names for those techniques.

For instance, in the book of Roberts and Spanos (1990) statistical linearization in "Kazakov's sense" is described in the section entitled "Nonlinear elements without memory". Similarly in the book of Soong and Grigoriu (1993) statistical linearization is introduced in section entitled "Memoryless Transformation" while equivalent linearization in section entitled "Transformation with memory". Also, in the survey paper by Socha and Soong Socha and Soong (1991) both approaches were separately reviewed. Therefore, we will discuss in sections 2 and 3 statistical and equivalent linearization approaches separately.

The material till 1990 was reviewed, for instance in the book by Roberts and Spanos (1990) and in the survey paper by Socha and Soong (1991). The development of linearization techniques in the study of stochastic models of dynamic systems as well in theoretical aspects as in application fields over the last decade was intensive. Over 200 papers in journals and conference proceedings have been appeared during last fifteen years. In several books and texts, for instance, Soong and Grigoriu (1993), Lin and Cai (1995), Lutes and Sarkani (1997), Solnes (1997) one can find that linearization techniques are already treated as standard mathematical tools in the analysis of stochastic dynamic systems. They have been partially reviewed in Elishakoff (1995a, 2000), Proppe et al. (2003) and Socha (2005).

Since the material is very wide we present only the basic approaches. In sections 2–4 we discuss some theoretical results connected with derivation of linearization coefficients, linearization approaches with moment criteria including energy criteria and with criteria in probability density functions. The considered nonlinear dynamic systems are under external or parametric excitation in the form of stationary or nonstationary stochastic processes. We discuss also the cases of Gaussian and non–Gaussian excitations.

In sections 5–7 we discuss applications of stochastic linearization in response analysis of structures under earthquake excitation and in control of nonlinear stochastic systems with examples of active control of vehicle suspension. In particular we show an application of the classical technique for linear systems with quadratic criterion under Gaussian excitation (LQG) and linearization methods with different criteria in joint iterative procedures that determine quasi–optimal controls for nonlinear stochastic dynamic systems.

2 Statistical linearization of stochastic dynamic systems under external excitations

2.1 Moment criteria

We start our consideration with presentation the standard "classical" results about statistical linearization presented by Kazakov (1956); Kazakow (1975).

Consider a dynamic system described by nonlinear vector Ito stochastic differential equation

$$d\mathbf{x}(t) = \Phi(\mathbf{x}, t)dt + \sum_{k=1}^{M} \mathbf{G}_{k0} d\xi_k(t), \qquad \mathbf{x}(t_0) = \mathbf{x}_0, \tag{2.1.1}$$

where $\mathbf{x} = [x_1, ..., x_n]^T$ is the vector state, $\Phi = [\phi_1, ..., \phi_n]^T$ is a nonlinear vector function, $\mathbf{G}_{k0} = [\sigma_{k0}^1, ..., \sigma_{k0}^n]^T$, $k = 1, ..., M$, are deterministic vectors of intensities of noise, ξ_k are independent standard Wiener processes, the initial condition $\mathbf{x}_0$ is a vector random variable independent of ξ_k, $k = 1, ..., M$. We assume that the solution of equation (2.1.1) exists.

As was mentioned in the Introduction the objective of statistical linearization is to find for nonlinear vector $\Phi(\mathbf{x}, t)$ an equivalent one "in some sense" but in a linear form, i.e., replacing

$$\mathbf{y} = \Phi(\mathbf{x}, t) \tag{2.1.2}$$

in equation (2.1.1) by a linearized form

$$\mathbf{y} = \Phi_0(\mathbf{m}_x, \Theta_x, t) + \mathbf{K}(\mathbf{m}_x, \Theta_x, t)\mathbf{x}^0, \tag{2.1.3}$$

where $\Phi_0 = [\phi_0^1, ..., \phi_0^n]^T$ is a nonlinear vector function of the moments of variables x_j, $\mathbf{x}^0 = [x_1^0, ..., x_n^0]^T$, $\mathbf{K} = [k_{ij}]$ is an $n \times n$ matrix of statistical linearization coefficients. $\mathbf{m}_x = E[\mathbf{x}] = [m_{x_1}, ..., m_{x_n}]^T$, $m_{x_i} = E[x_i]$, $\Theta_x = [\theta_{ij}] = E[x_i^0 x_j^0]$, $i, j = 1, ..., n$, and x_i^0 is the centralized coordinate of the vector stochastic process

$$x_i^0 = x_i - m_{x_i}. \tag{2.1.4}$$

Consider first the case of one -dimensional nonlinearity. Then Φ and Φ_0 are scalars and $\mathbf{K}$ is transpose of the vector, i.e. $\mathbf{K} = [k_1, ..., k_n]^T$, where k_i are scalars. Their determination depends upon the choice of an equivalence criterion. In what follows, a few standard equivalence criteria are presented. First we quote two basic criteria introduced by Kazakov (1956).

Criterion 1^0 *Equality of the First and Second Moments of Nonlinear and Linearized Variables*

$$E[y] = m_y = \phi_0, \tag{2.1.5}$$

$$E[(y - E[y])^2] = E[(\phi - \phi_0)^2] = \sum_{i=1}^{n}\sum_{j=1}^{n} k_i k_j \theta_{ij}. \tag{2.1.6}$$

Except for the one–dimensional case, equations (2.1.5) and (2.1.6) do not determine coefficients k_i uniquely. Hence, additional relations are needed. This can be done by introducing, for example, the equality of cross second moments, i.e.

$$E[(y - E[y])(x_j - m_{x_j})] = E\left[\sum_{i=1}^{n} k_i x_i^0 x_j^0\right], \quad j = 1, ..., n. \tag{2.1.7}$$

One then obtains a system of linear equations in the form

$$\theta_{\phi_j} = \sum_{i=1}^{n} k_i \theta_{ij}, \quad j = 1, ..., n\ , \tag{2.1.8}$$

where

$$\theta_{\phi_j} = E[\phi x_j^0]. \tag{2.1.9}$$

Equations (2.1.8) and (2.1.9) combined with equation (2.1.6) produce a system of $(n+1)$ equations for n unknowns k_i, $i = 1, ..., n$. One can now either eliminate one of the equations or use another approximate means to determine k_i. One of the procedures proposed by Kazakov (1956) suggest that the coefficients k_i be sought in the form

$$k_i = \left[\frac{E[\phi^2] - \phi_0^2}{\Theta_{ii}}\mu\right]^{1/2} sgn\left(\frac{\partial \phi_0}{\partial m_x}\right), \quad i = 1, ..., n\ , \tag{2.1.10}$$

where $sgn(.)$ denotes the signum function and μ is a parameter which is found by substituting equation (2.1.10) into equation (2.1.6). With the requirement that the signs for k_i should agree with the signs of the corresponding terms of $\partial\phi_0/\partial m_x$ for normally distributed variables x_i, one obtains

$$k_i = \left\{\frac{E[\phi^2] - \phi_0^2}{\theta_{ii}}\left[\sum_{r,l=1}^{n} \frac{\theta_{rl}}{\sqrt{\theta_{rr}\theta_{ll}}}\, sgn\left(\frac{\partial \phi_0}{\partial m_{x_r}}\right) sgn\left(\frac{\partial \phi_0}{\partial m_{x_l}}\right)\right]^{-1}\right\}^{1/2} sgn\left(\frac{\partial \phi_0}{\partial m_x}\right),$$

$$i = 1, ..., n. \tag{2.1.11}$$

In the particular case of one–dimensional nonlinearity with $\Phi = \phi(x,t)$ in equation (2.1.1) one finds that $n = 1$ and

$$k_1 = \sqrt{\frac{D\phi}{\theta_{11}}}\, sgn\left(\frac{\partial \phi_0}{\partial m_{x_1}}\right), \tag{2.1.12}$$

where

$$\phi_0 = \int_{-\infty}^{+\infty} \phi(x_1,t) g(x_1,t) dx_1, \tag{2.1.13}$$

$$D\phi = \int_{-\infty}^{+\infty} \phi^2(x_1,t) g(x_1,t) dx_1 - \phi_0^2, \tag{2.1.14}$$

$$\theta_{11} = \int_{-\infty}^{+\infty} x_1^2 g(x_1,t) dx_1 - m_{x_1}^2, \tag{2.1.15}$$

$$g(x_1,t) = \frac{1}{\sqrt{2\pi\theta_{11}}} \exp\left\{-\frac{(x_1 - m_{x_1})^2}{2\theta_{11}}\right\}. \tag{2.1.16}$$

The function ϕ_0 in equation (2.1.3) in this case usually takes the form

$$\phi_0 = k_0 m_x, \tag{2.1.17}$$

where k_0 is the linearization coefficient with respect to the mean value m_x .

Criterion 2^0 *Mean–square error of approximation*

Consider the mean–square error of the linear approximation of function $y = \phi(\mathbf{x}, t)$ by the linearized form (2.1.3), i.e.

$$\delta = E\left[\left(\phi(x_1, ..., x_n, t) - \phi_0 - \sum_{i=1}^{n} k_i x_i^0\right)^2\right]. \tag{2.1.18}$$

The necessary condition of the minimum of this criterion have the form

$$\frac{\partial \delta}{\partial \phi_0} = 0, \quad \frac{\partial \delta}{\partial k_i} = 0, \quad i = 1, ..., n. \tag{2.1.19}$$

From equations (2.1.19) we calculate the following relations

$$\phi_0 = E[\phi], \tag{2.1.20}$$

$$\theta_{\phi_j} = \sum_{i=1}^{n} k_i \theta_{ij}, \quad j = 1, ..., n, \tag{2.1.21}$$

where θ_{ij} and θ_{Φ_j} are defined by equations (2.1.3) and (2.1.9), respectively, i.e.

$$\theta_{ij} = E[x_i^0 x_j^0], \qquad \theta_{\phi_j} = E[\phi x_j^0]. \tag{2.1.22}$$

Solving the system of equations (2.1.21) we find the solutions

$$k_i = \sum_{j=1}^{n} (-1)^{i+j} \frac{\Delta_{ij}}{\Delta} \theta_{\Phi_j}, \quad i = 1, ..., n, \tag{2.1.23}$$

where $\Delta = det[\theta_{ij}]$, Δ_{ij} is the cofactor of the ith column and jth row of the determinant Δ.

In the case of uncorrelated variables, equation (2.1.23) leads

$$k_i = \frac{1}{D_i} \theta_{\phi_i}, \quad i = 1, ..., n, \tag{2.1.24}$$

where $D_i = \Theta_{ii}$ is the variance of x_i.

The calculations are greatly simplified when x_i are Gaussian. Kazakov (1956) has shown that in this case the coefficients k_i are

$$k_i = \frac{\partial \phi_0}{\partial m_{x_i}}, \quad i = 1, ..., n. \tag{2.1.25}$$

Equality (2.1.25) is of course a particular case of (2.1.23).

We note that although the linearization coefficients have been determined separately they are nonlinear functions of moments of solution of system (2.1.1). Since the moments arising in linearization coefficients formulas are unknown they are approximated by the corresponding moments obtained for linearized system, i.e. for system (2.1.1), where the nonlinear vector (2.1.2) is replaced by its linearized form (2.1.3)

$$d\mathbf{x}(t) = [\mathbf{A}_0(t) + \mathbf{A}(t)\mathbf{x}(t)]dt + \sum_{k=1}^{M} \mathbf{G}_k(t)d\xi_k(t), \qquad \mathbf{x}(t_0) = \mathbf{x}_0, \tag{2.1.26}$$

where $\mathbf{A}_0 = \Phi_0 - \mathbf{K}\mathbf{m}_x = [A_{01}, ..., A_{0n}]^T$, $\mathbf{A} = \mathbf{K} = [a_{ij}]$, $i, j = 1, ..., n$ are a vector and a matrix of linearization coefficients.

First, by using Ito formula and next by averaging one can obtain the differential equations for first and second order moments of vector state $\mathbf{x}(t)$

$$\frac{d\mathbf{m}_x(t)}{dt} = \mathbf{A}_0(t) + \mathbf{A}(t)\mathbf{m}_x(t), \qquad \mathbf{m}_x(t_0) = E[\mathbf{x}_0], \tag{2.1.27}$$

$$\frac{d\Gamma_L(t)}{dt} = \mathbf{m}_x(t)\mathbf{A}_0^T(t) + \mathbf{A}_0(t)\mathbf{m}_x{}^T(t) + \Gamma_L\mathbf{A}^T(t) + \mathbf{A}(t)\Gamma_L(t)$$

$$+ \sum_{k=1}^{M} \mathbf{G}_k(t)\mathbf{G}_k^T(t), \qquad \Gamma_L(t_0) = E[\mathbf{x}_0\mathbf{x}_0^T], \tag{2.1.28}$$

where $\mathbf{m}_x = E[\mathbf{x}]$, $\Gamma_L = E[\mathbf{x}\mathbf{x}^T]$

or in an equivalent form

$$\frac{dm_{xi}(t)}{dt} = \phi_0^i(\mathbf{m}_x, \Theta_x, t), \qquad m_{xi}(t_0) = E[x_{i0}], \quad i = 1, ..., n, \tag{2.1.29}$$

$$\frac{d\theta_{ij}}{dt} = \sum_{l=1}^{n}(k_{il}\theta_{lj}(t) + k_{jl}\theta_{il}(t)) + \sum_{k=1}^{M} \sigma_{k0}^i(t)\sigma_{k0}^j(t), \;\; \theta_{ij}(t_0) = E[x_{i0}^0 x_{j0}^0] = \Theta_{ij0},$$

$$i, j = 1, ..., n, \tag{2.1.30}$$

where m_{xi} are elements (components) of vector $\mathbf{m}_x$, i.e. $\mathbf{m}_x = [m_{x_1}, ..., m_{x_n}]^T$, θ_{ij} are elements of the matrix Θ_x, i.e. $\Theta_x = [\theta_{ij}]$.

Since the elements ϕ_0^i and k_{ij} are functions depending only on $\mathbf{m}_x$ and Θ_x, i.e. equations (2.1.29) and (2.1.30) are nonlinear differential equations and can be solved by standard methods, for instance by Runge–Kutta method. In the case of the determination of stationary solutions (for $t \to \infty$) right sides of equations (2.1.29) and (2.1.30) are equated to zero. Then we obtain a nonlinear system of algebraic equations. The determination of linearization coefficients and response characteristics for dynamic system can be done by an iterative procedure involving algorithms of solving Lyapunov equation. It will be illustrated further on an example.

Energy criteria

A new class of linearization criteria for nonlinear stochastic dynamic systems were developed by Elishakoff and his coworkers. The idea of these criteria was the replacement of the displacements in Kazakov's criteria by corresponding energies of displacements. Then the corresponding criterion of equality of first two order moments of potential energies of nonlinear and linearized displacement and mean–square error of potential energies have the following form

Criterion 3^0 *Criterion of equality of first two order moments of potential energies of nonlinear and linearized displacement* Elishakoff and Zhang (1991)

$$E[(U(x))^2] = E[(\frac{1}{2}k_{eq}x^2)^2], \tag{2.1.31}$$

where $U(x)$ is the potential energy of the nonlinear displacement (static element) $\Phi = f(x)$,

$$U(x) = \int_0^x f(s)ds. \tag{2.1.32}$$

Hence, we find

$$k_{eq} = 2\sqrt{\frac{E[U(x)^2]}{E[x^4]}}. \tag{2.1.33}$$

Criterion 4^0 *Mean–square criterion for potential energies* Zhang et al. (1991)

$$E\left[\left(U(x) - \frac{1}{2}k_{eq}x^2\right)^2\right] \longrightarrow min. \tag{2.1.34}$$

Assuming that the averaging operation arising in criterion (2.1.34) does not depend on the linearization coefficient k_{eq} one can find the necessary condition of minimum of criterion (2.1.34) and then calculate

$$k_{eq} = 2\frac{E[U(x)x^2]}{E[x^4]}. \tag{2.1.35}$$

Similarly to criteria 1^0 and 2^0 the linearization coefficients obtained by energy criteria also depend on moments of the solution of nonlinear dynamic system (higher order moments than in the case of criteria 1^0 and 2^0). To obtain required higher order moments one can solve a system of nonlinear algebraic equations or nonlinear differential equations in the case of stationary or nonstationary moments, respectively. Another possibility is the replacement of the moments of nonlinear dynamic system by the corresponding moments of the linearized system in an iterative procedure in a similar way as for criteria 1^0 and 2^0. Then the situation simplifies because the considered processes are Gaussian and all higher order moments can be expressed as polynomials of the first two order moments. We show the functioning of the iterative procedure for criteria 1^0, 2^0, 3^0 and 4^0 on the following example.

Example 2.1 Consider the nonlinear scalar dynamic system

$$dx = -(x + Ax^3)dt + \sqrt{q}d\xi, \tag{2.1.36}$$

where $A > 0$ and $q > 0$ are constant coefficients and ξ is a standard Wiener process.
First we calculate linearization coefficients for a nonlinear function $Y = Ax^3$. Statistically linearized function has the form

$$Y = k_0 m_x + k_1(x - m_x). \tag{2.1.37}$$

We assume that the variable $x(t)$ is approximately a Gaussian process $N(m_x, \sigma_x)$, and the probability distribution function is given by (2.1.16) for $\sigma_x^2 = \theta_{11}$. The coefficient k_0 that for all considered criteria is the same and is calculated as follows

$$k_0 = \frac{1}{\sqrt{2\pi}\sigma_x}\int_{-\infty}^{+\infty} A(\sigma_x z + m_x)^3 \exp\{-\frac{1}{2}z^2\}dz = A(m_x^2 + 3\sigma_x^2). \tag{2.1.38}$$

By calculation of the integrals of polynomials with respect to Gaussian measure we use the property that the integral $\int_{-\infty}^{+\infty}\phi(.)dx$ of an odd function is equal to zero and from the following properties

$$\int_0^{+\infty} z^{2k+1}\exp\{-az^2\}dz = \int_{-\infty}^{0} z^{2k+1}\exp\{-az^2\}dz = \frac{k!}{2a^{k+1}}, \tag{2.1.39}$$

$$\int_0^{+\infty} z^{2k}\exp\{-az^2\}dz = \int_{-\infty}^{0} z^{2k}\exp\{-az^2\}dz = \sqrt{\frac{\pi}{a}}\frac{(2k-1)!!}{2(2a)^k}. \tag{2.1.40}$$

Criterion 1^0 We use the relations (2.1.13)-(2.1.17)

$$k_0 = A(m_x^3 + 3\sigma_x^2 m_x), \tag{2.1.41}$$

$$k_1^{(1)} = \sqrt{\frac{D\Phi}{\sigma_x^2}} = A\sqrt{9m_x^4 + 36m_x^2\sigma_x^2 + 15\sigma_x^4}. \tag{2.1.42}$$

Criterion 2^0 We use the relation (2.1.25)

$$k_1^{(2)} = \frac{\partial\phi_0}{\partial m_x} = 3A(m_x^2 + \sigma_x^2). \tag{2.1.43}$$

Criterion 3^0 We use the relation (2.1.33)

$$k_1^{(3)} = 2\sqrt{\frac{E[U(x)^2]}{E[x^4]}} = \frac{\sqrt{35}}{2}A(m_x^2 + \sigma_x^2). \tag{2.1.44}$$

Criterion 4^0 We use the relation (2.1.35)

$$k_1^{(4)} = 2\frac{E[U(x)x^2]}{E[x^4]} = \frac{5}{2}A(m_x^2 + \sigma_x^2). \tag{2.1.45}$$

Now, we calculate the differential equations for the first and second order moments for system (2.1.36). They have the form

$$\frac{dm_x}{dt} = \phi_0, \tag{2.1.46}$$

$$\frac{dE[x^2]}{dt} = -2k_1^{(i)}E[x^2] + q, \qquad i = 1, ..., 4. \tag{2.1.47}$$

Hence, we find that the mean value of stationary solution of (2.1.46) is equal to zero $m_x = 0$ and for the stationary second order moment can be determined from the following iterative procedure

$$E[x^2]^{h+1} = \kappa^{h+1} = \frac{q}{2k_1^{(i)}(m_x^h, \sigma_x^h)} = \frac{q}{2[1 + \alpha_i A\kappa^h]}, \quad h \in \mathbf{N}, \tag{2.1.48}$$

where

$$\alpha_1 = \sqrt{15}, \quad \alpha_2 = 3, \quad \alpha_3 = \frac{\sqrt{35}}{2}, \quad \alpha_4 = 2.5. \tag{2.1.49}$$

Every of four procedures corresponding to four considered criteria one can rewrite in the form

$$\kappa^{h+1} = F(\kappa^h), \tag{2.1.50}$$

where

$$F(y) = \frac{q}{2[1 + \alpha_i Ay]}, \quad \alpha_i > 0, \ y > 0. \tag{2.1.51}$$

If $|F| = \sup_{|y|=1}|F(y)|$ is smaller than 1 for sufficiently small q, iterative procedures (2.1.48) for $i = 1, 2, 3, 4$ converge for $h \to \infty$. In stationary case the second order moment one can determine from the following algebraic equation

$$2\alpha_i(\kappa_i^\infty)^2 + 2(\kappa_i^\infty) + q = 0, \quad i = 1, 2, 3, 4, \tag{2.1.52}$$

then

$$(\kappa_i^\infty) = \frac{-1 + \sqrt{1 + 2A\alpha_i q}}{2A\alpha_i}, \quad i = 1, 2, 3, 4. \tag{2.1.53}$$

A comparison of characteristics of stationary responses obtained by statistical linearization with cited energy criteria and standard mean–square criterion for displacements (criterion 2^0) for Gaussian stationary excitations was done on simple examples in Elishakoff (1991), Elishakoff and Zhang (1992), Elishakoff and Falsone (1993), Elishakoff (1995b). The energy criterion 2.1.31 was also compared with statistical linearization with mean–square criterion on more complex examples such as n–story structure with installed tuned liquid dampers with crossed tube–like containers Zhang et al. (1993), nonlinear sliding structure Zhang et al. (1994), simply supported or clamped beam on elastic foundation Fang et al. (1995), Elishakoff et al. (1995). These linearization techniques were compared with the exact solutions available in those cases and with simulations. The comparison study shows that statistical linearization with energy criteria yield the results (response characteristics) in closer vicinity with the exact or simulation results than the standard statistical linearization (criteria 1^0 and 2^0).

2.2 Criteria in probability density functions space

As was mentioned in the Introduction in the case of moment criteria some information is lost, because not all order moments of the response are taken into account. Since the complete information about a continuou random variable is contained in its probability density function the present author proposed new criteria of linearization depending on difference between probability densities of responses of nonlinear and linearized systems and two approximate approaches, (see for instance Socha (2002)). One of these approaches is a method called *s*tatistical linearization with probability density criteria Socha (1999c). The objective of this method was to replace nonlinear elements in a model by corresponding linear forms, where the coefficients of linearization can be found separately for every element based on the criterion of linearization which is a probabilistic metric in probability density functions space. The elements of this space are found as probability density functions of random variables obtained by linear and nonlinear transformation of one-dimensional Gaussian variable. In this subsection we present the basic linearization criteria and procedures of the determination of linearization coefficients.

We consider nonlinear static elements in the form

$$Y_j = \psi_j(x_j), \qquad j = 1, ...n_y \tag{2.2.1}$$

and the corresponding linearized elements

$$Y_j = k_j x_j, \qquad j = 1, ..., n_y. \tag{2.2.2}$$

To determine the linearization coefficients k_j it was proposed in Socha (1999c) the following two criteria for scalar functions $\psi_j(x_j)$ for $j = 1, ..., n_y$.

Criterion 1_{PD}. Probabilistic square metric

$$I_{1_j} = \int_{-\infty}^{+\infty} (g_N(y_j) - g_L(y_j))^2 dy_j, \quad j = 1, ..., n_y, \tag{2.2.3}$$

where $g_N(y_j)$ and $g_L(y_j)$ are probability density functions of variables Y_j of nonlinear elements (2.2.1) and linearized elements (2.2.2), respectively.

Criterion 2_{PD}. Pseudo-moment metric

$$I_{2_j} = \int_{-\infty}^{+\infty} |y_j|^{2l} |g_N(y_j) - g_L(y_j)| dy_j, \quad j = 1, ..., n_y. \tag{2.2.4}$$

If we assume that the input processes acting on static elements are Gaussian processes with mean values $m_{x_j}(t) = 0$ for $j = 1, ..., n$ and probability density functions

$$g_I(x_j(t)) = \frac{1}{\sqrt{2\pi}\sigma_{x_j}(t)} \exp\left\{ - \frac{x_j^2}{2\sigma_{x_j}^2(t)} \right\}, \tag{2.2.5}$$

where $\sigma_{x_j}^2(t) = E[x_j^2(t)]$, then the output processes Y_j for $j = 1, ..., n_y$ from the static linear elements defined by equality (2.2.2) are also zero mean Gaussian processes with corresponding probability density functions

$$g_{L_j}(x_j(t)) = \frac{1}{\sqrt{2\pi}k_j\sigma_{x_j}(t)} \exp\left\{ - \frac{x_j^2}{2k_j^2\sigma_{x_j}^2(t)} \right\}. \tag{2.2.6}$$

To apply the proposed criteria 1_{PD} and 2_{PD} we have to find probability density functions $g_N(y_j)(t)$, $j = 1, ..., n_y$. Unfortunately, except for some special cases it is impossible to find them in analytical forms Pugacev and Sinicyn (1987). It is well known that one of these special cases is for a scalar strictly monotonically increasing or decreasing function

$$Y_j = \phi_j(x_j), \qquad j = 1, ..., n_y, \tag{2.2.7}$$

with continuous derivatives $\phi_j'(x_j)$ for all $x_j \in R$. Then the probability density functions of the output variables (2.2.7) are given by

$$g_N(y_j) = g_{Y_j}(y_j) = g_I(h(y_j))|h'(y_j)|, \qquad j = 1, ..., n_y, \tag{2.2.8}$$

where $g_I(x_j)$, $j = 1, ..., n$ are the probability density functions of the input variables and h_j are the inverse functions to $\phi_j(x_j)$, i.e.

$$x_j = h_j(Y_j) = \phi_j^{-1}(Y_j), \qquad j = 1, ..., n_y. \tag{2.2.9}$$

Criterion 1_{PD} can be also adjusted to energies of displacement and considered in the following form

Criterion 3_{PD}. Probabilistic square metric

$$I_{1_j} = \int_0^{+\infty} (g_{E_N}(y_j) - g_{E_L}(y_j))^2 dy_j, \quad j = 1, ..., n_y, \tag{2.2.10}$$

where $g_{E_N}(y_j)$ and $g_{E_L}(y_j)$ are probability density functions of stationary solutions of energies of nonlinear elements (2.2.1) and linearized elements (2.2.2), respectively.

Example 2.2. Consider the static nonlinear element defined by

$$Y = -\omega_0^2 x - \alpha x^3, \tag{2.2.11}$$

where ω_0 and $\alpha > 0$ are constant parameters.

For simplicity we limit our considerations to the stationary case and we assume that the input process $X(t)$ is a stationary zero mean Gaussian process described by the corresponding probability density function (2.2.5), i.e.

$$g_I(x) = \frac{1}{\sqrt{2\pi}\sigma_x} \exp\left\{-\frac{x^2}{2\sigma_x^2}\right\}, \tag{2.2.12}$$

where $\sigma_x^2 = E[X^2]$ is the variance of the process X.

Then the probability density function of stationary output process $Y(t)$ has the form

$$g_Y(y) = \frac{1}{\sqrt{2\pi}\sigma_x} \exp\left\{-\frac{(v_1+v_2)^2}{2\sigma_x^2}\right\} \frac{1}{6a\alpha}\left[\frac{a+y}{v_1^2} + \frac{a-y}{v_2^2}\right], \tag{2.2.13}$$

where

$$v_1 = \left[\frac{1}{2\alpha}\left(y + \sqrt{y^2 + 4\omega_0^6/27\alpha}\right)\right]^{\frac{1}{3}}, \quad v_2 = \left[\frac{1}{2\alpha}\left(y - \sqrt{y^2 + 4\omega_0^6/27\alpha}\right)\right]^{\frac{1}{3}},$$

$$a = \sqrt{y^2 + 4\omega_0^6/27\alpha}. \tag{2.2.14}$$

The probability density function of the linearized variable

$$Y = kx \tag{2.2.15}$$

has the form

$$g_L(y) = \frac{1}{\sqrt{2\pi}k\sigma_x(t)} \exp\left\{-\frac{x^2}{2k^2\sigma_x^2(t)}\right\}, \tag{2.2.16}$$

where k is the linearization coefficient.

In the case of Criterion 3_{PD} the energies of nonlinear and linearized systems are given by

$$E_N = \frac{\omega_0^2}{2}x^2 + \frac{\varepsilon}{4}x^4, \tag{2.2.17}$$

$$E_L = \frac{k}{2}x^2 \tag{2.2.18}$$

and the corresponding probability density functions $g_{E_N}(y_j)$ and $g_{E_L}(y_j)$ of energies of nonlinear and linearized elements have the forms

$$g_{E_N}(E_{N_j}) = \frac{2}{\sqrt{2\pi}\sigma_{x_j}(t)|\omega_0^2 z_j + \varepsilon z_j^3|} \exp\left\{-\frac{z_j^2}{2\sigma_{x_j}^2(t)}\right\}, \tag{2.2.19}$$

$$g_{E_L}(E_{L_j}) = \frac{2}{\sqrt{2\pi}\sigma_{x_j}(t)|\omega_0^2 E_{L_j}|} \exp\left\{-\frac{E_{L_j}}{2\sigma_{x_j}^2(t)}\right\}, \tag{2.2.20}$$

where

$$z_j = \sqrt{\frac{-\omega_0^2 + \sqrt{\omega_0^4 + 4\varepsilon E_{N_j}}}{\varepsilon}}. \tag{2.2.21}$$

In general case when the nonlinear functions $\psi_j(x)$, $j = 1, ..., n_y$ are not strongly monotonically increasing or decreasing or not differentiable everywhere the approximation methods have to be used.

To obtain approximate probability density functions of nonlinear random variables (2.2.7) one can use for instance the Gram-Charlier expansion (see Pugacev and Sinicyn (1987)). In particular case for a scalar function $Y_j = \phi_j(x_j)$ of a scalar random variable x_j the nonlinear variable has the probability density function

$$g_{Y_j}(y_j) = \frac{1}{\sqrt{2\pi}c_j\sigma_{Y_j}}\, exp\left\{-\frac{(y_j - m_{Y_j})^2}{2\sigma_{Y_j}^2}\right\}\left[1 + \sum_{\nu=3}^{N}\frac{c_{\nu j}}{\nu!}H_\nu\left(\frac{y_j - m_{Y_j}}{\sigma_{Y_j}}\right)\right], \tag{2.2.22}$$

where $m_{Y_j} = E[Y_j]$, $\sigma_{Y_j}^2 = E[(Y_j - m_{Y_j})^2]$; c_j are normalized constants, $c_{\nu j} = E[G_\nu(y_j - m_{Y_j})]$,$j = 1, ..., n$, $\nu = 3, 4, ..., N$ are quasi–moments, $H_\nu(x)$ and $G_\nu(x)$ are Hermite polynomials of one variable

$$H_\nu(x) = (-1)^\nu \exp\left\{\frac{x^2}{2\sigma^2}\right\}\frac{d^\nu}{dx^\nu}\exp\left\{-\frac{x^2}{2\sigma^2}\right\}, \tag{2.2.23}$$

$$G_\nu(x) = (-1)^\nu \exp\left\{\frac{x^2}{2\sigma^2}\right\}\left[\frac{d^\nu}{dy^\nu}\exp\left\{-\frac{y^2\sigma^2}{2}\right\}\right]_{y=(x/\sigma)^2}. \tag{2.2.24}$$

In contrast to statistical linearization with moment criteria in state space one can not find expressions for linearization coefficients in an analytical form. However, in some particular cases some analytical considerations can be done. For instance, for criterion 1_{PD} defined by (2.2.3) and for an input zero mean Gaussian process the necessary condition of minimum can be derived in the following form

$$\frac{\partial I_{1_j}}{\partial k_j} = \int_{-\infty}^{+\infty}(g_N(y_j,t) - g_L(y_j,t))\frac{1}{k_j}\left(1 - \frac{y_j^2}{k_j^2\sigma_{Y_j}^2}\right)g_L(y_j,t)dy_j, \quad j = 1, ..., n_y. \tag{2.2.25}$$

To apply the proposed linearization criteria in probability density functions space to the determination of the linearization coefficients k_j, $j = 1, ..., n_y$ and approximate characteristics of the stationary response of system (2.1.1) one can use moment equations for the linearized system (2.1.27) and (2.1.28), i.e.

$$\frac{d\mathbf{m}_x(t)}{dt} = \mathbf{A}_0(t) + \mathbf{A}(t)\mathbf{m}_x(t), \qquad \mathbf{m}_x(t_0) = E[\mathbf{x}_0] \; ,$$

$$\frac{d\Gamma_L(t)}{dt} = \mathbf{m}_x(t)\mathbf{A}_0^T(t) + \mathbf{A}_0(t)\mathbf{m}_x{}^T(t) + \Gamma_L\mathbf{A}^T(t) + \mathbf{A}(t)\Gamma_L(t) \; ,$$

$$+\sum_{k=1}^{M} \mathbf{G}_k(t)\mathbf{G}_k^T(t), \qquad \Gamma_L(t_0) = E[\mathbf{x}_0\mathbf{x}_0^T] \; ,$$

where $\mathbf{m}_x = E[\mathbf{x}]$, $\Gamma_L = E[\mathbf{x}\mathbf{x}^T]$.

These equations are solved in the following iterative procedure proposed by Socha (2002).

Procedure SL_{PD}

Step 1. Substitute initial values of linearization coefficients k_j, $j = 1, ..., n_y$ and calculate moments of the stationary response of linearized system by solving system (2.1.27) and (2.1.28).

Step 2. For all nonlinear and linearized elements (one dimensional) defined by (2.2.1) and (2.2.2) calculate probability density functions using m_j, σ_{x_j}, $j = 1, ..., n_y$ obtained from elements of response moments $\mathbf{m}_x$ and Γ_L and relations between moments of the stationary response obtained in *Step 1* and Hermite polynomials defined by (2.2.23) and (2.2.24).

Step 3. Consider a criterion, for instance, I_1 and apply it to all nonlinear functions separately, i.e.

$$I_{1_j} = \int_{-\infty}^{+\infty} (g_{Y_j}(y_j) - g_{L_j}(y_j))^2 dy_j, \quad j = 1, ..., n_y, \tag{2.2.26}$$

where $g_{Y_j}(y_j)$ and $g_{L_j}(y_j)$, $j = 1, ..., n_y$ are probability density functions defined by (2.2.22) and (2.2.6), respectively and find the coefficients $k_{j_{min}}$, $j = 1, ..., n_y$ which minimize criterion (2.2.26) separately for each nonlinear function. Next, substitute $k_j = k_{j_{min}}$, $j = 1, ..., n_y$.

Step 4. Calculate moments of the stationary response of the linearized system by solving system (2.1.27) and (2.1.28).

Step 5. For all nonlinear and linearized (one–dimensional) elements defined by (2.2.1) and (2.2.2) redefine probability density functions using new m_j, σ_{x_j}, $j = 1, ..., n_y$ obtained from elements of response moments $\mathbf{m}_x$ and Γ_L in *Step 4* and using the relations between moments and Hermite polynomials defined by (2.2.23) and (2.2.24).

Step 6. Repeat *Steps 3-5* until m_j, σ_{x_j}, k_j converge (with a given accuracy).

2.3 Stationary and nonstationary Gaussian excitations

In the considered models of stochastic dynamic systems it was assumed that the external excitations are Gaussian white noises. The derived relations for linearization coefficients and iterative procedures are also correct for the case when the excitations are assumed to be Gaussian stationary processes with a given correlation matrix or power spectral density matrix, i.e. for colored noise excitations. The only one difference is in calculation of stationary response moments. The following considerations will illustrate this fact.

Consider the nonlinear vector stochastic differential equation

$$\frac{d\mathbf{x}(t)}{dt} = \Phi(\mathbf{x}, t) + \sum_{k=1}^{M} \mathbf{G}_{k0}\eta_k(t), \qquad \mathbf{x}(t_0) = \mathbf{x}_0, \tag{2.3.1}$$

where $\mathbf{x} = [x_1, ..., x_n]^T$ is the vector state, $\Phi = [\phi_1, ..., \phi_n]^T$ is a nonlinear vector function, $\mathbf{G}_{k0} = [\sigma_{k0}^1, ..., \sigma_{k0}^n]^T$, $k = 1, ..., M$, are constant deterministic vectors of noise intensity, η_k are independent Gaussian processes; the initial condition $\mathbf{x}(t_0)$ is a vector random variable independent of η_k, $k = 1, ..., M$. We assume that η_k can be treated as outputs from linear filters which inputs are excited by Gaussian white noises, i.e. the filters satisfy the following algebraic equation and linear Ito stochastic differential equations

$$\eta_k(t) = \mathbf{C}_k^T \zeta_k(t), \tag{2.3.2}$$

$$d\zeta_k(t) = \mathbf{A}_{\zeta k}\zeta_k dt + \mathbf{G}_{\zeta k} d\xi_k, \tag{2.3.3}$$

where $\mathbf{A}_{\zeta k}$ are $n_k \times n_k$ dimensional negative definite matrices, $Re\lambda_{ki}(\mathbf{A}_{\zeta k}) < 0$, $\mathbf{C}_k^T, \mathbf{G}_{\zeta k}$ are n_k dimensional vectors of the real coefficients modeling the linear filter with single input and single output (SISO); ξ_k are independent standard Wiener processes.

Similarly to the case of Gaussian white noise excitation we find for system (2.3.1) the corresponding equations for linearized system

$$\frac{d\mathbf{x}(t)}{dt} = [\mathbf{A}_0(t) + \mathbf{A}(t)\mathbf{x}(t)] + \sum_{k=1}^{M} \mathbf{G}_k\eta_k(t), \qquad \mathbf{x}(t_0) = \mathbf{x}_0, \tag{2.3.4}$$

where $\mathbf{A}_0 = \Phi_0 - \mathbf{K}\mathbf{m}_x = [A_{01}, ..., A_{0n}]^T$, $\mathbf{A} = \mathbf{K} = [a_{ij}]$, $i, j = 1, ..., n$ are vector and matrix of linearization coefficients, respectively.

If we joint system (2.3.4) with filter equations (2.3.2)-(2.3.3), then we obtain a new linear vector Ito stochastic differential equation

$$d\mathbf{X}(t) = [\bar{\mathbf{A}}_0(t) + \bar{\mathbf{A}}(t)\mathbf{X}(t)]dt + \sum_{k=1}^{M} \bar{\mathbf{G}}_k d\xi_k(t), \qquad \mathbf{X}(t_0) = \mathbf{X}_0, \tag{2.3.5}$$

where

$$\mathbf{X} = \begin{bmatrix} \mathbf{x} \\ \zeta_1 \\ \zeta_2 \\ \vdots \\ \zeta_M \end{bmatrix}, \quad \bar{\mathbf{A}} = \begin{bmatrix} \mathbf{A} & \mathbf{G}_1\mathbf{C}_1^T & \mathbf{G}_2\mathbf{C}_2^T & \cdots & \mathbf{G}_M\mathbf{C}_M^T \\ \mathbf{0} & \mathbf{A}_{\zeta 1} & \mathbf{0} & \cdots & \mathbf{0} \\ \mathbf{0} & \mathbf{0} & \mathbf{A}_{\zeta 2} & \cdots & \mathbf{0} \\ \vdots & \vdots & \vdots & \ddots & \vdots \\ \mathbf{0} & \mathbf{0} & \mathbf{0} & \cdots & \mathbf{A}_{\zeta M} \end{bmatrix}, \quad \bar{\mathbf{A}}_0 = \begin{bmatrix} \mathbf{A}_0 \\ \mathbf{0} \\ \mathbf{0} \\ \vdots \\ \mathbf{0} \end{bmatrix},$$

$$\bar{\mathbf{G}}_k = \left[\mathbf{0}^T \cdots \mathbf{0}^T \ \mathbf{G}_{\zeta_k}^T \ \mathbf{0}^T \cdots \mathbf{0}^T \right]^T, \ k = 1, ..., M, \ \mathbf{X}_0 = [\mathbf{x}_0^T \mathbf{0}^T, ..., \mathbf{0}^T]^T]. \qquad (2.3.6)$$

Hence, we find moment equations that have the form similar to equations (2.1.27) and (2.1.28), i.e.

$$\frac{d\bar{\mathbf{m}}_x(t)}{dt} = \bar{\mathbf{A}}_0(t) + \bar{\mathbf{A}}(t)\bar{\mathbf{m}}_x(t), \qquad \bar{\mathbf{m}}_x(t_0) = E[\mathbf{X}_0] = E[[\mathbf{x}^T\mathbf{0}^T, ..., \mathbf{0}^T]^T], \qquad (2.3.7)$$

$$\frac{d\bar{\Gamma}_L(t)}{dt} = \bar{\mathbf{m}}_x(t)\bar{\mathbf{A}}_0^T(t) + \bar{\mathbf{A}}_0(t)\bar{\mathbf{m}}_x^T(t) + \bar{\Gamma}_L\bar{\mathbf{A}}^T(t) + \bar{\mathbf{A}}(t)\bar{\Gamma}_L(t)$$

$$+ \sum_{k=1}^{M} \bar{\mathbf{G}}_k\bar{\mathbf{G}}_k^T, \qquad \bar{\Gamma}_L(t_0) = E[\mathbf{X}_0\mathbf{X}_0^T], \qquad (2.3.8)$$

where $\bar{\mathbf{m}}_x = E[\mathbf{X}]$, $\bar{\Gamma}_L = E[\mathbf{X}\mathbf{X}^T]$. To determine linearization coefficients we use only a part of moments connected with the vector $\mathbf{x}$, i.e. $\mathbf{m}_x = E[\mathbf{x}]$, $\Gamma_L = E[\mathbf{x}\mathbf{x}^T]$.

An application of statistical linearization to the response analysis of stochastic dynamic systems under external nonstationary excitations is for some classes of excitations more complicated than for stationary excitations. The most common type of nonstationary inputs for which analytical solutions (exact or approximate) have been proposed is the uniformly (amplitude) modulated or separable random process which is defined as the product of a stationary process and a deterministic envelope function, also called the time modulating function. For the application of statistical linearization it is convenient to assume that the component stationary process is a Gaussian process. There are two basic approaches of application of statistical linearization to the response analysis of stochastic dynamic systems under external nonstationary excitations in the form product of a Gaussian colored noise and a deterministic envelope function called by Roberts and Spanos (1990) *pre–filters method* and *decomposition method.*

To present the first approach we consider the nonlinear vector stochastic differential equation

$$\frac{d\mathbf{x}(t)}{dt} = \Phi(\mathbf{x}, t) + \sum_{k=1}^{M} \mathbf{G}_{k0}(t)\eta_k(t), \qquad \mathbf{x}(t_0) = \mathbf{x}_0, \qquad (2.3.9)$$

where $\mathbf{x} = [x_1, ..., x_n]^T$ is the vector state, $\Phi = [\phi_1, ..., \phi_n]^T$ is a nonlinear vector function, $\mathbf{G}_{k0}(t) = [\sigma_{k0}^1(t), ..., \sigma_{k0}^n(t)]^T$, $k = 1, ..., M$, are time depending deterministic vectors of noise intensity, η_k are independent zero mean Gaussian stationary processes with a given correlation functions $K_{\eta_k}(\tau)$ or spectral density functions $S_{\eta_k}(\omega)$, the initial condition $\mathbf{x}_0$ is a vector random variable independent of η_k, $k = 1, ..., M$.

In pre–filter method it is assumed that η_k can be treated as outputs from linear filters which inputs are excited by Gaussian white noises, i.e. the filters satisfy the following linear algebraic equation and Ito stochastic differential equations

$$\eta_k(t) = \mathbf{C}_k^T \zeta_k(t), \tag{2.3.10}$$

$$d\zeta_k(t) = \mathbf{A}_{\zeta k} \zeta_k dt + \mathbf{G}_{\zeta k} d\xi_k, \tag{2.3.11}$$

where $\mathbf{A}_{\zeta k}$ are $n_k \times n_k$ dimensional negative definite matrices, $Re\lambda_{ki}(\mathbf{A}_{\zeta k}) < 0$, $\mathbf{C}_k^T, \mathbf{G}_{\zeta k}$ are n_k dimensional vectors of the real coefficients modeling the linear filter with single input and single output (SISO); ξ_k are independent standard Wiener processes.

Similarly to the case of Gaussian white or colored noise excitation we find for system (2.3.9) the corresponding equations for linearized system

$$\frac{d\mathbf{x}(t)}{dt} = [\mathbf{A}_0(t) + \mathbf{A}(t)\mathbf{x}(t)] + \sum_{k=1}^{M} \mathbf{G}_k(t)\eta_k(t), \qquad \mathbf{x}(t_0) = \mathbf{x}_0, \tag{2.3.12}$$

where $\mathbf{A}_0 = \Phi_0 - \mathbf{K}\mathbf{m}_x = [A_{01}, ..., A_{0n}]^T$, $\mathbf{A} = \mathbf{K} = [a_{ij}]$, $\quad i, j = 1, ..., n$ are vector and matrix of linearization coefficients, respectively.

Next, repeating considerations for stationary excitations case one can joint system equations (2.1.15) with filter equations (2.3.10)–(2.3.11) and rewrite them in the form of the linear vector Ito stochastic differential equation (2.3.5)–(2.3.6) for $\mathbf{G}_k = \mathbf{G}_k(t)$. Then the corresponding moment equations have the form (2.3.7)–(2.3.8).

We note that the problem of the determination of the coefficients of filter equations for a zero mean stationary process with a given correlation function or a power spectral density function was presented in many monographs, for instance Priestly (1981).

To determine linearization coefficients we have to solve moment equations (2.3.7)–(2.3.8) for $\mathbf{G}_k = \mathbf{G}_k(t)$ and for linearization matrices $\mathbf{A}_0(t)$ and $\mathbf{A}(t)$ depending on moments $\mathbf{m}_x(t) = E[\mathbf{x}(t)]$ and $\Gamma_L(t) = E[\mathbf{x}(t)\mathbf{x}^T(t)]$ with a given set of initial conditions for $\mathbf{m}_x(t)$, $\Gamma_L(t)$ and separately calculated initial conditions for the filter part of moment equations. These initial conditions for the filter part have to ensure the existence of stationary solution for all $t \geq 0$ for second order moments of filter variables (for details see, for instance Roberts and Spanos (1990)).

The second approach of the application of statistical linearization to stochastic dynamic systems under external nonstationary excitations in the form product of a Gaussian colored noise and a deterministic envelope function proposed by Sakata and Kimura (1980) and Roberts and Spanos (1990) and developed by Smyth and Masri (2002) called by Roberts and Spanos (1990) *decomposition method* is more complicated. The main idea of this approach is to calculate the differential equation for variance matrix $\mathbf{K}(t)$

and to approximate the convolution of Green function corresponding to linearized system and covariance matrix of nonstationary excitation appearing in this equation by a series. It is based on the spectral decomposition the random process by the orthogonal Karhunen–Loeve expansion. For details see Roberts and Spanos (1990), Smyth and Masri (2002).

2.4 Non–Gaussian excitations

The statistical linearization method was also applied to nonlinear systems with non–Gaussian excitations. First time it was applied by Tylikowski and Marowski (1986) and Grigoriu (1995a,b) for excitation in the form of stationary Poisson processes. This approach was generalized for stationary continuous non–Gaussian processes and compound Poisson processes by Sobiechowski and Socha (1998, 2000). The fundamental difference between statistical linearization methods for Gaussian and non–Gaussian excitations is the method of calculation of moments (stationary or nonstationary) for linearized systems. Since in the case of non–Gaussian excitations this method is more complex we illustrate it on an example of the Duffing oscillator described by the following differential equation

$$\ddot{X}(t) + 2\zeta\dot{X}(t) + X(t) + \epsilon X^3(t) = \eta(t), \tag{2.4.1}$$

where ζ and ϵ are constant parameters, $\eta(t)$ is a zero mean non–Gaussian stochastic process.

The objective of statistical linearization is to replace the nonlinear element $\phi = \epsilon X^3(t)$ by the linear form $\phi_L = k_0 + k_e(X(t) - E[X(t)])$, k_0 and k_e are linearization coefficients, such that a certain equivalence criterion is satisfied. Then, the linearized system has the form

$$\ddot{X}(t) + 2h\dot{X}(t) + \omega_e^2 X(t) + \epsilon_e = \eta(t), \tag{2.4.2}$$

where $\omega_e^2 = 1 + k_e$ and $\epsilon_e = k_0 - k_e E[X(t)]$.
We consider the following four equivalence criteria:

1. Criterion of equality of the first and second moments of nonlinear and linearized variables Kazakov (1956),

$$\epsilon^k E[X^{3k}(t)] = E[(\phi_L(t))^k], \quad \text{for } k = 1, 2. \tag{2.4.3}$$

2. Minimization of the mean square error of approximation Booton (1954), Kazakov (1956),

$$E[(\phi_L(t) - \epsilon X^3(t))^2]. \tag{2.4.4}$$

3. Criterion of equality of the first moments of variables and the corresponding potential energies,

$$\begin{aligned} \epsilon E[X^3(t)] &= E[(\phi_L(t))], \\ \epsilon E[\int_0^{X(t)} \xi^3 d\xi] &= E[\int_0^{X(t)} \phi_L d\xi]. \end{aligned} \tag{2.4.5}$$

4. Minimization of the mean square difference of the potential energies Zhang et al. (1991)

$$E[(\int_0^{X(t)} (\phi_L(t) - \epsilon\xi^3)d\xi)^2]. \tag{2.4.6}$$

If the stochastic process $\eta(t)$ has vanishing odd moments, the corresponding odd moments of the response are equal to zero. In this case the linearization coefficients are,

$$k_{0i} = 0, \quad \text{for } i = 1, 2, 3, 4, \tag{2.4.7}$$

$$k_{e1} = \epsilon\sqrt{\frac{E[X^6(t)]}{E[X^2(t)]}}, \; k_{e2} = \epsilon\frac{E[X^4(t)]}{E[X^2(t)]}, \; k_{e3} = \epsilon\frac{E[X^4(t)]}{2E[X^2(t)]}, \; k_{e4} = \epsilon\frac{E[X^6(t)]}{2E[X^4(t)]}. \tag{2.4.8}$$

We note that the linearization coefficients as well for moments criteria as for criteria in probability density functions space depend in a nonlinear way from moments of solutions of equation (2.4.1). Therefore, to calculate the linearization coefficients one have to calculate the moments of solutions of equation (2.4.1). However, such a method is too complicated and similarly to Gaussian excitations case we replace the moments of solution of equation (2.4.1) by the corresponding moments of solutions of linearized system (2.4.2). For convenience we rewrite equation (2.4.2) in the form of two dimensional vector equation

$$\begin{aligned} dX_1(t) &= X_2(t)dt, \\ dX_2(t) &= [-\omega_e^2 X_1(t) - 2hX_2(t) - \epsilon_1 + \eta(t)]dt. \end{aligned} \tag{2.4.9}$$

In this case, the following algorithm for the determination of the linearized system can be formulated:

1. Guess initial values for k_{ei}, $i = 1, 2, 3, 4$; for instance, $k_{ei} = 0$, $i = 1, 2, 3, 4$.
2. Calculate $E[X^k]$ for $k = 1, 2, 3, 4$ for the linear system (2.4.9).
3. Calculate coefficients k_{ei} from $i = 1, 2, 3, 4$ from corresponding relations (2.4.8).
4. Go back to step 2 and iterate until convergence.

Thus, the whole problem consists in the determination of the moments $E[X^k]$, $k = 1, \ldots, 4$, for the linear system (2.4.9). This is described below. As we only want to compare predictions for the variance of the stationary state, we limit ourselves to the stationary moments.

Calculation of the stationary moments of solutions of linearized system

We assume that the stochastic process $\eta(t)$ is non–Gaussian and can be represented by a polynomial form of a normal filtered process described by

$$\eta(t) = \sum_{i=1}^{M} \alpha_i Y^i(t) \tag{2.4.10}$$

and

$$dY(t) = -\alpha Y(t)dt + qd\xi(t), \tag{2.4.11}$$

where $\alpha, \alpha_i, (i = 1, ..., M)$ and q are positive constant parameters, $Y(t)$ is a one-dimensional colored Gaussian process, and $\xi(t)$ is a standard Wiener process.

The moment equations corresponding to the linearized system (2.4.9)– (2.4.11) have the form

$$\begin{aligned}\frac{dE[X_1^{p_1}X_2^{p_2}]}{dt} =& p_1E[X_1^{p_1-1}X_2^{p_2+1}] - \omega_e^2 p_2 E[X_1^{p_1+1}X_2^{p_2-1}] - 2hp_2E[X_1^{p_1}X_2^{p_2}] \\ &-\epsilon_1 p_2 E[X_1^{p_1}X_2^{p_2-1}] + \sum_{i=1}^{M}\alpha_i E[X_1^{p_1}X_2^{p_2-1}Y^i],\end{aligned} \tag{2.4.12}$$

for $p_1, p_2 = 0, 1, \ldots, p$, $p_1 + p_2 = p$, $p = 1, 2, \ldots, N_p$,

$$\begin{aligned}\frac{dE[X_1^{p_1}X_2^{p_2-1}Y^i]}{dt} &= p_1E[X_1^{p_1-1}X_2^{p_2}Y^i] - \omega_e^2(p_2-1)E[X_1^{p_1+1}X_2^{p_2-2}Y^i] \\ &- 2h(p_2-1)E[X_1^{p_1}X_2^{p_2-1}Y^i] - \epsilon_1(p_2-1)E[X_1^{p_1}X_2^{p_2-2}Y^i] \\ &+ (p_2-1)\sum_{j=1}^{M}\alpha_j E[X_1^{p_1}X_2^{p_2-2}Y^{i+j}] \\ &- \alpha i E[X_1^{p_1}X_2^{p_2-1}Y^i] + \frac{1}{2}i(i-1)q^2E[X_1^{p_1}X_2^{p_2-1}Y^{i-2}],\end{aligned} \tag{2.4.13}$$

for $p_1, p_2 - 1 = 0, 1, \ldots, p-1$, $p_1 + p_2 = p$, $p = 1, \ldots, N_p - 1$ and

$$\frac{dE[Y^i]}{dt} = -\alpha i E[Y^i] + \frac{1}{2}i(i-1)q^2E[Y^{i-2}], \tag{2.4.14}$$

for $i = 1, \ldots, MN_p$.

The number of equations is equal to $\frac{N_p(N_p+3)}{2} + M[N_p + 2(N_p-1) + ... + (N_p-1)2 + N_p]$. For example, if $M = 3$ and $N_p = 6$, we obtain 195 equations. We note that although system (2.4.9)–(2.4.11) is nonlinear, the moment equations (2.4.12)–(2.4.14) are in exact closed form and no closure technique has to be applied.

To illustrate the obtained results, a comparison of mean–square displacements $E[x_1^2]$ for four criteria of statistical linearization for system (2.4.9)– (2.4.11) is shown

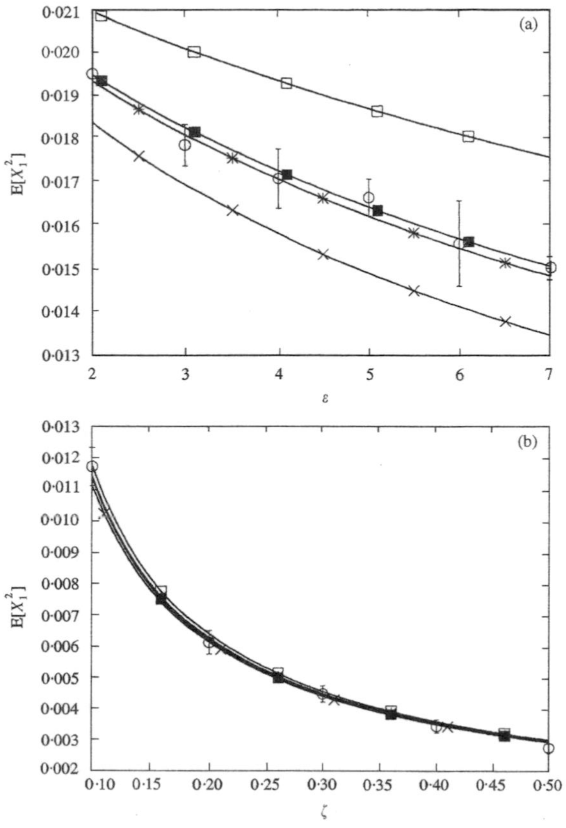

Figure 1. Prediction of $E[X_1^2]$ for the Duffing oscillator under continuous non-Gaussian external excitation Sobiechowski and Socha (2000). Equations (2.4.10) and (2.4.11) with $\alpha_1 = 0.25$, $\alpha_2 = 0.25$, $\alpha_3 = 0.25$, $\alpha = 1$, $q^2 = 0.1$, (a) $\zeta = 0.05$, (b)$\epsilon = 1$ $\times$, SL–1; $*$, SL–2; $\square$, SL–3; $\blacksquare$, SL–4; $-\circ-$, Simulation;.

in Figure 1. The notation SL–i, $i = 1, ..., 4$ is the key of the figures refers to the four linearization criteria.

Figures 1 show that there are no significant differences between response characteristics obtained by considered criteria and simulation results. However, with increasing values for ε the approximation error increases and the mean–square criteria for variables and their potential energies yield better approximations than the criterion of equality of the first and second moments of nonlinear and linearized variables and the criterion of equality of the first moments of variables and potential energies.

Another representation of continuous stationary non–Gaussian processes was proposed by Iyengar and Jaiswal (1993), and similar derivation of moment equations can be done. For details see Iyengar and Jaiswal (1993) and Sobiechowski and Socha (2000). The derivation for Poisson process excitation was also given by Sobiechowski and Socha (2000).

3 Equivalent linearization of stochastic dynamic systems under external excitation

3.1 Moment criteria

In this section we present the method of equivalent linearization and we show the differences between statistical and equivalent linearization. We consider again the nonlinear stochastic dynamic system with external excitation in the form of Gaussian white noise described by Ito vector stochastic differential equation

$$d\mathbf{x} = \Phi(\mathbf{x}, t)dt + \sum_{k=1}^{M} \mathbf{G}_k(t)d\xi_k, \qquad \mathbf{x}(t_0) = \mathbf{x}_0, \tag{3.1.1}$$

where $\mathbf{x} = [x_1, ..., x_n]^T$ is the vector state, $\Phi = [\phi_1, ..., \phi_n]^T$ is a nonlinear vector function, $\mathbf{G}_{k0} = [\sigma_{k0}^1, ..., \sigma_{k0}^n]^T$, $k = 1, ..., M$, are deterministic vectors of intensities of noise, ξ_k are independent standard Wiener processes, the initial condition $\mathbf{x}_0$ is a vector random variable independent of ξ_k, $k = 1, ..., M$.

The linearized system has the form

$$d\mathbf{x} = [\mathbf{A}(t)\mathbf{x} + \mathbf{C}(t)]dt + \sum_{k=1}^{M} \mathbf{G}_k(t)d\xi_k, \qquad \mathbf{x}(t_0) = \mathbf{x}_0, \tag{3.1.2}$$

where $\mathbf{A}(t) = [a_{ij}(t)], i, j = 1, ..., n$ and $\mathbf{C}(t) = [C_1(t), ..., C_n(t)]^T$ are time depending matrix and vector of linearization coefficients, respectively. As in the case of statistical linearization we consider two basic moment criteria

Criterion 1_{EL}^0 *Equality of mean values and second order moments of the response of nonlinear and linearized systems* Kozin (1989) (Kozin Criterion)

$$E[\mathbf{x}_N(t)] = E[\mathbf{x}_L(t)], \tag{3.1.3}$$

$$E[\mathbf{x}_N(t)\mathbf{x}_N(t)^T] = E[\mathbf{x}_L(t)\mathbf{x}_L(t)^T], \tag{3.1.4}$$

where $\mathbf{x}_N(t)$ and $\mathbf{x}_L(t)$ are solutions of equations (3.1.1) and (3.1.2), respectively.

Then we obtain a system of n^2+n equations with n^2+n unknowns a_{ij} and C_i, $i, j = 1, ..., n$), where the linearization coefficients depend on moments of solutions of nonlinear systems (3.1.1). As in statistical linearization, moments of solutions of nonlinear system are approximated by the corresponding moments of linearized system (3.1.2).

Criterion 2^0_{EL} *The mean-square error of approximation.*

Consider the mean–square error of approximation of the function $y = \Phi(\mathbf{x}, t)$ by its linearized form (3.1.2), i.e.

$$\delta_{RL} = E\left[[\Phi(x_1, ..., x_n, t) - \mathbf{A}(t)\mathbf{x} - \mathbf{C}(t)]^T[\Phi(x_1, ..., x_n, t) - \mathbf{A}(t)\mathbf{x} - \mathbf{C}(t)]\right]. \tag{3.1.5}$$

where the expectation appearing in equality (3.1.5) can be generated by the solution of nonlinear or linearized system. We denote them by $E_N[.]$ and $E_L[.]$, respectively.

The necessary conditions for minimum of Criterion 2^0_{EL} have the form

$$\frac{\partial \delta_{RL}}{\partial a_{ij}} = 0, \quad \frac{\partial \delta}{\partial C_i} = 0, \quad i, j = 1, ..., n. \tag{3.1.6}$$

The different definitions of averaging operations in Criterion 2^0_{EL} (3.1.5) give also different necessary conditions for minimum of criterion. Shortly, one can say that if the averaging operation is generated by the solution of nonlinear system or with respect to a measure independent of linearized system, then we obtain the same linearization coefficients as in the case of statistical linearization. We illustrate this idea on an example of a nonlinear oscillator.

Example 3.1. Consider a nonlinear oscillator described by Ito vector stochastic differential equation

$$d\mathbf{x} = \mathbf{F}(\mathbf{x})dt + \mathbf{G}d\xi(t), \tag{3.1.7}$$

where

$$\mathbf{x} = \begin{bmatrix} x_1 \\ x_2 \end{bmatrix}, \quad \mathbf{F}(\mathbf{x}) = \begin{bmatrix} x_2 \\ -2h\omega_0 x_2 - \omega_0^2 x_1 - f(x_1) \end{bmatrix}, \quad \mathbf{G} = \begin{bmatrix} 0 \\ q \end{bmatrix}, \tag{3.1.8}$$

$f(x)$ is a nonlinear function, $f(0) = 0$, ω_0^2, h and q are positive constants, ξ is the standard Wiener process.

The linearized stochastic differential equation has the form

$$d\mathbf{x} = \mathbf{B}x dt + \mathbf{G}d\xi(t), \tag{3.1.9}$$

where

$$\mathbf{B} = \begin{bmatrix} 0 & 1 \\ -k & -2h\omega_0 \end{bmatrix}. \tag{3.1.10}$$

First, we determine the linearization coefficient by statistical linearization. We limit our considerations to the mean–square criterion (Criterion 2^0). For nonlinear element $y = f(\eta)$ the equivalent linear element has the form $k\eta$, where η is a stationary zero mean input process $E[\eta] = 0$. Then Criterion 2^0 has the form

$$I_1 = E_{n1}[(f(\eta) - k\eta)^2], \tag{3.1.11}$$

where the averaging operation E_{n1} can be defined by the probability density function $g_{n1}(x)$ of a one dimensional non–Gaussian random variable, see Figure 2, i.e.

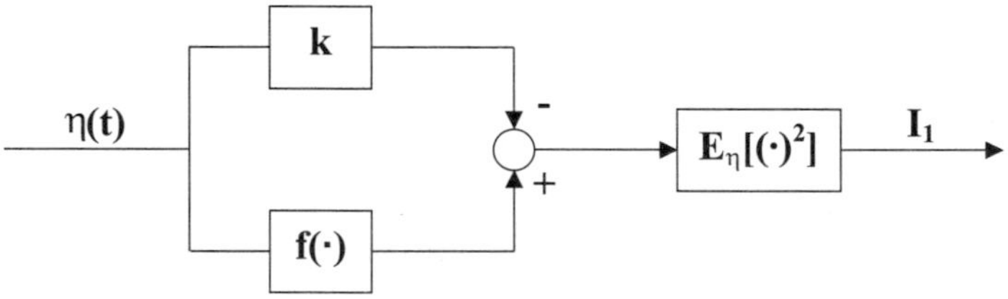

Figure 2. Schematic of statistical linearization

$$E_{n1}[.] = \int_{-\infty}^{+\infty} [.]\, g_{n1}(x_1)dx_1. \tag{3.1.12}$$

Hence, finding the necessary conditions for minimum of Criterion 2^0 we calculate the linearization coefficient

$$k_{2SL} = \omega_0^2 + \frac{E_{n1}[f(x_1)x_1]}{E_{n1}[x_1^2]}. \tag{3.1.13}$$

In the case of equivalent linearization as well the mean–square criterion as the linearization coefficient have the same form as in the case of statistical linearization, i.e. defined by relations (3.1.11) and (3.1.13), respectively, where the averaging operation is defined by the probability density function $g_{N1}(x)$ of the two dimensional random variable $\mathbf{x} = [x_1, x_2]^T$ being the stationary solution of nonlinear dynamic system (3.1.7). The averaging operation has the form

$$E_{N1}[.] = \int_{-\infty}^{+\infty} \int_{-\infty}^{+\infty} [.]\, g_{N1}(x_1, x_2)dx_1 dx_2, \tag{3.1.14}$$

where

$$g_{N1}(x_1, x_2) = \frac{1}{c_{N1}} \exp\left\{ -\frac{2h\omega_0}{q^2} \left(\omega_0^2 x_1^2 + \int_0^{x_1} f(s)ds + x_2^2 \right) \right\}, \tag{3.1.15}$$

c_{N1} is a normalized constant.

Hence, finding the necessary conditions for minimum of Criterion 2^0_{EL} we calculate the linearization coefficient

$$k_{2RL} = \omega_0^2 + \frac{E_{N1}[f(x_1)x_1]}{E_{N1}[x_1^2]}. \tag{3.1.16}$$

To obtain approximate moments of stationary solutions in both methods one should to find moment equations and apply iterative procedures, in which the averaging operations $E_{n1}[.]$ and $E_{N1}[.]$ with respect to non–Gaussian probability density functions defined by g_{n1} and g_{N1}, respectively, are replaced the averaging operation with respect to Gaussian probability density functions, i.e. in the case of statistical linearization by $E_{l1}[.]$ defined as

$$E_{l1}[.] = \int_{-\infty}^{+\infty} [.]\, g_{l1}(x_1) dx_1, \tag{3.1.17}$$

where the probability density function of Gaussian random variable has the form

$$g_{l1}(x_1) = \frac{1}{\sqrt{2\pi}\sigma_{x_1}} \exp\left\{ - \frac{x_1^2}{2\sigma_{x_1}^2} \right\}. \tag{3.1.18}$$

In the case of equivalent linearization the averaging operation E_{N1} is replaced by operation E_{L1} defined as

$$E_{L1}[.] = \int_{-\infty}^{+\infty} \int_{-\infty}^{+\infty} [.]\, g_{L1}(x_1, x_2, k) dx_1 dx_2, \tag{3.1.19}$$

where the probability density function of two dimensional Gaussian random variable has the form

$$g_{L1}(x_1, x_2, k) = \frac{1}{c_{L1}} \frac{4h\omega_0\sqrt{k}}{q^2} \exp\left\{ - \frac{2h\omega_0}{q^2}(kx_1^2 + x_2^2) \right\}, \tag{3.1.20}$$

see Figure 3 and 4

The moment equations for stationary solutions of linearized equation (3.1.9) are

$$2E_{L1}[x_1x_2] = 0, \tag{3.1.21}$$

$$E_{L1}[x_2^2] - kE_{L1}[x_1^2] - 2h\omega_0 E_{L1}[x_1x_2] = 0, \tag{3.1.22}$$

$$-2kE_{L1}[x_1x_2] - 4h\omega_0 E_{L1}[x_2^2] + q^2 = 0. \tag{3.1.23}$$

In the case of statistical linearization the iterative procedure has the form

Procedure A_{L1}

(1) Substitute $k = \omega_0^2$ in moment equations (3.1.21)-(3.1.23).

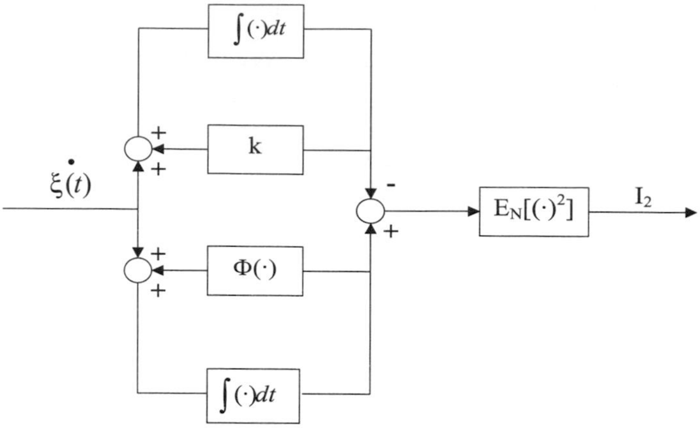

Figure 3. Schematic of equivalent linearization for E_{N1}.

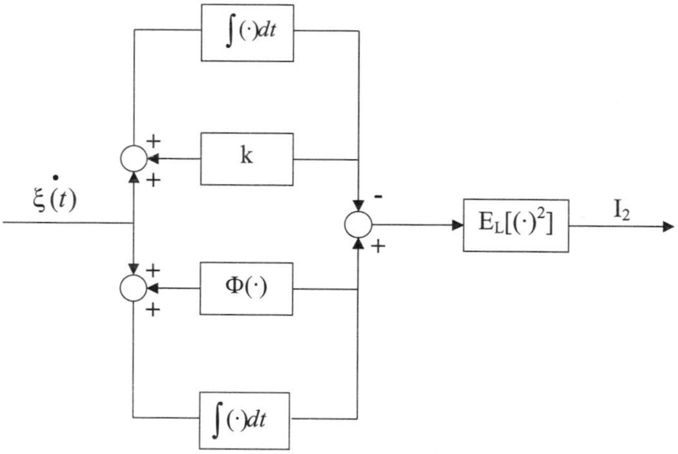

Figure 4. Schematic of equivalent linearization for E_{L1}.

(2) Replace the moments in equalities (3.1.13) by corresponding solutions of (3.1.21)-(3.1.23) and determine new linearization coefficient k using Gaussian closure.

(3) Substitute the coefficient k into moment equations (3.1.21)-(3.1.23) and solve them.

(4) Repeat steps (2) and (3) until convergence.

In the case of equivalent linearization the iterative procedure is the same except step (2), that should be replaced by the following one.

(2') Replace moments in equality (3.1.16) by corresponding moments being the solutions of moment equations (3.1.21)-(3.1.23) and determine the new linearization coefficient k using Gaussian closure.

At the end we consider the case when the averaging operation in mean–square criterion (3.1.11) is replaced by $E_{L1}[.]$ defined by (3.1.19) and (3.1.20), i.e. the mean–square criterion has the form

$$I_{L1} = E_{L1}[(f(\frac{y}{k}) - y)^2]$$

$$= \int_{-\infty}^{+\infty}\int_{-\infty}^{+\infty} \left[(f(\frac{y}{k}) - y)^2\right] \frac{1}{c_{L1}} \frac{4h\omega_0\sqrt{k}}{q^2} \exp\left\{-\frac{2h\omega_0}{q^2}(ky^2 + x_2^2)\right\} dy\, dx_2, \tag{3.1.24}$$

where c_{L1} is a normalized constant. Substituting formally $y = kx_1$ we obtain

$$I_{L1} = \int_{-\infty}^{+\infty}\int_{-\infty}^{+\infty} [(f(x_1) - kx_1)^2] \frac{1}{c_{L1}} \frac{4h\omega_0\sqrt{k^3}}{q^2} \exp\left\{-\frac{2h\omega_0}{q^2}(k^3x_1^2 + x_2^2)\right\} dx_1\, dx_2. \tag{3.1.25}$$

The necessary condition for minimum of functional I_{L1} has the form

$$\frac{\partial I_{L1}}{\partial k}$$

$$= 2\int_{-\infty}^{+\infty}\int_{-\infty}^{+\infty} [(f(x_1) - kx_1)]\,(-x_1)\frac{1}{c_{L1}} \frac{4h\omega_0\sqrt{k^3}}{q^2} \exp\left\{-\frac{2h\omega_0}{q^2}(k^3x_1^2 + x_2^2)\right\} dx_1\, dx_2$$

$$+ \int_{-\infty}^{+\infty}\int_{-\infty}^{+\infty} [(f(x_1) - kx_1)^2]\, \frac{\partial}{\partial k}\left(\frac{1}{c_{L1}} \frac{4h\omega_0\sqrt{k^3}}{q^2} \exp\left\{-\frac{2h\omega_0}{q^2}(k^3x_1^2 + x_2^2)\right\}\right) dx_1\, dx_2. \tag{3.1.26}$$

Since the integrating operation with respect to x_2 does not depend on linearization coefficient k the relation (3.1.26) reduces to the following one

$$\frac{\partial I_{L1}}{\partial k} = 2\bar{I}_L \int_{-\infty}^{+\infty} [(f(x_1) - kx_1)] (-x_1) k \bar{g}_L(x_1, k) dx_1$$

$$+\bar{I}_L \int_{-\infty}^{+\infty} [(f(x_1) - kx_1)^2] \left(\frac{3}{2} - \frac{6h\omega_0 k^3 x_1^2}{q^2} \right) \bar{g}_L(x_1, k) dx_1, \qquad (3.1.27)$$

where

$$\bar{I}_L = \int_{-\infty}^{+\infty} \frac{1}{c_{L1}} \frac{4h\omega_0}{q^2} \exp\left\{ - \frac{2h\omega_0}{q^2} x_2^2 \right\} dx_2, \quad \bar{g}_L(x_1, k) = \frac{\sqrt{k}}{c_{L1}} \exp\left\{ - \frac{2h\omega_0}{q^2} k^3 x_1^2 \right\}. \qquad (3.1.28)$$

The further simplifications are possible, for instance, for the function $\phi(x_1)$ in the form of the odd order polynomial and using properties of Gaussian random variables.

In a similar way one can show the differences between statistical and equivalent linearization methods for energy criteria.

3.2 Criteria in probability density space

An application of criteria in probability density space is also possible for equivalent linearization. It has been proposed first by the present author in Socha (1995) and further results were summarized in Socha (2002). In contrast to statistical linearization where two criteria (probabilistic metrics) were considered in equivalent linearization there are two groups of methods. In the first one, similar to methods in statistical linearization called *direct methods* probabilistic metrics are used while in the second one the Fokker–Planck–Kolmogorov equations. To discuss both approaches we consider again the nonlinear system (3.1.1) and linearized one (3.1.2).

The objective of the probability density equivalent linearization is to find the elements a_{ij} and C_i which minimize the following criterion

$$I = \int_{\mathbf{R}^n} w(\mathbf{x}) \Psi(g_N(\mathbf{x}) - g_L(\mathbf{x})) d\mathbf{x}, \qquad (3.2.1)$$

where Ψ is a convex function, $w(\mathbf{x})$ is a weight function, $g_N(\mathbf{x})$ and $g_L(\mathbf{x})$ are probability density functions of stationary solutions of nonlinear (3.1.1) and linearized (3.1.2) systems, respectively. As in the case of statistical linearization, in what follows two basic criteria are considered

(a) Probabilistic square metric

$$I_{1e} = \int_{\mathbf{R}^n} (g_N(\mathbf{x}) - g_L(\mathbf{x}))^2 d\mathbf{x}. \qquad (3.2.2)$$

(b) Pseudo–moment metric

$$I_{2e} = \int_{\mathbf{R}^n} |\mathbf{x}|^{2l} |g_N(\mathbf{x}) - g_L(\mathbf{x})| d\mathbf{x}. \tag{3.2.3}$$

It means that the discussed equivalent linearization method is also made in the space of probability density functions. To apply the proposed criterion (3.2.2) we have to find the probability density functions for stationary solutions of nonlinear system $g_N(\mathbf{x})$ and linearized one $g_L(\mathbf{x})$. Unfortunately, except some special cases it is impossible to find the both functions in analytical form. However, it can be done by approximation methods, for instance, by the Gramm–Charlier expansion or by simulations. For n-dimensional system the one dimensional density has the following truncated form Pugacev and Sinicyn (1987)

$$g_N(\mathbf{x}) = \frac{1}{c_N} g_G(\mathbf{x}) \left[1 + \sum_{k=3}^{N} \sum_{\sigma(\nu)=k} \frac{c_\nu H_\nu(\mathbf{x} - m)}{\nu_1!...\nu_n!} \right], \tag{3.2.4}$$

where c_N is a normalized constant, $g_G(\mathbf{x})$ is the probability density function of a Gaussian vector random variable

$$g_G(\mathbf{x}) = [(2\pi)^n |\mathbf{K}_G|]^{1/2} \exp\left\{ -\frac{1}{2}(\mathbf{x} - m)\mathbf{K}_G^{-1}(\mathbf{x} - m) \right\}, \tag{3.2.5}$$

where $\mathbf{m} = E[\mathbf{x}]$, $\mathbf{K}_G = E[(\mathbf{x} - m)(\mathbf{x} - m)^T]$ are the mean value and covariance matrix of vector variable $\mathbf{x}$; ν is the multiindex, $\nu = [\nu_1, ..., \nu_n]^T$, $|\mathbf{K}_G|$ is the determinant of the matrix $\mathbf{K}_G$, $\sigma(\nu) = \sum_{i=1}^n \nu_i$, N is a number of elements in truncation series, and $c_\nu = E[G_\nu(\mathbf{x} - m_x)]$, $\nu = 3, 4, ..., N$ are quasi moments.

$H_\nu(\mathbf{x})$ and $G_\nu(\mathbf{x})$ are Hermite's polynomials defined by

$$H_\nu(\mathbf{x}) = (-1)^{\sigma(\nu)} \exp\left\{ \frac{1}{2}\mathbf{x}^T K^{-1} x \right\} \frac{\partial^{\sigma(\nu)}}{\partial x_1^{\nu_1}...\partial x_n^{\nu_n}} \exp\left\{ -\frac{1}{2}\mathbf{x}^T K^{-1} x \right\}, \tag{3.2.6}$$

$$G_\nu(\mathbf{x}) = (-1)^{\sigma(\nu)} \exp\left\{ \frac{1}{2}\mathbf{x}^T K^{-1} x \right\} \left[\frac{\partial^{\sigma(\nu)}}{\partial y_1^{\nu_1}...\partial y_n^{\nu_n}} \exp\left\{ -\frac{1}{2}\mathbf{y}^T K^{-1} y \right\} \right]_{\mathbf{y}=K^{-1}x}, \tag{3.2.7}$$

where $\mathbf{K}$ is a real positive definite matrix. To obtain quasimoments c_ν first we derive moment equations for the system (3.1.1) which can be closed, for instance, by cumulant or quasi moment closure technique and next we use the algebraic relationships between quasi moments and moments.

The probability density function of the stationary solution of linearized system (3.1.2) is known and has the analytical form that can be expressed as follows:

$$g_L(\mathbf{x}) = [(2\pi)^n |\mathbf{K}_L|]^{1/2} \exp\left\{ -\frac{1}{2}(\mathbf{x} - m_x)\mathbf{K}_L^{-1}(\mathbf{x} - m_x) \right\}, \tag{3.2.8}$$

where $\mathbf{m}_x = E[\mathbf{x}]$, $\mathbf{K}_L = E[(\mathbf{x} - m_x)(\mathbf{x} - m_x)^T]$ are the mean and covariance matrix of the solution of linearized system (3.1.2). The vector $\mathbf{m}_x$ and matrix $\mathbf{K}_L$ satisfy equations

$$\frac{d\mathbf{m}_x(t)}{dt} = \mathbf{A}(t)\mathbf{m}_x(t) + \mathbf{C}(t), \qquad \mathbf{m}_x(t_0) = E[\mathbf{x}_0], \tag{3.2.9}$$

$$\begin{aligned} \frac{d\mathbf{K}_L(t)}{dt} &= \mathbf{K}_L\mathbf{A}^T(t) + \mathbf{A}(t)\mathbf{K}_L(t) + \sum_{k=1}^{M} \mathbf{G}_k(t)\mathbf{G}_k^T(t), \\ \mathbf{K}_L(t_0) &= E[(\mathbf{x}_0 - E[\mathbf{x}_0])(\mathbf{x}_0 - E[\mathbf{x}_0])^T]. \end{aligned} \tag{3.2.10}$$

Since the probability density function $g_L(\mathbf{x})$ of the linearized system (3.1.2) is a function of linearization coefficients a_{ij} and C_i then the necessary conditions of minimization of functionals I_{1e} and I_{2e} defined by (3.2.2), (3.2.3), respectively are

$$\frac{\partial I_{re}(t)}{\partial a_{ij}} = 0, \qquad \frac{\partial I_{re}(t)}{\partial C_i} = 0, \qquad i,j = 1,...,n, \;\; r = 1,2. \tag{3.2.11}$$

Using necessary conditions (3.2.11) and moment equations for nonlinear and linearized systems one can determine linearization coefficients by an iterative procedure. It is illustrated for the functional I_{1e}, rewritten as follows

$$I_{1e} = \int_{-\infty}^{+\infty} \ldots \int_{-\infty}^{+\infty} [g_N(x_1,...,x_n) - g_L(x_1,...,x_n)]^2 dx_1 ... dx_n. \tag{3.2.12}$$

The proposed iterative procedure has the following form

Procedure EL_{PD}

Step 1: Substitute initial values of linearization coefficients a_{ij}, C_i, $i,j = 1,...,n$ and calculate moments of the stationary response of linearized system (3.2.9) and (3.2.10).

Step 2: Calculate approximate moments of stationary response of nonlinear system (3.1.1) .

Step 3: Calculate probability density functions $g_L(x_1,...,x_n)$ and $g_N(x_1,...,x_n)]$ of the stationary response of nonlinear (3.1.1) and linearized (3.1.2) systems, respectively, using the moments obtained in *Step 1:* and *Step 2:*, respectively and relations between moments of the corresponding stationary responses and Hermite polynomials defined by (3.2.5) - (3.2.7).

Step 4: Choose any criterion, for instance, I_{1e} and find the linearization coefficients $a_{ij_{min}}, C_{i_{min}}$, $i,j = 1,...,n$, that minimize I_{1e} (jointly for whole nonlinear dynamic system) and substitute $a_{ij} = a_{ij_{min}}$, $C_i = C_{i_{min}}$,

Step 5: Calculate the stationary moments of linearized system (3.2.9) and (3.2.10).

Step 6: Redefine probability density functions for linearized system, by substituting the moments obtained in *Step 5* into relations defined by (3.2.4) - (3.2.7)

Step 7: Repeat *Steps 3-6* until a_{ij} and C_i, $i,j = 1,...,n$ converge.

When the probability density function of nonlinear system is unknown and for some reason the direct optimization technique can not be applied the present author proposed instead of state equations (3.1.1) - (3.1.2) to consider the corresponding reduced Fokker-Planck equations (for stationary probability density functions).

$$\frac{\partial g_N}{\partial t} = -\sum_{i=1}^{n} \frac{\partial}{\partial x_i}[\phi_i(\mathbf{x},t)g_N] + \frac{1}{2}\sum_{i=1}^{n}\sum_{j=1}^{n} \frac{\partial^2}{\partial x_i \partial x_j}[b_{Nij}g_N] = 0 \tag{3.2.13}$$

and

$$\frac{\partial g_L}{\partial t} = -\sum_{i=1}^{n} \frac{\partial}{\partial x_i}[(\mathbf{A}_i^T x + C_i)g_L] + \frac{1}{2}\sum_{i=1}^{n}\sum_{j=1}^{n} \frac{\partial^2}{\partial x_i \partial x_j}[b_{Lij}g_L] = 0, \tag{3.2.14}$$

where $\mathbf{A}_i^T$ is ith row of matrix $\mathbf{A}$; $\mathbf{B}_N = [b_{Nij}]$ and $\mathbf{B}_L = [b_{Lij}]$ are the diffusion matrices

$$b_{Nij} = b_{Lij} = \sum_{i=k}^{M} G_{ki}G_{kj}. \tag{3.2.15}$$

If we denote

$$p_1 = g_N, \quad p_{2i} = \frac{\partial g_N}{\partial x_i}, \quad q_1 = g_L, \quad q_{2i} = \frac{\partial g_L}{\partial x_i}, \tag{3.2.16}$$

then equations (3.2.13) and (3.2.13) can be transformed to the following two–dimensional vector systems

$$\frac{\partial p_1}{\partial x_i} = p_{2i},$$

$$\sum_{i=1}^{n}\left[\frac{\partial \phi_i}{\partial x_i}p_1 + \phi_i p_{2i}\right] - \frac{1}{2}\sum_{i=1}^{n}\sum_{j=1}^{n}\left[\frac{\partial^2 b_{ij}}{\partial x_i \partial x_j}p_1 + \frac{\partial b_{ij}}{\partial x_j}p_{2i} + \frac{\partial b_{ij}}{\partial x_i}p_{2j} + b_{ij}\frac{\partial p_{2j}}{\partial x_i}\right] = 0,$$

$$i,j = 1,...,n \tag{3.2.17}$$

and

$$\frac{\partial q_1}{\partial x_i} = q_{2i},$$

$$\sum_{i=1}^{n}\left[a_{ij}q_1 + (A_i^T\mathbf{x} + C_i)q_{2i}\right] - \frac{1}{2}\sum_{i=1}^{n}\sum_{j=1}^{n}\left[\frac{\partial^2 b_{ij}}{\partial x_i \partial x_j}q_1 + \frac{\partial b_{ij}}{\partial x_j}q_{2i} + \frac{\partial b_{ij}}{\partial x_i}q_{2j} + b_{ij}\frac{\partial q_{2j}}{\partial x_i}\right] = 0,$$

$$i,j = 1,...,n, \tag{3.2.18}$$

where $b_{ij} = b_{Nij}$ for system (3.2.17) and $b_{ij} = b_{Lij}$ for system (3.2.18), respectively. Comparing the system equations (3.2.17) with (3.2.18) we find that g_N and $\partial g_N/\partial x_i$ are approximated by g_L and $\partial g_L/\partial x_i$. Then the following joint criterion is proposed Socha (1999b)

$$I_3 = \int_{-\infty}^{+\infty} \epsilon_1(x)^2 dx, \qquad (3.2.19)$$

where

$$\epsilon_1(x) = \sum_{i=1}^{n} \frac{\partial}{\partial x_i} \left[(\Phi_i - A_i^T x - C_i) g_L \right]. \qquad (3.2.20)$$

Then the necessary conditions of minimum for criterion I_3 have the form (3.2.11), and the linearization coefficients are determined by an iterative procedure similar to Procedure EL_{PD}. An application of the direct method and Fokker–Planck–Kolmogorov equation method is illustrated on the following example.

Example 3.2. Consider a scalar nonlinear Ito stochastic differential equation with a nonlinear function $f(x) = \omega_0^2 x + \phi_1(x)$, i.e.

$$dx = -[\omega_0^2 x + \phi_1(x)]dt + qd\xi(t), \qquad (3.2.21)$$

where $\phi_1(x)$ is a nonlinear function, $\phi_1(0) = 0$, $q > 0$ is a constant, ξ is a standard Wiener process.

The linearized system corresponding to equation (3.2.21) is described by

$$dx = -kxdt + qd\xi(t), \qquad (3.2.22)$$

where k is a linearization coefficient.

The stationary solutions of equations (3.2.20) and (3.2.21) are characterized by corresponding stationary probability density functions

$$g_N(x) = \frac{1}{c_N} \exp\left\{ -\frac{2}{q^2} \int_0^x [\omega_0^2 s + \phi_1(s)] ds \right\}, \qquad (3.2.23)$$

$$g_L(x) = \frac{1}{c_L \sigma_L} \exp\left\{ -\frac{x^2}{2\sigma_L^2} \right\}, \qquad (3.2.24)$$

$$\sigma_L^2 = \frac{q^2}{2k}, \qquad (3.2.25)$$

where c_N and c_L are normalized constants, σ_L^2 is the variance of stationary solution of linearized system (3.2.21). Here, for both systems the mean values of stationary solutions are equal to zero $E[x] = 0$.

First we determine the linearization coefficient by the direct method for criterion I_{1e}. The necessary conditions of minimum for I_{1e} are

$$\frac{\partial}{\partial k}\int_{-\infty}^{+\infty}(g_N(x)-g_L(x))^2dx = -2\int_{-\infty}^{+\infty}(g_N(x)-g_L(x))\frac{\partial g_L(x)}{\partial k}\,dx = 0. \qquad (3.2.26)$$

Hence, we find

$$\int_{-\infty}^{+\infty}(g_N(x)-g_L(x))\left(\frac{1}{2k}-\frac{x^2}{q^2}\right)g_L(x)dx = 0, \qquad (3.2.27)$$

where $g_N(x)$ and $g_L(x)$ are defined by (3.2.23)-(3.2.25). The linearization coefficient k can be found numerically.

Next we consider the application of the Fokker-Planck equations approach to the system (3.2.21)-(3.2.22). Then the linearization criterion has the form (3.2.19), i.e.

$$I_3 = \int_{-\infty}^{+\infty}\left\{\frac{\partial}{\partial x}[\omega_0^2x+\phi_1(x)-kx]g_L(x)\right\}^2dx, \qquad (3.2.28)$$

where $g_L(x)$ is defined by (3.2.24)-(3.2.25).

The necessary condition of minimum for criterion I_3 one can find by equating to zero the derivative of I_3 with respect to the linearization coefficient k.

$$\frac{\partial}{\partial k}\int_{-\infty}^{+\infty}\left\{\frac{\partial}{\partial x}[\omega_0^2x+\phi_1(x)-kx]g_L(x)\right\}^2dx = 0. \qquad (3.2.29)$$

Hence, we calculate

$$\int_{-\infty}^{+\infty}\left\{\left(\frac{\partial f}{\partial x}-k\right)g_L+(f(x)-kx)\frac{\partial g_L}{\partial x}\right\}$$
$$\times\left\{\left(\frac{\partial f}{\partial x}-k\right)\frac{\partial g_L}{\partial k}+(f(x)-kx)\frac{\partial^2 g_L}{\partial x\partial k}-\frac{\partial}{\partial x}(xg_L)\right\}dx = 0, \qquad (3.2.30)$$

where

$$f(x)=\omega_0^2x+\phi_1(x), \qquad g_L=g_L(x,k), \qquad \frac{\partial g_L}{\partial x}=\frac{2kx}{q^2}g_L,$$

$$\frac{\partial}{\partial x}(xg_L)=\left(1-\frac{2kx^2}{q^2}\right)g_L, \qquad \frac{\partial g_L}{\partial k}=\left(\frac{1}{2k}-\frac{x^2}{q^2}\right)g_L, \qquad (3.2.31)$$

$$\frac{\partial}{\partial k}\left(\frac{\partial g_L}{\partial x}\right)=\left(-3+\frac{2kx^2}{q^2}\right)\frac{x}{q^2}g_L.$$

The further calculations are possible only for some class of nonlinear functions, for instance, if it is assumed that the function $f(x)$ is an odd polynomial of third order

$$f(x) = \omega_0^2 x + \varepsilon x^3 \tag{3.2.32}$$

Then condition (3.2.30) takes the form

$$\int_{-\infty}^{+\infty} \left\{ [3\varepsilon x^2 + (\omega_0^2 - k)] + [\varepsilon x^3 + (\omega_0^2 - k)x] \left(-\frac{2kx}{q^2} \right) \right\}$$

$$\times \left\{ [3\varepsilon x^2 + (\omega_0^2 - k)] \left(\frac{1}{2k} - \frac{x^2}{q^2} \right) + [\varepsilon x^3 + (\omega_0^2 - k)x] \left(-3 + \frac{2kx^2}{q^2} \right) \frac{x}{q^2} \right. \tag{3.2.33}$$

$$\left. - \left(1 - \frac{2kx^2}{q^2} \right) \right\} \frac{2k}{c_L q^2} \exp\left\{ -\frac{2kx^2}{q^2} \right\} dx = 0.$$

Next, using the properties of Gaussian process we calculate from equation (3.2.33) the integrals of polynomials with weight function $\exp\{-2kx^2/q^2\}$. In paper Socha (2002) it was shown that, the linearization coefficient k satisfies the following algebraic equation

$$48k^3(\omega_0^2 - k)^2 - 192k^4(\omega_0^2 - k) - 891k\varepsilon^2 q^4 - 48\varepsilon k^3 q^2 - 24(\omega_0^2 - k)k^2 q^2 = 0. \tag{3.2.34}$$

We note that, although statistical linearization and equivalent linearization with criteria in probability density function space were applied to dynamic systems under Gaussian white noise excitation both approaches can be extended to dynamic systems under Gaussian colored noise. The difference is in calculation of moments of stationary solutions.

3.3 Criteria in power spectral density space

As was mentioned in the Introduction, the idea of linearization of nonlinear elements in the frequency domain for deterministic systems and signals was called in the literature as *harmonic balance* or *describing function* and was used mainly by control engineers to the analysis of nonlinear automatic control systems. The mathematical background for this approach was given by Russian researchers Krylov in 1934 and Bogoliuboff in 1937 partially presented in Krilov and Bogoliuboff (1943) and developed by many authors (see for details and references, for instance in Kazakow and Dostupow (1962), Thaler and Pastel (1962), Naumov (1972)). The first attempts of application of describing function method to stochastic dynamic systems were done in 1980s and 1990s in Apetaur and Opicka (1983), Apetaur (1988), Apetaur (1991). The derivation of formulas for linearization coefficients in the frequency domain is much more complicated than in the time domain. The basic difficulty was the problem of the ("good approximation") of the power spectral density function of the stationary response of a nonlinear dynamic system. At the same time it has been found that in contrast to linearization criteria in time domain where one nonlinear element is replaced by a linear one in frequency domain such replacement leads always to incorrect approximation. It was necessary to approximate even a simple nonlinear dynamic system with stochastic external excitation by a multi–degree–of–freedom with deterministic coefficients and the same external excitation or by a simple linear dynamic system with random parameters.

The linearization procedure of stochastic dynamic systems in frequency domain consist with four basic stages:

(i) The determination of an approximate power spectral density function $S_N(\lambda)$ of the stationary solution of the considered nonlinear dynamic system for a given set of frequencies $\lambda \in \Lambda$.

(ii) The determination of an approximate operator transfer function $H(s)$ which defines a linear system such that his response power spectral density $S_L(\lambda)$ approximates the power spectral density function $S_N(\lambda)$ for $\lambda \in \Lambda$, for instance by relation $S_L(\lambda_k) = S_N(\lambda_k)$ for a given set of $\lambda_k \in \Lambda$. This step is called *Spectral decomposition.*

(iii) Identification by ARMA models, i.e. the determination of an approximate ARMA model for the function $H(e^{-\,i\lambda}) = \frac{Q}{P}(e^{-\,i\lambda})$, where Q, P are some polynomials.

(iv) The determination of an approximate ARMA model for a linear mechanical system, i.e. an approximate model for the function $\frac{Q}{P}(e^{-i\lambda})$ by ARMA model for a linear mechanical system.

In what follows we discuss shortly all proposed steps, starting from approximate methods of the determination of the power spectral density function of the response of a simple nonlinear dynamic system

The determination of the power spectral density function of the stationary response of a nonlinear dynamic system $S_N(\lambda)$

In the first works concerning the determination of power spectral density functions of the stationary response of simple nonlinear dynamic systems with external excitation in the form of Gaussian white noise the perturbation methods were used, for instance, by Crandall (1963, 1973), Morton and Corrisin (1970). Since the obtained results were not enough satisfying new approaches have been proposed, for instance, Roy and Spanos (1993), Krenk and Roberts (1999), Cai and Lin (1997). Roy and Spanos (1993) have proposed to expand in series the approximated power spectral density function, where the coefficients of this expansion are spectral moments that are determined by a recursive method. Krenk and Roberts (1999) used the stochastic averaging for the total energy to the determination of the conditional correlation function $R_x(\tau|E)$ and next the conditional power spectral density function $S_x(\lambda|E)$. It has the form of a series, that components represent oscillators with damping and different resonance frequencies.

Based on the knowledge of an exact probability density function for the total energy $p(E)$ that is known as an exact solution of the FPK equation corresponding to the averaged stochastic differential equation for the total energy one can calculate the sought power spectral density function of stationary response of nonlinear system in the form

$$S_x(\lambda) = \int_0^{+\infty} S_x(\lambda|E)p(E)dE. \tag{3.3.1}$$

One of the new proposed methods of the determination of an approximate power spectral density of the stationary solution of a nonlinear oscillator is an application of cumulant closure technique to the differential equations for correlation moments by Cai and Lin (1997) and next the application of Fourier transform to the obtained differential equations for correlation moments in the closed form.

Spectral decomposition

One of the simplest method of the approximation of a given power spectral density function $S_N(\lambda)$ by the power spectral density function of a stationary solution of a linear continuous time dynamic system $S_L(\lambda, \mathbf{r})$ depending on a vector of constant parameters $\mathbf{r}$ is the analysis of the minimum of a square criterion defined, for instance, by

$$I(\mathbf{r}) = \int_{-\infty}^{+\infty} (S_N(\lambda) - S_L(\lambda, \mathbf{r}))^2 d\lambda \tag{3.3.2}$$

or application of this criterion to the approximation of $S_N(\lambda)$ by the power spectral density function of a stationary solution of a linear discrete time dynamic system defined by an ARMA(p, q) model

$$x(t) + \sum_{i=1}^{p} a_i x(t-i) = \sum_{j=0}^{p} b_j v(t-j), \tag{3.3.3}$$

where $v(t)$ is a discrete time white noise. The power spectral density function of the stationary solution of such ARMA process is a periodic function over the interval $[-\pi, \pi]$ defined by

$$S_A(\lambda) = \left| \frac{MA(e^{-\ i\lambda})}{AR(e^{-\ i\lambda})} \right|^2, \tag{3.3.4}$$

where

$$AR(z) = 1 + \sum_{i=1}^{p} a_i z^i, \qquad MA(z) = 1 + \sum_{j=0}^{q} b_j z^j \tag{3.3.5}$$

are two polynomials characterizing ARMA model (AR – Auto Regressive, MA – Moving Average), a_i and b_j are parameters that should be found from conditions of minimization of functional (3.3.2), when instead of $S_L(\lambda, \mathbf{r})$ we substitute $S_A(\lambda, a_i, b_j)$ defined by (3.3.4), i.e.

$$I(a_i, b_j) = \int_{-\infty}^{+\infty} (S_N(\lambda) - S_L(\lambda, a_i, b_j))^2 d\lambda. \tag{3.3.6}$$

Unfortunately, the both criteria (3.3.2) and (3.3.6), are nonconvex with respect to parameters: elements of vector $\mathbf{r}$ for continuous time model and a_i, b_j for discrete time model, respectively. This fact causes that the optimization process is very complicated.

The mentioned difficulties were omitted by Bernard (1991) who proposed a modification of a linearization method in frequency domain. His idea was developed by Bernard and Taazount (1994) and Ismaili (1996). We discuss their approaches.

If the obtained power spectral density function for scalar stationary response process of a nonlinear system we restrict to the frequency interval $[-\lambda_c, +\lambda_c]$, where λ_c is the cut–off pulsation, then by a suitable change of variables $\lambda = \bar{\lambda}\pi/\lambda_c$ the range $[-\lambda_c, +\lambda_c]$ is transformed to $[-\pi, +\pi]$. Further we consider the obtained power spectral density function of the stationary solution of nonlinear system $S_N(\bar{\lambda})$ for $\bar{\lambda} \in [-\pi, +\pi]$.

If we assume that the function $\log S(\bar{\lambda})$ is Lebesgue integrable and is restricted to the functions that can be developed in Laurent series on the ring containing this unit circle then the power spectral density function can be represented as follows Priestly (1981), Ismaili (1996)

$$S(\bar{\lambda}) = e^{c_o} H(e^{-i\bar{\lambda}}) H^*(e^{-i\bar{\lambda}}) \quad \forall \bar{\lambda} \in [-\pi, +\pi], \tag{3.3.7}$$

where

$$H(z) = \exp\left\{\sum_{k=1}^{+\infty} c_k z^k\right\}, \quad c_k \in \mathbf{R}, \quad \sum_{k\geq 0} |c_k|^2 < \infty, \tag{3.3.8}$$

$$c_k = \frac{1}{2\pi} \int_{-\pi}^{+\pi} \log\left(S(\bar{\lambda})\right) e^{ik\bar{\lambda}} d\bar{\lambda}, \quad k \geq 0. \tag{3.3.9}$$

The coefficients c_k can be determined by Fast Fourier Transform (FFT) based on the knowledge of values of the function $S_N(\bar{\lambda})$ in $2N+1$ points in the range $[-\pi, \pi]$. Denoting

$$\bar{\lambda}_j = \frac{\pi j}{N} \quad for \quad j = 0, ..., N, \tag{3.3.10}$$

we obtain approximate coefficients c_k for $k = 0, ..., \frac{N}{2}$,

$$c_k = \frac{1}{N} \left(\sum_{j=0}^{N-1} \log\left(S(\bar{\lambda}_j)\right) \right) e^{ik\bar{\lambda}_j}. \tag{3.3.11}$$

The power spectral density function in the form (3.3.7) can be treated as a power spectral density function of an output process of a linear filter excited by a Gaussian white noise.

Identification by ARMA models

For the obtained in previous section function $H(e^{-i\bar{\lambda}})$ in (3.3.8) we find an approximate ARMA model by minimization the following square criterion

$$I(a_i, b_j) = \int_{-\pi}^{+\pi} |H(e^{-i\bar{\lambda}}) P(e^{-i\bar{\lambda}}) - Q(e^{-i\bar{\lambda}})|^2 d\bar{\lambda}, \tag{3.3.12}$$

where Q, P are polynomials

$$P(z) = 1 + \sum_{i=1}^{p} a_i z^i, \qquad Q(z) = 1 + \sum_{j=0}^{q} b_j z^j, \quad q < p. \tag{3.3.13}$$

After discretization of criterion (3.3.12) we obtain

$$I(a_i, b_j) = \sum_{k=1}^{N-1} |P(e^{-i\bar{\lambda}_k}) H(e^{-i\bar{\lambda}_k}) - Q(e^{-i\bar{\lambda}_k})|^2. \tag{3.3.14}$$

Using the necessary conditions for minimum of criterion $I(a_i, b_j)$

$$\frac{\partial I}{\partial a_i} = 0, \quad i = 1, ..., p; \qquad \frac{\partial I}{\partial b_j} = 0, \quad i = 1, ..., q; \tag{3.3.15}$$

we obtain the system of linear algebraic equations for coefficients a_i and b_j in the form

$$\mathbf{F}Y = \mathbf{G}, \tag{3.3.16}$$

where $\mathbf{Y}^T = [\mathbf{a}^T \mathbf{b}^T]$, $\mathbf{a}^T = [a_1, ..., a_p]$, $\mathbf{b}^T = [b_0, ..., b_q]$, elements of matrix $\mathbf{F} = [F_{ij}]$ and vector $\mathbf{G} = [G_1, ..., G_{p+q+1}$ are determined by elements of the matrix $\mathbf{H}$ Bernard and Taazount (1994). Hence, from equation (3.3.16) we find the linearization coefficients a_i, $i = 1, ..., p$ and b_j, $j = 0, ..., q$.

Mechanical ARMA model approximation

The approximation of the power spectral density function for scalar stationary response process of a nonlinear system presented above gives good results. However, the corresponding linear system does not represent any model of a real mechanical system. Therefore Ismaili (1996) proposed the following modification of the discussed procedure.

Consider again the function $H(e^{-i\lambda})$ in the form

$$H(e^{-i\lambda}) = \frac{Q}{P}(e^{-i\lambda}), \qquad \forall\lambda \in [-\pi, \pi]. \tag{3.3.17}$$

If some poles z_i of function H are real then they represent aperiodic terms and can not be used to modeling of mechanical oscillators. Therefore Ismaili (1996) has made a decomposition of polynomial P as a product of two polynomials

$$P = P_1 P_2, \tag{3.3.18}$$

where the roots of P_1 are non–real and the roots of P_2 are real. In further consideration we take into account only the polynomial P_1 and we are interested in approximation of function $H(e^{-i\lambda})$ by the ratio $\frac{Q_1}{P_1}(e^{-i\lambda})$, i.e. we have to find the coefficients of the polynomial Q_1 that minimize the following criterion

$$I = \int_{-\pi}^{+\pi} |H(e^{-i\lambda}) P_1(e^{-i\lambda}) - Q_1(e^{-i\lambda})|^2 d\lambda, \tag{3.3.19}$$

where the polynomial P_1 and the degree of the polynomial Q_1 are fixed, i.e. if the degree of the polynomial P_1 is equal to $2n_1$ then the degree of the polynomial Q_1 is equal to $2n_1 - 1$; the transfer function H has the form (3.3.17).

To obtain a spectrally equivalent linear mechanical model the denominator P_1 of the rational function $\frac{Q_1}{P_1}$ will be decomposed into irreducible quadratic factors. Then the function $\frac{Q_1}{P_1}$ can be decomposed as:

$$\frac{Q_1}{P_1}(e^{-i\lambda}) = \sum_{j=1}^{n_1} \alpha_j \frac{MA_j}{AR_j}(e^{-i\lambda}), \qquad \forall \lambda \in [-\pi, \pi], \tag{3.3.20}$$

where $[\alpha_1, ..., \alpha_{n_1}]^T \in R^{n_1}$, $\frac{MA_j}{AR_j}(e^{-i\lambda})$ is an output of the filter of an ARMA (2,1) process, the polynomials MA_j and AR_j define linear oscillators described by the following equations

$$\ddot{x}_j(t) + 2\eta_j\omega_{0j}\dot{x}_j(t) + \omega_{0j}^2 x_j(t) = q_j\dot{\xi}(t), \tag{3.3.21}$$

Ismaili (1996) has found the explicit formula for the coefficients η_j and ω_{0j} defining "j"–th oscillator. We note that the approximate transfer function of nonlinear dynamic system is approximated by the transfer function of mechanical ARMA$(2n, 2n-1)$ model, i.e. by a linear combination of n linear oscillators.

4 Linearization of dynamic systems with stochastic parametric excitations

4.1 Statistical linearization

The generalization of statistical linearization for dynamic systems under stochastic parametric excitations was proposed by Kazakow and Malczikow (1975). It consists of independent linearization procedures for deterministic and stochastic parts of a dynamic system that can be modeled by Stratonovich stochastic differential equation. We discus a modification of this approach using equivalent Ito stochastic differential equation.

Consider a nonlinear stochastic dynamic system described by Ito vector stochastic differential equation.

$$d\mathbf{x}(t) = \Phi(\mathbf{x}, t)dt + \sigma(\mathbf{x}, t)d\xi(t), \qquad \mathbf{x}(t_0) = \mathbf{x}_0, \tag{4.1.1}$$

where $\mathbf{x} = [x_1, ..., x_n]^T$ is the vector state, $\Phi = [\phi_1, ..., \phi_n]^T$ is a nonlinear vector function, $\sigma = [\sigma_{jk}]$ is a matrix with elements being nonlinear functions, $\sigma : R^n \times R^+ \rightarrow R^n \times R^M$, $\xi = [\xi_1, ..., \xi_M]^T$, ξ_k are independent standard Wiener processes, the initial condition $\mathbf{x}_0$ is a vector random variable independent of ξ_k, $k = 1, ..., M$.

We assume that the solution of equation (4.1.1) exists and the sufficient conditions of the stability of equilibrium of this equation are satisfied.

In the case of parametric excitations the objective of statistical linearization is the replacement as well the vector Φ as the matrix σ by the corresponding linearized forms

$$\Phi(\mathbf{x},t) \approx \Phi_0(\mathbf{m}_x,\Theta_x,t) + \mathbf{K}_\phi(\mathbf{m}_x,\Theta_x,t)\mathbf{x}^0, \tag{4.1.2}$$

$$\sigma(\mathbf{x},t) \approx \sigma_0(\mathbf{m}_x,\Theta_x,t) + \boldsymbol{K}_\sigma^T(\mathbf{m}_x,\Theta_x,t)\mathbf{x}^0 \tag{4.1.3}$$

or

$$\sigma(\mathbf{x},t) \approx \sigma_0(\mathbf{m}_x,\Theta_x,t) + \sum_{i=1}^{n} \boldsymbol{K}_{\sigma i}(\mathbf{m}_x,\Theta_x,t)\mathbf{x}_i^0, \tag{4.1.4}$$

where $\Phi_0(\mathbf{m}_x,\Theta_x,t)$, $\mathbf{K}_\phi(\mathbf{m}_x,\Theta_x,t)$, $\mathbf{m}_x$, Θ_x and $\mathbf{x}^0$ have the same meaning as in the case of statistical linearization for dynamic systems under external excitation,

$$\sigma_0(\mathbf{m}_x,\Theta_x,t) = [\sigma_{jk0}(\mathbf{m}_x,\Theta_x,t)] \tag{4.1.5}$$

is a $n \times m$ matrix,

$$\boldsymbol{K}_\sigma = [\ \boldsymbol{K}_{\sigma 1}, ...,\ \boldsymbol{K}_{\sigma n}]^T \tag{4.1.6}$$

is a block vector which elements are matrices defined by

$$\boldsymbol{K}_{\sigma i}(\mathbf{m}_x,\Theta_x,t) = [h_{jki}(\mathbf{m}_x,\Theta_x,t)] = \begin{bmatrix} h_{11i} & \cdots & h_{1mi} \\ \vdots & & \vdots \\ h_{n1i} & \cdots & h_{nki} \end{bmatrix}, \quad i,j = 1,...,n, \quad k = 1,...,M. \tag{4.1.7}$$

Linearization of the matrix $\sigma(\mathbf{x},t)$ is made similarly to the linearization of the vector $\Psi(\mathbf{x},t)$, i.e. each element of the matrix σ is linearized separately by one of the criterion presented in subsection 2.2. For instance, if we use the mean–square criterion defined by

$$\delta_{jk} = E\left[\left(\sigma_{jk}(x_1,...,x_n,t) - \sigma_{jk0} - \sum_{i=1}^{n} h_{jki}x_i^0\right)^2\right], \tag{4.1.8}$$

then linearization coefficients σ_{jk0} and h_{jki}, $i,j = 1,...,n$, $k = 1,...,M$, are determined from necessary conditions of minimum of considered criterion

$$\frac{\partial \delta_{jk}}{\partial \sigma_{jk0}} = 0, \qquad \frac{\partial \delta_{jk}}{\partial h_{jki}} = 0, \qquad i = 1,...,n. \tag{4.1.9}$$

Similarly to the derivation presented in subsection 2.2 we obtain under the assumption that variables x_i are Gaussian

$$\sigma_{jk0} = E[\sigma_{jk}] \quad j = 1,...,n, \quad k = 1,...,M. \tag{4.1.10}$$

$$h_{jki} = \frac{\partial \sigma_{jk0}}{\partial m_{x_i}} \quad i,j = 1,...,n, \quad k = 1,...,M. \tag{4.1.11}$$

As in the case of application of statistical linearization to the vector function $\Psi(\mathbf{x}, t)$ also the elements of linearized matrices σ_0 and $\boldsymbol{K}_\sigma$ are nonlinear functions of mean value and covariance matrix of state vector $\mathbf{x}$. These nonlinear relations one can find from equations (4.1.10) and (4.1.11). Since the determination of exact moments of first and second order is impossible in general case, they are replaced by the corresponding moments of linearized system. It is obtained by substitution of relations (4.1.2) and (4.1.3) to equation (4.1.1), i.e.

$$d\mathbf{x}(t) = [\Phi_0(\mathbf{m}_x, \Theta_x, t) + \mathbf{K}_\phi(\mathbf{m}_x, \Theta_x, t)\mathbf{x}^0]dt$$

$$+[\sigma_0(\mathbf{m}_x, \Theta_x, t) + \sum_{i=1}^{n} \boldsymbol{K}_{\sigma i}(\mathbf{m}_x, \Theta_x, t)\mathbf{x}_i^0]d\xi, \qquad \mathbf{x}(t_0) = \mathbf{x}_0, \tag{4.1.12}$$

or in an equivalent form

$$d\mathbf{x}(t) = [\mathbf{A}_0(t) + \mathbf{A}(t)\mathbf{x}(t)]dt + \sum_{k=1}^{m}[\mathbf{D}_k(t)\mathbf{x}(t) + \mathbf{G}_k(t)]d\xi_k(t), \qquad \mathbf{x}(t_0) = \mathbf{x}_0, \tag{4.1.13}$$

where $\mathbf{A}_0 = \Phi_0 - \mathbf{K}_\phi \mathbf{m}_x = [A_{01}, ..., A_{0n}]^T$; $\mathbf{G}_k = \sigma_{k0} - \mathbf{D}_k \mathbf{m}_x = [G_{k1}, ..., G_{kn}]^T$, $\mathbf{A} = \mathbf{K}_\phi = [a_{ij}]$, $i, j = 1, ..., n$; $\mathbf{D}_k = [h_{jki}]$, $i, j = 1, ..., n$, $k = 1, ..., M$ are vectors and matrices of linearization coefficients, respectively.

$$\sigma_{k0} = \begin{bmatrix} \sigma_{1k0} \\ \vdots \\ \sigma_{nk0} \end{bmatrix}, \quad \bar{\mathbf{K}}_{\sigma i} = \begin{bmatrix} h_{1ki} \\ \vdots \\ h_{nki} \end{bmatrix}, \quad \mathbf{D}_k = \begin{bmatrix} h_{1k1} & \cdots & h_{1kn} \\ \vdots & & \vdots \\ h_{nk1} & \cdots & h_{nkn} \end{bmatrix}, \tag{4.1.14}$$
$$i, j = 1, ..., n, \quad k = 1, ..., M.$$

Using Ito formula one can derive the moment equation of first and second order

$$\frac{d\mathbf{m}_x(t)}{dt} = \Phi_0(\mathbf{m}_x, \Theta_x, t), \qquad \mathbf{m}_x(t_0) = E[\mathbf{x}_0], \tag{4.1.15}$$

$$\frac{d\Theta_x(t)}{dt} = \mathbf{K}_\phi(\mathbf{m}_x, \Theta_x, t)\Theta_x + \Theta_x \mathbf{K}_\phi^T(\mathbf{m}_x, \Theta_x, t) + \sum_{k=1}^{m} \mathbf{D}_k(\mathbf{m}_x, \Theta_x, t)\Theta_x \mathbf{D}_k^T(\mathbf{m}_x, \Theta_x, t)$$

$$+ \sigma_0(\mathbf{m}_x, \Theta_x, t)\sigma_0^T(\mathbf{m}_x, \Theta_x, t), \qquad \theta_x(t_0) = E[\mathbf{x}_0^0(\mathbf{x}_0^0)^T], \tag{4.1.16}$$

or in an equivalent form

$$\frac{d\mathbf{m}_x(t)}{dt} = \mathbf{A}_0(t) + \mathbf{A}(t)\mathbf{m}_x(t), \qquad \mathbf{m}_x(t_0) = E[\mathbf{x}_0], \tag{4.1.17}$$

$$\frac{d\Gamma_{Lp}(t)}{dt} = \mathbf{m}_x(t)\mathbf{A}_0^T(t) + \mathbf{A}_0(t)\mathbf{m}_x{}^T(t) + \Gamma_{Lp}\mathbf{A}^T(t) + \mathbf{A}(t)\Gamma_{Lp}(t)$$

$$+\sum_{k=1}^{m}\left[\mathbf{G}_k(t)\mathbf{G}_k^T(t) + \mathbf{D}_k(t)\mathbf{m}(t)\mathbf{A}_0^T + \mathbf{A}_0(t)\mathbf{m}^T(t)\mathbf{D}_k^T + \mathbf{D}_k(t)\Gamma_{Lp}\mathbf{D}_k^T(t)\right], \quad (4.1.18)$$

$$\Gamma_{Lp}(t_0) = E[\mathbf{x}_0\mathbf{x}_0^T],$$

where $\mathbf{m}_x = E[\mathbf{x}]$, $\Gamma_{Lp} = E[\mathbf{x}\mathbf{x}^T]$.

The moment equations (4.1.15) and (4.1.16) are nonlinear differential equations that can be solved by standard methods (usually numerical methods). To obtain stationary solutions one should solve the system of algebraic equations obtained by equating to zero the right sides of the moment equations (4.1.15) and (4.1.16) or equations (4.1.17) and (4.1.18) for $t \to \infty$ taking into consideration the nonlinear dependence of linearization coefficients with respect to response moments. Then one can use the iterative procedure discussed in subsection 3.2.

We note that this approach is valid for the white noise excitation. In the case of colored noise excitations a modification of this approach has to be done. The problem is connected with the solution of moment equations for linear systems with colored noise parametric excitations. It was discussed in section 2.

In the case of statistical linearization with criteria in probability density functions space the procedure is the same as in the case of moment criteria, i.e. we linearize separately each nonlinear element and next we determine moments of solutions using, for instance a modified procedure SL_{PD}, where the solution of moment equations for nonlinear system (2.1.27) and (2.1.28) is replaced by solutions of corresponding moments for linearized system (4.1.17) and (4.1.18).

The described method of linearization for dynamic systems with stochastic parametric excitations one can generalized on dynamic systems with stochastic external and parametric excitations. Then linearization coefficients have the same form (they are calculated separately) and moment equations for linearized system depend on external and parametric excitations. We illustrate this approach on an example.

Example 4.1. Consider the vector nonlinear stochastic Ito equation

$$d\mathbf{x}(t) = [\mathbf{A}x + B(x)]dt + [\mathbf{C}(x) + D]d\xi, \qquad \mathbf{x}(t_0) = \mathbf{x}_0, \quad (4.1.19)$$

where

$$\mathbf{x} = \begin{bmatrix} x_1 \\ x_2 \end{bmatrix}, \quad \mathbf{A} = \begin{bmatrix} 0 & 1 \\ -\omega_0^2 & -2h \end{bmatrix}, \quad \mathbf{B}(x) = \begin{bmatrix} 0 \\ \varepsilon x_1^3 \end{bmatrix},$$

$$\mathbf{C}(\mathbf{x}) = \begin{bmatrix} 0 & 0 \\ \varepsilon x_1^3 & c|x_2|x_2 \end{bmatrix}, \quad \mathbf{D} = \begin{bmatrix} 0 & 0 \\ q_1 & q_2 \end{bmatrix}, \quad \xi = \begin{bmatrix} \xi_1 \\ \xi_2 \end{bmatrix}, \quad \mathbf{x}_0 = \begin{bmatrix} x_{10} \\ x_{20} \end{bmatrix}, \quad (4.1.20)$$

where $\omega_0, h, q_1, q_2, \varepsilon$ and c are positive constant parameters satisfying sufficient conditions of stability with probability one; ξ_1 and ξ_2 are independent standard Wiener processes, the initial condition $\mathbf{x}_0$ is a vector random variable independent of ξ_1 and ξ_2.

The nonlinear functions $\phi_1(x_1) = \varepsilon x_1^3$ and $\phi_2(x_2) = c|x_2|x_2$ we replace by the corresponding linearized forms, i.e.

$$\phi_1(x_1) = \varepsilon x_1^3 \approx k_0 m_{x_1} + k_1(x_1 - m_{x_1}), \tag{4.1.21}$$

$$\phi_2(x_2) = c|x_2|x_2 \approx \sigma_0 m_{x_2} + h_1(x_2 - m_{x_2}), \tag{4.1.22}$$

where k_0, k_1, σ_0 and h_1 are linearization coefficients. If we substitute the linearized functions to system (4.1.19) and (4.1.20) then we obtain a linear vector Ito stochastic differential equation

$$d\mathbf{x}(t) = [\mathbf{A}_1 x + A_0]dt + \sum_{k=1}^{2}[\mathbf{D}_k x + G_k]d\xi_k, \qquad \mathbf{x}(t_0) = \mathbf{x}_0, \tag{4.1.23}$$

where

$$\mathbf{A}_0 = \begin{bmatrix} 0 \\ (k_0 - k_1)m_{x_1} \end{bmatrix}, \quad \mathbf{A}_1 = \begin{bmatrix} 0 & 1 \\ -\omega_0^2 + k_1 & -2h \end{bmatrix}, \quad \mathbf{D}_1 = \begin{bmatrix} 0 & 0 \\ k_1 & 0 \end{bmatrix},$$

$$\mathbf{D}_2 = \begin{bmatrix} 0 & 0 \\ 0 & h_1 \end{bmatrix}, \mathbf{G}_1 = \begin{bmatrix} 0 \\ (k_0 - k_1)m_{x_1} + q_1 \end{bmatrix}, \quad \mathbf{G}_2 = \begin{bmatrix} 0 \\ (\sigma_0 - h_1)m_{x_2} + q_2 \end{bmatrix}. \tag{4.1.24}$$

Then we obtain

$$\frac{d\mathbf{m}_x(t)}{dt} = \mathbf{A}_1 m_x + A_0. \tag{4.1.25}$$

The stationary solution of equation (4.1.25) is equal to zero, i.e.

$$m_{x_1}(\infty) = m_{x_2}(\infty) = 0. \tag{4.1.26}$$

Hence it follows that also linearization coefficients for stationary solution are equal to zero

$$k_0 = \sigma_0 = 0, \tag{4.1.27}$$

while coefficients k_1 and h_1 in the case, for instance, Gaussian excitations and mean–square criterion have the form

$$k_1 = E\left[\frac{\partial(\varepsilon x_1^3)}{\partial x_1}\right] = 3\varepsilon E[x_1^2], \qquad h_1 = E\left[\frac{\partial(c|x_2|x_2)}{\partial x_2}\right] = c\sqrt{\frac{8}{\pi}}E[x_2^2]. \tag{4.1.28}$$

Deriving second order moment equations for linearized system (4.1.23) and taking into consideration equalities (4.1.27) we find relations for stationary solutions

$$2E[x_1 x_2] = 0, \tag{4.1.29}$$

$$E[x_2^2] + (-\omega_0^2 + k_1)E[x_1^2] - 2hE[x_1 x_2] = 0, \tag{4.1.30}$$

$$2(-\omega_0^2+k_1)E[x_1x_2]-4hE[x_2^2]+q_1^2+q_2^2+k_1^2E[x_1^2]+h_1^2E[x_2^2]+2k_1h_1E[x_1x_2]=0. \quad (4.1.31)$$

Hence, after substitution relations (4.1.28) to equations (4.1.29) - (4.1.31) we obtain

$$E[x_2^2]+(-\omega_0^2+3\varepsilon E[x_1^2])E[x_1^2]=0, \quad (4.1.32)$$

$$-4hE[x_2^2]+q_1^2+q_2^2+9\varepsilon^2E[x_1^2]^3+\frac{8c^2}{\pi}E[x_2^2]^3=0. \quad (4.1.33)$$

Using notations

$$y=E[x_1^2], \quad a_1=\frac{648\varepsilon^3c^2}{\pi}, \quad a_2=-\frac{216\varepsilon^2c^2\omega_0^2}{\pi}, \quad a_3=\frac{72\varepsilon c^2\omega_0^4}{\pi},$$

$$a_4=-\left(\frac{8c^2\omega_0^6}{\pi}+9\varepsilon^2\right), \quad a_5=-12h\varepsilon, \quad a_6=4h\omega_0^2, \quad a_7=-q_1^2-q_2^2$$

the system of equations (4.1.32) and (4.1.33) one can reduce to the following one equation

$$\sum_{i=1}^{7}a_iy^{7-i}=0. \quad (4.1.34)$$

Hence, it follows that to the determination of linearization coefficients k_1 and h_1 for stationary solutions it is necessary to solve numerically algebraic equation (4.1.34).

In the case of application of statistical linearization with a criterion in probability density functions space, for instance, Criterion 1_{PD} we assume that for each nonlinearity the input process is a stationary Gaussian process $x(t)$ with the probability density function given by

$$g_I(x)=\frac{1}{\sqrt{2\pi}\sigma_x}\exp\left\{-\frac{x^2}{2\sigma_x^2}\right\}, \quad (4.1.35)$$

where $\sigma_x^2=E[X^2]$ is the variance of the process $x(t)$.

Then, the probability density function for stationary output processes $Y_i(t)=\phi_i(X(t))$, $i=1,2$ have the form

$$g_{Y_1}(y)=\frac{1}{\sqrt{2\pi}\sigma_x}\exp\left\{-\frac{v_1^2}{2\sigma_x^2}\right\}\frac{1}{3\varepsilon v_1^2}, \quad (4.1.36)$$

$$g_{Y_2}(y)=\frac{1}{\sqrt{2\pi}\sigma_x}\exp\left\{-\frac{v^2}{2\sigma_x^2}\right\}\frac{1}{2cv_2}, \quad (4.1.37)$$

where

$$v_1=\left[\frac{y}{\varepsilon}\right]^{\frac{1}{3}}, \quad v_2=\left[\frac{|y|}{c}\right]^{\frac{1}{2}}.$$

The probability density functions for linearized variables

$$Y_1 = k_1 x, \qquad Y_2 = h_1 x \tag{4.1.38}$$

have the form

$$g_{L_1}(y) = \frac{1}{\sqrt{2\pi} k_1 \sigma_x(t)} \exp\left\{ - \frac{x^2}{2k_1^2 \sigma_x^2(t)} \right\}, \tag{4.1.39}$$

$$g_{L_2}(y) = \frac{1}{\sqrt{2\pi} h_1 \sigma_x(t)} \exp\left\{ - \frac{x^2}{2h_1^2 \sigma_x^2(t)} \right\}, \tag{4.1.40}$$

respectively.

To obtain the linearization coefficients and approximate response stationary moments we use a simplified of Procedure SL_{PD}.

Procedure SSL_{PD}

Step 1: Substitute the initial values of linearization coefficients $k_1 = -\omega_0^2$, $h_1 = 0$ and calculate the stationary moments of solutions of the linearized system (4.1.32) and (4.1.33), i.e.

$$E[x_1^2] = \frac{q_1^2 + q_2^2}{(4h - h_1^2)(\omega_0^2 - k_1) + k_1^2}, \tag{4.1.41}$$

$$E[x_2^2] = \frac{(q_1^2 + q_2^2)(\omega_0^2 - k_1)}{(4h - h_1^2)(\omega_0^2 - k_1) + k_1^2}, \tag{4.1.42}$$

under the stability with probability one assumptions

$$\omega_0^2 > k_1, \; 4h > h_1. \tag{4.1.43}$$

Step 2: For nonlinear and corresponding linearized elements (one dimensional) defined by (4.1.21) and (4.1.22), respectively, determine probability density functions using $E[x_i] = 0$ and $E[x_i^2]$, $i = 1,2$ calculated from moments obtained in *Step 1:*. They have the forms (4.1.36) and (4.1.37), for nonlinear elements and (4.1.39) and (4.1.40) for the corresponding linearized ones.

Step 3: Choose any criterion, for instance, I_1 and apply it to all nonlinear elements separately, i.e.

$$I_{1_j} = \int_{-\infty}^{+\infty} (g_{Y_j}(y_j) - g_{L_j}(y_j))^2 dy_j, \quad j = 1,2 \tag{4.1.44}$$

where $g_{Y_j}(y_j)$ and $g_{L_j}(y_j)$, $j = 1,2$ are probability density functions defined by (4.1.36), (4.1.37) and (4.1.39), (4.1.40) and next find the coefficients $k_{1_{min}}$ and $h_{1_{min}}$

that minimize criterion (4.1.44) separately for each nonlinear element and then substitute $k_1 = k_{1_{min}}$ and $h_1 = h_{1_{min}}$.

Step 4: If the stability conditions (4.1.43) are satisfied, then calculate the stationary moments of linearized system (4.1.41) and (4.1.42), otherwise stop.

Step 5: For all nonlinear and linearized elements (one dimensional) defined by (4.1.21) and (4.1.22) determine new stationary probability density functions using $E[x_i] = 0$, $E[x_i^2]$, $i = 1, 2$ calculated on the basis of stationary moments obtained in *Step 4:*.

*S*tep 6: Repeat *S*teps 3-5 until k_1, h_1, $E[x_i^2]$, $i = 1, 2$ converge with a given accuracy.

We note that if we treat equation (4.1.39) as the Stratonovich stochastic differential equation, then there are two possibility of linearization. In the first one, that is similar to the proposed by Kazakow and Malczikow (1975) first we linearize separately deterministic and stochastic part of Stratonovich stochastic differential equation and next we transform the obtained linearized Stratonovich equation to the corresponding Ito equation and then we calculate moment equations. In the second approach we first transform the Stratonovich equation to the corresponding Ito equation and next we linearize separately deterministic and stochastic part of Ito equation and we determine moment equations for linearized Ito equation. The results obtained by both approaches are different.

4.2 Equivalent linearization

Similar to statistical linearization, the method of equivalent linearization has been applied to models as well with parametric as external and parametric excitations. Two basic approaches of equivalent linearization have been proposed in the literature. The first one was given by Kottalam et al. (1986) and independently by Young and Chang (1987), while the second approach by Bruckner and Lin (1987). We illustrate the first approach.

Kottalam et al. (1986) considered the nonlinear system with parametric excitations described by the vector Ito stochastic differential equation

$$d\mathbf{x} = \Phi(x)dt + \sum_{k=1}^{M} \sigma_k(x)d\xi_k, \qquad \mathbf{x}(t_0) = \mathbf{x}_0, \tag{4.2.1}$$

where $\mathbf{x} = [x_1, ..., x_n]^T$ is a state vector, $\xi_k, k = 1, ..., M$ are standard Wiener processes, such that

$$E[d\xi_i(t)d\xi_j(t)] = \delta_{ij}dt, \tag{4.2.2}$$

$[\delta_{ij}]$ is the matrix of intensity, the initial condition $\mathbf{x}_0$ is a vector random variable independent of ξ_k, $k = 1, ..., M$, $\Phi(x)$ is a vector nonlinear function having components

$$\phi_i(\mathbf{x}) = f_i(\mathbf{x}) + \frac{1}{2}\sum_{j=1}^{n}\sum_{k=1}^{M}\sum_{l=1}^{M}\left[\frac{\partial}{\partial x_j}\sigma_{ik}(\mathbf{x})\right]\delta_{ij}\sigma_{jl}(\mathbf{x}), \quad i = 1, ..., n, \tag{4.2.3}$$

$\sigma_k(x) = [\sigma_{1k}(\mathbf{x}), ..., \sigma_{nk}(\mathbf{x})]^T$, $f_i(\mathbf{x})$ and $\sigma_{ik}(\mathbf{x})$ are nonlinear functions $i, j = 1, ..., n$, $k = 1, ..., M$.

Elements of the diffusion matrix $\mathbf{B}(x) = [b_{ij}(\mathbf{x})]$ corresponding to equation (4.2.1) have the form

$$b_{ij}(\mathbf{x}) = \frac{1}{2} \sum_{k=1}^{M} \sum_{l=1}^{M} \sigma_{ik}(\mathbf{x}) \delta_{kl} \sigma_{lj}(\mathbf{x}). \tag{4.2.4}$$

We replace equation (4.2.1) by a linear one with time dependent coefficients and additive nonstationary excitations, i.e.

$$d\mathbf{x} = [\mathbf{A}(t)\mathbf{x} + \mathbf{C}(t)]dt + \mathbf{G}(t) d\xi(t), \qquad \mathbf{x}(t_0) = \mathbf{x}_0, \tag{4.2.5}$$

where $\mathbf{A} = [a_{ij}]$, $\mathbf{G} = [G_{ij}]$ and $\mathbf{C} = [C_1, ..., C_n]^T$, are the matrices and vector of time depending linearization coefficients, $\xi(t) = [\xi_1(t), ..., \xi_M]^T$ is the standard vector Wiener process.

$$G_{ij}(t) = \sum_{k=1}^{M} \sum_{l=1}^{M} \gamma_{ik}(t) \delta_{kl} \gamma_{lj}(t). \tag{4.2.6}$$

The elements $a_{ij}, C_i, \gamma_{ij}, \quad i, j = 1, ..., n$ are determined from the necessary conditions of minimization of mean–square linearization criteria I_1 and I_2 defined by

$$I_1 = E\left[[\Phi(\mathbf{x}) - \mathbf{A}(t)\mathbf{x} - C(t)]^T [\Phi(\mathrm{x}) - \mathbf{A}(t)\mathbf{x} - C(t)] \right], \tag{4.2.7}$$

$$I_2 = E[\mathbf{G}(t)\mathbf{G}^T(t) - \mathbf{B}(\mathbf{x})]. \tag{4.2.8}$$

Deriving equations for first and second order moments for nonlinear system (4.2.1) and linearized one (4.2.5), and next equating the right hand sides of these equations we find

$$\mathbf{A}(t) = \left[E[\Phi(\mathbf{x})\mathbf{x}^T] - E[\Phi] E[\mathbf{x}^T] \right] \left[E[\mathbf{x}x^T] - E[\mathbf{x}] E[\mathbf{x}^T] \right]^{-1}, \tag{4.2.9}$$

$$\mathbf{C}(t) = E[\Phi(\mathbf{x})] - \mathbf{A}(t) E[\mathbf{x}]. \tag{4.2.10}$$

To calculate the linearization coefficients and response moments one can use, as in previous cases, an iterative procedure. The moment equations are linear with time depending coefficients and time depending excitations, i.e.

$$\frac{d\mathbf{m}_x}{dt} = \mathbf{A}(t)\mathbf{m}_x + \mathbf{C}(t), \tag{4.2.11}$$

$$\frac{d\Gamma_L}{dt} = \Gamma_L \mathbf{A}^T(t) + \mathbf{A}(t)\Gamma_L + \mathbf{C}(t)\mathbf{x}^T + \mathbf{x}\mathbf{C}^T(t) + \mu(t), \tag{4.2.12}$$

where

$$\mu(t) = E[\mathbf{B}(\mathbf{x})]. \tag{4.2.13}$$

This approach was illustrated by Kottalam, Lindberg and West for the Duffing oscillator subject to both additive and stiffness stochastic excitations

Example 4.2. Kottalam et al. (1986). Consider the Duffing oscillator wth external and parametric excitation described by the vector Ito stochastic differential equation

$$d\mathbf{x}(t) = [\mathbf{A}_1\mathbf{x} + \mathbf{F}(\mathbf{x})]dt + \sum_{k=1}^{2} \sigma_k(\mathbf{x})d\xi_k, \qquad \mathbf{x}(t_0) = \mathbf{x}_0, \tag{4.2.14}$$

where

$$\mathbf{x} = \begin{bmatrix} x_1 \\ x_2 \end{bmatrix}, \quad \mathbf{A}_1 = \begin{bmatrix} 0 & 1 \\ -\omega_0^2 & -\delta_{11} \end{bmatrix}, \quad \mathbf{F} = \begin{bmatrix} 0 \\ -\varepsilon x_1^3 - 2\delta_{12}x_1x_2 - \delta_{22}x_1^2x_2 \end{bmatrix},$$

$$\sigma_1 = \begin{bmatrix} 0 \\ 1 \end{bmatrix}, \quad \sigma_2 = \begin{bmatrix} 0 \\ x_1 \end{bmatrix}, \quad E[d\xi_i d\xi_j] = \delta_{ij}dt, \quad \mathbf{x}_0 = \begin{bmatrix} x_{10} \\ x_{20} \end{bmatrix} \tag{4.2.15}$$

The linearized system has the form (4.2.5), where

$$\mathbf{A} = [a_{ij}], \quad \mathbf{C} = [C_1, C_2]^T, \quad \mathbf{G} = [G_1, G_2]^T,$$

$$a_{11} = 0, \quad a_{12} = 1, \quad C_1 = 0, \quad G_1 = 0,$$

$$a_{21} = -\frac{1}{\Delta}\{[Q(\mathbf{x})(E[x_1]E[x_2^2] - E[x_2]E[x_1x_2])$$
$$-E[x_1Q(\mathbf{x})](E[x_2^2] - (E[x_2])^2) + E[x_2Q(\mathbf{x})](E[x_1x_2] - E[x_1]E[x_2])\},$$

$$a_{22} = -\frac{1}{\Delta}\{[Q(\mathbf{x})(E[x_1]E[x_1x_2] - E[x_2]E[x_1^2])$$
$$+E[x_1Q(\mathbf{x})](E[x_1x_2] - E[x_1]E[x_2]) - E[x_2Q(\mathbf{x})](E[x_1^2] - (E[x_1])^2\},$$

$$C_2 = -\frac{1}{\Delta}\{-[Q(\mathbf{x})(E[x_1^2]E[x_2^2] - (E[x_1x_2])^2)$$
$$+E[x_1Q(\mathbf{x})](E[x_1]E[x_2^2] - E[x_1x_2]E[x_2]) - E[x_2Q(\mathbf{x})](E[x_1]E[x_1x_2] - E[x_1^2]E[x_2])\},$$

$$G_2 = \left(\frac{E[(\delta_{11} + 2\delta_{12}x_1 + \delta_{22}x_1^2)]}{\delta_{11}}\right)^{\frac{1}{2}},$$

$$\Delta = E[x_1^2]E[x_2^2] - (E[x_1x_2])^2 - (E[x_1])^2E[x_2^2]$$
$$+ 2E[x_1]E[x_2]E[x_1x_2] - E[x_1^2](E[x_2])^2, \tag{4.2.16}$$

$$Q(\mathbf{x}) = -\omega_0^2x_1 - \varepsilon x_1^3 - (\delta_{11} + 2\delta_{12}x_1 + \delta_{22}x_1^2)x_2.$$

The solution of the first and second order moment equations for system (4.2.5) with the coefficients given by (4.2.15) is possible only by application of, for instance, cumulant closure technique of second order. Then we obtain a system of nonlinear differential equations that can be directly solved or by the iterative procedure discussed in previous section. In the case of the application of the iterative procedure one has to solve multiple times the system of linear differential equations.

5 Vibration of structures under earthquake excitations

A wide area of application of statistical and equivalent linearization is the response analysis of multistory structures under stochastic ground excitations. Wen (1980) and Baber and Wen (1981, 1982) considered a multi–degree of freedom shear beam model subjected to horizontal ground acceleration $\ddot{\xi}_g$ (see Figure 5).

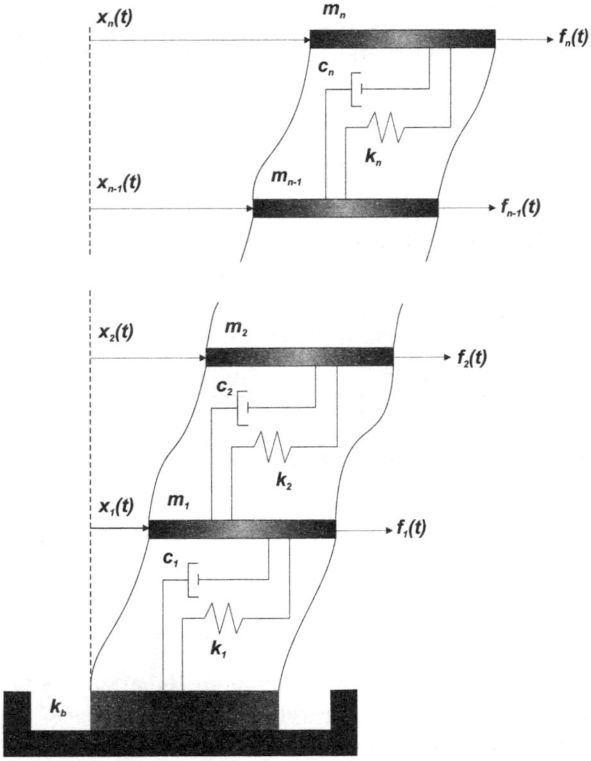

Figure 5. Model of multi–degree of freedom shear beam subjected to horizontal ground acceleration Baber and Wen (1981). Copyright from Random Vibration Hysteretic, Degrading Systems by T.Baber and Y.K.Wen. Reproduced by permission of American Society of Civil Engineers, License Number LS601.

Noting that the quantities v_i are the relative displacements of the ith and $(i+1)$th stories, the equation of motion may be written in the form

$$m_i\left(\sum_{j=1}^{i}\ddot{v}_j+\ddot{\xi}_g\right)+q_i-q_{i+1}=0, \quad i=1,...,n, \tag{5.0.1}$$

where m_i is the mass of the ith floor, $\ddot{\xi}_g$ is the ground acceleration and q_i is the ith restoring force, including viscous damping, given by

$$q_i=c_i\dot{v}_i+a_ik_iv_i+(1-\alpha_i)k_iz_i, \quad i=1,...,n, \tag{5.0.2}$$

where c_i is the viscous damping, k_i controls the initial tangent stiffness, α_i controls the ratio of post yield stiffness and z_i is the ith hysteresis. Equation (5.0.1) can be rewritten with the accelerations decoupled if the $(i-1)$ equation is subtracted from the ith equation, for $i=2,...,n$. This result may be summarized by

$$\ddot{v}_i-(1-\delta_{i1})\frac{q_{i-1}}{m_{i-1}}+\frac{q_i}{m_i}\left[1+(1-\delta_{i1})\frac{m_i}{m_{i-1}}\right]-(1-\delta_{in})\frac{q_{i+1}}{m_{i+1}}\left(\frac{m_{i+1}}{m_i}\right)=-\delta_{i1}\ddot{\xi}_g,$$

$$i=1,...,n, \tag{5.0.3}$$

in which δ_{i1},δ_{in} are Kronecker deltas.

The time history dependent hysteretic restoring force is modeled by the first order nonlinear differential equation

$$\dot{z}_i=\frac{A_i\dot{v}_i-\nu_i(\beta_i|\dot{v}_i||z_i|^{n_i-1}z_i+\gamma_i\dot{v}_i|z_i|^{n_i})}{\eta_i}, \quad i=1,...,n, \tag{5.0.4}$$

in which A_i,ν_i,β_i,η_i and n_i are parameters which control the hysteresis shape and model degradation.

Using equivalent linearization, Baber and Wen (1981) looked for an equivalent linear equation in the form

$$z_i=c_{e_i}\dot{v}_i+k_{e_i}z_i, \quad i=1,...,n. \tag{5.0.5}$$

It should be stressed that as seen above, equivalent linearization is not applied to the whole system (5.0.2)-(5.0.4) but only separately to each hysteresis.

Under the assumption that $\dot{v}_i$ and z_i are zero mean joint Gaussian process, coefficients c_{e_i} and k_{e_i} can be found from the conditions

$$c_{e_i}=E\left[\frac{\partial\dot{z}_i}{\partial\dot{v}_i}\right], \quad k_{e_i}=E\left[\frac{\partial\dot{z}_i}{\partial z_i}\right], \quad i=1,...,n \tag{5.0.6}$$

in closed form with the result

$$c_{e_i} = \frac{\mu_{A_i} - \mu_{\nu_i}(\beta_i F_{1_i} + \gamma_i F_{2_i})}{\mu_{\eta_i}}, \quad k_{e_i} = -\mu_{\nu_i}\frac{\beta_i F_{1_i} + \gamma_i F_{2_i}}{\mu_{\eta_i}}, \quad i = 1, ..., n, \tag{5.0.7}$$

where

$$\mu_{A_i} = E[A_i], \quad \mu_{\nu_i} = E[\nu_i], \quad \mu_{\eta_i} = E[\eta_i],$$

$$F_{1_i} = \frac{\sigma_{z_i}^{n_i}}{\pi}\,\Gamma\left(\frac{n_i+2}{2}\right)2^{n_i/2}(I_{s_{1i}} - I_{s_{2i}}), \quad F_{2_i} = \frac{\sigma_{z_i}^{n_i}}{\sqrt{\pi}}\,\Gamma\left(\frac{n_i+1}{2}\right)2^{n_i/2},$$

$$F_{3_i} = \frac{n_i\sigma_{\dot{v}_i}\sigma_{z_i}^{n_i-1}}{\pi}\,\Gamma\left(\frac{n_i+2}{2}\right)2^{n_i/2}\left[\frac{2(1-\rho_{\dot{v}_i z_i}^2)^{n_i+1)/2}}{n_i} + \rho_{\dot{v}_i z_i}(I_{s_{1i}} - I_{s_{2i}})\right], \tag{5.0.8}$$

$$F_{4_i} = \frac{n_i\rho_{\dot{v}_i z_i}\sigma_{\dot{v}_i}\sigma_{z_i}^{n_i-1}}{\sqrt{\pi}}\,\Gamma\left(\frac{n_i+1}{2}\right)2^{n_i/2}, \quad i = 1, ..., n,$$

In the above $\Gamma(.)$ is the gamma function,

$$I_{s_{1i}} = \int_0^{\alpha_i} \sin^{n_i}\theta d\theta, \quad I_{s_{2i}} = \int_{\alpha_i}^{\pi} \sin^{n_i}\theta d\theta, \quad i = 1, ..., n, \tag{5.0.9}$$

where

$$\alpha_i = \arctan\left(-\frac{\sqrt{1-\rho_{\dot{v}_i z_i}^2}}{\rho_{\dot{v}_i z_i}}\right), \quad i = 1, ..., n. \tag{5.0.10}$$

The integrals in equation (5.0.9) are widely tabulated. The coefficients of equation (5.0.7) are response dependent since

$$\sigma_{z_i}^2 = E[z_i^2(t)], \quad \sigma_{\dot{v}_i} = E[\dot{v}_i^2(t)], \quad \rho_{\dot{v}_i z_i} = \frac{E[\dot{v}_i(t) z_i(t)]}{\sigma_{\dot{v}_i}\sigma_{z_i}}, \quad i = 1, ..., n. \tag{5.0.11}$$

They can be found using standard iterative procedures Baber and Wen (1981).

6 Application of linearization techniques in control problems of stochastic dynamic systems

Stochastic linearization was used as mathematical tool in control theory. It was first time applied in control theory by Wonham and Cashman (1969) who combined statistical linearization and Linear Quadratic Gaussian (LQG) theory to the determination of a quasi–optimal control for a nonlinear stochastic system. This idea was later developed by many authors, for instance, Beaman (1984), Kazakov (1984), Yoshida (1984). In these papers systems with nonlinear plants, linear actuators and nonlinear criteria were studied and the linearization coefficients were determined from the mean square criterion. The quasi–optimal control was obtained from an iterative procedure where the linearization coefficients were derived using the mean–square linearization criterion and the optimal control for linear stochastic system with the mean–square control criterion was calculated.

To discuss the problem of application of linearization techniques in control problems of stochastic dynamic systems we introduce the basic results of LQG theory.

6.1 Linear quadratic Gaussian problems

Consider the following optimal control problem. The dynamic system is described by the following vector Ito stochastic differential equation

$$d\mathbf{x}(t) = [\mathbf{A}\mathbf{x}(t) + \mathbf{B}u(t)]dt + \sum_{k=1}^{M} \mathbf{G}_k d\xi_k(t), \qquad \mathbf{x}(t_0) = \mathbf{x}_0, \tag{6.1.1}$$

where $\mathbf{x}(t)$ is the state vector $\mathbf{x} \in \mathbf{R}^n$, $\mathbf{u}(t)$ the control vector $\mathbf{u} \in \mathbf{R}^m$, $\mathbf{G}_k$ are constant vectors of intensity of noise, $\mathbf{G}_k \in \mathbf{R}^n$, ξ_k, $k = 1, ..., M$ are correlated standard Wiener processes with covariance matrix $\mathbf{Q}_\xi$; the initial condition $\mathbf{x}_0$ is a vector random variable independent of ξ_k, $k = 1, ..., M$, $\mathbf{A}$, $\mathbf{B}$ and $\mathbf{Q}_\xi$ are constant matrices of $n \times n$, $n \times m$ and $M \times M$ dimensions, respectively.

The mean-square criterion is defined by

$$I = \lim_{T\to\infty} E\left\{\frac{1}{T}\int_{t_0}^{T} \left[\mathbf{x}^T(t)\mathbf{Q}\mathbf{x}(t) + 2\mathbf{x}^T(t)\mathbf{N}\mathbf{u}(t) + \mathbf{u}^T(t)\mathbf{R}\mathbf{u}(t)\right] dt\right\}, \tag{6.1.2}$$

where $\mathbf{Q}$, $\mathbf{N}$ and $\mathbf{R}$ are matrices of $n \times n$, $n \times m$ and $m \times m$ dimensions, respectively. $\mathbf{Q} \geq \mathbf{0}$ and $\mathbf{R} > \mathbf{0}$ are symmetric. It is assumed that the state vector $\mathbf{x}$ is complete measurable.

Using results of LQG theory (see Kwakernaak and Sivan (1972)) the optimal control is determined by

$$\mathbf{u}(t) = -\mathbf{K}\mathbf{x}(t), \tag{6.1.3}$$

where the control gain matrix $\mathbf{K}$ is constant and given by

$$\mathbf{K} = \mathbf{R}^{-1}(\mathbf{N}^T + \mathbf{B}^T\mathbf{P}), \tag{6.1.4}$$

where $\mathbf{P}$ is a symmetric, positive-definite solution of the algebraic Riccati equation

$$\mathbf{P}\mathbf{A}_N + \mathbf{A}_N^T\mathbf{P} - \mathbf{P}\mathbf{B}\mathbf{R}^{-1}\mathbf{B}^T\mathbf{P} + \mathbf{Q}_N = \mathbf{0}, \tag{6.1.5}$$

where

$$\mathbf{A}_N = \mathbf{A} - \mathbf{B}\mathbf{R}^{-1}\mathbf{N}^T \geq \mathbf{0}, \qquad \mathbf{Q}_N = \mathbf{Q} - \mathbf{N}\mathbf{R}^{-1}\mathbf{N}^T \geq \mathbf{0}. \tag{6.1.6}$$

The performance index for optimal control is determined by

$$I = \text{tr}(\mathbf{P}\mathbf{G}\mathbf{Q}_\xi \mathbf{G}^T) \tag{6.1.7}$$

or alternatively can be calculated from the algebraic Lyapunov equation

$$(\mathbf{A} - \mathbf{B}\mathbf{K})\mathbf{V} + \mathbf{V}(\mathbf{A} - \mathbf{B}\mathbf{K})^T + \mathbf{G}\mathbf{Q}_\xi \mathbf{G}^T = \mathbf{0}, \tag{6.1.8}$$

where $\mathbf{G} = [\mathbf{G}_1, ..., \mathbf{G}_M]$, $\mathbf{V}$ is the covariance matrix of vector state $\mathbf{x}$ i.e.

$$\mathbf{V}(t) = E[\mathbf{x}(t)\mathbf{x}^T(t)]. \tag{6.1.9}$$

6.2 Algorithms for quasi–optimal control problems for nonlinear system

Consider the following optimal control problem. The nonlinear stochastic system is described by

$$d\mathbf{x}(t) = [\mathbf{A}\mathbf{x}(t) + \Phi(\mathbf{x}) + \mathbf{B}\mathbf{u}(t)]dt + \sum_{k=1}^{M} \mathbf{G}_k d\xi_k, \qquad \mathbf{x}(t_0) = \mathbf{x}_0, \tag{6.2.1}$$

where $\mathbf{x} \in R^n$ and $\mathbf{u} \in R^m$ are the state vector and the control vector, respectively. $\mathbf{A}$ and $\mathbf{B}$ are time invariant matrices of appropriate dimensions, $\Phi = [\phi_1, ..., \phi_n]^T$ is a vector nonlinear function, such that, $\Phi(\mathbf{0}) = \mathbf{0}$, $\mathbf{G}_k$ are time invariant deterministic vectors, ξ_k, $k = 1, ..., M$ are correlated standard Wiener processes with correlation matrix $\mathbf{Q}_\xi$; the initial condition $\mathbf{x}_0$ is a vector random variable independent of ξ_k, $k = 1, ..., M$. We assume that the unique solution of equation (6.2.1) exists. The control strategy is designed to minimize the criterion (6.1.2)

We assume that the nonlinear vector $\Phi(\mathbf{x})$ can be substituted by a linearized form

$$\Phi(\mathbf{x}) = \mathbf{A}_e \mathbf{x}, \tag{6.2.2}$$

where $\mathbf{A}_e$ is a $n \times n$ matrix of linearization coefficients such that $(\mathbf{A}+\mathbf{A}_e, \mathbf{B})$ is stabilizable and detectable. Then the optimal control for the linearized system

$$d\mathbf{x}_L(t) = [(\mathbf{A} + \mathbf{A}_e)\mathbf{x}_L(t) + \mathbf{B}\mathbf{u}(t)]dt + \sum_{k=1}^{M} \mathbf{G}_k d\xi_k, \qquad \mathbf{x}(t_0) = \mathbf{x}_0 \tag{6.2.3}$$

can be found by LQG standard method presented in previous subsection in the linear feedback form (6.1.3), i.e.

$$\mathbf{u} = -\mathbf{K}x_L, \tag{6.2.4}$$

where the gain matrix $\mathbf{K}$ is equal

$$\mathbf{K} = \mathbf{R}^{-1}(\mathbf{N}^T + \mathbf{B}^T\mathbf{P}), \tag{6.2.5}$$

and $\mathbf{P}$ is a positive solution of the algebraic Riccati equation

$$\mathbf{P}(\mathbf{A}_N + \mathbf{A}_e) + (\mathbf{A}_N + \mathbf{A}_e)^T\mathbf{P} - \mathbf{P}BR^{-1}\mathbf{B}^T\mathbf{P} + \mathbf{Q}_N = \mathbf{0}, \tag{6.2.6}$$

where

$$\mathbf{A}_N = \mathbf{A} - \mathbf{B}\mathbf{R}^{-1}\mathbf{N}^T \geq \mathbf{0}, \qquad \mathbf{Q}_N = \mathbf{Q} - \mathbf{N}\mathbf{R}^{-1}\mathbf{N}^T \geq \mathbf{0}. \tag{6.2.7}$$

Substituting (6.2.5) into equation (6.2.3) we find

$$d\mathbf{x}_L(t) = [(\mathbf{A} + \mathbf{A}_e - \mathbf{B}\mathbf{K})\mathbf{x}_L(t)]dt + \sum_{k=1}^{M} \mathbf{G}_k d\xi_k, \qquad \mathbf{x}(t_0) = \mathbf{x}_0. \tag{6.2.8}$$

The corresponding covariance equation has the form

$$(\mathbf{A} + \mathbf{A}_e - \mathbf{B}K)\mathbf{V}_L + \mathbf{V}_L(\mathbf{A} + \mathbf{A}_e - \mathbf{B}\mathbf{K})^T + \sum_{k=1}^{M} \mathbf{G}_k\mathbf{Q}_\xi\mathbf{G}_k^T = \mathbf{0} \tag{6.2.9}$$

and the criterion is equal

$$I_L = E[\mathbf{x}_L^T(\mathbf{Q} + \mathbf{K}^T\mathbf{R}\ \mathbf{K})\mathbf{x}_L] = tr[(\mathbf{Q} + \mathbf{K}^T\mathbf{R}\ \mathbf{K})\mathbf{V}_L] \tag{6.2.10}$$

where the subindex L corresponds to the linearized problem, "tr" denotes the trace of matrix and

$$\mathbf{V}_L = E[\mathbf{x}_L\mathbf{x}_L^T] \tag{6.2.11}$$

To obtain quasi–optimal control usually the authors have used the mean–square criterion in statistical linearization. Since in the vibration analysis of stochastic systems in mechanical and structural engineering many criteria of linearization were considered we show only a comparison of application of moment criteria.

We consider three criteria of statistical linearization: the mean - square error of displacement, equality of the second order moments of nonlinear and linearized elements, the mean-square error of potential energy of displacement and true linearization method.

In statistical linearization the elements of the nonlinear vector have to be replaced by corresponding equivalent elements "in the sense of a given criterion" in a linear form. The following criteria of linearization for a scalar function $\varphi(x)$ are considered.

Criterion of linearization 1. Mean-square error of displacements.

$$E[(c_1x - \varphi(x))^2] \rightarrow min \tag{6.2.12}$$

Criterion of linearization 2. Equality of second order moments of nonlinear and linearized elements.

$$E[(c_2x)^2] = E[(\varphi(x))^2] \tag{6.2.13}$$

Criterion of linearization 3. Mean-square error of potential energies.

$$E[(\int_0^x (c_3x - \varphi(x))dx)^2] \rightarrow min \tag{6.2.14}$$

In true linearization Kozin (1989) the linearization coefficients are determined from the equality of covariance matrices of responses of nonlinear and linearized systems to random excitations. The main idea of this approach for multidimensional system can be formulated as a next criterion.

Criterion of linearization 4. True linearization.

$$E[\mathbf{x}_N\mathbf{x}_N^T] = E[\mathbf{x}_L\mathbf{x}_L^T] \tag{6.2.15}$$

where $\mathbf{x}_N$ and $\mathbf{x}_L$ are stationary solutions of nonlinear and linearized dynamic systems, respectively.

In the application of linear feedback gain obtained for linearized system to the nonlinear system the state equation and the corresponding mean–square criterion of control have the form

$$d\mathbf{x}_N(t) = [(\mathbf{A} - \mathbf{B}\ \mathbf{K})\mathbf{x}_N(t) + \Phi(\mathbf{x}_N(t))]dt + \sum_{k=1}^{M} \mathbf{G}_k d\xi_k, \qquad \mathbf{x}(t_0) = \mathbf{x}_0, \tag{6.2.16}$$

$$I_N = E[\mathbf{x}_N^T(\mathbf{Q} + \mathbf{K}^T\mathbf{R}\ \mathbf{K})\mathbf{x}_N] = tr[(\mathbf{Q} + \mathbf{K}^T\mathbf{R}\ \mathbf{K})\mathbf{V}_N], \tag{6.2.17}$$

where the subindex N corresponds to the original nonlinear problem and

$$\mathbf{V}_N = E[\mathbf{x}_N\mathbf{x}_N^T] \tag{6.2.18}$$

In general case the covariance matrix $\mathbf{V}_N$ can be found approximately. To obtain the linearization matrix $\mathbf{A}_e$ and quasi-optimal control one of the fourth proposed criteria in an iterative procedure can be applied. For the first three criteria of linearization the procedure given in Yoshida (1984) can be used while for Criterion of linearization 4 a procedure given bellow. We note that in the case of true linearization the determination of linearization coefficients from condition (6.2.15) for multidimensional systems is not always unique and additional conditions are required, for instance, the equality of even higher order linearized systems, i.e.

$$E[\mathbf{x}_N^{[p]}] = E[\mathbf{x}_L^{[p]}], \tag{6.2.19}$$

where $\mathbf{x}^{[p]}$ denotes pth forms of the of the components of vector $\mathbf{x}$ i.e.

$$\mathbf{x}^{[p]} = x_1^{p_1} x_2^{p_2} ... x_n^{p_n}, \qquad \sum_{j=1}^{n} p_j = p, \qquad p = 4, 6, ..., 2k \tag{6.2.20}$$

or to condition (6.2.15) could be added the following one

$$I_A = \min_{\mathbf{A}_{e_{ij}}} |E[\mathbf{x}_N^{[p]}(\mathbf{A}_e)] - E[\mathbf{x}_L^{[p]}(\mathbf{A}_e)]|. \tag{6.2.21}$$

The following procedure is a modified version of a standard one given in Yoshida (1984)

Control Procedure (CP) *S*tep 1. First put $\mathbf{A}_e = 0$ in (6.2.3) and then solve (6.2.6) and (6.2.9). The solutions of (6.2.6) and (6.2.9) are $\mathbf{P}$ and $\mathbf{V}_L$, respectively.

*S*tep 2. Substitute $\mathbf{P}$ obtained in *S*tep 1 into (6.2.5) and find $\mathbf{K}$. Next, substitute $\mathbf{K}$ into (6.2.4) and (6.2.10) find $\mathbf{u}$ and I_L , respectively.

*S*tep 3. Substitute $\mathbf{K}$ obtained in *S*tep 2 into (6.2.17) and (6.2.18) and then find exactly or approximately I_N and V_N .

*S*tep 4. Find linearization coefficients from Criterion 4 and if necessary from additional conditions (6.2.19) or apply a minimization procedure with respect to coefficients $\mathbf{A}_{e_{ij}}$.

*S*tep 5. If the accuracy is greater then a given parameter ε_1 then repeat solving equations (6.2.6) and (6.2.9) and steps 2 - 4 until $\mathbf{V}_L$ and $\mathbf{P}$ converge.

We illustrate the main results of this subsection by an example.

Example 5.1.(Duffing oscillator)

Consider the Duffing oscillator described by

$$\begin{aligned} dx_1 &= x_2 dt, \qquad x_1(t_0) = x_{10} \\ dx_2 &= [-\omega_0^2 x_1 - 2hx_2 - \alpha x_1^3 + bu]dt + gd\xi, \quad x_2(t_0) = x_{20}, \end{aligned} \tag{6.2.22}$$

where $\mathbf{x} = [x_1 x_2]^T$ is the state vector, u is the scalar control, ω_0, h, α, b and g are positive constants, ξ is the standard Wiener process, the initial conditions x_{10} and x_{20} are random variables independent of ξ and the mean-square criterion is

$$I = E[\bar{\mathbf{x}}^T \mathbf{Q} \bar{\mathbf{x}} + r\bar{u}^2] \tag{6.2.23}$$

where $\bar{\mathbf{x}} = [\bar{x}_1 \bar{x}_2]^T$ is a stationary solution of (6.2.22) and $\bar{u}$ is the corresponding stationary control, $\mathbf{Q} = diag\{Q_i\}$; Q_i and r are positive constants. The linearized system has the following form

$$\begin{aligned} dx_1 &= x_2 dt, \quad x_1(t_0) = x_{10} \\ dx_2 &= [-\omega_0^2 x_1 - 2hx_2 - \alpha k x_1 + bu]dt + g d\xi, \quad x_2(t_0) = x_{20} \end{aligned} \tag{6.2.24}$$

where k is a linearization coefficient. The coordinates of the solutions of algebraic Riccati and covariance equations denoted by $\mathbf{P} = [p_{ij}]$ and $\mathbf{V}_L = [v_{L_{ij}}]$, respectively, for i, j=1,2, are the following

$$\begin{aligned} p_{12} &= \frac{1}{\beta}(-\gamma + \sqrt{\gamma^2 + Q_1 \beta}), \quad p_{22} = \frac{1}{\beta}\sqrt{4h^2 + \beta(Q_2 + 2p_{12})}, \\ p_{11} &= 2hp_{12} + \gamma p_{22} + \beta p_{12} p_{22}, \end{aligned} \tag{6.2.25}$$

$$v_{L_{22}} = \frac{g^2}{2(2h + \beta p_{22})}, \quad v_{L_{12}} = 0, \quad v_{L_{11}} = \frac{v_{L_{22}}}{\gamma + \beta p_{12}}, \tag{6.2.26}$$

where $\beta = b^2/r$ and $\gamma = \omega_0^2 + \alpha k$. The optimal value of the criterion for linearized system is equal to

$$I_L = (Q_1 + \beta p_{12}^2) v_{L_{11}} + (Q_2 + \beta p_{22}^2) v_{L_{22}} \tag{6.2.27}$$

If we apply the obtained linear feedback control to nonlinear system we obtain the state equation and the corresponding criterion

$$\begin{aligned} dx_1 &= x_2 dt, \quad x_1(t_0) = x_{10} \\ dx_2 &= [-\omega_0^2 x_1 - 2hx_2 - \alpha x_1^3 - \beta(x_1 p_{12} + x_2 p_{22})]dt + g d\xi, \quad x_2(t_0) = x_{20} \end{aligned} \tag{6.2.28}$$

$$I_{N_{opt}} = (Q_1 + \beta p_{12}^2) v_{N_{11}} + (Q_2 + \beta p_{22}^2) v_{N_{22}}, \tag{6.2.29}$$

where the second order moments $v_{N_{11}}$ and $v_{N_{22}}$ can be found in analytical form.

$$v_{N_{ii}} = \int_{-\infty}^{+\infty} \int_{-\infty}^{+\infty} x_i^2 g_N(x_1, x_2) dx_1 dx_2, \quad i = 1, 2, \tag{6.2.30}$$

where

$$g_N(x_1, x_2) = \frac{1}{c_N} exp\left\{ - \frac{2h + \beta p_{22}}{g^2}((\omega_0^2 + \beta p_{12}) x_1^2 + \alpha \frac{x_1^4}{2} + x_2^2) \right\}, \tag{6.2.31}$$

c_N is a normalized constant. The linearization coefficients for first three criteria have the form

$$k_1 = 3E[x_L^2], \quad k_2 = \sqrt{15}E[x_L^2], \quad k_3 = 2.5E[x_L^2] \tag{6.2.32}$$

and for Criterion of linearization 4 it can be found from condition $v_{N_{ii}} = v_{L_{ii}}$, where $v_{N_{ii}}$ and $v_{L_{ii}}$ are defined by (6.2.30) and (6.2.26), respectively.

To obtain the quasi-optimal control for the Duffing oscillator the iterative procedure (CP) has been used. To illustrate the obtained results a comparison of the criterion $I_{N_{opt}}$ defined by (6.2.29) for different linearization methods versus parameters of system (6.2.22) and criterion (6.2.23) has been shown. In this comparison four criteria of linearization, namely mean-square error of displacements, equality of second order moments, mean–square error of potential energies and true linearization were considered. The numerical results denoted by circle, diamond, square lines and by stars, respectively are presented in Figures 6–7.

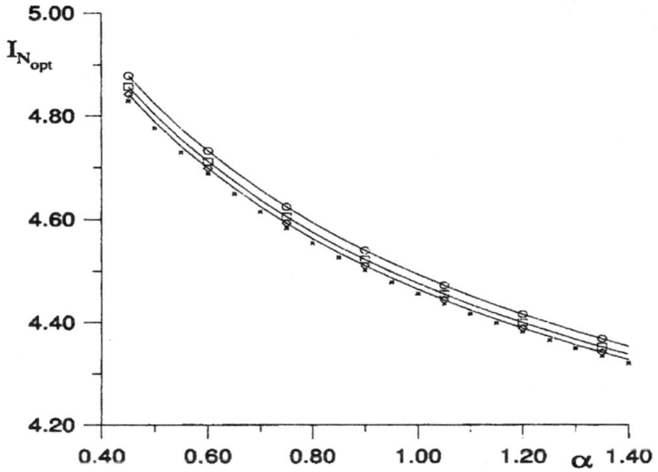

Figure 6. The comparison of optimization criteria obtained by application of different statistical linearization techniques vs. parameter α with $\omega_0^2 = 1,\ h = 0.05,\ b = 1,\ g = 1,$ $Q_1 = Q_2 = 1,\ \ r = 100.$

Since for the Duffing oscillator do not exist the optimal feedback control and the corresponding mean-square criterion in exact forms but they exist for the linear feedback control we use as a measure accuracy the relative error defined by

$$\Delta_{opt} = \Delta_{opt}(par) = \frac{|I_{N_{opt}}(par) - I_{lin}(par)|}{I_{lin}(par)}, \tag{6.2.33}$$

where $I_{N_{opt}}(par)$ and $I_{lin}(par)$ are criteria for nonlinear system with linear feedback and linear system with linear feedback, respectively. The argument "par" denotes a parameter of system (6.2.22). The relative error are presented in Figure 7 for (par=h)

Numerical studies show that for a given mean-square criterion of minimization (6.2.23) there are no significant differences between considered linearization methods. However,

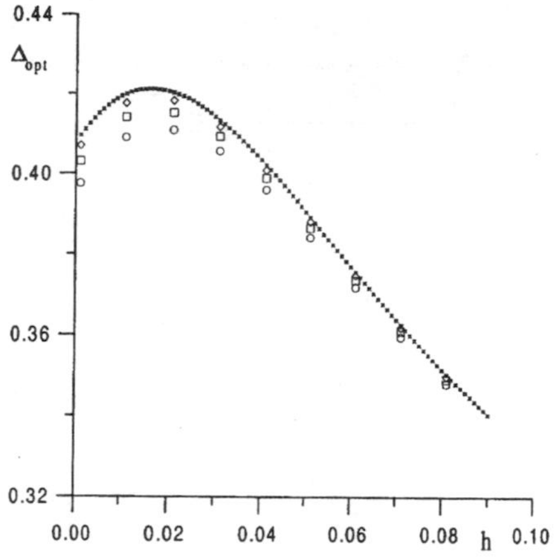

Figure 7. The comparison of relative errors of optimal criteria obtained by application of different statistical linearization techniques vs. parameter h with $\omega_0^2 = 1, \ \ \alpha = 1, \ \ b = 1, \ \ g = 1, \ \ Q_1 = Q_2 = 1, \ \ r = 100$.

one can show Socha and Błachuta (2000) that the relative errors obtained for linearization methods with criteria in probability density functions space are smaller than the corresponding ones for statistical linearization with moment criteria.

6.3 Applications in vehicle models

Linear stochastic models

The LQG optimal control theory was first applied to the determination of an active suspension by Hac (1985). He used the linear system and the mean square criterion defined by (6.1.1) and (6.1.2), respectively to a simple 2-DOF quarter car model (see Figure 8)

The vector equation of motion is

$$\frac{d\mathbf{z}(t)}{dt} = \mathbf{A}_z\mathbf{z}(t) + \mathbf{B}_z u(t) + \mathbf{G}_z w(t), \tag{6.3.1}$$

where $\mathbf{z} = \begin{bmatrix} z_1 & \dot{z}_1 & z_2 & \dot{z}_2 \end{bmatrix}^T$, u and w are the state vector, scalar active control and scalar stochastic disturbance, respectively, whereas

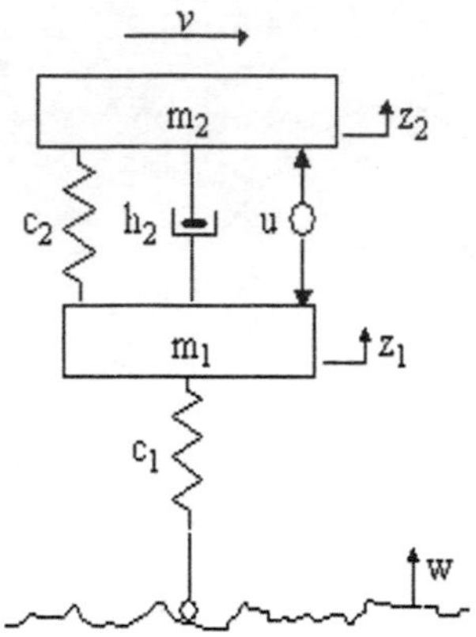

Figure 8. Model of linear suspension system (6.3.1)

$$\mathbf{A}_z = \begin{bmatrix} 0 & 1 & 0 & 0 \\ -(c_2+c_1)/m_1 & -h_2/m_1 & c_2/m_1 & h_2/m_1 \\ 0 & 0 & 0 & 1 \\ c_2/m_2 & h_2/m_2 & -c_2/m_2 & -h_2/m_2 \end{bmatrix},$$

$$\mathbf{B}_z = \begin{bmatrix} 0 \\ -1/m_1 \\ 0 \\ 1/m_2 \end{bmatrix}, \qquad \mathbf{G}_z = \begin{bmatrix} 0 \\ c_1/m_1 \\ 0 \\ 0 \end{bmatrix}, \tag{6.3.2}$$

where c_1, c_2 and h_1, h_2 are constant spring and damper parameters. The stochastic disturbance which describes the road irregularities is a stationary colored noise modeled as an output of first order linear filter with a white noise input process given by

$$\dot{w}(t) = -avw(t) + \xi(t), \tag{6.3.3}$$

where σ^2, a, v and ξ denote the variance of the road irregularities, a constant parameter describing the road surface, the vehicle velocity and the standard white noise process, respectively.

By introducing the notation $A_w = -av$, $G_w = \sigma\sqrt{2av}$, $Q_\xi = 2\sigma^2 av$ and $G_w = 1$ equations (6.3.1) and (6.3.3) can be rewritten in a joint vector form (6.1.1), where

$$\mathbf{x} = \begin{bmatrix} \mathbf{z} \\ \eta \end{bmatrix}, \qquad \mathbf{A} = \begin{bmatrix} \mathbf{A}_z & G_w \\ \mathbf{0} & A_w \end{bmatrix}, \qquad \mathbf{B} = \begin{bmatrix} \mathbf{B}_z \\ \mathbf{0} \end{bmatrix}, \qquad \mathbf{G} = \begin{bmatrix} \mathbf{0} \\ G_w \end{bmatrix}. \tag{6.3.4}$$

The performance index I is defined by stationary response characteristics of considered suspension system

$$I = I_1 + \rho_1 I_2 + \rho_2 I_3 + \rho_3 I_4. \tag{6.3.5}$$

where $I_1 = E[\ddot{z}_2]$ represents a measure of ride comfort, $I_2 = E[(z_2 - z_1)^2]$ limits the space required for the suspension, $I_3 = E[(z_1 - w)^2]$ avoids loosing contact between the wheel and the road, $I_4 = E[u^2]$ limits the control force; ρ_i $(i = 1, \ldots, 3)$ are weight coefficients.

This criterion is an extended version of a criterion given for linear deterministic model in A.G.Thompson (1976). In new state variables the performance index I has the form

$$I = E\Bigg[\left(-\frac{c_2}{m_2}\bar{x}_1 - \frac{h_2}{m_2}\bar{x}_2 + \frac{c_2}{m_2}\bar{x}_3 + \frac{h_2}{m_2}\bar{x}_4 + \frac{1}{m_2}\bar{u}\right)^2 + \rho_1(\bar{x}_1 - \bar{x}_3)^2 + \rho_2(\bar{x}_1 - \bar{x}_5)^2 + \rho_3\bar{u}^2\Bigg] = E\big[\bar{\mathbf{x}}^T\mathbf{Q}\bar{\mathbf{x}} + 2\bar{\mathbf{x}}^T\mathbf{N}\bar{\mathbf{u}} + \bar{\mathbf{u}}^T\mathbf{R}\bar{\mathbf{u}}\big]. \quad (6.3.6)$$

where $\bar{x}_i$ and $\bar{u}$ denotes the stationary values of corresponding x_i and u. Equation (6.3.6) defines matrices $\mathbf{Q}$, $\mathbf{N}$ and $\mathbf{R}$.

Nonlinear stochastic models

An application of LQG theory to the determination of quasi-optimal control for nonlinear models of active suspension systems was presented, for instance, by Gordon et al. (1991), Narayanan and Raju (1992); Raju and Narayanan (1995), Socha (1999a, 2000). Mainly the authors proposed iterative procedures where LQG technique and a linearization method were used. To present this approach we follow the results presented by Socha (1999a, 2000). We consider the linear 2-DOF vehicle model described in previous section with one nonlinear suspension spring between masses m_1 and m_2 (see Fig.9).

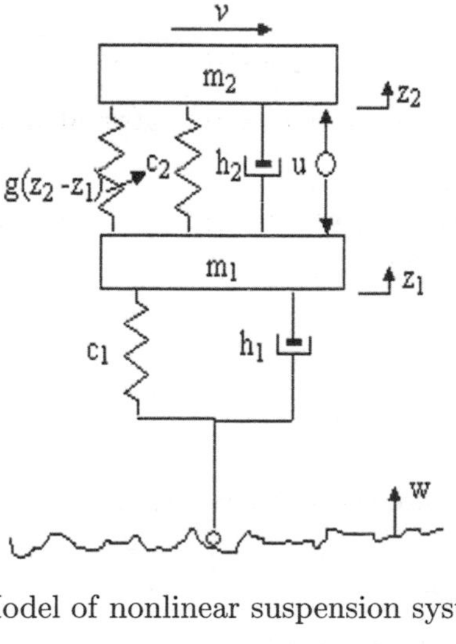

Figure 9. Model of nonlinear suspension system (6.3.7).

The equation of motion for new state variables are described by the following Ito stochastic differential equations

$$
\begin{aligned}
dx_1 &= x_3 dt, \\
dx_2 &= x_4 dt, \\
dx_3 &= \Bigg[-\frac{c_1}{m_1} x_1 - \frac{h_1}{m_1} x_3 + \frac{c_2}{m_1} x_2 + \frac{h_2}{m_1} x_4 + \frac{1}{m_1} g(x_2) \\
&\qquad - \frac{1}{m_1} u + a_1 a_2 x_5 + (a_1 + a_2) x_6 \Bigg] dt - q d\xi, \\
dx_4 &= \Bigg[\frac{c_1}{m_1} x_1 - \frac{m_1 + m_2}{m_1 m_2} c_2 x_2 + \frac{h_1}{m_1} x_3 - \frac{m_1 + m_2}{m_1 m_2} h_2 x_4 \\
&\qquad - \frac{m_1 + m_2}{m_1 m_2} g(x_2) + \frac{m_1 + m_2}{m_1 m_2} u \Bigg] dt, \\
dx_5 &= x_6 dt, \\
dx_6 &= [-a_1 a_2 x_5 - (a_1 + a_2) x_6] dt + q d\xi,
\end{aligned}
\tag{6.3.7}
$$

where the new state variables are defined by

$$
\begin{aligned}
x_1 &= z_1 - w, \qquad & x_2 &= z_2 - z_1, \qquad & x_3 &= \dot{z}_1 - \dot{w}, \\
x_4 &= \dot{z}_2 - \dot{z}_1, \qquad & x_5 &= w, \qquad & x_6 &= \dot{w},
\end{aligned}
\tag{6.3.8}
$$

c_1 and c_2 are stiffness constant parameters, h_1 and h_2 are damping constant parameters, ξ is a standard Wiener process, a_1, a_2 and q are constant parameters of linear filter defined by

$$
a_1 = a_1^* v, \qquad a_2 = a_2^* v, \qquad q = q^* \sqrt{a_1 a_2 v}, \tag{6.3.9}
$$

where a_1^*, a_2^* and q^* are constant parameters of random road profile, v is the constant speed of vehicle.

The objective of the use of the active control u is to minimize the modified performance index I defined by (6.3.6), namely

$$
I = \frac{1}{m_2} E \left[\left(\frac{c_2}{m_2} \bar{x}_2 + \frac{h_2}{m_2} \bar{x}_4 + \frac{1}{m_2} g(\bar{x}_2) - \frac{1}{m_2} \bar{u} \right)^2 + \rho_1 (\bar{x}_2)^2 + \rho_2 (\bar{x}_3)^2 + \rho_3 \bar{u}^2 \right]. \tag{6.3.10}
$$

Here $\bar{x}_i$, $i = 2, 3, 4$ and $\bar{u}$ are the stationary quantities. If the nonlinear stiffness

$$
Y = g(x_2) \tag{6.3.11}
$$

can substituted by the following linearized form

$$
Y = \alpha k x_2, \tag{6.3.12}
$$

where α is the constant parameter and k is the linearization coefficient, then using equations (6.3.7) and (6.3.12) the optimal control problem can be transformed to the standard one

$$d\mathbf{x} = [\mathbf{A}(k)\mathbf{x} + \mathbf{B}u]dt + \mathbf{G}d\xi, \tag{6.3.13}$$

$$I = E[\bar{\mathbf{x}}^T \mathbf{Q}(k)\bar{\mathbf{x}} + 2\bar{\mathbf{x}}^T \mathbf{N}(k)\bar{u} + r\bar{u}^2], \tag{6.3.14}$$

where matrices $\mathbf{A}$, $\mathbf{Q}$ and vectors $\mathbf{N}$, $\mathbf{N}$ are defined by equations (6.3.7) and (6.3.12); $\mathbf{A}(k)$, $\mathbf{Q}(k)$ and $\mathbf{N}(k)$ for a given linearization coefficient k are constant matrices and vector, respectively.

A comparison of statistical linearization methods for Gaussian excitations corresponding to the following moment criteria was done in Socha (1999a):

Criterion 1. Equality of second order moments Kazakov (1956):

$$E\left[(k_a x_2)^2\right] = E\left[(g(x_2))^2\right]. \tag{6.3.15}$$

Criterion 2. Mean-square error of displacements Kazakov (1956):

$$E\left[(k_b x_2 - g(x_2))^2\right] \to \min. \tag{6.3.16}$$

Criterion 3. Mean-square error of potential energies Elishakoff (1991):

$$E\left[\left(\int_0^{x_2} [k_c v - g(v)]dv\right)^2\right] \to \min, \tag{6.3.17}$$

where k_a, k_b and k_c are linearization coefficients,

and two criteria in the probability density functions space:

Criterion 4. Square probability metric Socha (2002):

$$\int_{-\infty}^{+\infty} (g_N(y) - g_L(y, k_d))^2 dy. \tag{6.3.18}$$

Criterion 5. Pseudo-moment metric Socha (2002):

$$\int_{-\infty}^{+\infty} |y|^{2l} |g_N(y) - g_L(y, k_e)| dy, \tag{6.3.19}$$

where $l = 1, 2, \ldots$, $g_N(y)$ and $g_L(y)$ are probability density functions of variables defined by

$$Y = \psi(x_2) = c_2 x_2 + g(x_2) \tag{6.3.20}$$

and

$$Y = k^* x_2, \tag{6.3.21}$$

respectively, where k^* is the linearization coefficient. Then the optimal control problem can be transformed to the modified version of (6.3.13) and (6.3.14). As an example it was considered a nonlinear function

$$g(x_2) = \alpha x_2^3. \tag{6.3.22}$$

In the case of moment criteria it was shown Kazakov (1956), Elishakoff (1991) that the corresponding linearization coefficients have the form

$$k_a = \sqrt{15} E[x_2^2], \qquad k_b = 3E[x_2^2], \qquad k_b = 2.5E[x_2^2], \tag{6.3.23}$$

and in the case of criteria 4 and 5 the nonlinear function $\psi(x_2)$ and the corresponding probability density function have the form

$$Y = \psi(x_2) = c_2 x_2 + \alpha x_2^3, \tag{6.3.24}$$

$$g_Y(y) = \frac{1}{\sqrt{2\pi}\sigma_L} \exp\left[-\frac{(v_1+v_2)^2}{2\sigma_L^2}\right] \frac{1}{6a\varepsilon}\left(\frac{a+y}{v_1^2} + \frac{a-y}{v_2^2}\right), \tag{6.3.25}$$

where

$$v_1 = \left(\frac{y}{2\alpha} + \sqrt{\frac{y^2}{4\alpha^2} + \frac{c_2^3}{27\alpha^3}}\right)^{1/3}, \qquad v_2 = \left(\frac{y}{2\alpha} - \sqrt{\frac{y^2}{4\alpha^2} + \frac{c_2^3}{27\alpha^3}}\right)^{1/3}, \tag{6.3.26}$$
$$a = \sqrt{y^2 + \frac{4c_2^3}{27\alpha}}.$$

The probability density of the linearized variable (6.3.21) has the form

$$g_L(y, k^*) = \frac{1}{\sqrt{2\pi} k^* \sigma_L} \exp\left\{-\frac{y^2}{2(k^*)^2\sigma_L^2}\right\}, \tag{6.3.27}$$

where $\sigma_L^2 = E[x^2]$ is the variance of the input Gaussian variable, k^* is equal to k_d or k_e in criterion 4 or 5, respectively.

To determine the quasi-optimal control for a nonlinear system with nonlinear criterion the idea proposed in the literature (see for instance Yoshida (1984) Beaman (1984)) consisting in application of the statistical linearization and LQG method was used. The following two iterative procedures were proposed in Socha and Błachuta (2000).

Iterative procedures

Procedure A (for criteria 1-3):

Step 1. Assume that one of the linearization coefficients is equal to zero, for instance, $k = k_a = 0$.

***S*tep 2.** Calculate $\mathbf{A} = \mathbf{A}(k)$, $\mathbf{Q} = \mathbf{Q}(k)$ and $\mathbf{N} = \mathbf{N}(k)$ in (6.3.13)-(6.3.14) and then solve the algebraic Riccati equation

$$\mathbf{P}\mathbf{A}_N + \mathbf{A}_N^T\mathbf{P} - \frac{1}{r}\mathbf{P}\mathbf{B}\mathbf{B}^T\mathbf{P} + \mathbf{Q}_N = \mathbf{0}, \tag{6.3.28}$$

where $\mathbf{A}_N = \mathbf{A} - \frac{1}{r}\mathbf{B}\mathbf{N}^T \geq \mathbf{0}$, $\mathbf{Q}_N = \mathbf{Q} - \frac{1}{r}\mathbf{N}\mathbf{N}^T \geq \mathbf{0}$. The solution is a symmetric positive definite matrix $\mathbf{P}$.

***S*tep 3.** Find the optimal control and the matrix $\mathbf{K}$.

$$\bar{\mathbf{u}}(t) = -\mathbf{K}\bar{\mathbf{x}}(t) = -\frac{1}{r}(\mathbf{N}^T + \mathbf{B}^T\mathbf{P})\bar{\mathbf{x}}(t). \tag{6.3.29}$$

Next, substitute $\mathbf{K}$, $\mathbf{A}(k)$, $\mathbf{Q}(k)$, $\mathbf{N}(k)$ into the covariance equation

$$(\mathbf{A}(k) - \mathbf{B}K)\mathbf{V}_L + \mathbf{V}_L(\mathbf{A}(k) - \mathbf{B}K)^T + \mathbf{G}G^T = \mathbf{0} \tag{6.3.30}$$

and solve the equation. The solution of equation (6.3.30) is $\mathbf{V}_L$.

***S*tep 4.** Substitute the element of covariance matrix $E[\bar{x}_2^2] = V_{L_{22}}$ obtained in *S*tep 3 into the linearization coefficient k_a defined by (6.3.23).

***S*tep 5.** Calculate $\mathbf{P}$, $\mathbf{u}$ and $\mathbf{V}_L$ using equations (6.3.28)- (6.3.30) and the linearization coefficient k_a obtained in the last step.

***S*tep 6.** Iterate *S*teps 2-5 until $\mathbf{V}_L$ and $\mathbf{P}$ converge.

***S*tep 7.** Calculate the optimal value of criterion I_{opt} using the solution of the Riccati equation obtained in *S*tep 5,

$$I = tr(\mathbf{P}\mathbf{G}\mathbf{G}^T). \tag{6.3.31}$$

Procedure B (for criteria 4-5):

***S*tep 1.** Assume $k^* = c_2$.

***S*tep 2.** Calculate modified matrices $\mathbf{A} = \mathbf{A}(k^*)$, $\mathbf{Q} = \mathbf{Q}(k^*)$ and $\mathbf{N} = \mathbf{N}(k^*)$ in (6.3.13)-(6.3.14) and then solve the algebraic Riccati equation (6.3.28). The solution is a symmetric positive definite matrix $\mathbf{P}$.

***S*tep 3.** Substitute $\mathbf{P}$ obtained in *S*tep 2 into equation (6.3.29) and find the matrix $\mathbf{K}$. Next, substitute $\mathbf{K}$, $\mathbf{A}(k)$, $\mathbf{Q}(k)$, $\mathbf{N}(k)$ into the covariance equation (6.3.30) and solve the equation. The solution of equation (6.3.30) is $\mathbf{V}_L$.

***S*tep 4.** For $\mathbf{C}$ obtained in previous step calculate for linearized element the variance $\sigma_{x_2}^2 = E[\bar{x}_2^2]$ of the input Gaussian variable and next the corresponding probability density functions given by (6.3.25) and (6.3.27), respectively.

***S*tep 5.** For nonlinear element find the linearization coefficient k^* which minimize, for instance, Criterion 4.

***S*tep 6.** Substitute the linearization coefficient k^* obtained in *S*tep 5 into equation (6.3.30) and then solve the equation.

***S*tep 7.** If the error of accuracy is greater then a given parameter ε_1 then repeat *S*teps 4-6 until $\mathbf{V}_L$ converges.

***S*tep 8.** Calculate new $\mathbf{A} = \mathbf{A}(k^*)$, $\mathbf{Q} = \mathbf{Q}(k^*)$ and $\mathbf{N} = \mathbf{N}(k^*)$ and next using these matrices calculate $\mathbf{P}, \mathbf{K}$ and $\mathbf{V}_L$ using equations (6.3.28)-(6.3.30).

***S*tep 9.** Iterate *S*teps 4-8 until $\mathbf{V}_L$ and $\mathbf{P}$ converge.

***S*tep 10.** Calculate the optimal value of criterion I_{opt} substituting the solution of the Riccati equation obtained in *S*tep 8 into relation (6.3.31).

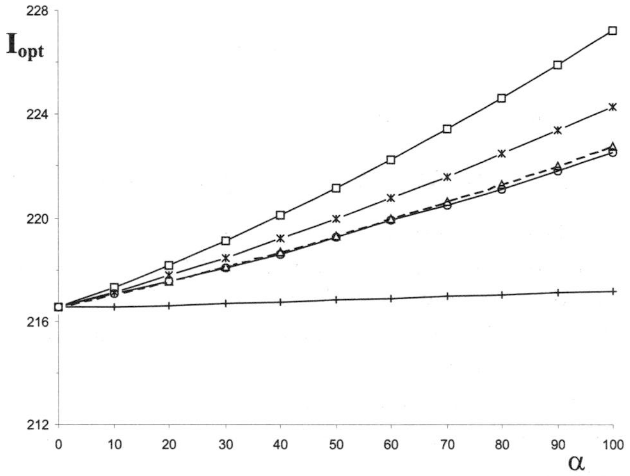

Figure 10. Comparison between optimal criteria obtained by application of different statistical linearization techniques versus parameter α.

An illustration of the obtained results in the form of a comparison of the criterion I_{opt} defined by (6.3.31) versus parameter alpha was presented by Socha (2000). In this comparison three moment criteria and two criteria in probability density functions space of linearization techniques, namely equality of second order moments of nonlinear and linearized elements, mean-square error of the displacement, mean-square error of the potential energies, square probability metric and pseudomoment probability metric are considered. The numerical results denoted by lines with circles stars, squares, triangles and crosses, respectively are presented in Fig.10. The parameters selected for calculations and further simulations are $m_1 = 100$, $m_2 = 500$, $c_1 = 100$, $c_2 = 50$, $h_1 = 1$, $h_2 = 5$, $a_1^* = 0.025$, $a_2^* = 0.075$, $q^* = \sqrt{0.0067}$, $v = 20$, $\rho_1 = 1$, $\rho_2 = 1000$, $\rho_3 = 10000$, $\rho_4 = 1$.

Figure 11 shows the dependence of the criterion I_{opt} upon the speed of vehicle as it changes from 10^0 to 10^2. The other parameters are the same except $\rho_2 = 100$, $\rho_3 = 1000$ and $\alpha = 20$.

From the numerical results it follows that for given sets of parameters there are no significant differences between mean-square criteria obtained by application considered

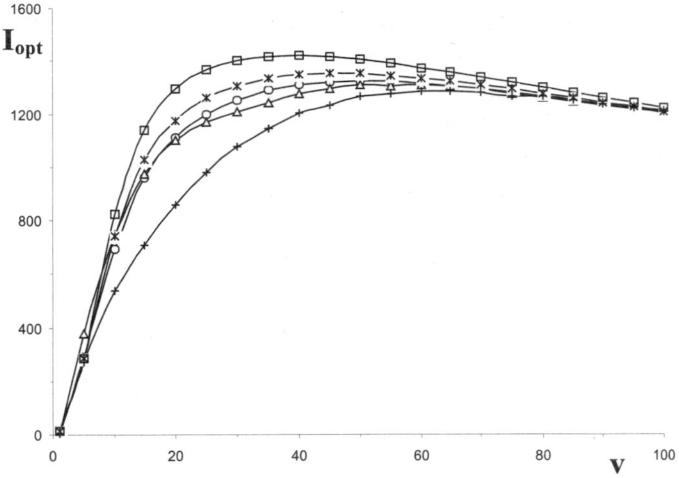

Figure 11. Comparison between optimal criteria obtained by application of different statistical linearization techniques versus parameter v.

linearization methods.

An extension of Wonham and Cashman's idea for quasi–optimal control for seismic–excited hysteretic structural systems without and with observation equation was done by Yang et al. (1994) and Suzuki (1995), respectively. Raju and Narayanan (1995) and Narayanan and Senthil (1998) applied this approach to the two–degree–of–freedom vehicle suspension model with fully observed state vector. In both cases equivalent linearization is applied to hysteretic parts of considered systems and LQG theory with or without filtering equation is used.

Acknowledgements
Computer programming assistance given by A.Wojak and M.Swiety is greatfully acknowledged.

Bibliography

A.G.Thompson. An active suspension with optimal lineartate feedback. *Veh.Syst.Dyn.*, 5:187–203, 1976.

M. Apetaur. Modified second order linearization procedure - problems encountered and their solution. *Veh. System Dynam.*, 17:255–265, 1988.

M. Apetaur. Linearization function of a complex nonlinear two–force–element. *Veh. System Dynam.*, 20:309–320, 1991.

M. Apetaur and F. Opicka. Linearization of nonlinear stochastically excited dynamic systems. *J. Sound Vib.*, 86:563–585, 1983.

T.T. Baber and Y.K. Wen. Random vibration of hysteretic degrading systems. *ASCE J. Engrg. Mech. Div.*, 107:1069–1087, 1981.

T.T. Baber and Y.K. Wen. Stochastic response of multi-story yielding frames random vibration of hysteretic degrading systems. *Earthquake Engrg. and Struct. Dyn.*, 10: 403–416, 1982.

J.J. Beaman. Non–linear quadratic Gaussian control. *Int. J. Cont.*, 39:343–361, 1984.

P. Bernard. About stochastic linearization. In N. Bellomo and F. Casciati, editors, *Nonlinear Stochastic Mechanics*, pages 61–70. Springer, 1991.

P. Bernard and M. Taazount. Random dynamics of structures with gaps: simulation and spectral linearization. *Nonlinear Dynamics*, 5:313–335, 1994.

R. Booton. The analysis of nonlinear central systems with random inputs. *IRE Transactions on Circuit Theory*, 1:32–34, 1954.

A. Bruckner and Y.K. Lin. Generalization of the equivalent linearization method for nonlinear random vibration problems. *Int. J. Nonlinear Mech.*, 22:227–235, 1987.

G.Q. Cai and Y.K. Lin. Response spectral densities of strongly nonlinear systems under random excitation. *Prob. Engng. Mech.*, 12:41–47, 1997.

T.K. Caughey. Response of van der pol's oscillator to random excitations. *Trans. ASME J. Appl. Mech.*, 26:345–348, 1959.

T.K. Caughey. Random excitation of a loaded nonlinear string. *Trans. ASME J. Appl. Mech.*, 27:575–578, 1960.

S.H. Crandall. Perturbation techniques for random vibration of nonlinear systems. *J. Acoust. Soc. Amer.*, 35:1700–1705, 1963.

S.H. Crandall. Correlations and spectra of nonlinear system response. *Problems of Nonlinear Vibrations*, 14:39–53, 1973.

I. Elishakoff. Method of stochastic linearization: revisited and improved. In P.D.Spanos and C.A. Brebbia, editors, *Computational Stochastic Mechanics*, pages 101–111. Computational Mechanics Publication and Elsevier Applied Science, 1991.

I. Elishakoff. Random vibration of structures: A personal perspective. *Appl.Mech. Rev.*, 48:809–825, 1995a.

I. Elishakoff. Some results in stochastic linearization of nonlinear systems. In W.H.Klieman and N. Namachchivaya, editors, *Nonlinear Dynamics and Stochastic Mechanics*, pages 259–281. CRC Press, 1995b.

I. Elishakoff. Stochastic linearization technique: A new interpretation and a selective review. *The Shock and Vib. Dig.*, 32:179–188, 2000.

I. Elishakoff and G. Falsone. Some recent developments in stochastic linearization technique. In A.H. Cheng and C.Y.Yang, editors, *Computational Stochastic Mechanics*, pages 175–194. Computational Mechanics Publication, 1993.

I. Elishakoff, J. Fang, and R. Caimi. Random vibration of a nonlinearly deformed beam by a new stochastic linearization technique. *Int. J. Solids Structures*, 32:1571–1584, 1995.

I. Elishakoff and R.C. Zhang. Comparison of the new energy–based version of the stochastic linearization technique. In N.Bellomo and F.Casciati, editors, *Nonlinear Stochastic Mechanics*, pages 201–212. Springer, 1991.

I. Elishakoff and X.T. Zhang. An appraisal of different stochastic linearization criteria. *J. Sound Vib.*, 153:370–375, 1992.

J. Fang, I. Elishakoff, and R. Caimi. Nonlinear response of a beam under stationary random excitation by improved stochastic linearization method. *Appl. Math. Modeling*, 19:106–111, 1995.

T. Gordon, C. March, and M. Milsted. A comparison of adaptive lqg and nonlinear controllers for vehicle suspension systems. *Veh. Syst. Dyn.*, 20:321–340, 1991.

M. Grigoriu. Equivalent linearization for Poisson white noise input. *Prob. Eng. Mech.*, 10:45–51, 1995a.

M. Grigoriu. Linear and nonlinear systems with non–Gaussian white noise input. *Prob. Eng.Mech.*, 10:171–180, 1995b.

A. Hac. Suspension optimization of a 2-dof vehicle model using a stochastic optimal control technique. *J. Sound Vib.*, 100:343–357, 1985.

M.A. Ismaili. Design of a system of linear oscillators spectrally equivalent to a non-linear one. *Int. J. Nonlinear Mech.*, 31:573–580, 1996.

R.N. Iyengar and O.R. Jaiswal. A new model for non–Gaussian random excitations. *Prob. Eng. Mech.*, 8:281–287, 1993.

I.E. Kazakov. An approximate method for the statistical investigation for nonlinear systems (in russian). *Trudy VVIA im Prof. N. E. Zhukovskogo*, 394:1–52, 1954.

I.E. Kazakov. Approximate probabilistic analysis of the accuracy of the operation of essentially nonlinear systems. *Avtomatika i Telemekhanika*, 17:423–450, 1956.

I.E. Kazakov. Analytical synthesis of a quasi–optimal additive control in a nonlinear stochastic system. *Avtomatika i Telemekhanika*, 45:34–46, 1984.

I.E. Kazakow. *Statistical Theory of Control Systems in State Space (in Russian)*. Nauka, 1975.

I.E. Kazakow and B.G. Dostupow. *Statistical Dynamics of Nonlinear Automatic Control Systems (in Russian)*. Fizmatgiz, 1962.

I.E. Kazakow and S.W. Malczikow. *Analysis of Stochastic Systems in State Space (in Russian)*. Nauka, 1975.

J. Kottalam, K. Lindberg, and B. West. Statistical replacement for systems with delta-correlated fluctuations. *J. Statist. Phys.*, 42:979–1008, 1986.

F. Kozin. The method of statistical linearizationfor non-linear stochastic vibration. In F.Ziegler and G.I. Schueller, editors, *Nonlinear Stochastic Dynamic Engineering Systems*, pages 45–56. Springer, 1989.

S. Krenk and J.B. Roberts. Local similarity in nonlinear random vibration. *Trans. ASME J.Appl. Mech.*, 66:225–235, 1999.

N. Krilov and N. Bogoliuboff. *Introduction to Non-linear Mechanics*. Princeton University Press, 1943.

H. Kwakernaak and R. Sivan. *Linear Optimal Control Systems*. J. Wiley, 1972.

Y.K. Lin and G.Q. Cai. *Probabilistic structural dynamics*. McGraw Hill, 1995.

L.D. Lutes and S. Sarkani. *Stochastic Analysis of Structural and Mechanical Vibrations.* Prentice Hall, 1997.

J.B. Morton and S. Corrisin. Consolidated expansions for estimating the response of a randomly driven nonlinear oscillator. *J. Statist. Phys.*, 2:153–194, 1970.

S. Narayanan and G.V. Raju. Active control of non–stationary response of vehicles with nonlinear suspensions. *Veh.Syst.Dyn.*, 21:73–87, 1992.

S. Narayanan and S. Senthil. Stochastic optimal active control of a 2-DOF quarter car model with non-linear passive suspension elements. *J. Sound Vib.*, 211:495–506, 1998.

B.N. Naumov. *The Theory of Nonlinear Automatic Control Systems. Frequency methods (In Russian).* Nauka, 1972.

M.B. Priestly. *Spectral Analysis and Time Series.* Academic Press, 1981.

C. Proppe, H.J. Pradlwarter, and G.I. Schueller. Equivalent linearization and Monte Carlo simulation in stochastic dynamics. *Prob. Engng. Mech.*, 18:1–15, 2003.

W.S. Pugacev and I.N. Sinicyn. *Stochastic Differential Systems.* Willey, 1987.

G.V. Raju and S. Narayanan. Active control of nonstationary response of a 2-degree of freedom vehicle model with nonlinear suspension. *Sadhana-Acad. P. Engng. S.*, 20: 489–499, 1995.

J.B. Roberts and P.D. Spanos. *Random Vibration and Statistical Linearization.* John Wiley and Sons, 1990.

R.V. Roy and P.D. Spanos. Power spectral density of nonlinear system response: the recursive method. *Trans. ASME J.Appl. Mech.*, 60:358–365, 1993.

M. Sakata and K. Kimura. Calculation of the non–stationary mean square response of a non–linear system subjected to non–white excitation. *J. Sound Vib.*, 73:333–343, 1980.

A.W. Smyth and S.F. Masri. Nonstationary response of nonlinear systems using equivalent linearization with a compact analytical form of the excitation process. *Prob. Eng. Mech.*, 17:97–108, 2002.

C. Sobiechowski and L. Socha. Statistical linearization of the Duffing oscillator under non–Gaussian external excitation. In P.D.Spanos, editor, *Proceedings of the International Conference on Computational Stochastic Mechanics*, pages 125–133. Balkema, 1998.

C. Sobiechowski and L. Socha. Statistical linearization of the Duffing oscillator under non–Gaussian external excitation. *J. Sound Vib.*, 231:19–35, 2000.

L. Socha. Application of probability metrics to the linearization and sensitivity analysis of stochastic dynamic systems. In *Proceedings of the International Conference on Nonlinear Stochastic Dynamics*, pages 193–202, 1995.

L. Socha. Active control of nonlinear 2-degrees-of-freedom vehicle suspension under stochastic excitations. In J. Holnicki-Szulc and J.Rodelalar, editors, *Smart Structures*, pages 321–327. Kluwer Academic Publishers, 1999a.

L. Socha. Probability density equivalent linearization and non–linearization techniques. *Archive Mechanics*, 51:487–507, 1999b.

L. Socha. Statistical and equivalent linearization techniques with probability density criteria. *J. Theor. Appl. Mech.*, 37:369–382, 1999c.

L. Socha. Application of statistical linearization techniques to design of quasi–optimal active control of nonlinear systems. *J. Theor. Appl.Mech.*, 38:591–605, 2000.

L. Socha. Probability density statistical and equivalent linearization techniques. *Int.J. System Sci.*, 33:107–127, 2002.

L. Socha. Linearization in analysis of nonlinear stochastic systems: Recent results–part i:theory. *Appl.Mech. Rev.*, 58:178–205, 2005.

L. Socha and M. Błachuta. Application of linearization methods with probability density criteria in control problems. In *Proceedings of the American Control Conference, Chicago*, pages 2775–2779, 2000.

L. Socha and T.T. Soong. Linearization in analysis of nonlinear stochastic systems. *Appl.Mech. Rev.*, 44:399–422, 1991.

J. Solnes. *Stochastic Processes and Random Vibrations Theory and Practice.* John Wiley & Sons, 1997.

T.T. Soong and M. Grigoriu. *Random Vibration of Mechanical and Structural Systems.* PTR Prentice Hall, 1993.

Y. Suzuki. Stochastic control of hysteretic structural systems. *Sadhana*, 20:475–488, 1995.

G.J. Thaler and M.P. Pastel. *Analysis and Design of Nonlinear Feedback Control Systems.* MacGraw-Hill, 1962.

A. Tylikowski and W. Marowski. Vibration of a non–linear single degree of freedom system due to Poissonian impulse excitation. *Int. J. of Non–Linear Mech.*, 21:229–238, 1986.

Y.K. Wen. Equivalent linearization for hysteretic systems under random excitation. *ASME J. Appl. Mech.*, 47:150–154, 1980.

W.M. Wonham and W.F. Cashman. A computational approach to optimal control of stochastic saturating systems. *Int. J. Cont.*, 10(1):77–98, 1969.

J.N. Yang, Z. Li, and S. Vongchavalitkul. Stochastic hybrid control of hysteretic structures. *Prob. Engng. Mech.*, 9:125–133, 1994.

K. Yoshida. A method of optimal control of non–linear stochastic systems with non–quadratic criteria. *Int. J. Cont.*, 39:279–291, 1984.

G.E. Young and R.J. Chang. Prediction of the response of nonlinear oscillators under stochastic parametric and external excitations. *Int. J. Nonlinear Mech.*, 22:151–160, 1987.

R. Zhang, I. Elishakoff, and M. Shinozuka. Analysis of nonlinear sliding structures by modified stochastic linearization methods. *Nonlinear Dynamics*, 5:299–312, 1994.

X.T. Zhang, I. Elishakoff, and R.C. Zhang. A new stochastic linearization technique based on minimum mean-square deviation of potential energies. In Y.K. Lin and I. Elishakoff, editors, *Stochastic Structural Dynamics - New Theoretical Developments*, pages 327–338. Springer Verlag, 1991.

X.T. Zhang, R.C. Zhang, and Y.L. Xu. Analysis on control of flow–induced vibration by tuned liquid damper with crossed tube–like containers. *J. Wind Engr. Indust. Aerod.*, 50:351–360, 1993.

Models, Verification, Validation, Identification and Stochastic Eigenvalue Problems

Sondipon Adhikari

Department of Aerospace Engineering, University of Bristol, Bristol BS10 5BL, U. K.
Email: S.Adhikari@bristol.ac.uk

1 What is a Model? Why do We Need a Model? How We Use a Model?

1.1 Introduction

Problems involving vibration occur in many areas of mechanical, civil and aerospace engineering: wave loading of offshore platforms, cabin noise in aircrafts, earthquake and wind loading of cable stayed bridges and high rise buildings, performance of machine tools – to pick only few examples. Human beings usually regard noise and vibration as uncomfortable. Beside this, an engineering structure can fail due to excessive vibration - the devastating effects of earthquakes on our society is a prime example of this fact. Due to this reasons over the years the aim of the vibration engineers has been to reduce vibration. In order to achieve this in an efficient and economic manner a good understanding of the physics of vibration phenomena in complex engineering structures is needed. In the last few decades, the sophistication of modern design methods together with the development of improved composite structural materials instilled a trend towards lighter structures. At the same time, there is also a constant demand for larger structures, capable of carrying more loads at higher speeds with minimum noise and vibration level as the safety/workability and environmental criteria become more stringent. Unfortunately, these two demands are conflicting and the problem cannot be solved without proper understanding of the vibration phenomena.

Broadly speaking, as with other branches of science and engineering, there are two ways to understand the dynamics of complex structures. The first is the experimental approach. A carefully conducted experiment can yield high quality data which can provide crucial information regarding the dynamics of a system. However, the experimental process is time consuming, expensive and it may not be possible to dynamically test an extremely complex structure under various loading conditions which the structure might experience during its service period. The alternative is to 'replace' the real structure by a *mathematical model* and perform numerical experiments in a computer. Therefore, a model is a mathematical representation of the true structure. A model can be created

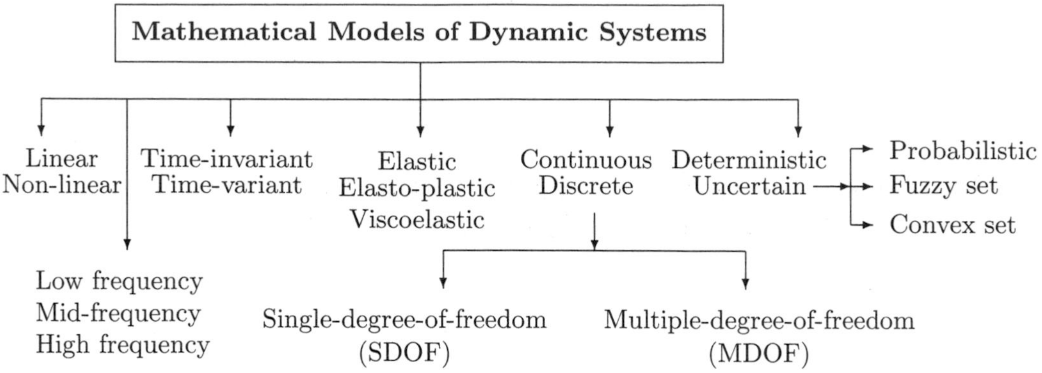

Figure 1.1. Different types of mathematical models for dynamic systems

with various levels of sophistication. Figure 1.1 shows different types of models of dynamic systems used by today's engineers. The choice of a particular model depends on the physics of the system, accuracy required for the problem and also on the nature of the forces the structure is expected to withstand in practice. The quality of a model of a dynamic system depends on the following three factors (see for example Hemez and Ben-Haim, 2004):

- *Fidelity to (experimental) data:*
  The results obtained from a numerical or mathematical model undergoing a given excitation force should be close to the results obtained from the vibration testing of the same structure undergoing the same excitation.
- *Robustness with respect to (random) errors:*
  Errors in estimating the system parameters, boundary conditions and dynamic loads are unavoidable in practice. The output of the model should not be very sensitive to such errors.
- *Predictive capability*
  In general it is not possible to experimentally validate a model over the entire domain of its scope of application. The model should predict the response well beyond its validation domain.

Here attention will be paid to the above equalities of a model. Linear system behaviour will be assumed throughout this paper.

1.2 Equation of Motion of Linear Mechanical Systems

A key step in the modeling of dynamical systems is to obtain the equation of motion. Lord Rayleigh (1877) in his classic monograph 'Theory of Sound' outlined the fundamental concepts of modeling and analysis of linear dynamic systems. Current methods for modeling and analysis of complex engineering systems are largely based on Rayleigh's approach. The books by Géradin and Rixen (1997), Meirovitch (1997) give excellent account of the basic concepts of structural dynamics. From a mathematical point of view, models of vibrating systems are commonly divided into two broad classes – discrete, or lumped-parameter models, and continuous, or distributed-parameter models (see figure 1.1). However, systems can contain both distributed and lumped parameter models, for example, a beam with a tip mass. Distributed-parameter modeling of vibrating systems leads to *partial-differential equations* as the equation of motion. Exact solutions of such equations were historically possible only for a limited number of problems with simple geometry, boundary conditions, and material properties (such as constant mass density). Recent works by Elishakoff (2005) however show that exact closed-form solutions can be obtained for a wide range of distributed parameter systems. In spite of these developments, for general complex engineering structures, such as an aircraft, normally we need some kind of approximate methods for dynamic analysis. Such approaches are generally based on spatial discretization of the displacement field (for example, the Finite Element Method, Zienkiewicz and Taylor, 1991), which amounts to approximating distributed-parameter systems by lumped-parameter systems. Equations of motion of lumped-parameter systems can be shown to be expressed by a set of coupled *ordinary-differential equations*. In this paper we will mostly deal with such lumped-parameter systems. We also restrict our attention to the linear system behavior only.

Suppose that a system with N degrees of freedom is executing small oscillations around equilibrium points. The theory of small oscillations was studied in detail by Rayleigh (1877). Considering the vector of *generalized coordinates*

$$\mathbf{q} = \{q_1(t), q_2(t), \cdots, q_N(t)\}^T \in \mathbb{R}^N \tag{1.1}$$

it can be shown that the equation of motion of an undamped non-gyroscopic system is given by

$$\mathbf{M}\ddot{\mathbf{q}}(t) + \mathbf{K}\mathbf{q}(t) = \mathfrak{f}(t) \tag{1.2}$$

where $\mathbf{K} \in \mathbb{R}^{N\times N}$ is the (linear) stiffness matrix, $\mathbf{M} \in \mathbb{R}^{N\times N}$ is the *mass matrix* and $\mathfrak{f}(t) \in \mathbb{R}^N$ is the forcing vector. Equation (1.2) represents a set of coupled second-order ordinary-differential equations. The solution of this equation also requires knowledge of the initial conditions in terms of the displacements and velocities of all the coordinates.

1.3 Classical Modal Analysis

Rayleigh (1877) showed that the undamped linear systems (1.2) are capable of so-called *natural motions*. This essentially implies that all the system coordinates execute harmonic oscillation at a given frequency and form a certain displacement pattern. The oscillation frequency and displacement pattern are called *natural frequencies* and *normal modes*, respectively. The natural frequencies (ω_j) and the mode shapes ($\mathbf{x}_j$) are intrinsic characteristic of a system and can be obtained by solving the associated matrix eigenvalue problem

$$\mathbf{K}\mathbf{x}_j = \omega_j^2\mathbf{M}\mathbf{x}_j, \quad \forall j = 1, \cdots, N. \tag{1.3}$$

Since the above eigenvalue problem is in terms of real symmetric matrices $\mathbf{M}$ and $\mathbf{K}$, the eigenvalues and consequently the eigenvectors are real, that is $\omega_j \in \mathbb{R}$ and $\mathbf{x}_j \in \mathbb{R}^N$. In addition to this, it was also shown by Rayleigh that the undamped eigenvectors satisfy an orthogonality relationship over the mass and stiffness matrices, that is

$$\mathbf{x}_l^T \mathbf{M} \mathbf{x}_j = \delta_{lj} \tag{1.4}$$

$$\text{and} \quad \mathbf{x}_l^T \mathbf{K} \mathbf{x}_j = \omega_j^2 \delta_{lj}, \quad \forall\, l, j = 1, \cdots, N \tag{1.5}$$

where δ_{lj} is the Kroneker delta function. In the above equations the eigenvectors are unity mass normalized, a convention often used in practice. This orthogonality property of the undamped modes is very powerful as it allows to transform a set of coupled differential equations to a set of independent equations. For convenience, we construct the matrices

$$\boldsymbol{\Omega} = \text{diag}\left[\omega_1, \omega_2, \cdots, \omega_N\right] \in \mathbb{R}^{N \times N} \tag{1.6}$$

$$\text{and} \quad \mathbf{X} = \left[\mathbf{x}_1, \mathbf{x}_2, \cdots, \mathbf{x}_N\right] \in \mathbb{R}^{N \times N} \tag{1.7}$$

where the eigenvalues are arranged such that $\omega_1 < \omega_2$, $\omega_2 < \omega_3, \cdots, \omega_k < \omega_{k+1}$. Use a coordinate transformation

$$\mathbf{q}(t) = \mathbf{X}\mathbf{y}(t). \tag{1.8}$$

Substituting $\mathbf{q}(t)$ in equation (1.2), premultiplying by $\mathbf{X}^T$ and using the orthogonality relationships in (1.6) and (1.7), the equation of motion in the modal coordinates may be obtained as

$$\ddot{\mathbf{y}}(t) + \boldsymbol{\Omega}^2 \mathbf{y}(t) = \tilde{\mathfrak{f}}(t) \tag{1.9}$$

where $\tilde{\mathfrak{f}}(t) = \mathbf{X}^T \mathfrak{f}(t)$ is the forcing function in modal coordinates. Clearly, this method significantly simplifies the dynamic analysis because complex multiple degrees of freedom systems can be treated as collections of single-degree-of-freedom oscillators. This approach of analyzing linear undamped systems is known as *modal analysis*, possibly the most efficient tool for vibration analysis of complex engineering structures.

1.4 Models of Damping

Viscous Damping. Damping is the dissipation of energy from a vibrating structure. The most popular approach to model damping is to assume viscous damping. By analogy with the potential energy and the kinetic energy, Rayleigh (1877) introduced the *dissipation function*, which can be expressed as

$$\mathcal{F}\left(\mathbf{q}\right) = \frac{1}{2} \sum_{j=1}^{N} \sum_{k=1}^{N} C_{jk} \dot{q}_j \dot{q}_k = \frac{1}{2} \dot{\mathbf{q}}^T \mathbf{C} \dot{\mathbf{q}}. \tag{1.10}$$

In the above expression $\mathbf{C} \in \mathbb{R}^{N \times N}$ is a non-negative definite symmetric matrix, known as the viscous damping matrix. It should be noted that not all forms of the viscous damping matrix can be handled within the scope of classical modal analysis. Based on the solution method, viscous damping matrices can be further divided into classical and non-classical damping. Further discussions on viscous damping will follow in section 1.5.

Non-viscous Damping Models. It is important to avoid the widespread misconception that viscous damping is the *only* linear model of vibration damping. Any causal model which makes the energy dissipation functional non-negative is a possible candidate for a damping model. There have been several efforts to incorporate non-viscous damping models in linear systems. One popular approach is to model damping in terms of fractional derivatives of the displacements. The damping force using such models can be expressed by

$$\mathbf{F}_d = \sum_{j=1}^{l} \mathbf{g}_j D^{\nu_j}[\mathbf{q}(t)]. \tag{1.11}$$

Here $\mathbf{g}_j$ are complex constant matrices and the fractional derivative operator

$$D^{\nu_j}[\mathbf{q}(t)] = \frac{\mathrm{d}^{\nu_j}\mathbf{q}(t)}{\mathrm{d}t^{\nu_j}} = \frac{1}{\Gamma(1-\nu_j)}\frac{\mathrm{d}}{\mathrm{d}t}\int_0^t \frac{\mathbf{q}(t)}{(t-\tau)^{\nu_j}}\mathrm{d}\tau \tag{1.12}$$

where ν_j is a fraction and $\Gamma(\bullet)$ is the Gamma function. The familiar viscous damping appears as a special case when $\nu_j = 1$. Although this model might fit experimental data quite well, the physical justification for such models, however, is far from clear at the present time.

Possibly the most general way to model damping within the linear range is to consider non-viscous damping models which depend on the past history of motion via convolution integrals over some kernel functions. A *modified dissipation function* for such damping model can be defined as

$$\mathcal{F}(\mathbf{q}) = \frac{1}{2}\sum_{j=1}^{N}\sum_{k=1}^{N}\dot{q}_k\int_0^t \mathcal{G}_{jk}(t-\tau)\dot{q}_j(\tau)\mathrm{d}\tau = \frac{1}{2}\dot{\mathbf{q}}^T\int_0^t \boldsymbol{\mathcal{G}}(t-\tau)\dot{\mathbf{q}}(\tau)\mathrm{d}\tau. \tag{1.13}$$

Here $\boldsymbol{\mathcal{G}}(t) \in \mathbb{R}^{N\times N}$ is a symmetric matrix of the damping kernel functions, $\mathcal{G}_{jk}(t)$. The kernel functions, or others closely related to them, are described under many different names in the literature of different subjects: for example, retardation functions, heredity functions, after-effect functions and relaxation functions. In the special case when $\boldsymbol{\mathcal{G}}(t-\tau) = \mathbf{C}\,\delta(t-\tau)$, where $\delta(t)$ is the Dirac-delta function, equation (1.13) reduces to the case of viscous damping as in equation (1.10). The damping model of this kind is a further generalization of the familiar viscous damping. By choosing suitable kernel functions, it can also be shown that the fractional derivative model discussed before is also a special case of this damping model. Thus, as pointed by Woodhouse (1998), this damping model is the most general damping model within the scope of a linear analysis. For further discussions on non-viscously damped system see Adhikari (2002, 2000). Damping model of the form (1.13) is also used in the context of viscoelastic structures. The damping kernel functions are commonly defined in the frequency/Laplace domain. Conditions which $\mathbf{G}(s)$, the Laplace transform of $\boldsymbol{\mathcal{G}}(t)$, must satisfy in order to produce dissipative motion were given by Golla and Hughes (1985) and Adhikari (2000). Several authors have proposed different damping models and some of them are summarized in Table 1.1.

1.5 The Assumption of Proportional Damping: Classical Normal Modes and Complex Modes

Equations of motion of a viscously damped system can be obtained from the Lagrange's equation and using the Rayleigh's dissipation function given by (1.10). The equation of motion can expressed as

$$\mathbf{M}\ddot{\mathbf{q}}(t) + \mathbf{C}\dot{\mathbf{q}}(t) + \mathbf{K}\mathbf{q}(t) = \mathbf{f}(t). \tag{1.14}$$

Table 1.1. Summary of damping functions in the Laplace domain

Damping functions	Author, Year
$G(s) = \sum_{k=1}^{n} \frac{a_k s}{s + b_k}$	Biot (1955, 1958)
$G(s) = \frac{E_1 s^\alpha - E_0 b s^\beta}{1 + b s^\beta}$ $0 < \alpha < 1, \quad 0 < \beta < 1$	Bagley and Torvik (1983)
$sG(s) = G^\infty \left[1 + \sum_k \alpha_k \frac{s^2 + 2\xi_k \omega_k s}{s^2 + 2\xi_k \omega_k s + \omega_k^2} \right]$	Golla and Hughes (1985) and McTavish and Hughes (1993)
$G(s) = 1 + \sum_{k=1}^{n} \frac{\Delta_k s}{s + \beta_k}$	Lesieutre and Mingori (1990)
$G(s) = c \frac{1 - e^{-st_0}}{st_0}$	Adhikari (2000)
$G(s) = c \frac{1 + 2(st_0/\pi)^2 - e^{-st_0}}{1 + 2(st_0/\pi)^2}$	Adhikari (2000)

The aim is to solve this equation (together with the initial conditions) using the modal analysis as described in section 1.3. Utilizing the modal transformation in (1.8), premultiplying equation (1.14) by $\mathbf{X}^T$ and using the orthogonality relationships (1.6) and (1.7), equation of motion of a damped system in the modal coordinates may be obtained as

$$\ddot{\mathbf{y}}(t) + \mathbf{X}^T \mathbf{C} \mathbf{X} \dot{\mathbf{y}}(t) + \mathbf{\Omega}^2 \mathbf{y}(t) = \tilde{\mathbf{f}}(t). \tag{1.15}$$

Clearly, unless $\mathbf{X}^T \mathbf{C} \mathbf{X}$ is a diagonal matrix, no advantage can be gained by the employing modal analysis because the equation of motion will still be coupled. To solve this problem, it is common to assume *proportional damping* which we will discuss in some detail.

With the proportional damping assumption, the damping matrix $\mathbf{C}$ is simultaneously diagonalizable with $\mathbf{M}$ and $\mathbf{K}$. This implies that the damping matrix in the modal coordinates

$$\mathbf{C}' = \mathbf{X}^T \mathbf{C} \mathbf{X} \tag{1.16}$$

will be a diagonal matrix. The damping ratios ζ_j are defined from the diagonal elements of the modal damping matrix as

$$C'_{jj} = 2\zeta_j \omega_j \quad \forall j = 1, \cdots, N. \tag{1.17}$$

Such a damping model, introduced by Rayleigh (1877), allows us to analyze damped systems in very much the same manner as undamped systems since the equation of motion in the modal coordinate can be decoupled as

$$\ddot{y}_j(t) + 2\zeta_j \omega_j \dot{y}_j(t) + \omega_j^2 y_j(t) = \tilde{f}_j(t) \quad \forall j = 1, \cdots, N. \tag{1.18}$$

The proportional damping model expresses the damping matrix as a linear combination of the mass and stiffness matrices, that is

$$\mathbf{C} = \alpha_1 \mathbf{M} + \alpha_2 \mathbf{K} \tag{1.19}$$

where α_1, α_2 are real scalars. This damping model is also known as 'Rayleigh damping' or 'classical damping'. Modes of classically damped systems preserve the simplicity of the real normal modes as in the undamped case. Caughey (1960) proposed that a *sufficient* condition for the existence of classical normal modes is: if $\mathbf{M}^{-1}\mathbf{C}$ can be expressed in a series involving powers of $\mathbf{M}^{-1}\mathbf{K}$. Later, Caughey and O'Kelly (1965) proved that the series representation of damping

$$\mathbf{C} = \mathbf{M} \sum_{j=0}^{N-1} \alpha_j \left(\mathbf{M}^{-1}\mathbf{K}\right)^j \tag{1.20}$$

is the *necessary and sufficient* condition for existence of classical normal modes for systems without any repeated roots. This result generalized Rayleigh's result, which turns out to be the first two terms of the series. Assuming that the system is positive definite, a further generalized and useful form of proportional damping has been derived by Adhikari (2006). It was proved that viscously damped positive definite linear systems will have classical normal modes if and only if the damping matrix can be represented as

$$\mathbf{C} = \mathbf{M}\,\beta_1\left(\mathbf{M}^{-1}\mathbf{K}\right) + \mathbf{K}\,\beta_2\left(\mathbf{K}^{-1}\mathbf{M}\right) \tag{1.21}$$

or

$$\mathbf{C} = \beta_3\left(\mathbf{K}\mathbf{M}^{-1}\right)\mathbf{M} + \beta_4\left(\mathbf{M}\mathbf{K}^{-1}\right)\mathbf{K} \tag{1.22}$$

where $\beta_i(\bullet)$ are smooth analytic functions in the neighborhood of all the eigenvalues of their argument matrices. Rayleigh's result (1.19) can be obtained directly from equation (1.21) or (1.22) as a special case by choosing each matrix function $\beta_i(\bullet)$ as a real scalar times an identity matrix, that is

$$\beta_i(\bullet) = \alpha_i \mathbf{I}. \tag{1.23}$$

The damping matrix expressed in equation (1.21) or (1.22) provides a new way of interpreting the 'Rayleigh damping' or 'proportional damping' where the scalar constants α_i associated with $\mathbf{M}$ and $\mathbf{K}$ are replaced by arbitrary matrix functions $\beta_i(\bullet)$ with proper arguments. This kind of damping model will be called *generalized proportional damping.* We call the representation in equation (1.21) *right-functional form* and that in equation (1.22) *left-functional form.* The functions $\beta_i(\bullet)$ will be called as *proportional damping functions* which are consistent with the definition of proportional damping constants (α_i) in Rayleigh's model.

1.6 Forced Dynamic Response of Generally Damped Linear Systems

In the previous section, the concept of proportional damping was critically examined. It is clear that the conditions for existence of proportional damping are purely mathematical in nature and there is no reason why a general system should obey such conditions. Thus, in general a linear system will have non-proportional as well as non-viscous damping. The central theme of this section is to analyze non-viscously damped multiple-degrees-of-freedom linear systems with non-proportional damping. The equation of motion of forced vibration of an N-degrees-of-freedom linear system with non-viscous damping of the form (1.13) can be given by

$$\mathbf{M}\ddot{\mathbf{q}}(t) + \int_0^t \boldsymbol{\mathcal{G}}(t-\tau)\dot{\mathbf{q}}(\tau)\mathrm{d}\tau + \mathbf{K}\mathbf{q}(t) = \mathbf{f}(t). \tag{1.24}$$

The *initial conditions* associated with the above equation are

$$\mathbf{q}(0) = \mathbf{q}_0 \in \mathbb{R}^N \text{and} \quad \dot{\mathbf{q}}(0) = \dot{\mathbf{q}}_0 \in \mathbb{R}^N. \tag{1.25}$$

The nature of eigenvalues and eigenvectors is discussed under certain simplified but physically realistic assumptions on the system matrices and kernel functions. A first-order perturbation based method for the determination of complex eigensolutions is given. The transfer function matrix of the system is derived in terms of these eigenvectors. Exact closed-form expressions are derived for the transient response and the response due to non-zero initial conditions. Applications of the proposed method are discussed using a non-viscously damped three-degrees-of-freedom system.

Eigenvalues and Eigenvectors. Considering the free vibration and taking the Laplace transform of the equation of motion (1.24) one has

$$s^2\mathbf{M}\bar{\mathbf{q}} + s\mathbf{G}(s)\bar{\mathbf{q}} + \mathbf{K}\bar{\mathbf{q}} = \mathbf{0}. \tag{1.26}$$

Here $\bar{\mathbf{q}}(s) = \mathcal{L}[\mathbf{q}(t)] \in \mathbb{C}^N$, $\mathbf{G}(s) = \mathcal{L}[\boldsymbol{\mathcal{G}}(t)] \in \mathbb{C}^{N\times N}$ and $\mathcal{L}[\bullet]$ denotes the Laplace transform. In the context of structural dynamics, $s = \mathrm{i}\omega$, where $\mathrm{i} = \sqrt{-1}$ and $\omega \in \mathbb{R}^+$ denotes the frequency. It is assumed that: (a) $\mathbf{M}^{-1}$ exists, and (b) *all* the eigenvalues of $\mathbf{M}^{-1}\mathbf{K}$ are distinct and positive. Because $\boldsymbol{\mathcal{G}}(t)$ is a real function, $\mathbf{G}(s)$ is also a real function of the parameter s. We assume that $\mathbf{G}(s)$ is such that the motion is dissipative. Several physically realistic mathematical forms of the elements of $\mathbf{G}(s)$ were given in Table 1.1. For the linear viscoelastic case it can be shown that (see Bland, 1960, Muravyov, 1997), in general, the elements of $\mathbf{G}(s)$ can be represented as

$$G_{jk}(s) = \frac{p_{jk}(s)}{q_{jk}(s)} \tag{1.27}$$

where $p_{jk}(s)$ and $q_{jk}(s)$ are finite-order polynomials in s. Here, we do not assume any specific functional form of $G_{jk}(s)$ but assume that $|G_{jk}(s)| < \infty$ when $s \to \infty$. This in turn implies that the elements of $\mathbf{G}(s)$ are at the most of order $1/s$ in s or constant, as in the case of viscous damping. The eigenvalues, s_j, associated with equation (1.26) are roots of the characteristic equation

$$\det\left[s^2\mathbf{M} + s\mathbf{G}(s) + \mathbf{K}\right] = 0. \tag{1.28}$$

If the elements of $\mathbf{G}(s)$ have simple forms as in equation (1.27), then the characteristic equation becomes a polynomial equation of finite order. In other cases the characteristic equation can be expressed as a polynomial equation by expanding $\mathbf{G}(s)$ in a Taylor series. However, the order of the equation will be infinite in those cases. For practical purposes the Taylor expansion of $\mathbf{G}(s)$ can be truncated to a finite series to make the characteristic equation a polynomial equation of finite order. Such equations can be solved using standard numerical methods (see Press et al., 1992, Chapter 9). Suppose the order of the characteristic polynomial is m. In general m is more than $2N$, that is $m = 2N + p$; $p \geq 0$. Thus, although the system has N degrees of freedom, the number of eigenvalues is more than $2N$. This is a major difference between the non-viscously damped systems and the viscously damped systems where the number of eigenvalues is exactly $2N$, including any multiplicities.

A general analysis on the nature of the eigenvalues of non-viscously damped systems has not been reported in literature. Discussion on the nature of the eigenvalues of a single-degree-of-freedom system has been given recently by Adhikari (2005). It is assumed that *all* m eigenvalues are distinct. Because $\boldsymbol{\mathcal{G}}(t)$ is real, $\mathbf{G}^*(s) = \mathbf{G}(s^*)$, where $(\bullet)^*$ denotes complex conjugation. Using this and taking complex conjugation of (1.26) it is clear that ${s^*}^2\mathbf{M}\bar{\mathbf{q}}+s^*\mathbf{G}(s^*)\bar{\mathbf{q}}+\mathbf{K}\bar{\mathbf{q}}=\mathbf{0}$, *i.e.*, if s satisfies equation (1.26) then so does s^*. Therefore the roots of the characteristic equation (1.28) are either real, or if complex, then must appear in conjugate pairs. We restrict our attention to a special case when, among the m eigenvalues, $2N$ appear in complex conjugate pairs. Under such assumptions the remaining p eigenvalues become purely real. The mathematical conditions which $\mathbf{M}$, $\mathbf{K}$ and $\mathbf{G}(s)$ must satisfy in order to produce such eigenvalues will not be obtained but a physical justification will follow shortly. For convenience the eigenvalues are arranged as

$$s_1, s_2, \cdots, s_N, s_1^*, s_2^*, \cdots, s_N^*, s_{2N+1}, \cdots, s_m \tag{1.29}$$

The eigenvalue problem associated with equation (1.24) can be defined from (1.26) as

$$\mathbf{D}(s_j)\mathbf{z}_j=\mathbf{0}, \quad \text{for}\, j=1,\cdots,m \tag{1.30}$$

$$\text{where} \quad \mathbf{D}(s_j)=s_j^2\mathbf{M}+s_j\,\mathbf{G}(s_j)+\mathbf{K}\in\mathbb{C}^{N\times N} \tag{1.31}$$

is the *dynamic stiffness matrix* corresponding to the j-th eigenvalue and $\mathbf{z}_j$ is the j-th eigenvector. From equation (1.30) it is clear that, when s_j appear in complex conjugate pairs, $\mathbf{z}_j$ also appear in complex conjugate pairs, and when s_j is real $\mathbf{z}_j$ is also real. Corresponding to the $2N$ complex conjugate pairs of eigenvalues, the N eigenvectors together with their complex conjugates will be called *elastic modes.* These modes are related to the N modes of vibration of the structural system. Physically, the assumption of '$2N$ complex conjugate pairs of eigenvalues' implies that all the elastic modes are oscillatory in nature, that is, they are sub-critically damped. The modes corresponding to the 'additional' p eigenvalues will be called *non-viscous modes.* These modes are induced by the non-viscous effect of the damping mechanism. For stable passive systems the non-viscous modes are over-critically damped (i.e., negative real eigenvalues) and non-oscillatory in nature. Non-viscous modes, or similar to these, are known by different names in the literature of different subjects, for example, 'wet modes' in the context of ship dynamics (Bishop and Price, 1979) and 'damping modes' in the context of viscoelastic structures (McTavish and Hughes, 1993). Determination of the eigenvectors is considered next.

Elastic Modes. Once the eigenvalues are known, $\mathbf{z}_j, \forall j=1,\cdots,2N$ can be obtained from equation (1.30) by fixing any one element and inverting the matrix $\mathbf{D}(s_j)$. Clearly, the inversion of an $(N-1)\times(N-1)$ complex matrix is required to calculate each $\mathbf{z}_j$. Alternatively the eigenvalue problem can be formed in terms of of an augmented matrix eigenvalue problem (Wagner and Adhikari, 2003, see). Here we outline an approximate method based on the perturbation analysis.

If the damping is small, then the eigenvalues and the eigenvectors can be obtained using perturbation analysis as

$$\begin{aligned} &s_j \approx \pm \mathrm{i}\omega_j - G'_{jj}(\pm\mathrm{i}\omega_j)/2 \\ &\text{that is, } s_j \approx \mathrm{i}\omega_j - G'_{jj}(\mathrm{i}\omega_j)/2 \quad \text{or} \quad s_j \approx -\mathrm{i}\omega_j - G'_{jj}(-\mathrm{i}\omega_j)/2. \end{aligned} \tag{1.32}$$

and

$$\mathbf{z}_j \approx \mathbf{x}_j - \sum_{\substack{k=1\\k\neq j}}^{N} \frac{s_j G'_{kj}(s_j)\mathbf{x}_k}{\omega_k^2+s_j^2+s_j G'_{kk}(s_j)}. \tag{1.33}$$

This is the approximate first-order expression of the complex modes obtained by Woodhouse (1998). If a system has light non-proportional damping then equations (1.32) and (1.33) may not introduce significant error. A more accurate expression of the complex modes can be obtained using the approach proposed by Adhikari (2002).

Non-viscous Modes. When $2N < j \leq m$, the eigenvalues become real and consequently from equation (1.31) we observe that $\mathbf{D}(s_j) \in \mathbb{R}^{N\times N}$ and $\mathbf{z}_j \in \mathbb{R}^N$. The non-viscous modes can be obtained from equation (1.30) by fixing any one element of the eigenvectors. Partition $\mathbf{z}_j$ as

$$\mathbf{z}_j = \begin{Bmatrix} \mathbf{z}_{1j} \\ \mathbf{z}_{2j} \end{Bmatrix}. \tag{1.34}$$

We select $\mathbf{z}_{1j} = 1$ so that $\mathbf{z}_{2j} \in \mathbb{R}^{(N-1)}$ has to be determined from equations (1.30). Further, partition $\mathbf{D}(s_j)$ as

$$\mathbf{D}(s_j) = \begin{bmatrix} \mathbf{D}_{11}(s_j) & \mathbf{D}_{12}(s_j) \\ \mathbf{D}_{21}(s_j) & \mathbf{D}_{22}(s_j) \end{bmatrix} \tag{1.35}$$

where $\mathbf{D}_{11}(s_j) \in \mathbb{R}$, $\mathbf{D}_{12}(s_j) \in \mathbb{R}^{1\times(N-1)}$, $\mathbf{D}_{21}(s_j) \in \mathbb{R}^{(N-1)\times 1}$ and $\mathbf{D}_{22}{}^{(j)} \in \mathbb{R}^{(N-1)\times(N-1)}$. In view of (1.35) and recalling that $\mathbf{z}_{1j} = 1$, from equation (1.30) we can have

$$\mathbf{D}_{22}(s_j)\mathbf{z}_{2j} = -\mathbf{D}_{21}(s_j) \text{ or } \mathbf{z}_{2j} = -\left[\mathbf{D}_{22}(s_j)\right]^{-1}\mathbf{D}_{21}(s_j). \tag{1.36}$$

The determination of non-viscous modes is computationally more demanding than the elastic modes because inversion of an $(N-1)\times(N-1)$ real matrix is associated with each eigenvector. However, for most physically realistic non-viscous damping models it appears that the number of non-viscous modes is not very high and also their contribution to the global dynamic response is not very significant (see Adhikari, 2002). For this reason, the calculation of first few non-viscous modes is sufficient from a practical point view.

Transfer Function. The transfer function (matrix) of a system completely defines its input-output relationship in *steady-state*. It is well known that for any linear system, if the forcing function is harmonic, that is $\mathbf{f}(t) = \bar{\mathbf{f}}\exp[st]$ with $s = \mathrm{i}\omega$ and amplitude vector $\bar{\mathbf{f}} \in \mathbb{R}^N$, the steady-state response will also be harmonic at frequency ω. So we seek a solution of the form $\mathbf{q}(t) = \bar{\mathbf{q}}\exp[st]$, where $\bar{\mathbf{q}} \in \mathbb{C}^N$ is the response vector in the frequency domain. Substitution of $\mathbf{q}(t)$ and $\mathbf{f}(t)$ in equation (1.24) gives

$$s^2\mathbf{M}\bar{\mathbf{q}} + s\,\mathbf{G}(s)\bar{\mathbf{q}} + \mathbf{K}\bar{\mathbf{q}} = \bar{\mathbf{f}} \quad \text{or} \quad \mathbf{D}(s)\bar{\mathbf{q}} = \bar{\mathbf{f}} \tag{1.37}$$

where

$$\mathbf{D}(s) = s^2\mathbf{M} + s\,\mathbf{G}(s) + \mathbf{K} \in \mathbb{C}^{N\times N}. \tag{1.38}$$

is known as the *dynamic stiffness matrix*. From equation (1.37) the response vector $\bar{\mathbf{q}}$ can be obtained as

$$\bar{\mathbf{q}} = \mathbf{D}^{-1}(s)\bar{\mathbf{f}} = \mathbf{H}(s)\bar{\mathbf{f}} \tag{1.39}$$

where

$$\mathbf{H}(s) = \mathbf{D}^{-1}(s) \in \mathbb{C}^{N\times N} \tag{1.40}$$

is the transfer function matrix. From this equation one further has

$$\mathbf{H}(s) = \frac{\operatorname{adj}[\mathbf{D}(s)]}{\det[\mathbf{D}(s)]}. \tag{1.41}$$

The *poles* of $\mathbf{H}(s)$, denoted by s_j, are the eigenvalues of the system. Because it is assumed that *all* the m eigenvalues are distinct, each pole is a *simple pole*. The matrix inversion in (1.39) is difficult to carry out in practice because of the singularities associated with the poles. Moreover, such an approach would be an expensive numerical exercise and may not offer much physical insight. For these reasons, we seek a solution analogous to the classical modal series solution of the undamped or proportionally damped systems.

From the residue theorem it is known that any analytic complex function can be expressed in terms of the poles and *residues*, that is, the transfer function has the form

$$\mathbf{H}(s) = \sum_{j=1}^{m} \frac{\mathbf{R}_j}{s - s_j}. \tag{1.42}$$

Here

$$\mathbf{R}_j = \operatorname*{res}_{s=s_j} [\mathbf{H}(s)] \stackrel{\text{def}}{=} \lim_{s \to s_j} (s - s_j) [\mathbf{H}(s)] \tag{1.43}$$

is the residue of the transfer function matrix at the pole s_j. Equation (1.42) is equivalent to expressing the right hand side of equation (1.41) in the partial-fraction form. The residues, that is the coefficients in the partial-fraction form, can be obtained exactly (see Adhikari, 2002, for details) in terms of the system eigenvectors as

$$\mathbf{R}_j = \frac{\mathbf{z}_j \mathbf{z}_j^T}{\mathbf{z}_j^T \frac{\partial \mathbf{D}(s_j)}{\partial s_j} \mathbf{z}_j}. \tag{1.44}$$

Recalling that, among the m eigenvalues $2N$ appear in complex conjugate pairs, from equation (1.42) the transfer function matrix is obtained as

$$\mathbf{H}(\mathrm{i}\omega) = \sum_{j=1}^{N} \left[\frac{\gamma_j \mathbf{z}_j \mathbf{z}_j^T}{\mathrm{i}\omega - s_j} + \frac{\gamma_j^* \mathbf{z}_j^* \mathbf{z}_j^{*T}}{\mathrm{i}\omega - s_j^*} \right] + \sum_{j=2N+1}^{m} \frac{\gamma_j \mathbf{z}_j \mathbf{z}_j^T}{\mathrm{i}\omega - s_j}, \tag{1.45}$$

where

$$\gamma_j = \frac{1}{\mathbf{z}_j^T \frac{\partial \mathbf{D}(s_j)}{\partial s_j} \mathbf{z}_j}. \tag{1.46}$$

The transfer function matrix has two parts, the first part is due to the elastic modes, and the second part is due to the non-viscous modes. This expression is a natural generalization of the familiar transfer function matrices of undamped or viscously damped systems and they can be obtained as special cases of (1.45) as follows:

1. *Undamped systems*: In this case $\mathbf{G}(s) = 0$ results the order of the characteristic polynomial $m = 2N$; s_j is purely imaginary so that $s_j = \mathrm{i}\omega_j$ where $\omega_j \in \mathbb{R}$ are the undamped natural frequencies and $\mathbf{z}_j = \mathbf{x}_j \in \mathbb{R}^N$. In view of the mass normalization relationship in (1.4), $\gamma_j = \frac{1}{2\mathrm{i}\omega_j}$ and equation (1.45) leads to

$$\mathbf{H}(\mathrm{i}\omega) = \sum_{j=1}^{N} \frac{1}{2\mathrm{i}\omega_j} \left[\frac{1}{\mathrm{i}\omega - \mathrm{i}\omega_j} - \frac{1}{\mathrm{i}\omega + \mathrm{i}\omega_j} \right] \mathbf{x}_j \mathbf{x}_j^T = \sum_{j=1}^{N} \frac{\mathbf{x}_j \mathbf{x}_j^T}{\omega_j^2 - \omega^2}. \tag{1.47}$$

2. *Viscously-damped systems with non-proportional damping* (see for example, Géradin and Rixen, 1997): In this case $m = 2N$ and $\gamma_j = \frac{1}{\mathbf{z}_j^T[2s_j\mathbf{M}+\mathbf{C}]\mathbf{z}_j}$. These reduce expression (1.45) to

$$\mathbf{H}(\mathrm{i}\omega) = \sum_{j=1}^{N}\left[\frac{\gamma_j\mathbf{z}_j\mathbf{z}_j^T}{\mathrm{i}\omega - s_j} + \frac{\gamma_j^*\mathbf{z}_j^*\mathbf{z}_j^{*^T}}{\mathrm{i}\omega - s_j^*}\right]. \tag{1.48}$$

Dynamic Response. The steady-state response due to harmonic loads or the response due to broad-band random excitation can be obtained directly from the expression of the transfer function matrix in equation (1.45). In this section we consider the system response due to transient loads and initial conditions in the time and frequency domains. Taking the Laplace transform of equation (1.24) and considering the initial conditions in (1.25) we have

$$\begin{aligned} s^2\mathbf{M}\bar{\mathbf{q}} - s\mathbf{M}\mathbf{q}_0 - \mathbf{M}\dot{\mathbf{q}}_0 + s\,\mathbf{G}(s)\bar{\mathbf{q}} - \mathbf{G}(s)\mathbf{q}_0 + \mathbf{K}\bar{\mathbf{q}} = \bar{\mathbf{f}}(s) \\ \text{or}\quad \left[s^2\mathbf{M} + s\,\mathbf{G}(s) + \mathbf{K}\right]\bar{\mathbf{q}} = \bar{\mathbf{f}}(s) + \mathbf{M}\dot{\mathbf{q}}_0 + \left[s\mathbf{M} + \mathbf{G}(s)\right]\mathbf{q}_0. \end{aligned} \tag{1.49}$$

Using the expression for the transfer function derived before, the response vector $\bar{\mathbf{q}}$ may be obtained as

$$\bar{\mathbf{q}} = \sum_{j=1}^{m}\frac{\gamma_j\mathbf{z}_j\mathbf{z}_j^T}{s - s_j}\left\{\bar{\mathbf{f}}(s) + \mathbf{M}\dot{\mathbf{q}}_0 + \left[s\mathbf{M} + \mathbf{G}(s)\right]\mathbf{q}_0\right\}. \tag{1.50}$$

This can be simplified further to

$$\bar{\mathbf{q}}(\mathrm{i}\omega) = \sum_{j=1}^{m}\frac{\gamma_j A_j(\mathrm{i}\omega)}{\mathrm{i}\omega - s_j}\mathbf{z}_j \tag{1.51}$$

where the frequency-dependent complex scalar $A_j(\mathrm{i}\omega)$ is given by

$$A_j(\mathrm{i}\omega) = \mathbf{z}_j^T\bar{\mathbf{f}}(\mathrm{i}\omega) + \mathbf{z}_j^T\mathbf{M}\dot{\mathbf{q}}_0 + \mathrm{i}\omega\mathbf{z}_j^T\mathbf{M}\mathbf{q}_0 + \mathbf{z}_j^T\mathbf{G}(\mathrm{i}\omega)\mathbf{q}_0. \tag{1.52}$$

The summation in equation (1.51) may be split into two different parts – the first part would correspond to the $2N$ complex conjugate pairs of elastic modes and the second part would be the contribution of the non-viscous modes.

The response in the time domain due to any forcing function can be obtained using a convolution integral over the *impulse response function*. From the expression of the transfer function in equation (1.45), the impulse response function matrix $\mathbf{h}(t) \in \mathbb{R}^{N\times N}$ may be obtained as

$$\mathbf{h}(t) = \sum_{j=1}^{N}\left[\gamma_j\mathbf{z}_j\mathbf{z}_j^T e^{s_j t} + \gamma_j^*\mathbf{z}_j^*\mathbf{z}_j^{*^T} e^{s_j^* t}\right] + \sum_{j=2N+1}^{m}\gamma_j\mathbf{z}_j\mathbf{z}_j^T e^{s_j t}. \tag{1.53}$$

The response due to the initial conditions may also be obtained by taking the inverse transform of equation (1.50). First, simplify equation (1.50) to obtain

$$\bar{\mathbf{q}}(s) = \sum_{j=1}^{m}\gamma_j\left[\frac{\mathbf{z}_j^T\bar{\mathbf{f}}(s) + \mathbf{z}_j^T\mathbf{G}(s)\mathbf{q}_0}{s - s_j} + \frac{\mathbf{z}_j^T\mathbf{M}\dot{\mathbf{q}}_0}{s - s_j} + \left(1 + \frac{s_j}{s - s_j}\right)\mathbf{z}_j^T\mathbf{M}\mathbf{q}_0\right]\mathbf{z}_j. \tag{1.54}$$

From the above, one has

$$\mathbf{q}(t) = \mathcal{L}^{-1}[\bar{\mathbf{q}}(s)] = \sum_{j=1}^{N} \left[\gamma_j a_j(t)\mathbf{z}_j + \gamma_j^* a_j^*(t)\mathbf{z}_j^*\right] + \sum_{j=2N+1}^{m} \gamma_j a_j(t)\mathbf{z}_j \tag{1.55}$$

where the time-dependent scalar coefficients

$$a_j(t) = \int_0^t e^{s_j(t-\tau)} \left\{\mathbf{z}_j^T \mathbf{f}(\tau) + \mathbf{z}_j^T \boldsymbol{\mathcal{G}}(\tau)\mathbf{q}_0\right\} \mathrm{d}\tau + e^{s_j t} \left\{\mathbf{z}_j^T \mathbf{M}\dot{\mathbf{q}}_0 + s_j \mathbf{z}_j^T \mathbf{M}\mathbf{q}_0\right\}; \quad \forall t > 0. \tag{1.56}$$

The expression of the system response, either the frequency-domain description in equation (1.51) or the time-domain description in equation (1.55), is similar to the classical modal superposition result for undamped or proportionally damped systems usually obtained using the mode-orthogonality relationships. The formulation presented here is a generalization of the classical result where the real normal modes are appropriately replaced by the elastic modes and the non-viscous modes. Also note that we have not used any orthogonality relationship – the expression of the transfer function residue in equation (1.44) allows us to express the response in terms of superposition of individual modes even when the equation of motion cannot be decoupled.

Numerical Example. We consider a three degree-of-freedom system to illustrate the proposed method. Figure 1.2 shows the example taken together with the numerical values considered for mass and stiffness properties. Damping is associated only with

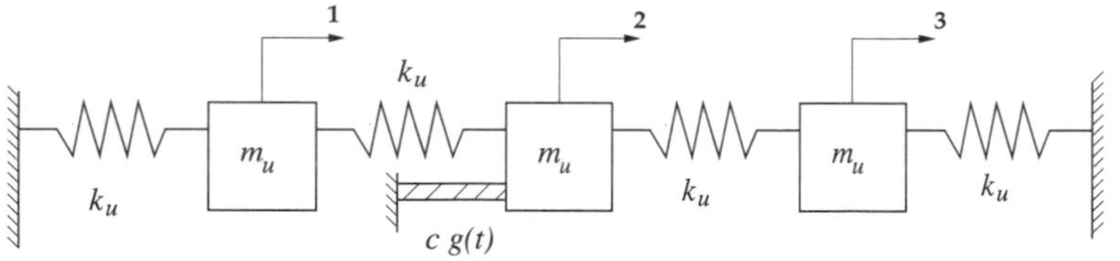

Figure 1.2. Three degree-of-freedom non-viscously damped system, $m_u = 1$ kg, $k_u = 1$ N/m, $c = 0.3$ Ns/m

the middle mass, and the kernel function corresponding to this damper has the form

$$\mathcal{G}_{22}(t) = c\,g(t), \text{ where } g(t) = \mu e^{-\mu t}; \quad \mu, t \geq 0. \tag{1.57}$$

The exponential damping function is possibly the simplest physically realistic non-viscous damping model. This function, often known as a 'relaxation function', was introduced by Biot (1955). It has been used extensively in the context of viscoelastic systems. The damping function has been scaled so as to have unit area when integrated to infinity. This makes it directly comparable with the viscous model in which the corresponding damping function would be a unit delta function, $g(t) = \delta(t)$, and the coefficient c would be the usual viscous damping coefficient. The difference between a delta function and $g(t)$ given by equation (1.57) is that at $t = 0$ it starts with a finite value of μ. Thus, the value of μ give a notion of *non-viscousness* – if it is large the damping behavior will be near-viscous, and vice versa. The mass and stiffness matrices and the damping matrix in the Laplace domain for the problem can be obtained as:

$$\mathbf{M} = \begin{bmatrix} m_u & 0 & 0 \\ 0 & m_u & 0 \\ 0 & 0 & m_u \end{bmatrix}, \quad \mathbf{K} = \begin{bmatrix} 2k_u & -k_u & 0 \\ -k_u & 2k_u & -k_u \\ 0 & -k_u & 2k_u \end{bmatrix}, \quad \mathbf{G}(s) = \begin{bmatrix} 0 & 0 & 0 \\ 0 & c\,G(s) & 0 \\ 0 & 0 & 0 \end{bmatrix}. \tag{1.58}$$

where $G(s) = \frac{\mu}{s+\mu}$. Using these expressions, the characteristic equation can be simplified as

$$m_u^3 s^7 + m_u^3 \mu\, s^6 + \left(2\, m_u^2 k_u + m_u \left(\mu\, c m_u + 4\, m_u k_u\right)\right) s^5 + 6\, k_u m_u^2 \mu\, s^4 \\ + \left(2\, k_u \left(\mu\, c m_u + 4\, m_u k_u\right) + m_u \left(2\, \mu\, c k_u + 2\, {k_u}^2\right)\right) s^3 + 10\, {k_u}^2 m_u \mu\, s^2 \\ + 2\, k_u \left(2\, \mu\, c k_u + 2\, {k_u}^2\right) s + 4\, {k_u}^3 \mu = 0. \tag{1.59}$$

The order of the above polynomial, $m = 7$. Since the system has three degrees of freedom there are three elastic modes corresponding to the three modes of vibration. The number of the non-viscous modes, $p = m - 2N = 1$. It is of interest to us to understand the

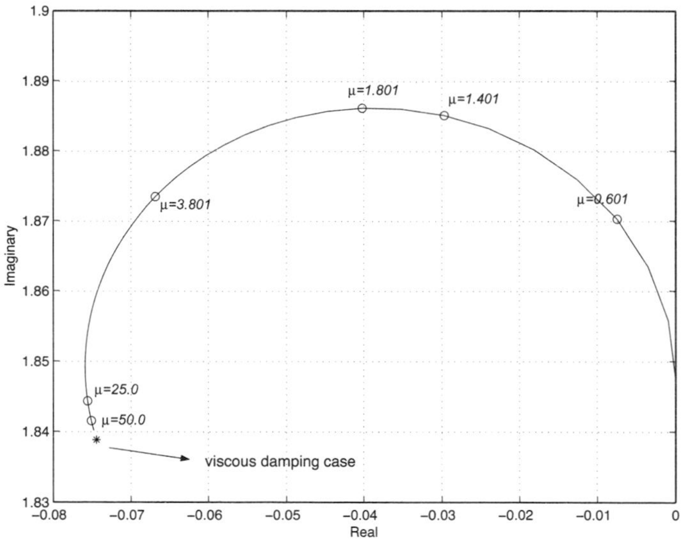

Figure 1.3. Root-locus plot of the third eigenvalue (s_3) as a function of μ.

effect of 'non-viscousness' on the eigensolutions. Figure 1.3 shows the locus of the third eigenvalue, that is s_3, plotted as a function of μ. It is interesting to observe that the locus is much more sensitive in the region of lower values of μ (*i.e.*, when damping is significantly non-viscous) compared to that in the region of higher values. The eigenvalue of the corresponding viscously damped system is also plotted (marked by *) in the same diagram. The non-viscous damping model approaches the viscous damping model when $\mu > 50.0$. Similar behaviour has been observed (results not shown here) for the locus of s_1 also. The second mode, in which the middle mass remains stationary, is not affected by damping.

To illustrate the dynamic response analysis, the problem of stationary random vibration analysis of the system is considered here. Suppose the system is subjected to a band-limited Gaussian *white noise* at the third DOF. We are interested in the resulting displacement of the system at the third DOF (i.e., $\mathbf{z}_3$). The power spectral density (PSD) of the response (see Nigam, 1983, for details) can be given by

$$S_{uu}(\mathrm{i}\omega) = |H_{33}(\mathrm{i}\omega)|^2 S_{ff}(\mathrm{i}\omega) \tag{1.60}$$

where $S_{ff}(\mathrm{i}\omega) = 1$ if $0 < \omega \leq 2.5$ and $S_{ff}(\mathrm{i}\omega) = 0$ elsewhere. In figure 1.4 the PSD

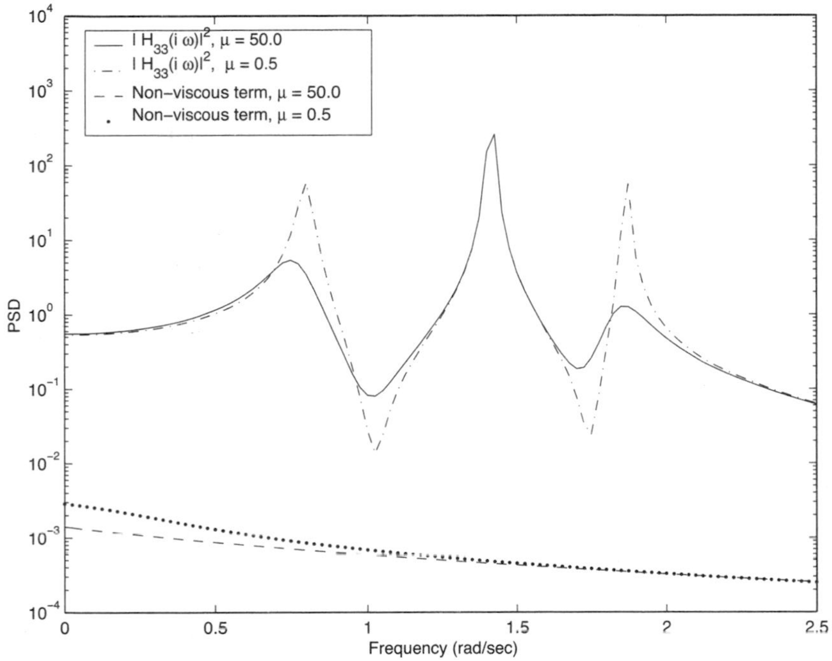

Figure 1.4. Power spectral density function of the displacement at the third DOF ($\mathbf{z}_3$)

of $\mathbf{z}_3$ that is $|H_{33}(\mathrm{i}\omega)|^2$ is plotted for the cases when $\mu = 50.0$ and $\mu = 0.5$. These results are obtained by direct application of equation (1.45). From the diagram observe that the damping is less for the case when $\mu = 0.5$ compared to when $\mu = 50.0$. Also note the (horizontal) shift in the position of the natural frequencies. These features may also be observed in the loot locus diagram as shown in figure 1.3. To understand the effect of 'non-viscosity', in the same diagram we have plotted the non-viscous term (the second term) appearing in equation (1.45) for both values of μ. For this problem the non-viscous part is quite small and becomes smaller at higher frequencies. When $\mu = 0.5$, that is when damping is significantly non-viscous, the value of the non-viscous part of the response is more than that when $\mu = 50.0$. This plot also clearly demonstrates that the non-viscous part of the response is *not* oscillatory in nature.

1.7 Conclusions

In this chapter different mathematical models used for dynamic analysis of engineering structures were introduced. Basics of the dynamics of undamped linear systems were discussed. It was mentioned that the undamped dynamics is characterized by the natural frequencies and the mode shapes. Different models of damping used in the literature have been reviewed. The assumption of proportional damping was critically examined and the concept of generalized proportional damping was introduced. The generalized proportional damping expresses the damping matrix in terms of *any* non-linear function involving specially arranged mass and stiffness matrices so that the system still possesses classical normal modes.

The problem of dynamic analysis of general non-viscously damped multiple-degrees-of-freedom linear systems was considered in details. The non-viscous damping model is such that the damping forces depend on the past history of motion via convolution integrals over some kernel functions. The transfer function matrix of the system was derived in terms of complex eigenvalues and eigenvectors of the system. Approximate expressions of the complex eigensolutions were derived. Exact closed-form expressions of the dynamic response due to arbitrary forcing functions and initial conditions were obtained and the results were illustrated using numerical examples.

2 What is Model Validation and Model Verification? How it is Done?

2.1 Introduction

In the previous section some commonly used approaches for the modeling and analysis of linear dynamic systems have been discussed. Once the modeling and the simulation have been performed, it is required to 'judge' the model and simulation against the three criteria mentioned before, namely, fidelity to experimental data, robustness with respect to random errors and predictive capability. This process broadly falls under model validation and verification. More precisely, we have the following definitions (Schlesinger, 1979):

Model Verification: Model verification is defined as ensuring that the computer program of the computerized model and its implementation are correct.

Model Validation: Model validation is defined to mean substantiation that a computerized model within its domain of applicability possesses a satisfactory range of accuracy consistent with the intended application of the model.

Historically the model validation and verification (will be referred as V&V) have been conducted in conjunction with physical experiments and measurements. One of the key reason behind the modeling and simulation in the context of structural dynamics is to perform engineering design. Before 1960 or so, the models were simple - often involved either some closed-form expressions or simple computations. The role of a model within the design activity was more 'passive' than 'active'. They were often used to provide an 'initial guess' or 'guideline'. The final design and certification of an engineering product was usually based on physical experiments and measurements.

With the development of the finite-element method together with easily available computational hardware, within the past three decades the models have become more complex and simulations have become more computationally intensive. These, combined with the development of powerful computing languages, software packages, realistic pre and post processing visualization software and hardware, development of sophisticated information technology (IT) systems and easily available trained software professionals have pushed the engineering community to rely more on their models than ever before. This fact is also fuelled by increasing cost of conducting full-scale experiments compared to simulations due to tough environmental regulations, stringent health and safety conditions and various social factors. As a direct result of all these factors, the developers and users of the models, the decision makers using information derived from the results of the models, and people affected by decisions based on such models are highly concerned with the accuracy, predictive capability and credibility of the modeling and simulation process as a whole. The objective of V&V activities is to provide measures by which these qualities can be judged in a scientific, methodical, rigorous, consistent, generalized and possibly simple manner. An essential step of V&V in structural dynamics is the experimental modal analysis (EMA). An overview of this method is given in the next section.

2.2 Brief Overview of Experimental Modal Analysis

Experimental modal analysis (EMA) is a widely used tool to analyze structures dynamically. It is beyond the scope of this paper to review all the aspects of this vast and mature subject. Readers are referred to the books by Ewins (2000), Maia and Silva (1997) and Silva and Maia (1998) for further details. Here we briefly outline the essential steps of experimental modal analysis.

Step 1: Measurement hardware setup:
The measurement hardware essentially consist of four components: (1) structural mounting system, (2) excitation/actuation system, (3) transducers/sensors system and (4) data collection and processing system.

The mounting system should ideally replicate the conditions under which the structure performs in real life. For example, if a plate is fixed along one edge, the mounting system should represent it as closely as possible. The mounting system should be selected carefully because it can have significant effect of the test results. Excitations are usually given by a shaker or an impulse hammer. Responses due to the excitations are normally recorded by piezoelectric accelerometers or laser vibrometers. These analogue signals are then converted to digital signals using an analogue to digital converter card (ADC) in a computer. Once the data is available in a digital format, several software are available to post-process the data and extract useful information. Figure 2.1 shows schematic representation of a typical modal testing setup. In this case a damped beam is hanged by a cord to simulate a free-free boundary condition.

Step 2: Excitation of the structure:
There are mainly two methods to excite a structure dynamically. The first is to connect a shaker to the structure which produces a force proportional to a specified input voltage. The second is to hit the structure by an instrumented hammer. Shaker and hammer excitations each have their advantages and disadvantages. Which method to

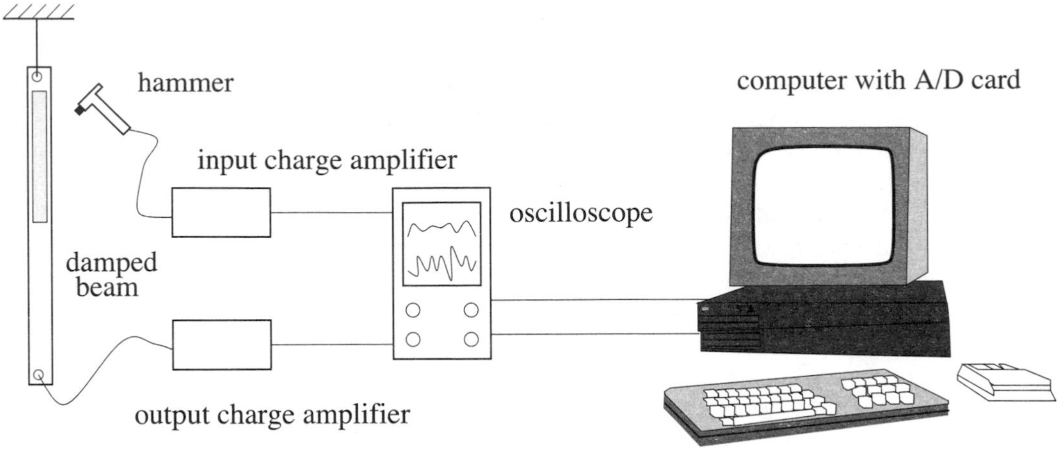

Figure 2.1. Schematic representation of a typical modal testing setup

be used depends on the equipments available, the type of structure to be analyzed and the future use of the measurement signals. Shakers are able to put more energy into a system, and more than one shaker may be used together to produce a more even distribution of energy producing more data which increases the accuracy of the subsequent analysis. Unfortunately, the shakers have to be fixed to the structure by some means. This is difficult to do in practice for all structures and moreover by doing so, the original structure gets modified in the form of local stiffening and mass loading. The extent of such modifications will depend on the details of the connection mechanism. Nevertheless, this will introduce some errors unless they are taken into account while modeling the structure. Impact excitation by a hammer is easy to apply and does not produce mass loading or local stiffening. By changing the tip of the hammer it is also possible to tailor the frequency range of excitation. However, sometimes it is difficult to impart sufficient energy to the structure to produce satisfactory response. A large impact may damage the structure or drive the structure into the non-linear range. Also the direction and the point of impact may be difficult to control or reproduce in practice.

Step 3: Measurement and post-processing of data:

There are mainly two approaches for the dynamic response measurements: piezoelectric materials based transducers and laser vibrometers. Piezoelectric accelerometers are used widely because they are not very expensive and relatively simple to use. They can be mounted onto the structures by a variety of methods, for example, using glue, magnets, bolts and beeswax. Depending on the resources available for the data processing, more than one accelerometers can be used simultaneously. The accelerometers together with the mounting system must be light so that the structure under test does not get modified to a great extent. This problem does not arise with the laser vibrometers. They also prove a clutter free environment for doing experiments since the cables associated with each accelerometer are not present. Moreover the laser beam can be pointed on any point of the structure, which may be physically inaccessible and where mounting a conventional accelerometers may be difficult or impossible. The main difficulties with a laser vibrometer based approach is that they are still quite expensive and one needs some training to use them successfully and efficiently as they are not used widely at present.

The analysis of vibration data is usually done digitally. The analogue signals from the force transducers and accelerometers are converted to digital signals using an analogue to digital card (ADC), usually attached to a personal computer. Time-domain data are often transformed to the frequency-domain using fast fourier transform (FFT). Data flow in a typical experimental modal analysis procedure is shown in figure 2.2. There are

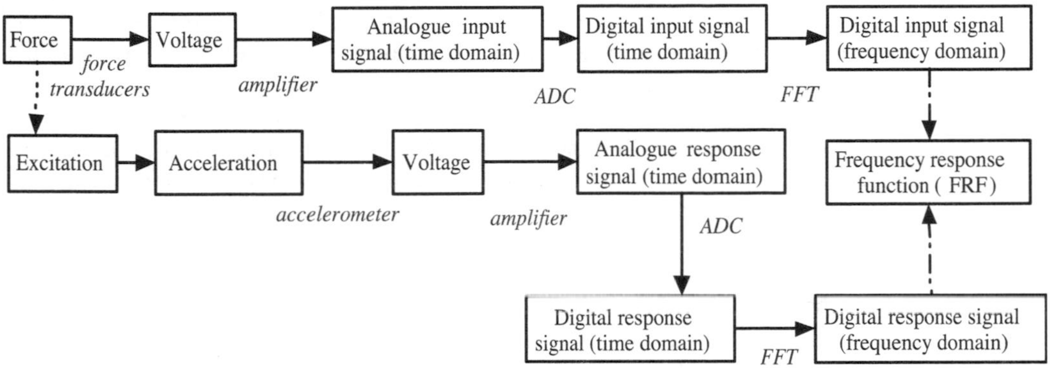

Figure 2.2. Data flow in a typical experimental modal analysis procedure

several important technical issues, for example aliasing, filtering, leakage etc, associated with this process. Another issue with the data is the presence of noise. Over the years several techniques have been developed to address these issues and many software are now available to obtain high quality data from modal testing. Once the dynamic response in the time or frequency domain is available, it can be used to obtain the modal parameters. The modal parameters in turn can be used for model verification and validation.

2.3 Measures of Model Correlations

A crucial step for dynamic model validation is to compare experimental results from the test structure with predicted results from the corresponding finite element model. The basic questions surrounding this process are

1. What quantities shall we compare?
2. How do we actually compare them?
3. How many modes shall we (or can we) compare?
4. The FE model has more degrees-of-freedom compared to the number of transducers in the experimental set up. How can we take account of this fact?
5. The FE model is normally undamped. How to deal with damping?
6. What about the (random) errors introduced by experimental hardware which have not been modeled in FE?

7. Finally, does a good comparison result always means a good analytical model? Various methods are available in literature to address these issues. It is beyond the scope of this paper to cover all of them. The methods addressing the first four points will be

briefly discussed while the methods addressing the last three points will be discussed in some details.

In the context of experimental modal analysis, the 'original' test data usually mean the transfer functions. From figure 2.2 it is however clear that the transfer functions are obtained using a sequence of processes involving several hardware and software. There are uncertainties and errors associated with each step and it is therefore required to perform some basic quality checks of the 'measured' transfer functions. The standard checks are reciprocity check, repeatability check and amplitude dependence (for non-linearity) check. In critical applications it is often useful to perform the repeatability check using different shakers/hammers, accelerometers, time-steps and excitation signals to ensure that the test data is independent of the hardware and excitation methods. Once a satisfactory set of transfer functions are obtained, the next step is to compare them with the results from the FE model. The comparison between the test results and analytical results can be performed in the following three levels:

(a) comparison of the modal properties
(b) comparison of the spatial properties (the mass and stiffness matrices)
(c) comparison of response (transfer functions)

Brief discussion on these topics are given next.

Comparison of the modal properties is the most widely used approach for dynamic model validation. This process consists of comparing the mode shapes and natural frequencies. Suppose $\mathbf{X}_e \in \mathbb{R}^{N_e \times m}$ is the matrix of modes shapes and $\boldsymbol{\Omega}_e \in \mathbb{R}^{m \times m}$ is the diagonal matrix of the natural frequencies obtained from the experiment. Comparing the natural frequencies obtained from the experiment with those from the FE model is the easiest task. Once can simply plot them together in a single graph and can visually feel their differences. From visual inspection one can identify which mode of the experiment correlates to which mode of the FE analysis. For systems with well separated modes it is also possible to compare the mode shapes graphically. A typical case is shown in figure 2.3 where first four experimental and analytical mode shapes of a free-free beam in compared. For structures with closely spaced modes or structures with complex geometry (such as an automobile body) this type of comparison is less simple to do. Here the main challenge is to identify an experimental mode which correlates to an FE mode or vice versa. Several researchers have developed different methods to quantify the correlation between experimental and FE mode shapes. Among all, most popular is the Modal Assurance Criterion (MAC) (Allemang and Brown, 1982). The MAC between an experimentally obtained mode $\mathbf{x}_{e_j}$ and an analytical mode $\mathbf{x}_{a_k}$ is defined as

$$\mathrm{MAC}_{jk} = \frac{\left|\mathbf{x}_{e_j}^T \mathbf{x}_{a_k}\right|^2}{\left|\mathbf{x}_{a_k}^T \mathbf{x}_{a_k}\right| \left|\mathbf{x}_{e_j}^T \mathbf{x}_{e_j}\right|}. \tag{2.1}$$

The value of MAC is between 0 and 1. A value of 1 indicates one mode shape is fully correlated to the other. To use equation (2.1), the experimental and analytical modes must contain the same number of elements, although their normalization can be different. Complex mode shapes can also be used provided a complex conjugate transpose is taken in place of the normal matrix transpose. Usually analytical modes are correlated with the measured modes and the values of MAC are placed in a matrix. For a well correlated

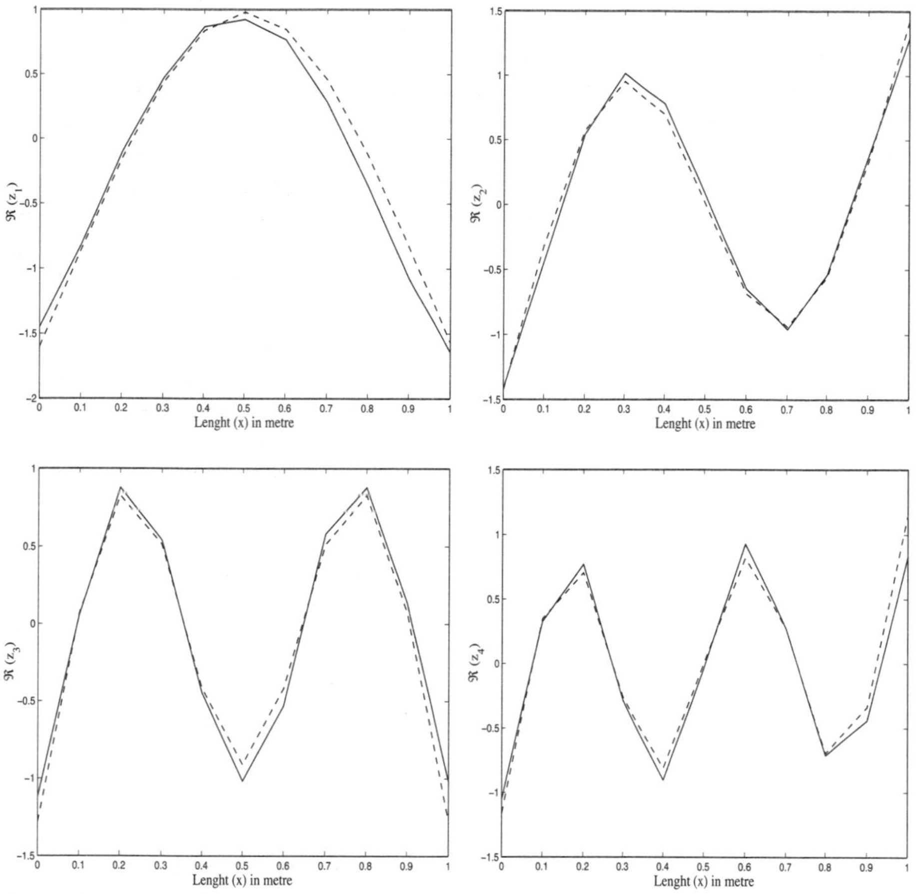

Figure 2.3. Comparison of the first four mode shapes of a free-free beam,'—' experiment, '−−' theory

model one expects the diagonal of the MAC matrix close to one and zero elsewhere.

Using the orthogonality properties of the mode shapes it can be shown that the mass and stiffness matrices of the experimental system are given by

$$\mathbf{M}_e^{-1} = \mathbf{X}_e \mathbf{X}_e^T \in \mathbb{R}^{N_e \times N_e} \tag{2.2}$$

$$\text{and} \quad \mathbf{K}_e^{-1} = \mathbf{X}_e \boldsymbol{\Omega}_e^{-2} \mathbf{X}_e^T \in \mathbb{R}^{N_e \times N_e} \tag{2.3}$$

Recall that N_e, the number of degrees-of-freedom of the experimental system (i.e., the number of response measurement point), is much smaller compared to the number of degrees-of-freedom of the FE model N. Since $N \gg N_e$, the comparison of $\mathbf{M}_e$ and $\mathbf{K}_e$ with $\mathbf{M}$ and $\mathbf{K}$ is not very meaningful. There are two approaches one could adopt at this stage. The first is to perform a *model reduction* of the FE system matrices so that they comparable to that of the experimental matrices. The second is to perform a *model expansion* of the experimental matrices using FE mode shapes. We refer the readers to the book by Friswell and Mottershead (1995) for further discussions on model reduction and model expansion methods.

The raw experimental data is in the form of dynamic response in the time domain. Although it is the most direct form of measurement, in practice it is difficult to compare them with the time histories obtained from the analytical model. The main reason behind this is the damping. The time histories are very sensitive to damping and since the initial numerical model normally has no any damping it is difficult to compare them directly. The usual way to compare the response is to look at the frequency domain response. Individual frequency response functions can be compared graphically by simply plotting the test and analytical results together. There are four issues to be kept in mind while comparing the frequency response functions: (a) the amount and nature of damping, (b) range of frequencies (c) number of modes and (d) separation between the modes. The amount of damping will control the sharpness of the individual peaks and troughs. The frequency range will control the amount of noise in the data, for example if one uses a soft-tip hammer for the excitation, it will be difficult to compare the FRFs beyond 1KHz or so. The number of modes and spacing between the modes will have an overall effect on the frequency response function. In some application it is required to compare the complete set of transfer functions. This is a difficult task simply due to overwhelming volume of data that one has to deal with. Extending the analogy of MAC, in this case a frequency response assurance criteria (FRAC) can be defined in a similar way

$$\mathrm{FRAC}_{jk} = \frac{\left|\mathbf{H}_e(\omega_j)^T \mathbf{H}_a(\omega_k)\right|^2}{\left|\mathbf{H}_a(\omega_k)^T \mathbf{H}_a(\omega_k)\right| \left|\mathbf{H}_e(\omega_j)^T \mathbf{H}_e(\omega_j)\right|}. \tag{2.4}$$

A diagram similar to MAC can be used to represent FRAC. The only difference will be that a FRAC diagram will be very dense since there are thousands of frequency points in each FRFs.

2.4 Conclusions

The concept of validation and verification of a numerical models of dynamic systems was introduced in chapter 2. The importance and scope of validation and verification activities were emphasized. The experimental modal analysis (EMA) plays a central role in the V&V of structural dynamic models. A brief overview of the experimental modal analysis procedure was presented. Several approaches to measure the correlation between experimental results and output of numerical models in dynamics, such as MAC and FRAC, were discussed. Experimental results of a damped beam were used to illustrate the model correlation process.

After an initial validation exercise one usually finds that there are discrepancies between the results from the FE models and that from the experiments. Once such differences are quantified using the methods outlined here, the next step is to update the FE model so that these differences can be reduced. Model updating is an essential step for building a credible numerical model. Model updating is an active area of research and it is beyond the scope of this paper to discuss them. Readers are referred to the book by Friswell and Mottershead (1995) for details.

3 Identification of Damping

3.1 Introduction

As mentioned in the previous section, the comparison between experimental modal analysis data and FE results hinges crucially on the damping. In general obtaining a damping matrix from the first-principle is difficult. Normally the damping matrix is constructed from modal testing data. The methods of damping identification can be divided into two basic categories: (a) damping identification from modal testing and analysis (Alvin et al., 1997, Hasselsman, 1972, Ibrahim, 1983, Minas and Inman, 1991) and (b) direct damping identification from the forced response measurements (Baruch, 1997, Chen et al., 1996, Mottershead, 1990). Here we will outline damping identification methods belonging the first category.

3.2 Identification of Non-proportional Damping Matrix

In section 1.6 and chapter 2 commonly used analytical and experimental methods for damped linear systems have been outlined. This gives us a firm basis for further analysis, that is, to use the details of the measured vibration data to learn more about the underlying damping mechanisms. Recall that complex modes arise in viscously damped systems provided the damping is non-proportional. There is no physical reason why a general system should follow the mathematical conditions for existence of proportional damping. In fact practical experience in modal testing shows that most real-life structures do not do so, as they possess complex modes instead of real normal modes. As Sestieri and Ibrahim (1994) have put it ' ... it is ironic that the real modes are in fact not real at all, in that in practice they do not exist, while complex modes are those practically identifiable from experimental tests. This implies that real modes are pure abstraction, in contrast with complex modes that are, therefore, the only reality!' For this reason it is legitimate to consider complex modes for the identification of damping.

The results derived for the elastic modes can be applied to viscously damped systems by considering the fact that the matrix of the damping functions, $\mathbf{G}(s)$, is a constant matrix. Say $\mathbf{G}(s) = \mathbf{C}$, $\forall s$, where $\mathbf{C}$ is the viscous damping matrix. Here we consider a further special case when the damping is light so that the first-order perturbation method can be applied. Suppose $\lambda_j, \mathbf{z}_j$ is the j-th complex natural frequency and complex mode shape. In the context of the notations used in section 1.6, $s_j = \mathrm{i}\lambda_j$. Using this, from equation (1.32) an approximate expression for the complex natural frequencies can be given by

$$\lambda_j \approx \pm\omega_j + \mathrm{i}C'_{jj}/2. \tag{3.1}$$

Using the light damping assumption, from equation (1.33) the first-order approximate expression of complex eigenvectors can be obtained as

$$\mathbf{z}_j \approx \mathbf{x}_j + \mathrm{i}\sum_{\substack{k=1\\k\neq j}}^{N} \frac{\omega_j C'_{kj}}{(\omega_j^2 - \omega_k^2)}\mathbf{x}_k. \tag{3.2}$$

In the above expressions $C'_{kl} = \mathbf{x}_k^T \mathbf{C} \mathbf{x}_l$ are the elements of the damping matrix in the modal coordinates. These results were obtained by Rayleigh (1877, see Section 102, equation 5 and 6). The above equation shows (up to first-order approximation) that the real parts of the complex modes are the same as the undamped modes and that the off-diagonal terms of the modal damping matrix are responsible for the imaginary parts. This fact will be exploited to obtain the non-proportional damping matrix from experimentally identified complex modes.

A General Complex Mode Based Approach. Consider $\hat{\lambda}_j$ and $\hat{\mathbf{z}}_j$ for all $j = 1, 2, \cdots m$ to be the *measured* complex natural frequencies and modes. Here $\hat{\mathbf{z}}_j \in \mathbb{C}^N$ where N denotes the number of measurement points on the structure and the number of modes considered in the study is m. In general $m \neq N$, usually $N \geq m$. Separating the real and imaginary parts of $\hat{\mathbf{z}}_j$ we have

$$\hat{\mathbf{z}}_j = \hat{\mathbf{u}}_j + \mathrm{i}\hat{\mathbf{v}}_j. \tag{3.3}$$

If the measured complex mode shapes are consistent with a viscous damping model then from equation (3.1) the real part of each complex natural frequency gives the undamped natural frequency:

$$\hat{\omega}_j = \Re\left(\hat{\lambda}_j\right), \tag{3.4}$$

where $\hat{\lambda}_j$ denotes the j-th complex natural frequency measured from the experiment. Similarly from equation (3.2), the real part of each complex mode $\hat{\mathbf{u}}_j$ immediately gives the corresponding undamped mode and the mass orthogonality relationship (1.4) will be automatically satisfied. Now from equation (3.2), expand the imaginary part of $\hat{\mathbf{z}}_j$ as a linear combination of $\hat{\mathbf{u}}_j$:

$$\hat{\mathbf{v}}_j = \sum_{k=1}^{m} B_{kj}\hat{\mathbf{u}}_k; \quad \text{where } B_{kj} = \frac{\hat{\omega}_j C'_{kj}}{\hat{\omega}_j^2 - \hat{\omega}_k^2}. \tag{3.5}$$

With $N \geq m$ this relation cannot be satisfied exactly in general. Then the constants B_{kj} should be calculated such that the error in representing $\hat{\mathbf{v}}_j$ by such a sum is minimized. Note that in the above sum we have included the $k = j$ term although in the original sum in equation (3.2) this term was absent. This is done to simplify the mathematical formulation to be followed, and has no effect on the result. Our interest lies in calculating C'_{kj} from B_{kj} through the relationship given by the second part of the equation (3.5), and indeed for $k = j$ we would obtain $C'_{kj} = 0$. The diagonal terms C'_{jj} are instead obtained from the imaginary part of the complex natural frequencies:

$$C'_{jj} = 2\Im(\hat{\lambda}_j). \tag{3.6}$$

The error from representing $\hat{\mathbf{v}}_j$ by the series sum (3.5) can be expressed as

$$\boldsymbol{\varepsilon}_j = \hat{\mathbf{v}}_j - \sum_{k=1}^{m} B_{kj}\hat{\mathbf{u}}_k. \tag{3.7}$$

To minimize the error a Galerkin approach can be adopted. The undamped mode shapes $\hat{\mathbf{u}}_l, \forall l = 1, \cdots, m$, are taken as 'weighting functions'. Using the Galerkin method on $\boldsymbol{\varepsilon}_j \in \mathbb{R}^N$ for a fixed j one obtains

$$\hat{\mathbf{u}}_l^T \boldsymbol{\varepsilon}_j = 0; \qquad \forall l = 1, \cdots m. \tag{3.8}$$

Combining equations (3.7) and (3.8) yields

$$\hat{\mathbf{u}}_l^T \left\{ \hat{\mathbf{v}}_j - \sum_{k=1}^{m} B_{kj} \hat{\mathbf{u}}_k \right\} = 0 \quad \text{or} \quad \sum_{k=1}^{m} W_{lk} \, B_{kj} = D_{lj}; \qquad l = 1, \cdots, m \tag{3.9}$$

with $W_{lk} = \hat{\mathbf{u}}_l^T \hat{\mathbf{u}}_k$ and $D_{lj} = \hat{\mathbf{u}}_l^T \hat{\mathbf{v}}_j$. Since W_{kl} is $j-$independent, for all $j = 1, \cdots m$ the above equations can be combined in matrix form

$$\mathbf{W}\,\mathbf{B} = \mathbf{D} \tag{3.10}$$

where $\mathbf{B} \in \mathbb{R}^{m \times m}$ is the matrix of unknown coefficients to be found, $\mathbf{W} = \hat{\mathbf{U}}^T \hat{\mathbf{U}} \in \mathbb{R}^{m \times m}$ and $\mathbf{D} = \hat{\mathbf{U}}^T \hat{\mathbf{V}} \in \mathbb{R}^{m \times m}$ with

$$\begin{aligned} \hat{\mathbf{U}} &= [\hat{\mathbf{u}}_1, \hat{\mathbf{u}}_2, \cdots \hat{\mathbf{u}}_m] \in \mathbb{R}^{N \times m} \\ \hat{\mathbf{V}} &= [\hat{\mathbf{v}}_1, \hat{\mathbf{v}}_2, \cdots \hat{\mathbf{v}}_m] \in \mathbb{R}^{N \times m}. \end{aligned} \tag{3.11}$$

Now $\mathbf{B}$ can be obtained by carrying out the matrix inversion associated with equation (3.10) as

$$\mathbf{B} = \mathbf{W}^{-1}\mathbf{D} = \left[\hat{\mathbf{U}}^T \hat{\mathbf{U}}\right]^{-1} \hat{\mathbf{U}}^T \hat{\mathbf{V}}. \tag{3.12}$$

Using the $\mathbf{B}$ matrix, the coefficients of the modal damping matrix can be derived as

$$C'_{kj} = \frac{(\hat{\omega}_j^2 - \hat{\omega}_k^2) B_{kj}}{\hat{\omega}_j}; \qquad k \neq j. \tag{3.13}$$

The above two equations together with equation (3.6) completely define the modal damping matrix $\mathbf{C}' \in \mathbb{R}^{m \times m}$. If $\hat{\mathbf{U}} \in \mathbb{R}^{N \times N}$ is the *complete* undamped modal matrix then the damping matrices in the modal coordinates and original coordinates are related by $\mathbf{C}' = \hat{\mathbf{U}}^T \mathbf{C} \hat{\mathbf{U}}$. Thus given $\mathbf{C}'$, the damping matrix in the original coordinates can be easily obtained by the inverse transformation as $\mathbf{C} = \mathbf{U}^{T^{-1}} \mathbf{C}' \hat{\mathbf{U}}^{-1}$. For the case when the full modal matrix is not available, that is $\hat{\mathbf{U}} \in \mathbb{R}^{N \times m}$ is not a square matrix, a pseudoinverse is required in order to obtain the damping matrix in the original coordinates. The damping matrix in the original coordinates is then given by

$$\mathbf{C} = \left[\left(\hat{\mathbf{U}}^T \hat{\mathbf{U}} \right)^{-1} \hat{\mathbf{U}}^T \right]^T \mathbf{C}' \left[\left(\hat{\mathbf{U}}^T \hat{\mathbf{U}} \right)^{-1} \hat{\mathbf{U}}^T \right]. \tag{3.14}$$

It is clear from the above equations that we need only the complex natural frequencies and mode shapes to obtain $\mathbf{C}$. The method is simple and does not require significant computational effort. Another advantage is that neither the estimation of mass and stiffness matrices nor the full set of modal data is required to obtain an estimate of the full damping matrix. Using a larger number of modes will of course produce better results with higher spatial resolution. In summary, this procedure can be described by the following steps:

1. Measure a set of transfer functions $H_{ij}(\omega)$.
2. Choose the number m of modes to be retained in the study. Determine the complex natural frequencies $\hat{\lambda}_j$ and complex mode shapes $\hat{\mathbf{z}}_j$ from the transfer functions, for all $j = 1, \cdots m$. Obtain the complex mode shape matrix $\hat{\mathbf{Z}} = [\hat{\mathbf{z}}_1, \hat{\mathbf{z}}_2, \cdots \hat{\mathbf{z}}_m] \in \mathbb{C}^{N \times m}$.
3. Estimate the 'undamped natural frequencies' as $\hat{\omega}_j = \Re(\hat{\lambda}_j)$.

4. Set $\hat{\mathbf{U}} = \Re\left[\hat{\mathbf{Z}}\right]$ and $\hat{\mathbf{V}} = \Im\left[\hat{\mathbf{Z}}\right]$, from these obtain $\mathbf{W} = \hat{\mathbf{U}}^T\hat{\mathbf{U}}$ and $\mathbf{D} = \hat{\mathbf{U}}^T\hat{\mathbf{V}}$. Now denote $\mathbf{B} = \mathbf{W}^{-1}\mathbf{D}$.
5. From the $\mathbf{B}$ matrix get $C'_{kj} = \dfrac{(\hat{\omega}_j^2 - \hat{\omega}_k^2)B_{kj}}{\hat{\omega}_j}$ for $k \neq j$ and $C'_{jj} = 2\Im(\hat{\lambda}_j)$.
6. Finally, carry out the transformation $\mathbf{C} = \left[\left(\hat{\mathbf{U}}^T\mathbf{U}\right)^{-1}\hat{\mathbf{U}}^T\right]^T \mathbf{C}' \left[\left(\hat{\mathbf{U}}^T\hat{\mathbf{U}}\right)^{-1}\hat{\mathbf{U}}^T\right]$ to get the damping matrix in physical coordinates.

Even if the measured transfer functions are reciprocal, this procedure does not necessarily yield a symmetric damping matrix. If we indeed obtain a non-symmetric damping matrix then it may be deduced that the physical law behind the damping mechanism in the structure is not viscous. This fact will be illustrated by an example in the next section. Under those circumstances, if an accurate model for the damping in the structure is needed then a non-viscous model of some kind must be fitted to the measured data. Identification of non-viscous damping model is difficult not only due to the complexity with the identification procedure itself, but also, due to the wide choice of non-viscous model is available in the literature. Adhikari and Woodhouse (2001) have proposed some methods for the identification of non-viscous damping. Here we take an alternative route in the next section by fitting a symmetric non-proportional viscous damping matrix.

Symmetry Preserving Damping Identification Method. An asymmetric fitted damping matrix is a non-physical result because the original system is reciprocal. There are good arguments to support the principle of reciprocity when the physical mechanism of damping arises from linear behaviour within some or all of the material of which the structure is built. Therefore, we describe a method to identify a viscous damping model which will not violate the principle of reciprocity.

Consider $\hat{\lambda}_j$ and $\hat{\mathbf{z}}_j$ for all $j = 1, 2, \cdots m$ to be the *measured* complex natural frequencies and modes. In view of equation (3.2), again expand the imaginary part of $\hat{\mathbf{z}}_j$ as a linear combination of $\hat{\mathbf{u}}_j$ as in equation (3.5). Our interest lies in calculating C'_{kj} from B'_{kj} through the relationship given by the second part of the equation (3.5) so that the resulting $\mathbf{C}$ matrix is symmetric. For symmetry of the identified damping matrix $\mathbf{C}$, it is required that $\mathbf{C}'$ is symmetric, that is

$$C'_{kj} = C'_{jk}. \tag{3.15}$$

Using the relationship given by the second part of the equation (3.5) the above condition reads

$$B_{kj}\frac{\hat{\omega}_j^2 - \hat{\omega}_k^2}{\hat{\omega}_j} = B_{jk}\frac{\hat{\omega}_k^2 - \hat{\omega}_j^2}{\hat{\omega}_k}. \tag{3.16}$$

Simplification of equation (3.16) yields

$$\frac{B_{kj}}{\hat{\omega}_j} = -\frac{B_{jk}}{\hat{\omega}_k} \quad \text{or} \quad B_{kj}\hat{\omega}_k + B_{jk}\hat{\omega}_j = 0; \quad \forall k \neq j. \tag{3.17}$$

For further calculations it is convenient to cast the above set of equations in a matrix form. Consider $\mathbf{B} \in \mathbb{R}^{m\times m}$ to be the matrix of unknown constants B_{kj} and define

$$\hat{\boldsymbol{\Omega}} = \mathrm{diag}\left[\hat{\omega}_1, \hat{\omega}_2, \cdots, \hat{\omega}_m\right] \in \mathbb{R}^{m\times m} \tag{3.18}$$

to be the diagonal matrix of the measured undamped natural frequencies. From equation (3.17) for all $k, j = 1, 2, \cdots, m$ (including $k = j$ for mathematical convenience) we have

$$\hat{\boldsymbol{\Omega}}\mathbf{B} + \mathbf{B}^T\hat{\boldsymbol{\Omega}} = \mathbf{0} \tag{3.19}$$

This equation must be satisfied by the matrix $\mathbf{B}$ in order to make the identified viscous damping matrix $\mathbf{C}$ symmetric. The error from representing $\hat{\mathbf{v}}_j$ by the series sum (3.5) can be expressed as

$$\boldsymbol{\varepsilon}_j = \hat{\mathbf{v}}_j - \sum_{k=1}^{m} B_{kj}\hat{\mathbf{u}}_k \in \mathbb{R}^N \tag{3.20}$$

We need to minimize the above error together with the constraints given by equation (3.17). The standard inner product norm of $\boldsymbol{\varepsilon}_j$ is selected to minimize the error. Considering the Lagrange multipliers ϕ_{kj} the objective function may be constructed as

$$\chi^2 = \sum_{j=1}^{m} \boldsymbol{\varepsilon}_j^T\boldsymbol{\varepsilon}_j + \sum_{j=1}^{m}\sum_{k=1}^{m} \left(B_{kj}\hat{\omega}_k + B_{jk}\hat{\omega}_j\right)\phi_{kj} \tag{3.21}$$

To obtain B_{jk} by the error minimization approach set

$$\frac{\partial \chi^2}{\partial B_{rs}} = 0; \quad \forall r, s = 1, \cdots, m. \tag{3.22}$$

Substituting $\boldsymbol{\varepsilon}_j$ from equation (3.20) one has

$$\begin{aligned} & -2\hat{\mathbf{u}}_r^T\left(\hat{\mathbf{v}}_s - \sum_{k=1}^{m} B_{ks}\hat{\mathbf{u}}_k\right) + \left[\phi_{rs} + \phi_{sr}\right]\hat{\omega}_r = 0 \\ \text{or} \quad & \sum_{k=1}^{m}\left(\hat{\mathbf{u}}_r^T\hat{\mathbf{u}}_k\right)B_{ks} + \frac{1}{2}\left[\hat{\omega}_r\phi_{rs} + \hat{\omega}_r\phi_{sr}\right] = \hat{\mathbf{u}}_r^T\hat{\mathbf{v}}_s; \quad \forall r, s = 1, \cdots, m. \end{aligned} \tag{3.23}$$

The above set of equations can be represented in a matrix form as

$$\mathbf{W}\mathbf{B} + \frac{1}{2}\left[\hat{\boldsymbol{\Omega}}\boldsymbol{\Phi} + \hat{\boldsymbol{\Omega}}\boldsymbol{\Phi}^T\right] = \mathbf{D} \tag{3.24}$$

where

$$\begin{aligned} \mathbf{W} &= \hat{\mathbf{U}}^T\hat{\mathbf{U}} \in \mathbb{R}^{m\times m} \\ \mathbf{D} &= \hat{\mathbf{U}}^T\hat{\mathbf{V}} \in \mathbb{R}^{m\times m} \end{aligned} \tag{3.25}$$

and $\boldsymbol{\Phi} \in \mathbb{R}^{m\times m}$ is the matrix of ϕ_{rs}. Note that both $\mathbf{B}$ and $\boldsymbol{\Phi}$ are unknown, so there are in total $2m^2$ unknowns. Equation (3.24) together with the symmetry condition (3.19) provides $2m^2$ equations. Thus both $\mathbf{B}$ and $\boldsymbol{\Phi}$ can be solved exactly provided their coefficient matrix is not singular or badly scaled.

Because in this study $\boldsymbol{\Phi}$ is not a quantity of interest, we try to eliminate it. Recalling that $\hat{\boldsymbol{\Omega}}$ is a diagonal matrix taking transpose of (3.24) one has

$$\mathbf{B}^T\mathbf{W}^T + \frac{1}{2}\left[\boldsymbol{\Phi}^T\hat{\boldsymbol{\Omega}} + \boldsymbol{\Phi}\hat{\boldsymbol{\Omega}}\right] = \mathbf{D}^T \tag{3.26}$$

Now postmultiplying equation (3.24) by $\hat{\boldsymbol{\Omega}}$ and premultiplying equation (3.26) by $\hat{\boldsymbol{\Omega}}$ and subtracting one has

$$\begin{aligned}&\mathbf{W}\mathbf{B}\hat{\boldsymbol{\Omega}} + \frac{1}{2}\hat{\boldsymbol{\Omega}}\boldsymbol{\Phi}\hat{\boldsymbol{\Omega}} + \frac{1}{2}\hat{\boldsymbol{\Omega}}\boldsymbol{\Phi}^T\hat{\boldsymbol{\Omega}} - \hat{\boldsymbol{\Omega}}\mathbf{B}^T\mathbf{W}^T - \frac{1}{2}\hat{\boldsymbol{\Omega}}\boldsymbol{\Phi}^T\hat{\boldsymbol{\Omega}} - \frac{1}{2}\hat{\boldsymbol{\Omega}}\boldsymbol{\Phi}\hat{\boldsymbol{\Omega}} = \mathbf{D}\hat{\boldsymbol{\Omega}} - \hat{\boldsymbol{\Omega}}\mathbf{D}^T\\ \text{or}\quad &\mathbf{W}\mathbf{B}\hat{\boldsymbol{\Omega}} - \hat{\boldsymbol{\Omega}}\mathbf{B}^T\mathbf{W}^T = \mathbf{D}\hat{\boldsymbol{\Omega}} - \hat{\boldsymbol{\Omega}}\mathbf{D}^T.\end{aligned} \tag{3.27}$$

This way $\boldsymbol{\Phi}$ has been eliminated. However, note that since the above is a rank deficient system of equations it cannot be used to obtain $\mathbf{B}$ and here we need to use the symmetry condition (3.19). Rearranging equation (3.19) we have

$$\mathbf{B}^T = -\hat{\boldsymbol{\Omega}}\mathbf{B}\hat{\boldsymbol{\Omega}}^{-1} \tag{3.28}$$

Substituting $\mathbf{B}^T$ in equation (3.27) and premultiplying by $\hat{\boldsymbol{\Omega}}^{-1}$ results in

$$\hat{\boldsymbol{\Omega}}^{-1}\mathbf{W}\mathbf{B}\hat{\boldsymbol{\Omega}} + \hat{\boldsymbol{\Omega}}\mathbf{B}\hat{\boldsymbol{\Omega}}^{-1}\mathbf{W}^T = \hat{\boldsymbol{\Omega}}^{-1}\mathbf{D}\hat{\boldsymbol{\Omega}} - \mathbf{D}^T. \tag{3.29}$$

Observe from equation (3.25) that $\mathbf{W}$ is a symmetric matrix. Now denote

$$\begin{aligned}\mathbf{Q} &= \hat{\boldsymbol{\Omega}}^{-1}\mathbf{W} = \hat{\boldsymbol{\Omega}}^{-1}\mathbf{W}^T\\ \mathbf{P} &= \hat{\boldsymbol{\Omega}}^{-1}\mathbf{D}\hat{\boldsymbol{\Omega}} - \mathbf{D}^T.\end{aligned} \tag{3.30}$$

Using the above definitions, equation (3.29) reads

$$\mathbf{Q}\mathbf{B}\hat{\boldsymbol{\Omega}} + \hat{\boldsymbol{\Omega}}\mathbf{B}\mathbf{Q} = \mathbf{P}. \tag{3.31}$$

This matrix equation represents a set of m^2 equations and can be solved to obtain $\mathbf{B}$ (m^2 unknowns) uniquely. To ease the solution procedure let us define the operation vec: $\mathbb{R}^{m\times n} \to \mathbb{R}^{mn}$ which transforms a matrix to a long vector formed by stacking the columns of the matrix in a sequence one below the other. It is known that (see Zhou et al., 1995, page 25) for any three matrices $\mathbf{A} \in \mathbb{C}^{k\times m}$, $\mathbf{B} \in \mathbb{C}^{m\times n}$, and $\mathbf{C} \in \mathbb{C}^{n\times l}$, we have $\text{vec}\,(\mathbf{ABC}) = \left(\mathbf{C}^T \otimes \mathbf{A}\right)\text{vec}(\mathbf{B})$ where $\otimes$ denotes the *Kronecker product*. Using this relationship and taking *vec* of both side of equation (3.31) one obtains

$$\begin{aligned}&\left(\hat{\boldsymbol{\Omega}} \otimes \mathbf{Q}\right)\text{vec}(\mathbf{B}) + \left(\mathbf{Q}^T \otimes \hat{\boldsymbol{\Omega}}\right)\text{vec}\,(\mathbf{B}) = \text{vec}\,(\mathbf{P})\\ \text{or}\quad &[\mathbf{R}]\,\text{vec}\,(\mathbf{B}) = \text{vec}\,(\mathbf{P})\end{aligned} \tag{3.32}$$

where

$$\mathbf{R} = \left(\hat{\boldsymbol{\Omega}} \otimes \mathbf{Q}\right) + \left(\mathbf{Q}^T \otimes \hat{\boldsymbol{\Omega}}\right) \in \mathbb{R}^{m^2\times m^2}. \tag{3.33}$$

Since $\mathbf{R}$ is square matrix equation (3.32) can be solved to obtain

$$\text{vec}\,(\mathbf{B}) = [\mathbf{R}]^{-1}\,\text{vec}\,(\mathbf{P}). \tag{3.34}$$

From $\text{vec}\,(\mathbf{B})$ the matrix $\mathbf{B}$ can be easily obtained by the inverse operation. Obtaining $\mathbf{B}$ in such a way will always make the identified damping matrix symmetric. The coefficients of the modal damping matrix can be derived from

$$C'_{kj} = \frac{(\hat{\omega}_j^2 - \hat{\omega}_k^2)B_{kj}}{\hat{\omega}_j}; \qquad k \neq j \tag{3.35}$$

Once $\mathbf{C}'$ is obtained, the damping matrix in the original coordinates can be obtained from equation (3.14). In summary, this procedure can be described by the following steps:

1. Measure a set of transfer functions $H_{ij}(\omega)$ at a set of N grid points. Fix the number of modes to be retained in the study, say m. Determine the complex natural frequencies $\hat{\lambda}_j$ and complex mode shapes $\hat{\mathbf{z}}_j$ from the transfer function, for all $j = 1, \cdots m$. Denote $\hat{\mathbf{Z}} = [\hat{\mathbf{z}}_1, \hat{\mathbf{z}}_2, \cdots \hat{\mathbf{z}}_m] \in \mathbb{C}^{N \times m}$ the complex mode shape matrix.
2. Set the 'undamped natural frequencies' to $\hat{\omega}_j = \Re(\hat{\lambda}_j)$. Denote the diagonal matrix $\hat{\mathbf{\Omega}} = \mathrm{diag}\left[\hat{\omega}_1, \hat{\omega}_2, \cdots, \hat{\omega}_m\right] \in \mathbb{R}^{m \times m}$.
3. Separate the real and imaginary parts of $\hat{\mathbf{Z}}$ to obtain $\hat{\mathbf{U}} = \Re\left[\hat{\mathbf{Z}}\right]$ and $\hat{\mathbf{V}} = \Im\left[\hat{\mathbf{Z}}\right]$.
4. From these obtain the $m \times m$ matrices $\mathbf{W} = \hat{\mathbf{U}}^T \hat{\mathbf{U}}$, $\mathbf{D} = \hat{\mathbf{U}}^T \hat{\mathbf{V}}$, $\mathbf{Q} = \hat{\mathbf{\Omega}}^{-1} \mathbf{W}$ and $\mathbf{P} = \hat{\mathbf{\Omega}}^{-1} \mathbf{D} \hat{\mathbf{\Omega}} - \mathbf{D}^T$.
5. Now denote $\mathbf{p} = \mathrm{vec}\,(\mathbf{P}) \in \mathbb{R}^{m^2}$ and calculate $\mathbf{R} = \left(\hat{\mathbf{\Omega}} \otimes \mathbf{Q}\right) + \left(\mathbf{Q}^T \otimes \hat{\mathbf{\Omega}}\right) \in \mathbb{R}^{m^2 \times m^2}$ (MATLAB™ command *kron* can be used to calculate Kronecker products).
6. Evaluate $\mathrm{vec}\,(\mathbf{B}) = [\mathbf{R}]^{-1} \mathbf{p}$ and obtain the matrix $\mathbf{B}$.
7. From the $\mathbf{B}$ matrix get $C'_{kj} = \dfrac{(\hat{\omega}_j^2 - \hat{\omega}_k^2) B_{kj}}{\hat{\omega}_j}$ for $k \neq j$ and $C'_{jj} = 2\Im(\hat{\lambda}_j)$.
8. Finally, carry out the transformation $\mathbf{C} = \left[\left(\hat{\mathbf{U}}^T \mathbf{U}\right)^{-1} \hat{\mathbf{U}}^T\right]^T \mathbf{C}' \left[\left(\hat{\mathbf{U}}^T \hat{\mathbf{U}}\right)^{-1} \hat{\mathbf{U}}^T\right]$ to get the damping matrix in physical coordinates.

A numerical illustration of the damping identification methods is considered next.

Numerical Examples. There is a major difference in emphasis between this example and other related studies on damping identification reported in the literature. Most of the methods assume from the outset that the system is viscously damped (see the review paper by Pilkey and Inman, 1998) and then formulate the theory to identify a viscous damping matrix. Here, we wish to investigate how much one can learn by fitting a viscous damping model when the actual system is non-viscously damped, as one must expect to be the case for most practical systems. It is far from clear in practice what kind of non-viscous damping behaviour a system might exhibit. We defer that question for the moment, and instead study by simulation a system which has a known non-viscous damping model. This simple system gives us a useful basis to carry out numerical investigations. Complex natural frequencies and modes can be calculated for the model system using the procedure outlined in 1.6, then treated like experimental data obtained from a modal testing procedure, and used for identifying a viscous damping model by the procedures described in the previous section. Note that in a true experimental environment the measured complex natural frequencies and mode shapes will be contaminated by noise. Since the simulation data are noise-free the results obtained using them are 'ideal', the best one can hope using this approach. Once promising algorithms have been identified in this way, the influence of noise in degrading the performance will have to be addressed.

The system considered is shown in figure 3.1. N masses, each of mass m_u, are connected by springs of stiffness k_u. Certain of the masses of the system shown in figure 3.1 have dissipative elements connecting them to the ground. For the numerical

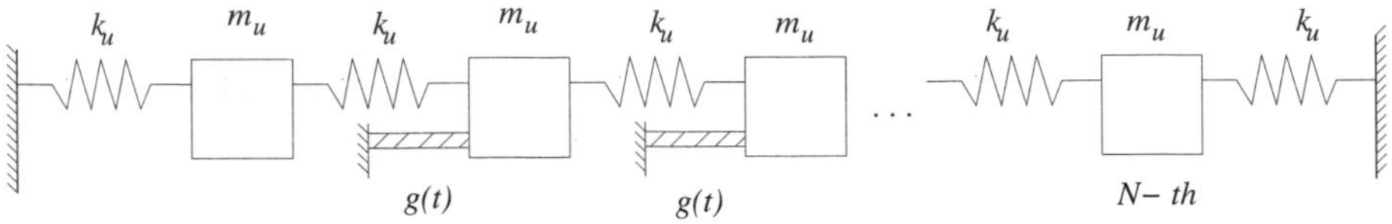

Figure 3.1. Linear array of N spring-mass oscillators, $N = 30$, $m_u = 1\,Kg$, $k_u = 4 \times 10^3 N/m$

calculations considered here, we have taken $N = 30$, and masses from $8th$ to $17th$ have dampers associated with them. The mass matrix of the system has the form $\mathbf{M} = m_u \mathbf{I}_N$ where $\mathbf{I}_N$ is the $N \times N$ identity matrix. The stiffness matrix of the system is given by

$$\mathbf{K} = k_u \begin{bmatrix} 2 & -1 & & & & \\ -1 & 2 & -1 & & & \\ & \ddots & \ddots & \ddots & & \\ & & -1 & 2 & -1 & \\ & & & \ddots & \ddots & -1 \\ & & & & -1 & 2 \end{bmatrix}. \tag{3.36}$$

The dissipative elements shown in figure 3.1 will be taken to be linear, but not to be simple viscous dashpots. For any such element, the force developed between the two ends will depend on the history of the relative motion of the two ends. The dependence can be written in terms of a convolution integral. Using the mass and the stiffness matrices described before, the equation of motion of free vibration can thus be expressed in the form

$$\mathbf{M}\ddot{\mathbf{q}}(t) + \bar{\mathbf{C}} \int_{-\infty}^{t} g(t-\tau)\, \dot{\mathbf{q}}(\tau)\, d\tau + \mathbf{K}\mathbf{q}(t) = \mathbf{0} \tag{3.37}$$

where $g(t)$ is the damping function (assumed to have the same form for all the damping elements in the system) and $\bar{\mathbf{C}}$ is the associated coefficient matrix which depends on the distribution of the dampers. Here $\bar{\mathbf{C}} = c\bar{\mathbf{I}}$ where c is a constant and $\bar{\mathbf{I}}$ is a block identity matrix which is non-zero only corresponding to the masses with the dampers. The function $g(t)$ is assumed to be

$$g(t) = \mu_1 e^{-\mu_1 t}; \quad t \geq 0 \tag{3.38}$$

where μ_1 is a constant. To quantify the amount of non-viscousness we define a characteristic time constant θ, via the first moment of $g(t)$ as

$$\theta = \int_0^{\infty} t\, g(t)\, dt = \frac{1}{\mu_1} \tag{3.39}$$

For viscous damping, $g(t) = c\delta(t)$ and therefore $\theta = 0$. So the characteristic time constant of a damping function gives a convenient measure of non-viscousness: if it is close to zero the damping behaviour will be near-viscous, and vice versa. For the purpose of numerical examples, the values $m_u = 1$ kg, $k_u = 4 \times 10^5$ N/m have been used. The resulting undamped natural frequencies then range from near zero to approximately 200 Hz. For damping models, the value $c = 25$ has been used, and various values of the time constant θ have been tested. These are conveniently expressed as a fraction of the period of the highest undamped natural frequency:

$$\theta = \gamma T_{min} \tag{3.40}$$

When γ is small compared with unity the damping behaviour can be expected to be essentially viscous. When γ is in the order of unity, non-viscous effects should become significant.

The complex natural frequencies and mode shapes can now be calculated from the analysis presented in section 1.6. We can then follow the steps outlined in the previous sections to obtain the viscous damping matrices which represents these 'measured' data most accurately. Two values of γ, namely $\gamma = 0.02$ and $\gamma = 0.5$ have been considered. When $\gamma = 0.02$ the non-viscous damping model is actually show near-viscous behaviour. Figure 3.2 shows the identified viscous damping matrix for this case. The fitted matrix identifies the damping in the system very well. The high portion of the plot corresponds to the spatial location of the dampers. The off-diagonal terms of the identified damping matrix are very small compared to the diagonal terms, indicating correctly that there is no damping corresponding to those masses.

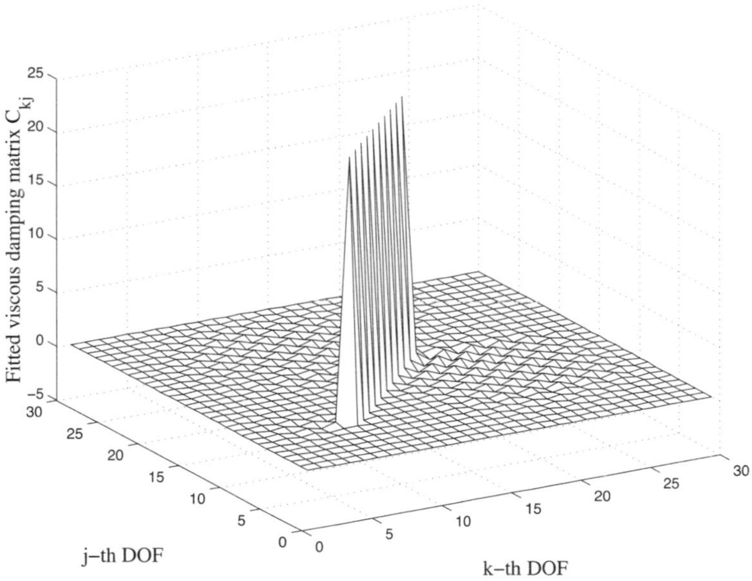

Figure 3.2. Fitted viscous damping matrix for $\gamma = 0.02$

When γ is larger the non-viscous damping model depart from the viscous damping model. For the value $\gamma = 0.5$, figure 3.3 shows the result of running the fitting procedure identical to that in figure 3.2. Although the identification method shows the position of the damper with reasonable accuracy, the fitted damping matrix is not symmetric. This is, in some sense, a non-physical result. This symmetry breaking phenomenon associated with the identification algorithm tells us that the original damping model (exponential in this case) and the fitted damping model (viscous) are significantly different. On the one

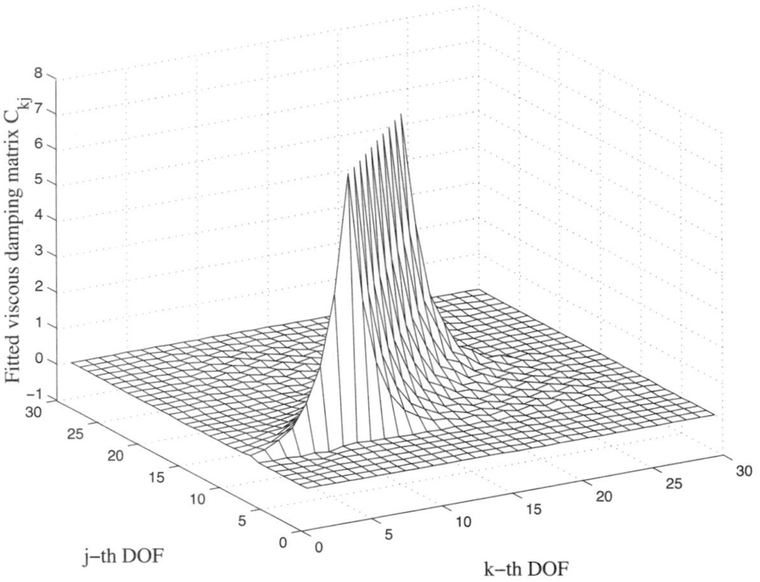

Figure 3.3. Fitted viscous damping matrix for $\gamma = 0.5$

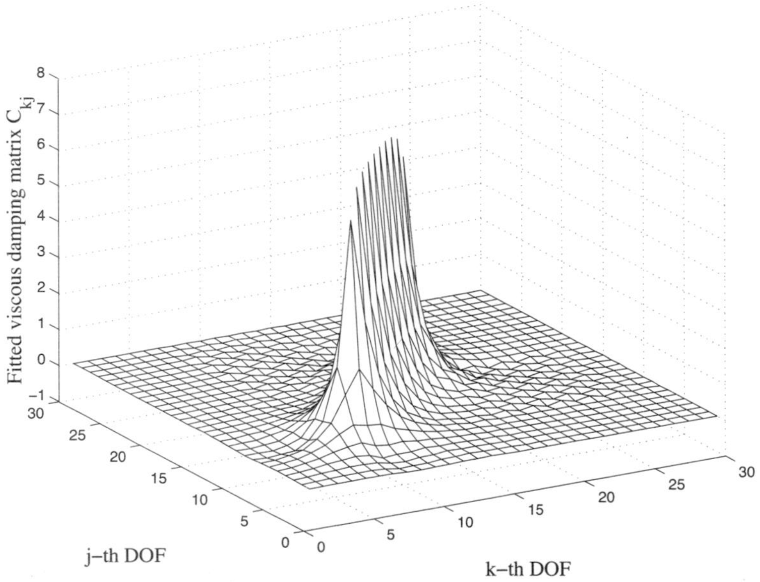

Figure 3.4. Symmetrically fitted viscous damping matrix for $\gamma = 0.5$

hand this is useful information since it indicates that we are fitting a wrong model, but on the other hand, the fitted asymmetric damping matrix is not very useful for further analysis as it violates the reciprocity principle. This problem can be 'solved' by applying the symmetry preserving damping identification method. Figure 3.4 shows the result of running the symmetry preserving fitting procedure. The resulting fitted viscous damping matrix is clearly physically more appealing. However, it should be remembered that although the symmetry is a desirable property in linear structural dynamics, enforcing the symmetry condition actually hides the true physics of the underlying damping behavior of the system.

3.3 Conclusions

The identification of the damping matrix from experimental measurements is a key component in model validation and verification. In this chapter two methods for the identification of damping matrix based on experimentally identified complex modes were described. The general approach may yield a non-symmetric damping matrix if the damping in the original system is not viscous. On the one hand this is useful information since it indicates that we are fitting a wrong model, but on the other hand, the fitted asymmetric damping matrix is not very useful for further analysis as it violates the reciprocity principle. This problem was resolved by a symmetry preserving damping identification method. So far we have not considered the effect of uncertainties in the modeling and simulation of dynamic systems. The next two chapters will explore some fundamental issues regarding uncertainty.

4 Model Uncertainty: Quantification and Propagation

4.1 Model Uncertainty: Why and How a Model Turns Uncertain?

Uncertainties are unavoidable in the description of real-life engineering systems. There are several sources of uncertainties, both in the mathematical models and in the experimental results. Uncertainties can be broadly divided into following three categories. The first type of uncertainty is due to the inherent variability in the system parameters, for example, different cars manufactured from a single production line are not exactly the same. This type of uncertainty is often referred to as *aleatoric uncertainty*. If enough samples are present, it is possible to characterize the variability using well established statistical methods and the probably density functions of the parameters can be obtained. The second type is uncertainty due to lack of knowledge regarding a system. This type of uncertainty is often referred to as *epistemic uncertainty* and generally arise in the modelling of complex systems, for example the problem of predicting cabin noise in helicopters. Due to its very nature, it is difficult to quantify or model this type uncertainties. Unlike aleatoric uncertainties, it is recognized that probabilistic models are not quite suitable for epistemic uncertainties. Several possibilistic approaches based on interval algebra, convex sets, Fuzzy sets and generalized Dempster-Schafer theory have been proposed to characterize this type of uncertainties. The third type of uncertainty is similar to the first type except that the corresponding variability characterization is not available, in which case work can be directed to gain better knowledge. This type of uncertainty often termed as *prejudicial uncertainty*, may consist of systematic and/or random errors, bias or other prejudices. An example of this type of uncertainty is the use of viscous damping model in spite of knowing that the true damping model is not viscous. The total uncertainty of a system is the combination of these three types of uncertainties. Different sources of uncertainties in the modeling and simulation of dynamic systems may be attributed, but not limited, to the following factors:

- Mathematical models: equations (linear, non-linear), geometry, damping model (viscous, non-viscous, fractional derivative), boundary conditions/initial conditions, input forces;
- Model parameters: Young's modulus, mass density, Poisson's ratio, damping model parameters (damping coefficient, relaxation modulus, fractional derivative order)
- Numerical algorithms: weak formulations, discretisation of displacement fields (in finite element method), discretisation of stochastic fields (in stochastic finite element method), approximate solution algorithms, truncation and roundoff errors, tolerances in the optimization and iterative methods, artificial intelligent (AI) method (choice of neural networks)
- Measurements: noise, resolution (number of sensors and actuators), experimental hardware, excitation method (nature of shakers and hammers), excitation and measurement point, data processing (amplification, number of data points, FFT), calibration

It is beyond the scope of this paper to discuss the quantification and propagation methods of the uncertainties arising from the above sources. We will focus our attention to the modeling and propagation of parametric uncertainties only. Moreover, only probabilistic models of uncertainty will be considered. In what follows next, basic understanding of the probability theory (see Papoulis and Pillai, 2002, for details) will be assumed.

4.2 Parametric Uncertainty in Structural Dynamics

The governing equation of motion of a linear structural system with stochastic parameter uncertainties, subjected to external excitations is most often a set of linear differential equation with random coefficients. The problem can be stated as finding the solution of the equation

$$\mathbf{L}(\Omega, \mathbf{r}, t)\mathbf{u}(\Omega, \mathbf{r}, t) = \mathbf{f}(\Omega, \mathbf{r}, t) \tag{4.1}$$

with prescribed boundary conditions and initial conditions. In the above equation $\mathbf{L}$ is a linear stochastic differential operator, $\mathbf{u}$ is the random system response to be determined, $\mathbf{f}$ is the dynamic excitation which can be random, $\mathbf{r}$ is the special coordinate vector, t is the time and Ω is the sample space denoting the stochastic nature of the problem. Equation (4.1) with $\mathbf{L}$ as a deterministic operator and $\mathbf{f}$ as a random forcing function, has been studied extensively within the scope of random vibration theory (Nigam, 1983). Here our interest is when the operator $\mathbf{L}$ itself is random. There are mainly two methods to model parametric uncertainty using the probabilistic approach: (a) uncertainty modeling using random variables, and (b) uncertainty modeling using stochastic processes. Since we often encounter distributed systems (such as beams, plates and shells) during the modeling of real-life systems, stochastic process models should be used for a realistic representation of the uncertainties in the system properties. This in turn leads to the operator $\mathbf{L}$ as a function of stochastic processes. Exact solutions of such stochastic differential equations, even for a simple system such as a standard Euler-Bernoulli beam, are very difficult to obtain. The problem becomes even more intractable when one has to reckon with a real-life engineering dynamics problems, where a set of stochastic boundary value problems defined on irregular spatial domain arises. This has motivated the engineers to seek approximate numerical methods for the solution of governing stochastic differential equations. The methods for solving structural dynamic problems with statistical uncertainties can be broadly grouped under Stochastic Finite Element Method (SFEM) and Statistical Energy Analysis (SEA). SEA was developed during 1960s (Lyon and Dejong, 1995) to analyze high frequency vibration problems where non-parametric uncertainties plays a key role. The stochastic finite element method is ideally suitable for low-frequency vibration problems where parametric uncertainties plays a key role. Here the stochastic finite element method is explained by applying it to an Euler-Bernoulli beam with stochastic parameter distributions.

4.3 Uncertainty Propagation Using Stochastic Finite Element Method

Stochastic finite element method (SFEM) is a generalization of the deterministic finite element method (FEM) to incorporate the random field models for the elastic, mass and damping properties (see the monographs by Elishakoff and Ren, 2003, Ghanem and Spanos, 1991, Kleiber and Hien, 1992). Application of the stochastic finite element method to linear structural dynamics problems typically consists of the following steps:

1. Selection of appropriate probabilistic models for parameter uncertainties and boundary conditions (such as Gaussian/non-Gaussian models).
2. Discretization of random fields, that is, replacement of the element property random fields by an equivalent finite set of random variables.
3. Formulation of the system equations of motion using stochastic generalization of standard methods such as variational method, energy method, virtual work method or weighted residual method. As a result of this process, the elements of $\mathbf{M}, \mathbf{C}$ and $\mathbf{K}$ will be random variables.
4. At this point one can take two routes. The first, and the most common approach, is to solve the free vibration problem, which in this case turns out to be a random matrix eigenvalue problem. The aim is to obtain the joint statistics of the mode shapes and natural frequencies. Once they are obtained, the next step is the characterization of response variability for the forced vibration problem.
5. The second route to solve the problem is using the dynamic stiffness method. The main challenge here is to invert the global dynamic stiffness matrix, which in general is a random complex symmetric matrix.

Extensive research works have been done in all of the above mentioned areas during the last few decades. In this section we will discuss the stochastic dynamic stiffness matrix method. The random eigenvalue problems will be discussed in the next chapter.

4.4 Stochastic Dynamic Stiffness Matrix Method

We will develop a finite element based formulation to obtain the dynamic stiffness matrix of a general beam element having randomly inhomogeneous mass density, flexural and axial rigidities and elastic foundation modulus. The beam element considered in this study is shown in figure 4.1. The governing field equation of motion, assuming linear

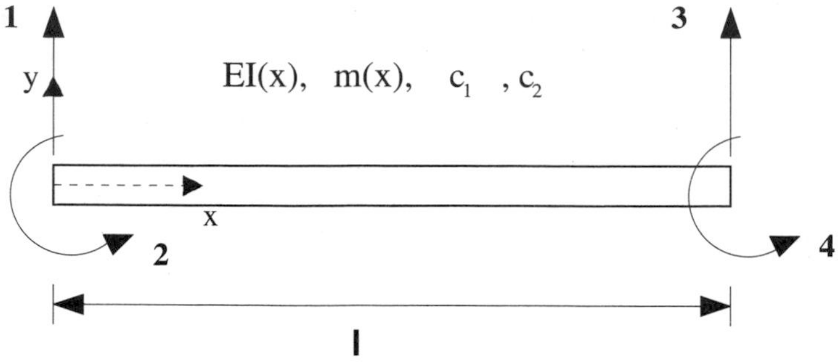

Figure 4.1. Random beam element in the local coordinate.

behavior and the validity of the Euler-Bernoulli hypotheses, is given by

$$\frac{\partial^2}{\partial x^2}\left[EI(x)\frac{\partial^2 Y(x,t)}{\partial x^2}+c_1\frac{\partial^3 Y(x,t)}{\partial x^2 \partial t}\right]+m(x)\frac{\partial^2 Y(x,t)}{\partial t^2}+c_2\frac{\partial Y(x,t)}{\partial t}=0 \qquad (4.2)$$

Here $Y(x,t)$ is the transverse flexural displacement, $EI(x)$ is the flexural rigidity, $m(x)$ is the mass per unit length, c_1 is the strain rate dependent viscous damping coefficient and c_2 is the velocity dependent viscous damping coefficient. The quantities $EI(x)$ and $m(x)$, in this study, are modeled as meansquare bounded, homogeneous random fields and are taken to have the following form

$$m(x)=m_0\left[1+\epsilon_1 f_1(x)\right], \quad \text{and} \quad EI(x)=EI_0\left[1+\epsilon_2 f_2(x)\right]. \qquad (4.3)$$

The subscript 0 indicates the mean values, ϵ_1 and ϵ_2 are deterministic constants which are usually small compared to the unity. The random fields $f_1(x)$ and $f_2(x)$ are taken to have zero mean, unit standard deviation with covariance $R_{ij}(\xi) =< f_i(x)f_j(x) >$ $(i,j=1,2)$; where $< \bullet >$ denotes the mathematical expectation operator. Since the linear system behavior is assumed, in the steady state the solution to the field equation can be expressed as

$$Y(x,t)=y(x)\exp\left[\mathrm{i}\omega t\right]. \qquad (4.4)$$

Consequently, the equation governing $y(x)$ has the form

$$\frac{\mathrm{d}^2}{\mathrm{d}x^2}\left[EI(x)\frac{\mathrm{d}^2 y}{\mathrm{d}x^2}+\mathrm{i}\omega c_1\frac{\mathrm{d}^2 y}{\mathrm{d}x^2}\right]+\left[-m(x)\omega^2+c_2\mathrm{i}\omega\right]y=0 \qquad (4.5)$$

An exact solution to this type of equation is in general not possible due to the presence of the random fields. The deterministic finite element method will be extended to obtain approximate solutions.

Instead of obtaining the standard mass and stiffness matrices, the problem will be solved using dynamic stiffness method. The dynamic stiffness method is an useful alternative to the more popular mode superposition method of vibration analysis. A vast amount of literature is available on the development of dynamic stiffness method in deterministic context, see, for example, Doyle (1989), Ferguson and Pilkey (1993a,b), Paz (1985). The dynamic stiffness coefficients are, by definition, frequency dependent. When system properties are modeled as random fields, these coefficients become random variables for a fixed value of the driving frequency. As the driving frequency is varied, these stiffness coefficients become random processes evolving in the frequency parameter. Furthermore, the presence of damping in the system makes these random processes complex in nature. The solution procedure is based on the application of finite element method which uses frequency dependent shape functions. This offers a powerful means to discretize the system property random fields and relaxes the dependence of finite element mesh size with respect to the frequency of excitation. The dynamic stiffness matrix can be achieved using the following steps.

Step 1 *Derivation of the shape functions:*

Obtain the shape functions using the deterministic undamped field equation. That is, solve the equation

$$\frac{\mathrm{d}^4 y}{\mathrm{d}x^4} - b^4 y = 0 \quad \text{where} \quad b^4 = \frac{m_0 \omega^2}{EI_0} \tag{4.6}$$

under a set of 'binary' boundary conditions. The shape functions $\{\mathbf{N}(x,\omega)\}$ can be shown to be given by

$$\{\mathbf{N}(x,\omega)\} = [\mathbf{\Gamma}(\omega)]\{\mathbf{s}(x,\omega)\}, \tag{4.7}$$

where

$$[\mathbf{\Gamma}(\omega)] = \begin{bmatrix} \frac{1}{2}\frac{cS+Cs}{cC-1} & -\frac{1}{2}\frac{1+sS-cC}{cC-1} & -\frac{1}{2}\frac{cS+Cs}{cC-1} & \frac{1}{2}\frac{cC+sS-1}{cC-1} \\ \frac{1}{2}\frac{cC+sS-1}{b(cC-1)} & \frac{1}{2}\frac{-Cs+cS}{b(cC-1)} & -\frac{1}{2}\frac{1+sS-cC}{b(cC-1)} & -\frac{1}{2}\frac{-Cs+cS}{b(cC-1)} \\ -\frac{1}{2}\frac{S+s}{cC-1} & \frac{1}{2}\frac{C-c}{cC-1} & \frac{1}{2}\frac{S+s}{cC-1} & -\frac{1}{2}\frac{C-c}{cC-1} \\ \frac{1}{2}\frac{C-c}{b(cC-1)} & -\frac{1}{2}\frac{S-s}{b(cC-1)} & -\frac{1}{2}\frac{C-c}{b(cC-1)} & -\frac{1}{2}\frac{S-s}{b(cC-1)} \end{bmatrix} \tag{4.8}$$

and

$$\{\mathbf{s}(x,\omega)\} = [\sin bx,\ \cos bx,\ \sinh bx,\ \cosh bx]^T \tag{4.9}$$

is the array of basis functions. Here $C = \cosh bl$, $c = \cos bl$, $S = \sinh bl$ and $s = \sin bl$. Plots of the first two shape functions, for a typical beam element to be considered later in numerical illustrations, are shown in figure 4.2. From equation (4.7) It can be derived that, for the static case (that is when $\omega = 0$), the shape functions are cubic polynomials in x and they agree with the well known beam shape functions. This feature is also observable from figure 4.2. With increasing value of ω, the shape functions adapt themselves and herein lies their major advantage.

Step 2 *Derivation of the Element Equation of Motion:*

The displacement field within the element is expressed in the form

$$Y(x,t) = \sum_{j=1}^{4} d_j(t) N_j(x,\omega) \tag{4.10}$$

Here $d_j(t), j = 1, \cdots, 4$ are the generalized coordinates representing the nodal displacements. Defining the Lagrangian $\mathcal{L}(t) = T(t) - V(t)$, the governing equations for the generalized coordinates $d_j(t)$ can be obtained from

$$\frac{\mathrm{d}}{\mathrm{d}t}\left[\frac{\partial \mathcal{L}}{\partial \dot{d}_j}\right] - \frac{\partial \mathcal{L}}{\partial d_j} = 0; \quad j = 1, \cdots, 4. \tag{4.11}$$

Step 3 *Undamped Element Stiffness Matrix:*

Since the motion is harmonic at frequency ω, it follows that $d_j(t) = A_j \exp[\mathrm{i}\omega t]$. Consequently, the undamped dynamic stiffness matrix $\mathbf{D}_u$ can be shown to be given by

$$\mathbf{D}_u(\omega) = \left[-\omega^2 I_{ij}(\omega) + J_{ij}(\omega)\right]_{(4\times 4)} \tag{4.12}$$

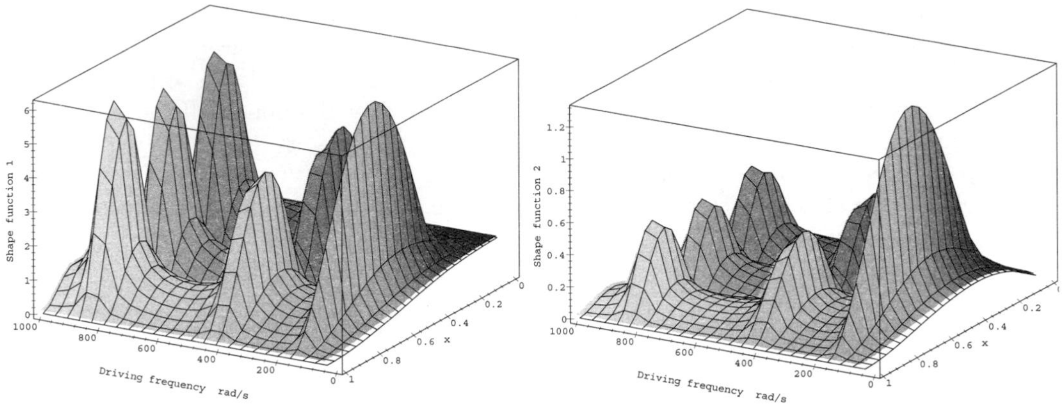

Figure 4.2. Shape functions for element number 2 of the portal frame in figure 4.3.

with

$$\begin{aligned} I_{ij}(\omega) &= \int_0^L m(x) N_i(x,\omega) N_j(x,\omega) \mathrm{d}x; \\ J_{ij}(\omega) &= \int_0^L \left[EI(x) \frac{\mathrm{d}^2}{\mathrm{d}x^2} N_i(x,\omega) \frac{\mathrm{d}^2}{\mathrm{d}x^2} N_j(x,\omega) \right] \mathrm{d}x. \end{aligned} \tag{4.13}$$

Substituting the expression of $m(x)$ and $EI(x)$ from equation (4.3) in equations (4.13) and after separating the random and deterministic parts, the undamped element dynamic stiffness matrix can be written as

$$\mathbf{D}_u(\omega) = \bar{\mathbf{D}}_u(\omega) + \sum_{l=1}^{10} \left[\boldsymbol{\alpha}^l(\omega) \right] X_l(\omega) \tag{4.14}$$

In the above expression $\bar{\mathbf{D}}_u(\omega)$ is the deterministic undamped element stiffness matrix which can be shown to be given by

$$\bar{\mathbf{D}}_u(\omega) = \begin{bmatrix} \frac{EI(cS+Cs)b^3}{-1+cC} & -\frac{EIsb^2S}{-1+cC} & \frac{EI(S+s)b^3}{-1+cC} & -\frac{EIb^2(C-c)}{-1+cC} \\ -\frac{EIsb^2S}{-1+cC} & \frac{EI(-Cs+cS)b}{-1+cC} & \frac{EIb^2(C-c)}{-1+cC} & -\frac{EI(S-s)b}{-1+cC} \\ \frac{EI(S+s)b^3}{-1+cC} & \frac{EIb^2(C-c)}{-1+cC} & -\frac{EI(cS+Cs)b^3}{-1+cC} & \frac{EIsb^2S}{-1+cC} \\ -\frac{EIb^2(C-c)}{-1+cC} & -\frac{EI(S-s)b}{-1+cC} & \frac{EIsb^2S}{-1+cC} & \frac{EI(-Cs+cS)b}{-1+cC} \end{bmatrix} \tag{4.15}$$

Furthermore, $X_l, l = 1, \cdots, 10$ are random in nature and are given by

$$\begin{aligned} X_1 = W_{11};\ X_2 = W_{12};\ X_3 = W_{13};\ X_4 = W_{14};\ X_5 = W_{22}; \\ X_6 = W_{23};\ X_7 = W_{24};\ X_8 = W_{33};\ X_9 = W_{34};\ X_{10} = W_{44} \end{aligned} \tag{4.16}$$

with

$$W_{kr} = \int_o^L \left[\{-m_0\omega^2\epsilon_1 f_1(x)\} s_k(x) s_r(x) + EI_0\epsilon_2 f_2(x) \frac{\mathrm{d}^2 s_k(x)}{\mathrm{d}x^2} \frac{\mathrm{d}^2 s_r(x)}{\mathrm{d}x^2} \right] \mathrm{d}x. \tag{4.17}$$

Also, $\left[\boldsymbol{\alpha}^l(\omega)\right], l = 1, \cdots, 10$ are 4×4 symmetric matrices of deterministic functions of ω. It may be noted that $X_l(\omega), l = 1, \cdots, 10$, are random processes evolving in the frequency parameter ω. Thus, for a fixed value of driving frequency ω, these quantities are random variables. The random processes $X_l(\omega)$ are known as the *dynamic weighted integrals*, since they arise as 'weighted integrals' of the random fields $f_1(x)$ and $f_2(x)$. The ransom variables X_l are linear functions of the random fields $f_1(x)$ and $f_2(x)$ and, therefore, if $f_1(x)$ and $f_2(x)$ are modeled as jointly Gaussian random fields, it follows that X_l are also jointly Gaussian. Furthermore, since the dynamic stiffness coefficients are linear function of X_l, it follows that these coefficients in turn are also Gaussian distributed.

Step 4 *Damped Element Stiffness Matrix:*

To allow for the effect of damping terms present in the field equation (4.2) the following steps are adopted:

(a) determine the damped dynamic stiffness matrix $\mathbf{D}_1(\omega)$ for the deterministic homogeneous beam element, that is, $\mathbf{D}_u(\omega)$ given by equation (4.14) with $\epsilon_1 = \epsilon_2 = 0$ and

$$b^4 = \frac{m_0\omega^2 + \mathrm{i}\omega c_2}{EI_0 + \mathrm{i}\omega c_1}; \tag{4.18}$$

(b) determine undamped dynamic stiffness matrix $\mathbf{D}_2(\omega)$ for the deterministic homogeneous beam element given again by equation (4.14) with $\epsilon_1 = \epsilon_2 = 0$ and $b = \frac{m_0\omega^2}{EI_0}$;

(c) compute the contribution to the dynamic stiffness $\mathbf{D}_d(\omega)$ from damping terms using $\mathbf{D}_d(\omega) = \mathbf{D}_1(\omega) - \mathbf{D}_2(\omega)$ and

(d) finally, the damped stochastic dynamic stiffness matrix is obtained as

$$\mathbf{D}(\omega) = \mathbf{D}_u(\omega) + \mathbf{D}_d(\omega) \tag{4.19}$$

Now substituting $\mathbf{D}(\omega)$ in place of $\mathbf{D}_u(\omega)$ in equation (12) the complete element dynamic stiffness matrix for an element 'e' can be expressed as

$$\mathbf{D}^e(\omega) = \bar{\mathbf{D}}^e(\omega) + \sum_{l=1}^{10} \left[\boldsymbol{\alpha}^l(\omega)\right]^e X_l^e \tag{4.20}$$

Here $\bar{\mathbf{D}}^e(\omega)$ represents the damped deterministic element dynamics stiffness matrix of element 'e'.

Step 5 *Element Stiffness Matrix in the Global Coordinates:*

The element stiffness matrices in the global coordinates can be written as

$$\mathbf{D}_G^e(\omega) = \mathbf{T}^{e^T} \mathbf{D}^e(\omega) \mathbf{T}^e \tag{4.21}$$

where $\mathbf{T}^e$ denotes the transformation matrix for the element concerned. Now substituting expression of $\mathbf{D}^e(\omega)$ from equation (4.20) into the above equation and separating the deterministic and random parts, $\mathbf{D}_G^e(\omega)$ can be cast in the form

$$\mathbf{D}_G^e(\omega) = \bar{\mathbf{D}}_G^e(\omega) + \mathbf{\Delta D}_G^e(\omega) \tag{4.22}$$

Here $\bar{\mathbf{D}}_G^e(\omega) = \mathbf{T}^{e^T} \bar{\mathbf{D}}^e(\omega)\mathbf{T}^e$ is the deterministic part of the element stiffness matrix in the global co-ordinates and $\mathbf{\Delta D}_G^e(\omega)$ is the corresponding random part. The covariance of the elements of $\mathbf{\Delta D}_G^e(\omega)$ associated with two distinct elements e_u and e_v can be expressed as

$$< \Delta D_{G_{pq}}^{e_u}(\omega)\Delta D_{G_{rs}}^{e_v}(\omega) >=$$

$$\sum_{i=1}^{10}\sum_{j=1}^{10}\sum_{k=1}^{4}\sum_{l=1}^{4}\sum_{m=1}^{4}\sum_{n=1}^{4} T_{kp}^{e_u} T_{lq}^{e_u} T_{mr}^{e_v} T_{ns}^{e_v} \left[\alpha_{kl}^{i}\right]^{e_u} \left[\alpha_{mn}^{j}\right]^{e_v} Cov(X_i^{e_u}, X_j^{e_v}) \tag{4.23}$$

where $p, q, r, s = 1, \cdots, 4$ and e_u, e_v runs over the number of elements in the structure. The term $Cov(X_i^{e_u}, X_j^{e_v})$ represents the statistics of the dynamic weighted integrals between two beam elements and is defined in the following section.

Step 6 *Statistics of the Dynamic Weighted Integrals*:

The covariance of the dynamic weighted integrals can be obtained by using equation (4.16) for two different elements of the structure, and can be expressed as

$$Cov(X_i^{e_u}, X_j^{e_v}) \quad = \; < W_{kr}^{e_u} W_{pq}^{e_v} > \; = In_{11} + In_{12} + In_{22} \tag{4.24}$$

where

$$In_{11} = m_0^{e_u} m_0^{e_v} \omega^4 \epsilon_1^{e_u} \epsilon_1^{e_v} \int_0^{L_u}\int_0^{L_v} \left\{ s_k^{e_u}(x_1) s_r^{e_u}(x_1) s_p^{e_v}(x_2) s_q^{e_v}(x_2) R_{11}^{e_u e_v}(x_1, x_2)\right\} \mathrm{d}x_1 \mathrm{d}x_2 \tag{4.25}$$

$$In_{12} = -\omega^2 \epsilon_1^{e_u} \epsilon_2^{e_v} \int_0^{L_u}\int_0^{L_v} \{ m_0^{e_u} EI_0^{e_v} s_k^{e_u}(x_1) s_r^{e_u}(x_1) \frac{\mathrm{d}^2 s_p^{e_v}(x_2)}{\mathrm{d}x_2^2}\frac{\mathrm{d}^2 s_q^{e_v}(x_2)}{\mathrm{d}x_2^2}$$
$$+ m_0^{e_v} EI_0^{e_u} \frac{\mathrm{d}^2 s_k^{e_u}(x_1)}{\mathrm{d}x_1^2}\frac{\mathrm{d}^2 s_r^{e_u}(x_1)}{\mathrm{d}x_1^2} s_p^{e_v}(x_2) s_q^{e_v}(x_2)\} \, R_{12}^{e_u e_v}(x_1, x_2)\mathrm{d}x_1\mathrm{d}x_2 \tag{4.26}$$

$$In_{22} = EI_0^{e_u} EI_0^{e_v} \epsilon_2^{e_u} \epsilon_2^{e_v}$$
$$\times \int_0^{L_u}\int_0^{L_v} \left\{ \frac{\mathrm{d}^2 s_k^{e_u}(x_1)}{\mathrm{d}x_1^2}\frac{\mathrm{d}^2 s_r^{e_u}(x_1)}{\mathrm{d}x_1^2}\frac{\mathrm{d}^2 s_p^{e_v}(x_2)}{\mathrm{d}x_2^2}\frac{\mathrm{d}^2 s_q^{e_v}(x_2)}{\mathrm{d}x_2^2} R_{22}^{e_u e_v}(x_1, x_2)\right\} \mathrm{d}x_1\mathrm{d}x_2 \tag{4.27}$$

In the above equations, $i, j = 1, \cdots, 10$; the relationship between i, j and k, r, p, q is according to equation (14), L_u, L_v are lengths of the elements e_u and e_v respectively and $(\bullet)^{e_u}$ represents the properties $(\bullet)$ corresponding to the element e_u. The expression $R_{lm}^{e_u e_v}(x_1, x_2)$, $(l, m = 1, 2)$ appearing in the above equations denotes the covariance function of the random processes $f_l(x_1)$ and $f_m(x_2)$ between the elements e_u and e_v, that is $R_{lm}^{e_u e_v}(x_1, x_2) =< f_l^{e_u}(x_1) f_m^{e_v}(x_2) >$. The random variability in the dynamic stiffness coefficients of the beam element is completely characterized in terms of the dynamic weighted integrals.

Global Dynamic Stiffness Matrix. After having obtained the element dynamic stiffness matrices in the global coordinate system, these matrices can be assembled to derive the the global dynamic stiffness matrix. The rules for assembling the element stiffness matrices are identical to those used in the traditional deterministic finite element analysis. This leads to the expression

$$\mathbf{K}_G(\omega) = \sum_{e=1}^{ne} \mathbf{D}_G^e(\omega) \tag{4.28}$$

where $\mathbf{K}_G(\omega)$ is the global dynamic stiffness matrix, ne is the number of elements and $\mathbf{D}_G^e(\omega)$ is the element dynamic stiffness matrix in the global coordinate system. The summation here implies the addition of appropriate element stiffness matrices at appropriate locations within the global stiffness matrix. The reduced global stiffness matrix $\mathbf{K}(\omega)$ can be obtained by deleting the rows and columns of $\mathbf{K}_G(\omega)$ corresponding to the fixed degrees of freedom. The equation of equilibrium is given by

$$\mathbf{K}(\omega)\mathbf{Z}(\omega) = \mathbf{F} \tag{4.29}$$

where $\mathbf{Z}(\omega)$ is the amplitude of the nodal harmonic displacement vector to be determined and $\mathbf{F}$ is the amplitude of nodal harmonic force vector and ω is the driving frequency. The reduced global dynamic stiffness matrix can be written as

$$\mathbf{K}(\omega) = \mathbf{K}^0(\omega) + \mathbf{\Delta K}(\omega) \tag{4.30}$$

where $\mathbf{K}^0(\omega)$ is the deterministic part and $\mathbf{\Delta K}(\omega)$ is the stochastic part. The deterministic part of the matrix is observed to be complex valued and symmetric in nature while the stochastic part is real valued and symmetric. The latter feature arises because (a) the element damping terms are deterministic, and (b) the shape functions are independent of any damping terms. The deterministic part is further represented by

$$\mathbf{K}^0(\omega) = \left[\mathbf{K}_R^0(\omega) + \mathrm{i}\mathbf{K}_I^0(\omega)\right] \tag{4.31}$$

where $\mathbf{K}_R^0(\omega)$ and $\mathbf{K}_I^0(\omega)$ are respectively, the real and imaginary parts of $\mathbf{K}^0(\omega)$. When element property random fields arise as Gaussian fields, the elements of stochastic part of the dynamic element stiffness matrix also become Gaussian distributed. Furthermore, since the global dynamic stiffness matrix is obtained by linear superpositioning of element stiffness matrices, it follows that elements of $\mathbf{\Delta K}(\omega)$ are also Gaussian distributed. Thus, the complete description of $\mathbf{K}(\omega)$ is given by its mean $\mathbf{K}^0(\omega)$ and the covariance matrix of the weighted integrals associated with the ne number of finite elements.

Inversion of the Global Dynamic Stiffness Matrix. The problem of determination of $\mathbf{Z}(\omega)$ requires the inversion of the global dynamic stiffness matrix. Although for a fixed value of ω the elements of $\mathbf{K}(\omega)$ are complex valued *Gaussian* random variables, upon inversion, the elements of $\mathbf{K}^{-1}(\omega)$ and consequently $\mathbf{Z}(\omega)$ will in general be non-Gaussian. Since $\mathbf{Z}(\omega)$ is complex valued, it can be described either in terms of its real and imaginary parts, or alternatively, by its amplitude and phase vectors. In the latter case a further nonlinear transformation of real and imaginary parts of $\mathbf{Z}(\omega)$ is implied. Three alternative strategies to find the inverse of $\mathbf{K}(\omega)$ will be presented. The first two approaches are analytical in nature while the last employs simulation techniques.

Random Eigenfunction Expansion Method

The equation of equilibrium (4.29) represents a set of linear random algebraic equations in $\mathbf{Z}(\omega)$. We seek an approximate solution to this equation of the form

$$\hat{\mathbf{Z}} = [\mathbf{\Phi}]\{\mathbf{a}\} \tag{4.32}$$

where $\mathbf{\Phi}$ represents a set of random basis vectors with known joint statistics and $\mathbf{a}$ is a vector of unknown complex valued deterministic constants. These unknown constants are determined by adopting a Galerkin type of error minimization scheme which leads to the expressions for the unknown $\mathbf{a}$ in terms of the statistics of elements of $\mathbf{\Phi}$. In principle any set of basis functions can be chosen to represent the unknown $\mathbf{Z}(\omega)$: in this study, these functions are taken to be the eigenvectors of the real part of the reduced global dynamic stiffness matrix. That is, the basis vectors are taken to be the eigenvectors of the matrix $\mathbf{K}_R(\omega) = \mathbf{K}_R^0(\omega) + \Delta\mathbf{K}(\omega)$. It may be noted that this matrix is symmetric, real valued, positive definite and its elements form a set of Gaussian random variables. The statistics of the eigenvectors of this matrix can be determined by solving the random eigenvalue problem given by

$$\left[\mathbf{K}_R^0(\omega) + \Delta\mathbf{K}(\omega)\right]\boldsymbol{\phi} = \lambda\boldsymbol{\phi} \tag{4.33}$$

This type of random eigenvalue problems typically arise in the determination of natural frequencies and buckling loads of randomly parametered structures and will be discussed in more details in chapter 5.

To determine the unknown constant $\mathbf{a}$, we begin by substituting $\hat{\mathbf{Z}}$ in equation (4.29) and define the error vector $\{\boldsymbol{\xi}\}$ as

$$\{\boldsymbol{\xi}\}_{(n\times 1)} = \mathbf{K}(\omega)[\mathbf{\Phi}]\{\mathbf{a}\} - \mathbf{F} = \left[\mathbf{K}^r(\omega) + \mathrm{i}\mathbf{K}_I^0(\omega)\right][\mathbf{\Phi}]\{\mathbf{a}\} - \mathbf{F} \tag{4.34}$$

Adopting the Galerkin weighted residual method, we impose the conditions

$$\left\langle \boldsymbol{\phi}_j, \boldsymbol{\xi}\right\rangle = 0; \quad \text{for} \quad j = 1, \cdots, n \tag{4.35}$$

where $\boldsymbol{\phi}_j$ is the jth column of the matrix $\mathbf{\Phi}$ which plays the role of the weighting function. This equation can be recast in a matrix form as

$$\left\langle [\mathbf{\Phi}]^T \{\boldsymbol{\xi}\}\right\rangle = 0 \tag{4.36}$$

Substituting $\{\boldsymbol{\xi}\}$ from equation (4.34) into the above equation we obtain

$$\left[< \mathbf{\Phi}^T\mathbf{K}^r(\omega)\mathbf{\Phi} > +\mathrm{i} < \mathbf{\Phi}^T\mathbf{K}_I^0(\omega)\mathbf{\Phi} >\right]\{\mathbf{a}\} =< \mathbf{\Phi}^T\mathbf{F} >$$

which leads to

$$\left[\left[\bar{\boldsymbol{\lambda}}\right] + \mathrm{i} < \mathbf{\Phi}^T\mathbf{K}_I^0(\omega)\mathbf{\Phi} >\right]\{\mathbf{a}\} =< \mathbf{\Phi}^T > \mathbf{F} \tag{4.37}$$

The above equation can further be written concisely as

$$\mathbf{G}\{\mathbf{a}\} = \mathbf{P} \tag{4.38}$$

where

$$\mathbf{G} = \left[\bar{\boldsymbol{\lambda}}\right] + \mathrm{i} < \mathbf{\Phi}^T\mathbf{K}_I^0(\omega)\mathbf{\Phi} > \quad \text{and} \quad \mathbf{P} =< \mathbf{\Phi}^T > \mathbf{F}. \tag{4.39}$$

Here $\left[\bar{\boldsymbol{\lambda}}\right]$ is a diagonal matrix containing the mean eigenvalues. The matrix $\mathbf{G}$ in the above equation can be written in the index form as

$$G_{ij} = \delta_{ij}\bar{\lambda}_i + \mathrm{i}\bar{w}_{ij}$$

where δ_{ij} is the Kronecker delta function and $\bar{w}_{ij}$ is given by

$$\bar{w}_{ij} =< w_{ij} >= \sum_{r=1}^{n}\sum_{s=1}^{n} < \phi_{ri}\phi_{sj} > {K^0}_{I_{rs}} \tag{4.40}$$

The term $< \phi_{ri}\phi_{sj} >$ in the above equation represents the correlation between rth element of ith eigenvector and sth element of jth eigenvector. This term is given by

$$< \phi_{ri}\phi_{sj} >= \sum_{l=1}^{n}\sum_{m=1}^{n} C_{rl}C_{sm}$$
$$\times \left[(1-\delta_{li})(1-\delta_{mj})\frac{1}{\mu_{li}\mu_{mj}} \sum_{p=1}^{n}\sum_{q=1}^{n}\sum_{t=1}^{n}\sum_{k=1}^{n} C_{pl}C_{qi}C_{tm}C_{kj} < \Delta K_{pq}(\omega)\Delta K_{tk}(\omega) > \right] \tag{4.41}$$

Here $\mu_{ij} = \mu_i - \mu_j$ and μ_i denotes the eigenvalues of the real deterministic matrix $\mathbf{K}_R^0(\omega)$, C_{ij} are the component of $\mathbf{C}$ which is the matrix containing the unity normalized eigenvectors of $\mathbf{K}_R^0(\omega)$. The quantity $< \Delta K_{pq}(\omega)\Delta K_{tk}(\omega) >$ denotes the correlations between the stiffness coefficients which in turn are expressible in terms of the statistics of the weighted integrals. Thus, having determined the matrix $\mathbf{G}$, the unknown constants $\mathbf{a}$ can now be determined using equation (4.38) which further leads to

$$\mathbf{Z}(\omega) = [\mathbf{\Phi}]\{\mathbf{a}\} \tag{4.42}$$

Using the first-order perturbation it can be shown that the elements of $[\mathbf{\Phi}]$, that is, $\boldsymbol{\phi}_j, j = 1, \cdots, n$, are Gaussian distributed. From equation (4.42) it follows that the solution vector $\mathbf{Z}(\omega)$, for a fixed value of ω, is also a vector of Gaussian random variables. This feature facilitates the evaluation of the statistics of the amplitude and phase vectors associate with $\mathbf{Z}(\omega)$. To achieve this, we separate $\mathbf{Z}(\omega)$ and $\mathbf{a}$ into their respective real and imaginary parts and write

$$z_i = {z^R}_i + \mathrm{i}{z^I}_i \quad \text{where} \quad {z^R}_i = \sum_{j=1}^{n} \phi_{ij}{a^R}_j, \quad \text{and} \quad {z^I}_i = \sum_{j=1}^{n} \phi_{ij}{a^I}_j \tag{4.43}$$

with

$$< {z^R}_i >= \sum_{j=1}^{n} C_{ij}{a^R}_j; \qquad < {z^I}_i >= \sum_{j=1}^{n} C_{ij}{a^I}_j$$

$$< {z^R_i}^2 >= \sum_{j=1}^{n}\sum_{k=1}^{n} < \phi_{ij}\phi_{ik} > {a^R}_j{a^R}_k; \qquad < {z^I_i}^2 >= \sum_{j=1}^{n}\sum_{k=1}^{n} < \phi_{ij}\phi_{ik} > {a^I}_j{a^I}_k$$

$$< {z^R}_i{z^I}_i >= \sum_{j=1}^{n}\sum_{k=1}^{n} < \phi_{ij}\phi_{ik} > {a^R}_j{a^I}_k \tag{4.44}$$

In the above expressions $(\bullet)^R$ and $(\bullet)^I$ are respectively the real and imaginary part of $(\bullet)$. The second order statistics of ${z^R}_i$ and ${z^I}_i$ are dependent upon joint statistics of the basis random vectors. Subsequently the moments of amplitude and phase of the elements of $\mathbf{Z}(\omega)$ are given by

$$< |z_i| >= \int_{-\infty}^{\infty}\int_{-\infty}^{\infty} \sqrt{{z^R_i}^2 + {z^I_i}^2} p_{{z^R}_i{z^I}_i}({z^R}_i, {z^I}_i)\mathrm{d}{z^R}_i\mathrm{d}{z^I}_i$$

$$< |z_i|^2 >= \int_{-\infty}^{\infty}\int_{-\infty}^{\infty} \left({z^R}_i^2 + {z^I}_i^2\right) p_{{z^R}_i {z^I}_i}({z^R}_i, {z^I}_i) \mathrm{d}{z^R}_i \mathrm{d}{z^I}_i \tag{4.45}$$

$$< arg(z_i) >= \int_{-\infty}^{\infty}\int_{-\infty}^{\infty} \{\tan^{-1}\left(\frac{z_i^I}{z_i^R}\right)\} p_{{z^R}_i {z^I}_i}({z^R}_i, {z^I}_i) \mathrm{d}{z^R}_i \mathrm{d}{z^I}_i$$

$$< (arg(z_i))^2 >= \int_{-\infty}^{\infty}\int_{-\infty}^{\infty} \{\tan^{-1}\left(\frac{z_i^I}{z_i^R}\right)\}^2 p_{{z^R}_i {z^I}_i}({z^R}_i, {z^I}_i) \mathrm{d}{z^R}_i \mathrm{d}{z^I}_i \tag{4.46}$$

In the above equations $p_{{z^R}_i {z^I}_i}({z^R}_i, {z^I}_i)$ is the two dimensional joint probability density function of ${z^R}_i$ and ${z^I}_i$. This is completely characterized by the mean, standard deviation and correlation coefficient of ${z^R}_i$ and ${z^I}_i$ which can be obtained from equations (4.44).

Complex Neumann Expansion Method

Shinozuka and Yamazaki (1998) have applied the Neumann expansion method to the inversion of static stiffness matrix in stochastic finite element applications. We begin by writing the equilibrium equation in the following form:

$$\left[\mathbf{K}_R^0(\omega) + \mathrm{i}\mathbf{K}_I^0(\omega) + \mathbf{\Delta K}(\omega)\right]\mathbf{Z}(\omega) = \mathbf{F} \tag{4.47}$$

Let $\mathbf{Z}^0(\omega)$ denote the solution in absence of system randomness. According to the Neumann expansion

$$\mathbf{Z}(\omega) = \left[\mathbf{K}^0(\omega) + \mathbf{\Delta K}(\omega)\right]^{-1}\mathbf{F} = \left[\mathbf{I} - \mathbf{R} + \mathbf{R}^2 - \mathbf{R}^3 + \cdots\right]\mathbf{Z}^0(\omega) \tag{4.48}$$

with

$$\mathbf{R} = \mathbf{K}^{0^{-1}}(\omega)\mathbf{\Delta K}(\omega) = \left[\mathbf{K}_R^0(\omega) + i\mathbf{K}_I^0(\omega)\right]^{-1}\mathbf{\Delta K}(\omega) \tag{4.49}$$

and $\mathbf{I}$ is the unit matrix. Here $\mathbf{R}$ is complex valued and is random in nature. Since the elements of $\mathbf{\Delta K}(\omega)$ are Gaussian distributed, it follows that elements of real and imaginary parts of $\mathbf{R}$ are also Gaussian distributed. If only the first-order terms are retained in the series expansion (4.48), the response vector $\mathbf{Z}(\omega)$ will have Gaussian distributed elements. Thus, retaining only the linear terms in $\mathbf{R}$, the response vector $\mathbf{Z}(\omega)$ is given by

$$\mathbf{Z}(\omega) = \left[\mathbf{I} - R\right]\mathbf{Z}^0(\omega) \quad \text{or} \quad z_i = {z^o}_i - \sum_{j=1}^{n} R_{ij}{z^o}_j \tag{4.50}$$

For notational convenience writing $\mathbf{Q} = \mathbf{K}_0^{-1}(\omega)$ equation (4.49) can be rewritten as

$$\mathbf{R} = \mathbf{Q}\mathbf{\Delta K}(\omega) \quad \text{or} \quad R_{ij} = \sum_{l=1}^{n} Q_{il}\Delta K_{lj} \tag{4.51}$$

Substituting R_{ij} from equation (4.51) into equation (4.50), the elements of the solution vector can be obtained as

$$z_i = z_i^0 - \sum_{j=1}^{n}\sum_{l=1}^{n} Q_{il} z_j \Delta K_{lj} \tag{4.52}$$

These elements are complex quantities with real and imaginary parts being Gaussian distributed. Denoting by ${z^R}_i$ and ${z^I}_i$ the real and imaginary parts of z_i respectively, we get

$$z^R{}_i = z^{0^R}{}_i - \sum_{j_1=1}^{n}\sum_{l_1=1}^{n}\Big(Q^R{}_{il_1} z^{0^R}{}_{j_1} - Q^I{}_{il_1} z^{0^I}{}_{j_1}\Big)\Delta K_{l_1 j_1} \tag{4.53}$$

$$z^I{}_i = z^{0^I}{}_i - \sum_{j_2=1}^{n}\sum_{l_2=1}^{n}\Big(Q^I{}_{il_2} z^{0^R}{}_{j_2} + Q^R{}_{il_2} z^{0^I}{}_{j_2}\Big)\Delta K_{l_2 j_2} \tag{4.54}$$

These expressions are further simplified to get

$$z^R{}_i = z^{0^R}{}_i - U^R{}_i; \qquad z^I{}_i = z^{0^I}{}_i - U^I{}_i \tag{4.55}$$

with

$$< z^R{}_i >= z^{0^R}{}_i, \qquad < z^I{}_i >= z^{0^I}{}_i, \tag{4.56}$$

$$\begin{aligned}&\mathrm{Var}(z^R{}_i) =< U^R{}_i{}^2 >= \\ &\sum_{j_1=1}^{n}\sum_{l_1=1}^{n}\sum_{j_2=1}^{n}\sum_{l_2=1}^{n}\Big(Q^R{}_{il_1} z^{0^R}{}_{j_1} - Q^I{}_{il_1} z^{0^I}{}_{j_1}\Big)\Big(Q^R{}_{il_2} z^{0^R}{}_{j_2} - Q^I{}_{il_2} z^{0^I}{}_{j_2}\Big) < \Delta K_{l_1 j_1}\Delta K_{l_2 j_2} >,\end{aligned} \tag{4.57}$$

$$\begin{aligned}&\mathrm{Var}(z^I{}_i) =< U^I{}_i{}^2 >= \\ &\sum_{j_1=1}^{n}\sum_{l_1=1}^{n}\sum_{j_2=1}^{n}\sum_{l_2=1}^{n}\Big(Q^I{}_{il_1} z^{0^R}{}_{j_1} + Q^R{}_{il_1} z^{0^I}{}_{j_1}\Big)\Big(Q^I{}_{il_2} z^{0^R}{}_{j_2} + Q^R{}_{il_2} z^{0^I}{}_{j_2}\Big) < \Delta K_{l_1 j_1}\Delta K_{l_2 j_2} >,\end{aligned} \tag{4.58}$$

and

$$\begin{aligned}&\mathrm{Cov}(z^R{}_i z^I{}_i) =< U^R{}_i U^I{}_i >= \\ &\sum_{j_1=1}^{n}\sum_{l_1=1}^{n}\sum_{j_2=1}^{n}\sum_{l_2=1}^{n}\Big(Q^R{}_{il_1} z^{0^R}{}_{j_1} - Q^I{}_{il_1} z^{0^I}{}_{j_1}\Big)\Big(Q^I{}_{il_2} z^{0^R}{}_{j_2} + Q^R{}_{il_2} z^{0^I}{}_{j_2}\Big) < \Delta K_{l_1 j_1}\Delta K_{l_2 j_2} > .\end{aligned} \tag{4.59}$$

Using the above expressions, the joint probability density function of $z^R{}_i$ and $z^I{}_i$ can be derived. Consequently, the moments of amplitude and phase of z_i can be evaluated using equations (4.45,4.46).

Combined Analytical and Simulation Method

The analytical methods presented in the preceding two sections introduce approximations at the stage of inverting the random dynamic stiffness matrix. These approximations are in addition to those involved in discretizing the random fields to formulate the element stiffness matrices and in handling damping properties. The approximations associated with inverting the matrix can be avoided if one adopts Monte Carlo simulation strategy to invert the reduced global stiffness matrix. This would require digital simulation of samples of $\mathbf{\Delta K}(\omega)$ and obtaining $\mathbf{Z}(\omega)$ by numerically inverting $\mathbf{K}(\omega)$. This method treats the inversion of the stiffness matrix in an exact manner while the other steps in the solution continue to be approximate in nature. The source of approximation here is associated with the discretization of the random fields, treatment of damping and also with the use of limited number of samples for estimating the response statistics. An advantage of this method over the other two methods presented so far is that it leads to non-Gaussian estimates for elements of $\mathbf{Z}(\omega)$.

Numerical Results And Discussions. To illustrate the relative performance of the different formulations, the harmonic response analysis of a portal frame shown in figure 4.3 is considered. The flexural rigidity and mass density for each element are taken to be independent, homogeneous, Gaussian random fields. It is assumed that $\epsilon_1 = \epsilon_2 = 0.05$, and the autocovariance of the processes $f_1(x)$ and $f_2(x)$ for all the three beam elements are taken to be of the form

$$R_{ii}(\xi) = cos\lambda_i\xi; \quad i = 1, 2 \qquad (4.60)$$

with $\lambda_i = 10\pi$ per unit length. The particular choice of autocovariance function and its parameters in this example is made only for purpose of illustrations and the theoretical results developed are expected to be valid for other forms of these functions also. It is also assumed that the random properties of distinct elements are mutually uncorrelated.

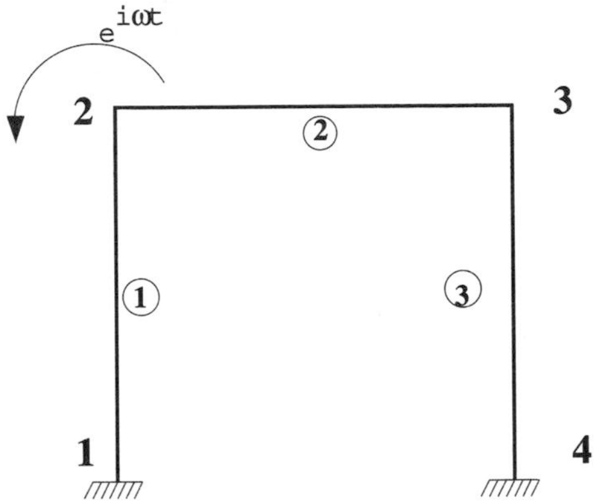

Figure 4.3. Finite element model for dynamic stiffness analysis of portal frame; EI_0,=10.0, L=1.0, m_0=0.2, ϵ_1=0.05, ϵ_2=0.05, $c_1 = 0$, c_2=1.0 (same for all the three elements).

It may be recalled that the harmonic displacement amplitudes for damped structural systems are complex valued. When the structural element is random, the displacement amplitudes can be interpreted as complex valued random processes evolving in the frequency parameter ω. Let us focus our attention on evaluating the mean and standard deviation of the amplitude and phase of the side sway as a function of the driving frequency ω. The results of theoretical analysis are compared with a 1000 samples full scale Monte Carlo simulation in figures 4.4. The algorithm used in the simulation work to simulate samples of the dynamic element stiffness matrix is as outlined by Manohar and Adhikari (1998), Sarkar and Manohar (1996). The simulation work does not involve discretization of the random fields, it treats the damping terms appearing in equation (4.5) exactly and inverts the random complex matrix in an exact way. Therefore, the simulation results serve to evaluate several aspects of the approximate analytical procedures.

The theoretical predictions generally compare well with the simulations results over the entire frequency range considered. This supports the approximations made in the treatment of system randomness and damping in deriving the element dynamic stiffness matrix and also in inverting the random global dynamic stiffness matrix to calculate the displacement amplitudes. At resonance points, the theory and simulations compare better than at the anti-resonance points. The numerical results on response statistics indicate that the amplitude and phase processes evolve in a nonstationary manner in ω. The mean results are found to closely follow the deterministic results. However,

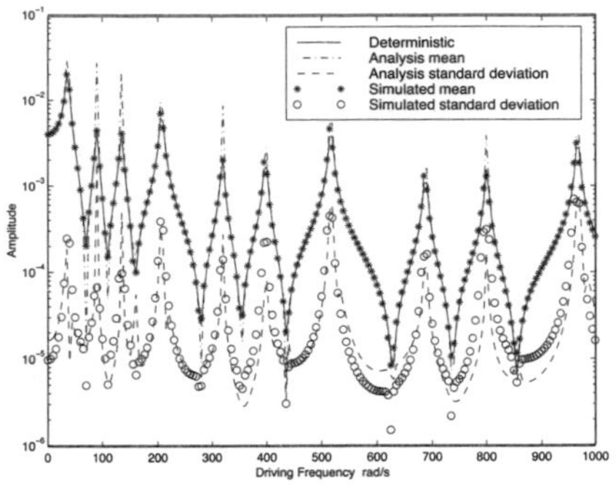

(a) Amplitude: Eigenfunction expansion method

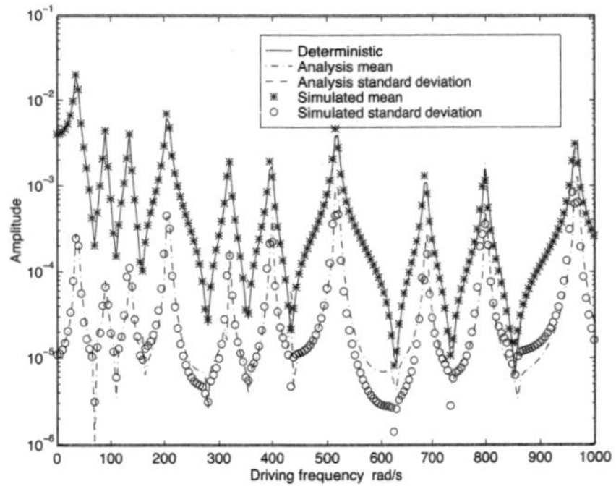

(b) Amplitude: Neumann expansion method

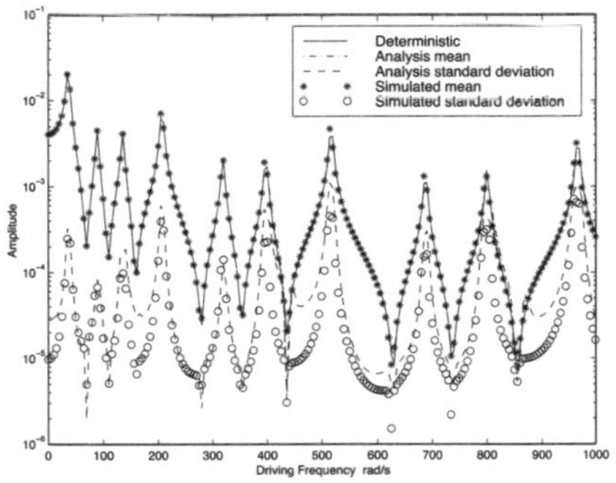

(c) Amplitude: Combined analytical-simulation method

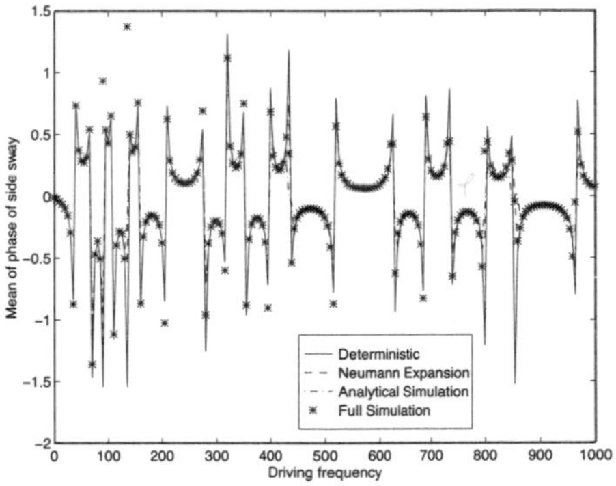

(d) Mean of phase

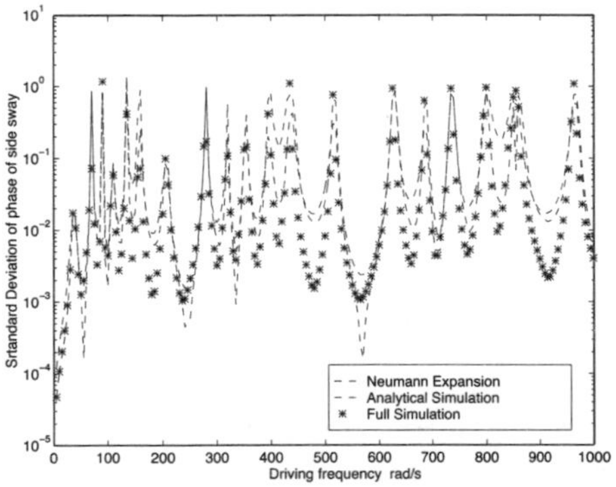

(e) Standard deviation of phase

Figure 4.4. Statistics of amplitude and phage angle of the side sway of the portal frame.

in high frequency ranges the response variability increases, which gets characterized by relatively higher values of standard deviation, especially near the resonant frequency points, where the standard deviation sometimes becomes comparable to the mean value.

This is significant, since, the standard deviations of the beam property random fields are only 5% of the corresponding mean values, which would mean that the system dynamics magnifies the structural uncertainties considerably in the high frequency ranges. This fact emphasizes the relevance of considering systems uncertainties, especially in higher driving frequency ranges.

4.5 Conclusions

The origins, quantification and propagation of uncertainties in structural dynamic models were discussed in chapter 4. The differences between aleatoric and epistemic uncertainties were pointed out and their potential sources were listed. Probabilistic model of uncertainty and its propagation methods using stochastic finite element method were discussed. As an example, harmonic vibration analysis of framed structures consisting of Euler-Bernoulli beam elements with stochastically inhomogeneous properties were considered. The element property random fields were discretized using frequency dependent shape functions. This results in a set of complex random linear algebraic equations as the governing equations of motion after the discretisation. Two analytical procedures and one combined analytical and simulation method were discussed for solutions of such equations and their performances were compared with the results from a full scale Monte Carlo simulation scheme. These comparisons were demonstrated to be reasonably good over a wide range of driving frequency. The methods discussed herein are particularly advantageous if operating frequency range lies in higher system natural frequency ranges and also when several structural modes contribute to the response.

The approach presented here bypasses the need to solve the random eigenvalue problem. However, this method is limited to one dimensional elements with simple boundary conditions for which suitable frequency-dependent shape functions can be obtained. For more general class of problems, solution using the random eigenvalues and eigenvectors is desirable. This topic is considered in the next section.

5 Random Eigenvalue Problems in Structural Dynamics

5.1 Introduction

Characterization of the natural frequencies and mode shapes play a key role in the analysis and design of engineering dynamic systems. The determination of natural frequency and mode shapes requires the solution of an eigenvalue problem. Eigenvalue problems also arise in the context of the stability analysis of structures. This problem could either be a differential eigenvalue problem or a matrix eigenvalue problem, depending on whether a continuous model or a discrete model is used to describe the given vibrating system. As discussed earlier, the descriptions of real-life engineering structural systems are inevitably associated with some amount of uncertainties in specifying material properties, geometric parameters, boundary conditions and applied loads. When we take account of these uncertainties, it is necessary to consider *random eigenvalue problems*. Several studies have been conducted on this topic since the mid-sixties. The study of probabilistic characterization of the eigensolutions of random matrix and differential operators is now an important research topic in the field of stochastic structural mechanics. The paper by Boyce (1968) and the book by Scheidt and Purkert (1983) are useful sources of information on early work in this area of research and also provide a systematic account of different approaches to random eigenvalue problems. Several review papers, for example, by Benaroya (1992), Ibrahim (1987), Manohar and Ibrahim (1999) and Manohar and Gupta (2003) have appeared in this field which summarize the current as well as the earlier works.

Here discrete linear systems or discretized continuous systems are considered. The random eigenvalue problem of undamped or proportionally damped systems can be expressed by

$$\mathbf{K}(\mathbf{x})\boldsymbol{\phi}_j = \lambda_j \mathbf{M}(\mathbf{x})\boldsymbol{\phi}_j \tag{5.1}$$

In the above equation λ_j and $\boldsymbol{\phi}_j$ are the eigenvalues and the eigenvectors of the dynamic system. $\mathbf{M}(\mathbf{x}) : \mathbb{R}^m \mapsto \mathbb{R}^{N\times N}$ and $\mathbf{K}(\mathbf{x}) : \mathbb{R}^m \mapsto \mathbb{R}^{N\times N}$, the mass and stiffness matrices, are assumed to be smooth, continuous and at least twice differentiable functions of a random parameter vector $\mathbf{x} \in \mathbb{R}^m$. The vector $\mathbf{x}$ may consist of material properties, e. g., mass density, Poisson's ratio, Young's modulus; geometric properties, *eg.*, length, thickness, and boundary conditions. The matrices $\mathbf{K}(\mathbf{x})$ and $\mathbf{M}(\mathbf{x})$ can be obtained using the stochastic finite element method (Kleiber and Hien, 1992). The statistical properties of the system are completely described by the joint probability density function $p_{\mathbf{X}}(\mathbf{x}) : \mathbb{R}^m \mapsto \mathbb{R}$. For mathematical convenience we express

$$p_{\mathbf{X}}(\mathbf{x}) = \exp\{-L(\mathbf{x})\} \tag{5.2}$$

where $-L(\mathbf{x})$ is often known as the log-likelihood function. For example, if $\mathbf{x}$ is a m-dimensional multivariate Gaussian random vector with mean $\boldsymbol{\mu} \in \mathbb{R}^m$ and covariance matrix $\boldsymbol{\Sigma} \in \mathbb{R}^{m\times m}$ then

$$L(\mathbf{x}) = \frac{m}{2}\ln(2\pi) + \frac{1}{2}\ln\|\boldsymbol{\Sigma}\| + \frac{1}{2}(\mathbf{x}-\boldsymbol{\mu})^T \boldsymbol{\Sigma}^{-1}(\mathbf{x}-\boldsymbol{\mu}) \tag{5.3}$$

It is assumed that in general the random parameters are non-Gaussian and correlated, *i.e.*, $L(\mathbf{x})$ can have any general form provided it is a smooth, continuous and at least twice differentiable function. It is further assumed that $\mathbf{M}$ and $\mathbf{K}$ are symmetric and positive definite random matrices so that all the eigenvalues are real and positive.

The central aim of studying random eigenvalue problems is to obtain the joint probability density function of the eigenvalues and the eigenvectors. The current literature on random eigenvalue problems arising in engineering systems is dominated by the mean-centered perturbation methods. These methods work well when the uncertainties are small and the parameter distribution is Gaussian. In theory, any set of non-Gaussian random variables can be transformed into a set of standard Gaussian random variables by using numerical transformations such as the Rosenblatt transformation or the Nataf transformation. These transformations are however often complicated, numerically expensive and inconvenient to implement in practice. Methods which are *not* based on mean-centered perturbation but still have the generality and computational efficiency to be applicable for engineering dynamic systems are rare. Here two new approaches will be described to obtain the moments and probability density functions of the eigenvalues. The first approach is based on a perturbation expansion of the eigenvalues about a point in the $\mathbf{x}$-space which is 'optimal' in some sense (in general different from the mean). The second approach is based on the asymptotic approximation of multidimensional integrals. These methods do not require the 'small randomness' assumption or Gaussian pdf assumption of the basic random variables often employed in literature. Moreover, they also eliminate the need to use intermediate numerical transformations of the basic random variables. In section 5.2 mean-centered perturbation methods are discussed. The perturbation method based on an optimal point is discussed in section 5.3. In section 5.4 a new method to obtain arbitrary order moments of the eigenvalues is proposed. Using these moments, some closed-form expressions of approximate pdf of the eigenvalues are derived in section 5.5. In section 5.6 the proposed analytical methods are applied to two problems and the results are compared with Monte Carlo simulations.

5.2 Classical Perturbation Methods

Perturbation Expansion. The mass and the stiffness matrices are in general non-linear functions of the random vector $\mathbf{x}$. Denote the mean of $\mathbf{x}$ as $\boldsymbol{\mu} \in \mathbb{R}$, and consider that

$$\mathbf{M}(\boldsymbol{\mu}) = \overline{\mathbf{M}}, \quad \text{and} \quad \mathbf{K}(\boldsymbol{\mu}) = \overline{\mathbf{K}} \tag{5.4}$$

are the 'deterministic parts' of the mass and stiffness matrices respectively. In general $\overline{\mathbf{M}}$ and $\overline{\mathbf{K}}$ are different from the mean matrices. The deterministic part of the eigenvalues:

$$\overline{\lambda}_j = \lambda_j(\boldsymbol{\mu}) \tag{5.5}$$

is obtained from the deterministic eigenvalue problem:

$$\overline{\mathbf{K}}\,\overline{\boldsymbol{\phi}}_j = \overline{\lambda}_j\,\overline{\mathbf{M}}\,\overline{\boldsymbol{\phi}}_j \tag{5.6}$$

The eigenvalues, $\lambda_j(\mathbf{x}) : \mathbb{R}^m \mapsto \mathbb{R}$ are non-linear functions of the parameter vector $\mathbf{x}$. If the eigenvalues are not repeated, then each $\lambda_j(\mathbf{x})$ is expected to be a smooth and twice differentiable function since the mass and stiffness matrices are smooth and twice differentiable functions of the random parameter vector. In the mean-centered perturbation approach the function $\lambda_j(\mathbf{x})$ is expanded by its Taylor series about the point $\mathbf{x} = \boldsymbol{\mu}$ as

$$\lambda_j(\mathbf{x}) \approx \lambda_j(\boldsymbol{\mu}) + \mathbf{d}_{\lambda_j}^T(\boldsymbol{\mu})\,(\mathbf{x}-\boldsymbol{\mu}) + \frac{1}{2}\,(\mathbf{x}-\boldsymbol{\mu})^T\,\mathbf{D}_{\lambda_j}(\boldsymbol{\mu})\,(\mathbf{x}-\boldsymbol{\mu}) \tag{5.7}$$

Here $\mathbf{d}_{\lambda_j}(\boldsymbol{\mu}) \in \mathbb{R}^m$ and $\mathbf{D}_{\lambda_j}(\boldsymbol{\mu}) \in \mathbb{R}^{m\times m}$ are respectively the gradient vector and the Hessian matrix of $\lambda_j(\mathbf{x})$ evaluated at $\mathbf{x} = \boldsymbol{\mu}$, that is

$$\left\{\mathbf{d}_{\lambda_j}(\boldsymbol{\mu})\right\}_k = \frac{\partial \lambda_j(\mathbf{x})}{\partial x_k}|_{\mathbf{x}=\boldsymbol{\mu}} \tag{5.8}$$

$$\text{and} \quad \left\{\mathbf{D}_{\lambda_j}(\boldsymbol{\mu})\right\}_{kl} = \frac{\partial^2 \lambda_j(\mathbf{x})}{\partial x_k\,\partial x_l}|_{\mathbf{x}=\boldsymbol{\mu}} \tag{5.9}$$

The expressions of the elements of the gradient vector and the Hessian matrix can be obtained from Plaut and Huseyin (1973). Due to equation (5.5), equation (5.7) implies that the eigenvalues are effectively expanded about their corresponding deterministic value $\overline{\lambda}_j$.

Equation (5.7) represents a quadratic form in the basic non-Gaussian random variables. The first-order perturbation, which is often used in practice, is obtained from equation (5.7) by neglecting the Hessian matrix. In this case the eigenvalues are simple linear functions of the basic random variables. This formulation is expected to produce acceptable results when the random variation in $\mathbf{x}$ is small. If the basic random variables are Gaussian then first-order perturbation results in a Gaussian distribution of the eigenvalues. In this case a closed-form expression for their joint probability density function can be obtained easily, see for example, Collins and Thomson (1969), Ramu and Ganesan (1993) and Sankar et al. (1993). However, if the elements of $\mathbf{x}$ are non-Gaussian then even the first-order perturbation method is not helpful because there is no general method to obtain the resulting pdf in a simple manner.

When the second-order terms are retained in equation (5.7) each $\lambda_j(\mathbf{x})$ results in a quadratic form in $\mathbf{x}$. If the elements of $\mathbf{x}$ are Gaussian then it is possible to obtain descriptive statistics using theory of quadratic forms as discussed next.

Eigenvalue Statistics Using the Theory of Quadratic Forms. Extensive discussions on quadratic forms in Gaussian random variables can be found in the books by Johnson and Kotz (1970)(Chapter 29) and Mathai and Provost (1992). Using the methods outlined in these references moments/cumulants of the eigenvalues are obtained in this section. Considering $\mathbf{x}$ as a multivariate Gaussian random vector, the moment generating function of $\lambda_j(\mathbf{x})$, for any $s \in \mathbb{C}$, can be obtained from (5.7) as

$$\begin{aligned} M_{\lambda_j}(s) &= \mathrm{E}\left[\exp\left\{s\lambda_j(\mathbf{x})\right\}\right] \\ &= \int_{\mathbb{R}^m} \exp\left\{s\lambda_j(\boldsymbol{\mu}) + s\mathbf{d}_{\lambda_j}^T(\boldsymbol{\mu})(\mathbf{x}-\boldsymbol{\mu}) + \frac{s}{2}(\mathbf{x}-\boldsymbol{\mu})^T \mathbf{D}_{\lambda_j}(\boldsymbol{\mu})(\mathbf{x}-\boldsymbol{\mu}) - L(\mathbf{x})\right\} d\mathbf{x} \end{aligned} \tag{5.10}$$

This integral can be evaluated exactly as

$$M_{\lambda_j}(s) = \frac{\exp\left\{s\overline{\lambda}_j + \frac{s^2}{2}\mathbf{d}_{\lambda_j}^T(\boldsymbol{\mu})\boldsymbol{\Sigma}\left[\mathbf{I} - s\boldsymbol{\Sigma}\,\mathbf{D}_{\lambda_j}(\boldsymbol{\mu})\right]^{-1}\mathbf{d}_{\lambda_j}(\boldsymbol{\mu})\right\}}{\sqrt{\left\|\mathbf{I} - s\boldsymbol{\Sigma}\,\mathbf{D}_{\lambda_j}(\boldsymbol{\mu})\right\|}} \tag{5.11}$$

To obtain the pdf of $\lambda_j(\mathbf{x})$, the inverse Laplace transform of (5.11) is required. Exact closed-form expression of the pdf can be obtained for few special cases only. Some approximate methods to obtain the pdf of $\lambda_j(\mathbf{x})$ will be discussed in section 5.5.

If mean-centered first-order perturbation is used then $\mathbf{D}_{\lambda_j}(\boldsymbol{\mu}) = \mathbf{O}$ and from equation (5.11) we obtain

$$M_{\lambda_j}(s) \approx \exp\left\{s\overline{\lambda}_j + \frac{s^2}{2}\mathbf{d}_{\lambda_j}^T(\boldsymbol{\mu})\boldsymbol{\Sigma}\,\mathbf{d}_{\lambda_j}(\boldsymbol{\mu})\right\} \tag{5.12}$$

This implies that $\lambda_j(\mathbf{x})$ is a Gaussian random variable with mean $\overline{\lambda}_j$ and variance $\mathbf{d}_{\lambda_j}^T(\boldsymbol{\mu})\boldsymbol{\Sigma}\,\mathbf{d}_{\lambda_j}(\boldsymbol{\mu})$. However, for second-order perturbation in general the mean of the eigenvalues is not the deterministic value. The cumulants of $\lambda_j(\mathbf{x})$ can be obtained from

$$\kappa_j^{(r)} = \frac{d^r}{ds^r} \ln M_{\lambda_j}(s)|_{s=0} \tag{5.13}$$

Here $\kappa_j^{(r)}$ is the rth order cumulant of jth eigenvalue and from equation (5.11) we have

$$\ln M_{\lambda_j}(s) = s\overline{\lambda}_j + \frac{s^2}{2}\mathbf{d}_{\lambda_j}^T(\boldsymbol{\mu})\boldsymbol{\Sigma}\left[\mathbf{I} - s\boldsymbol{\Sigma}\,\mathbf{D}_{\lambda_j}(\boldsymbol{\mu})\right]^{-1}\mathbf{d}_{\lambda_j}(\boldsymbol{\mu}) - \frac{1}{2}\ln\left\|\mathbf{I} - s\boldsymbol{\Sigma}\,\mathbf{D}_{\lambda_j}(\boldsymbol{\mu})\right\| \tag{5.14}$$

Using this expression and after some simplifications it can be shown that

$$\kappa_j^{(r)} = \overline{\lambda}_j + \frac{1}{2}\text{Trace}\left(\mathbf{D}_{\lambda_j}(\boldsymbol{\mu})\boldsymbol{\Sigma}\right) \quad \text{if} \quad r=1, \tag{5.15}$$

$$\kappa_j^{(r)} = \frac{r!}{2}\mathbf{d}_{\lambda_j}^T(\boldsymbol{\mu})\left[\boldsymbol{\Sigma}\,\mathbf{D}_{\lambda_j}(\boldsymbol{\mu})\right]^{r-2}\boldsymbol{\Sigma}\,\mathbf{d}_{\lambda_j}(\boldsymbol{\mu}) + \frac{(r-1)!}{2}\text{Trace}\left(\left[\mathbf{D}_{\lambda_j}(\boldsymbol{\mu})\boldsymbol{\Sigma}\right]^r\right) \quad \text{if} \quad r \geq 2 \tag{5.16}$$

The mean and first few cumulants of the eigenvalues can be explicitly obtained as

$$\widehat{\lambda}_j = \kappa_j^{(1)} = \overline{\lambda}_j + \frac{1}{2}\text{Trace}\left(\mathbf{D}_{\lambda_j}(\boldsymbol{\mu})\boldsymbol{\Sigma}\right) \tag{5.17}$$

$$\text{Var}\left[\lambda_j\right] = \kappa_j^{(2)} = \mathbf{d}_{\lambda_j}^T(\boldsymbol{\mu})\boldsymbol{\Sigma}\,\mathbf{d}_{\lambda_j}(\boldsymbol{\mu}) + \frac{1}{2}\text{Trace}\left(\left[\mathbf{D}_{\lambda_j}(\boldsymbol{\mu})\boldsymbol{\Sigma}\right]^2\right), \tag{5.18}$$

$$\kappa_j^{(3)} = 3\mathbf{d}_{\lambda_j}^T(\boldsymbol{\mu})\left[\boldsymbol{\Sigma}\,\mathbf{D}_{\lambda_j}(\boldsymbol{\mu})\right]\boldsymbol{\Sigma}\,\mathbf{d}_{\lambda_j}(\boldsymbol{\mu}) + \text{Trace}\left(\left[\mathbf{D}_{\lambda_j}(\boldsymbol{\mu})\boldsymbol{\Sigma}\right]^3\right), \tag{5.19}$$

$$\text{and } \kappa_j^{(4)} = 12\mathbf{d}_{\lambda_j}^T(\boldsymbol{\mu})\left[\boldsymbol{\Sigma}\,\mathbf{D}_{\lambda_j}(\boldsymbol{\mu})\right]^2\boldsymbol{\Sigma}\,\mathbf{d}_{\lambda_j}(\boldsymbol{\mu}) + 3\text{Trace}\left(\left[\mathbf{D}_{\lambda_j}(\boldsymbol{\mu})\boldsymbol{\Sigma}\right]^4\right) \tag{5.20}$$

From the cumulants, the raw moments $\mu_j^{(r)} = \text{E}\left[\lambda_j^r\right]$ and the central moments $\mu_j^{\prime(r)} = \text{E}\left[(\lambda_j - \bar{\lambda}_j)^r\right]$ can be obtained.

5.3 Perturbation Method Based on an Optimal Point

Perturbation Expansion. In the mean-centered perturbation method, $\lambda_j(\mathbf{x})$ is expanded in a Taylor series about $\mathbf{x} = \boldsymbol{\mu}$. This approach may not be suitable for all problems, especially if $\mathbf{x}$ is non-Gaussian then $p_{\mathbf{X}}(\mathbf{x})$ may not be centered around the mean. Here we are looking for a point $\mathbf{x} = \boldsymbol{\alpha}$ in the $\mathbf{x}$-space such that the Taylor series expansion of $\lambda_j(\mathbf{x})$ about this point

$$\lambda_j(\mathbf{x}) \approx \lambda_j(\boldsymbol{\alpha}) + \mathbf{d}_{\lambda_j}^T(\boldsymbol{\alpha})(\mathbf{x}-\boldsymbol{\alpha}) + \frac{1}{2}(\mathbf{x}-\boldsymbol{\alpha})^T\mathbf{D}_{\lambda_j}(\boldsymbol{\alpha})(\mathbf{x}-\boldsymbol{\alpha}) \tag{5.21}$$

is optimal in some sense. The optimal point $\boldsymbol{\alpha}$ can be selected in various ways. For practical applications the mean of the eigenvalues is often the most important. For this reason, the optimal point $\boldsymbol{\alpha}$ is selected such that the mean or the first moment of each eigenvalue is calculated most accurately. The mathematical formalism presented here is not restricted to this specific criteria and can be easily modified if any moment other than the first moment is required to be obtained more accurately. Using equation (5.2) the mean of $\lambda_j(\mathbf{x})$ can be obtained as

$$\widehat{\lambda}_j = \text{E}\left[\lambda_j(\mathbf{x})\right] = \int_{\mathbb{R}^m}\lambda_j(\mathbf{x})p_{\mathbf{X}}(\mathbf{x})\,d\mathbf{x} = \int_{\mathbb{R}^m}\lambda_j(\mathbf{x})e^{-L(\mathbf{x})}\,d\mathbf{x} \tag{5.22}$$

$$\text{or} \quad \widehat{\lambda}_j = \int_{\mathbb{R}^m} e^{-h_j(\mathbf{x})}\,d\mathbf{x} \tag{5.23}$$

$$\text{where} \quad h_j(\mathbf{x}) = L(\mathbf{x}) - \ln\lambda_j(\mathbf{x}) \tag{5.24}$$

Evaluation of the integral (5.23), either analytically or numerically, is in general difficult because (a) $\lambda_j(\mathbf{x})$ and $L(\mathbf{x})$ are complicated nonlinear functions of $\mathbf{x}$, (b) an explicit functional form $\lambda_j(\mathbf{x})$ is not easy to obtain except for very simple problems (usually an FE run is required to obtain λ_j for every $\mathbf{x}$), and (c) the dimension of the integral m is large. For these reasons some kind of approximation is required. From equation (5.23) note that the maximum contribution to the integral comes from the neighborhood where $h_j(\mathbf{x})$ is minimum. Therefore, expand the function $h_j(\mathbf{x})$ in a Taylor series about a point where $h_j(\mathbf{x})$ has its global minimum. By doing so the error in evaluating the integral (5.23) would be minimized. Thus, the optimal point can be obtained from

$$\frac{\partial h_j(\mathbf{x})}{\partial x_k} = 0 \quad \text{or} \quad \frac{\partial L(\mathbf{x})}{\partial x_k} = \frac{1}{\lambda_j(\mathbf{x})} \frac{\partial \lambda_j(\mathbf{x})}{\partial x_k}, \quad \forall k \tag{5.25}$$

Combining the above equations for all k, at $\mathbf{x} = \boldsymbol{\alpha}$ we have

$$\mathbf{d}_{\lambda_j}(\boldsymbol{\alpha}) = \lambda_j(\boldsymbol{\alpha})\mathbf{d}_L(\boldsymbol{\alpha}) \tag{5.26}$$

Equation (5.26) implies that at the optimal point the gradient vectors of the eigenvalues and log-likelihood function are parallel. The non-linear set of equations (5.26) have to be solved numerically. Due to the explicit analytical expression of $\mathbf{d}_{\lambda_j}$ in terms of the derivative of the mass and stiffness matrices, expensive numerical differentiation of $\lambda_j(\mathbf{x})$ at each step is not needed. Moreover, for most $p_{\mathbf{X}}(\mathbf{x})$, a closed-form expression of $\mathbf{d}_L(\mathbf{x})$ is available. For example, when $\mathbf{x}$ has multivariate Gaussian distribution, $L(\mathbf{x})$ is given by equation (5.3). By differentiating this we obtain

$$\mathbf{d}_L(\mathbf{x}) = \boldsymbol{\Sigma}^{-1}(\mathbf{x} - \boldsymbol{\mu}) \tag{5.27}$$

Substituting this in equation (5.26), the optimal point $\boldsymbol{\alpha}$ can be obtained as

$$\boldsymbol{\alpha} = \boldsymbol{\mu} + \frac{1}{\lambda_j(\boldsymbol{\alpha})}\boldsymbol{\Sigma}\,\mathbf{d}_{\lambda_j}(\boldsymbol{\alpha}) \tag{5.28}$$

This equation also gives a recipe for an iterative algorithm to obtain $\boldsymbol{\alpha}$. One starts with an initial $\boldsymbol{\alpha}$ in the right-hand side and obtains an updated $\boldsymbol{\alpha}$ in the left-hand side. This procedure can be continued till the difference between the values of $\boldsymbol{\alpha}$ obtained from both sides of (5.28) is less than (L_2 vector norm can be used to measure the difference) a predefined small value. A good value to start the iteration process is $\boldsymbol{\alpha} = \boldsymbol{\mu}$, as in the case of mean-centered approach. The form of equation (5.21) is similar to that of equation (5.7). As mentioned before, when the basic random variables are non-Gaussian, determination of moments and pdf is not straightforward. However, when $\mathbf{x}$ is Gaussian, some useful statistics of the eigenvalues can be obtained in closed-form.

Eigenvalue Statistics Using the Theory of Quadratic Forms. For notational convenience we rewrite the optimal perturbation expansion (5.21) as

$$\lambda_j(\mathbf{x}) \approx c_j + \mathbf{a}_j^T\mathbf{x} + \frac{1}{2}\mathbf{x}^T\mathbf{A}_j\mathbf{x} \tag{5.29}$$

where the constants $c_j \in \mathbb{R}$, $\mathbf{a}_j \in \mathbb{R}^m$ and $\mathbf{A}_j \in \mathbb{R}^{m\times m}$ are given by

$$c_j = \lambda_j(\boldsymbol{\alpha}) - \mathbf{d}_{\lambda_j}^T(\boldsymbol{\alpha})\boldsymbol{\alpha} + \frac{1}{2}\boldsymbol{\alpha}^T\mathbf{D}_{\lambda_j}(\boldsymbol{\alpha})\boldsymbol{\alpha} \tag{5.30}$$

$$\mathbf{a}_j = \mathbf{d}_{\lambda_j}(\boldsymbol{\alpha}) - \mathbf{D}_{\lambda_j}(\boldsymbol{\alpha})\boldsymbol{\alpha} \tag{5.31}$$

$$\mathbf{A}_j = \mathbf{D}_{\lambda_j}(\boldsymbol{\alpha}) \tag{5.32}$$

From equation (5.29), the closed-form expression of the moment generating function of $\lambda_j(\mathbf{x})$ can be obtained exactly in a way similar to what discussed for the case of mean-centered perturbation method. Using equation (5.3) it can be shown that

$$M_{\lambda_j}(s) = \mathrm{E}\left[\exp\left\{s\lambda_j(\mathbf{x})\right\}\right]$$
$$= (2\pi)^{-m/2}\|\boldsymbol{\Sigma}\|^{-1/2} \int_{\mathbb{R}^m} \exp\left\{ s\left(c_j + \mathbf{a}_j^T\mathbf{x} + \frac{1}{2}\mathbf{x}^T\mathbf{A}_j\mathbf{x}\right) - \frac{1}{2}\left(\mathbf{x}-\boldsymbol{\mu}\right)^T\boldsymbol{\Sigma}^{-1}\left(\mathbf{x}-\boldsymbol{\mu}\right)\right\} d\mathbf{x} \tag{5.33}$$

This m-dimensional integral can be evaluated exactly to obtain

$$M_{\lambda_j}(s) = \frac{\exp\left\{sc_j - \frac{1}{2}\boldsymbol{\mu}^T\boldsymbol{\Sigma}^{-1}\boldsymbol{\mu} + \frac{1}{2}\left(\boldsymbol{\mu}+s\boldsymbol{\Sigma}\,\mathbf{a}_j\right)^T\left[\mathbf{I}-s\boldsymbol{\Sigma}\,\mathbf{A}_j\right]^{-1}\left(\boldsymbol{\mu}+s\boldsymbol{\Sigma}\,\mathbf{a}_j\right)\right\}}{\sqrt{\|\mathbf{I}-s\boldsymbol{\Sigma}\,\mathbf{A}_j\|}} \tag{5.34}$$

From the preceding expression, the cumulants of the eigenvalues can be evaluated using equation (5.13) as

$$\kappa_j^{(r)} = c_j + \frac{1}{2}\mathrm{Trace}\left(\mathbf{A}_j\boldsymbol{\Sigma}\right) + \frac{1}{2}\boldsymbol{\mu}^T\mathbf{A}_j\,\boldsymbol{\mu} + \mathbf{a}_j^T\boldsymbol{\mu} \quad \text{if} \quad r=1, \tag{5.35}$$

$$\text{and} \quad \kappa_j^{(r)} = \frac{(r-1)!}{2}\mathrm{Trace}\left(\left[\mathbf{A}_j\boldsymbol{\Sigma}\right]^r\right) + \frac{r!}{2}\mathbf{a}_j^T\left[\boldsymbol{\Sigma}\,\mathbf{A}_j\right]^{r-2}\boldsymbol{\Sigma}\,\mathbf{a}_j$$
$$+\, r!\,\boldsymbol{\mu}^T\left[\mathbf{A}_j\boldsymbol{\Sigma}\right]^{r-1}\mathbf{A}_j\,\boldsymbol{\mu} + r!\,\mathbf{a}_j^T\left[\boldsymbol{\Sigma}\,\mathbf{A}_j\right]^{r-1}\mathbf{A}_j\,\boldsymbol{\mu} \quad \text{if} \quad r \geq 2 \tag{5.36}$$

The mean and first few cumulants of the eigenvalues can be explicitly obtained as

$$\widehat{\lambda}_j = \kappa_j^{(1)} = c_j + \frac{1}{2}\mathrm{Trace}\left(\mathbf{A}_j\boldsymbol{\Sigma}\right) + \frac{1}{2}\boldsymbol{\mu}^T\mathbf{A}_j\,\boldsymbol{\mu} + \mathbf{a}_j^T\boldsymbol{\mu} \tag{5.37}$$

$$\mathrm{Var}\left[\lambda_j\right] = \kappa_j^{(2)} = \frac{1}{2}\mathrm{Trace}\left(\left[\mathbf{A}_j\boldsymbol{\Sigma}\right]^2\right) + \mathbf{a}_j^T\boldsymbol{\Sigma}\,\mathbf{a}_j + +2\,\boldsymbol{\mu}^T\left[\mathbf{A}_j\boldsymbol{\Sigma}\right]\mathbf{A}_j\,\boldsymbol{\mu} + 2\,\mathbf{a}_j^T\left[\boldsymbol{\Sigma}\,\mathbf{A}_j\right]\mathbf{A}_j\,\boldsymbol{\mu}, \tag{5.38}$$

$$\kappa_j^{(3)} = \mathrm{Trace}\left(\left[\mathbf{A}_j\boldsymbol{\Sigma}\right]^3\right) + 3\mathbf{a}_j^T\left[\boldsymbol{\Sigma}\,\mathbf{A}_j\right]\mathbf{a}_j + 6\,\boldsymbol{\mu}^T\left[\mathbf{A}_j\boldsymbol{\Sigma}\right]^2\mathbf{A}_j\,\boldsymbol{\mu} + 6\,\mathbf{a}_j^T\left[\boldsymbol{\Sigma}\,\mathbf{A}_j\right]^2\mathbf{A}_j\,\boldsymbol{\mu}, \tag{5.39}$$

$$\text{and}\ \kappa_j^{(4)} = 3\mathrm{Trace}\left(\left[\mathbf{A}_j\boldsymbol{\Sigma}\right]^4\right) + 12\mathbf{a}_j^T\left[\boldsymbol{\Sigma}\,\mathbf{A}_j\right]^2\mathbf{a}_j + 24\,\boldsymbol{\mu}^T\left[\mathbf{A}_j\boldsymbol{\Sigma}\right]^3\mathbf{A}_j\,\boldsymbol{\mu} + 24\,\mathbf{a}_j^T\left[\boldsymbol{\Sigma}\,\mathbf{A}_j\right]^3\mathbf{A}_j\,\boldsymbol{\mu} \tag{5.40}$$

Since equations (5.15), (5.16) and (5.35), (5.36) give cumulants of arbitrary order, it is possible to construct the pdf of the eigenvalues from them. However, when the elements of $\mathbf{x}$ are non-Gaussian then neither the first-order perturbation nor the second-order perturbation methods are helpful because there is no general method to obtain the resulting statistics in a simple manner. In such cases the method outlined in the next section might be more useful.

5.4 Method Based on the Asymptotic Integral

Multidimensional Integrals in Unbounded Domains. In this section the moments of the eigenvalues are obtained based on the asymptotic approximation of the multidimensional integral. Consider a function $f(\mathbf{x}) : \mathbb{R}^m \mapsto \mathbb{R}$ which is smooth and at least twice differentiable. Suppose we want to evaluate an integral of the following form:

$$\mathcal{J} = \int_{\mathbb{R}^m} \exp\left\{-f(\mathbf{x})\right\} d\mathbf{x} \tag{5.41}$$

This is a m-dimensional integral over the unbounded domain $\mathbb{R}^m$. The maximum contribution to this integral comes from the neighborhood where $f(\mathbf{x})$ reaches its global minimum. Suppose that $f(\mathbf{x})$ reaches its global minimum at an *unique* point $\boldsymbol{\theta} \in \mathbb{R}^m$. Therefore, at $\mathbf{x} = \boldsymbol{\theta}$

$$\frac{\partial f(\mathbf{x})}{\partial x_k} = 0, \forall k \quad \text{or} \quad \mathbf{d}_f(\boldsymbol{\theta}) = \mathbf{0} \tag{5.42}$$

Using this, expand $f(\mathbf{x})$ in a Taylor series about $\boldsymbol{\theta}$ and rewrite equation (5.41) as

$$\begin{aligned}\mathcal{J} &= \int_{\mathbb{R}^m} \exp\left\{-\left\{f(\boldsymbol{\theta}) + \frac{1}{2}(\mathbf{x}-\boldsymbol{\theta})^T \mathbf{D}_f(\boldsymbol{\theta})(\mathbf{x}-\boldsymbol{\theta}) + \varepsilon(\mathbf{x},\boldsymbol{\theta})\right\}\right\} d\mathbf{x} \\ &= \exp\{-f(\boldsymbol{\theta})\} \int_{\mathbb{R}^m} \exp\left\{-\frac{1}{2}(\mathbf{x}-\boldsymbol{\theta})^T \mathbf{D}_f(\boldsymbol{\theta})(\mathbf{x}-\boldsymbol{\theta}) - \varepsilon(\mathbf{x},\boldsymbol{\theta})\right\} d\mathbf{x}\end{aligned} \tag{5.43}$$

where $\varepsilon(\mathbf{x},\boldsymbol{\theta})$ is the error if only the terms up to second-order were retained in the Taylor series expansion. With suitable scaling of $\mathbf{x}$ the integral in (5.41) can be transformed to the so called 'Laplace integral'. Under special conditions such integrals can be well approximated using asymptotic methods. The relevant mathematical methods and formal derivations are covered in detail in the books by Bleistein and Handelsman (1994) and Wong (2001). Here we propose a somewhat different version of asymptotic integrals. The error $\varepsilon(\mathbf{x},\boldsymbol{\theta})$ depends on higher order derivatives of $f(\mathbf{x})$ at $\mathbf{x} = \boldsymbol{\theta}$. If they are small compared to $f(\boldsymbol{\theta})$ and the elements of $\mathbf{D}_f(\boldsymbol{\theta})$, their contribution will be negligible to the value of the integral. Therefore, we assume $f(\boldsymbol{\theta})$ and the elements of $\mathbf{D}_f(\boldsymbol{\theta})$ are large so that

$$\left|\frac{1}{f(\boldsymbol{\theta})}\mathcal{D}^{(j)}(f(\boldsymbol{\theta}))\right| \to 0 \quad \text{and} \quad \forall k,l, \left|\frac{1}{[\mathbf{D}_f(\boldsymbol{\theta})]_{kl}}\mathcal{D}^{(j)}(f(\boldsymbol{\theta}))\right| \to 0 \quad \text{for} \quad j > 2 \tag{5.44}$$

where $\mathcal{D}^{(j)}(f(\boldsymbol{\theta}))$ is jth order derivative of $f(\mathbf{x})$ evaluated at $\mathbf{x} = \boldsymbol{\theta}$. Under such assumptions $\varepsilon(\mathbf{x},\boldsymbol{\theta}) \to 0$. Therefore, the integral in (5.43) can be approximated as

$$\mathcal{J} \approx \exp\{-f(\boldsymbol{\theta})\} \int_{\mathbb{R}^m} \exp\left\{-\frac{1}{2}(\mathbf{x}-\boldsymbol{\theta})^T \mathbf{D}_f(\boldsymbol{\theta})(\mathbf{x}-\boldsymbol{\theta})\right\} d\mathbf{x} \tag{5.45}$$

If $\boldsymbol{\theta}$ is the global minimum of $f(\mathbf{x})$ in $\mathbb{R}^m$, the symmetric Hessian matrix $\mathbf{D}_f(\boldsymbol{\theta}) \in \mathbb{R}^{m\times m}$ is also expected to be positive definite. The integral in equation (5.45) can be evaluated as

$$\mathcal{J} \approx (2\pi)^{m/2} \exp\{-f(\boldsymbol{\theta})\} \|\mathbf{D}_f(\boldsymbol{\theta})\|^{-1/2} \tag{5.46}$$

This approximation is expected to yield good result if the minima of $f(\mathbf{x})$ around $\mathbf{x} = \boldsymbol{\theta}$ is sharp. If $\boldsymbol{\theta}$ is not unique then it is required to sum the contributions arising from all such optimal points separately. Equation (5.46) will now be used to obtain moments of the eigenvalues.

Calculation of Arbitrary Moments of The Eigenvalues. An arbitrary rth order moment of the eigenvalues can be obtained from

$$\begin{aligned}\mu_j^{(r)} &= \mathrm{E}\left[\lambda_j^r(\mathbf{x})\right] = \int_{\mathbb{R}^m} \lambda_j^r(\mathbf{x}) p_{\mathbf{x}}(\mathbf{x})\, d\mathbf{x} \\ &= \int_{\mathbb{R}^m} \exp\{-(L(\mathbf{x}) - r\ln\lambda_j(\mathbf{x}))\}\, d\mathbf{x}, \quad r = 1,2,3\cdots\end{aligned} \tag{5.47}$$

The equation can be expressed in the form of equation (5.41) by choosing

$$f(\mathbf{x}) = L(\mathbf{x}) - r \ln \lambda_j(\mathbf{x}) \tag{5.48}$$

Differentiating the above equation with respect to x_k we obtain

$$\frac{\partial f(\mathbf{x})}{\partial x_k} = \frac{\partial L(\mathbf{x})}{\partial x_k} - \frac{r}{\lambda_j(\mathbf{x})} \frac{\partial \lambda_j(\mathbf{x})}{\partial x_k} \tag{5.49}$$

The optimal point $\boldsymbol{\theta}$ can be obtained from (5.42) by equating the above expression to zero. Therefore at $\mathbf{x} = \boldsymbol{\theta}$

$$\frac{\partial f(\mathbf{x})}{\partial x_k} = 0, \quad \forall k \tag{5.50}$$

$$\text{or} \quad \frac{r}{\lambda_j(\boldsymbol{\theta})} \frac{\partial \lambda_j(\boldsymbol{\theta})}{\partial x_k} = \frac{\partial L(\boldsymbol{\theta})}{\partial x_k}, \quad \forall k \tag{5.51}$$

$$\text{or} \quad \mathbf{d}_{\lambda_j}(\boldsymbol{\theta}) r = \lambda_j(\boldsymbol{\theta}) \mathbf{d}_L(\boldsymbol{\theta}) \tag{5.52}$$

Equation (5.52) is similar to equation (5.26) and needs to be solved numerically to obtain $\boldsymbol{\theta}$. The elements of the Hessian matrix $\mathbf{D}_f(\boldsymbol{\theta})$ can be obtained by differentiating equation (5.49) with respect to x_l:

$$\begin{aligned} \frac{\partial^2 f(\mathbf{x})}{\partial x_k \, \partial x_l} &= \frac{\partial^2 L(\mathbf{x})}{\partial x_k \, \partial x_l} - r \left(-\frac{1}{\lambda_j^2(\mathbf{x})} \frac{\partial \lambda_j(\mathbf{x})}{\partial x_l} \frac{\partial \lambda_j(\mathbf{x})}{\partial x_k} + \frac{1}{\lambda_j(\mathbf{x})} \frac{\partial^2 \lambda_j(\mathbf{x})}{\partial x_k \, \partial x_l} \right) \\ &= \frac{\partial^2 L(\mathbf{x})}{\partial x_k \, \partial x_l} + \frac{1}{r} \left\{ \frac{r}{\lambda_j(\mathbf{x})} \frac{\partial \lambda_j(\mathbf{x})}{\partial x_k} \right\} \left\{ \frac{r}{\lambda_j(\mathbf{x})} \frac{\partial \lambda_j(\mathbf{x})}{\partial x_l} \right\} - \frac{r}{\lambda_j(\mathbf{x})} \frac{\partial^2 \lambda_j(\mathbf{x})}{\partial x_k \, \partial x_l} \end{aligned} \tag{5.53}$$

At $\mathbf{x} = \boldsymbol{\theta}$ we can use equation (5.51) so that equation (5.53) reads

$$\frac{\partial^2 f(\mathbf{x})}{\partial x_k \, \partial x_l}\Big|_{\mathbf{x}=\boldsymbol{\theta}} = \frac{\partial^2 L(\boldsymbol{\theta})}{\partial x_k \, \partial x_l} + \frac{1}{r} \frac{\partial L(\boldsymbol{\theta})}{\partial x_k} \frac{\partial L(\boldsymbol{\theta})}{\partial x_l} - \frac{r}{\lambda_j(\boldsymbol{\theta})} \frac{\partial^2 \lambda_j(\boldsymbol{\theta})}{\partial x_k \, \partial x_l} \tag{5.54}$$

Combining this equation for all k and l we have

$$\mathbf{D}_f(\boldsymbol{\theta}) = \mathbf{D}_L(\boldsymbol{\theta}) + \frac{1}{r} \mathbf{d}_L(\boldsymbol{\theta}) \mathbf{d}_L(\boldsymbol{\theta})^T - \frac{r}{\lambda_j(\boldsymbol{\theta})} \mathbf{D}_{\lambda_j}(\boldsymbol{\theta}) \tag{5.55}$$

where $\mathbf{D}_{\lambda_j}(\bullet)$ is defined in equation (5.9). Using the asymptotic approximation (5.46), the rth moment of the eigenvalues can be obtained as

$$\mu_j^{(r)} \approx (2\pi)^{m/2} \lambda_j^r(\boldsymbol{\theta}) \exp\left\{-L(\boldsymbol{\theta})\right\} \left\| \mathbf{D}_L(\boldsymbol{\theta}) + \frac{1}{r} \mathbf{d}_L(\boldsymbol{\theta}) \mathbf{d}_L(\boldsymbol{\theta})^T - \frac{r}{\lambda_j(\boldsymbol{\theta})} \mathbf{D}_{\lambda_j}(\boldsymbol{\theta}) \right\|^{-1/2} \tag{5.56}$$

This is perhaps the most general formula to obtain the moments of the eigenvalues of linear stochastic dynamic systems. The optimal point $\boldsymbol{\theta}$ needs to be calculated by solving non-linear set of equations (5.52) for each λ_j and r. Several special cases arising from equation (5.56) are of practical interest:

- *Mean of the eigenvalues:* The mean of the eigenvalues can be obtained by substituting $r = 1$ in equation (5.56), that is

$$\widehat{\lambda}_j = \mu_j^{(1)} = (2\pi)^{m/2} \lambda_j(\boldsymbol{\theta}) \exp\left\{-L(\boldsymbol{\theta})\right\} \left\| \mathbf{D}_L(\boldsymbol{\theta}) + \mathbf{d}_L(\boldsymbol{\theta}) \mathbf{d}_L(\boldsymbol{\theta})^T - \mathbf{D}_{\lambda_j}(\boldsymbol{\theta}) / \lambda_j(\boldsymbol{\theta}) \right\|^{-1/2} \tag{5.57}$$

- *Central moments of the eigenvalues:* Once the mean in known, the central moments can be expressed in terms of the raw moments $\mu_j^{(r)}$ using the binomial transform

$$\mu_j^{\prime(r)} = \mathrm{E}\left[\left(\lambda_j - \widehat{\lambda}_j\right)^r\right] = \sum_{k=0}^{r} \binom{r}{k} (-1)^{r-k} \mu_j^{(k)} \widehat{\lambda}_j^{r-k} \tag{5.58}$$

- *Random vector* $\boldsymbol{x}$ *has multivariate Gaussian distribution:* In this case $L(\mathbf{x})$ is given by equation (5.3) and by differentiating equation (5.27) we obtain

$$\text{and} \quad \mathbf{D}_L(\mathbf{x}) = \boldsymbol{\Sigma}^{-1} \tag{5.59}$$

The optimal point $\boldsymbol{\theta}$ can be obtained from equation (5.52) as

$$\boldsymbol{\theta} = \boldsymbol{\mu} + \frac{r}{\lambda_j(\boldsymbol{\theta})} \boldsymbol{\Sigma}\, \mathbf{d}_{\lambda_j}(\boldsymbol{\theta}) \tag{5.60}$$

Using equation (5.27) and equation (5.59), the Hessian matrix can be derived from equation (5.55) as

$$\begin{aligned} \mathbf{D}_f(\boldsymbol{\theta}) &= \boldsymbol{\Sigma}^{-1} + \frac{1}{r} \boldsymbol{\Sigma}^{-1} (\boldsymbol{\theta} - \boldsymbol{\mu}) (\boldsymbol{\theta} - \boldsymbol{\mu})^T \boldsymbol{\Sigma}^{-1} - \frac{r}{\lambda_j(\boldsymbol{\theta})} \mathbf{D}_{\lambda_j}(\boldsymbol{\theta}) \\ &= \boldsymbol{\Sigma}^{-1} \left(\mathbf{I} + \frac{1}{r} (\boldsymbol{\theta} - \boldsymbol{\mu}) (\boldsymbol{\theta} - \boldsymbol{\mu})^T \boldsymbol{\Sigma}^{-1}\right) - \frac{r}{\lambda_j(\boldsymbol{\theta})} \mathbf{D}_{\lambda_j}(\boldsymbol{\theta}) \end{aligned} \tag{5.61}$$

Therefore, the rth moment of the eigenvalues can be obtained from equation (5.56) as

$$\mu_j^{(r)} \approx \lambda_j^r(\boldsymbol{\theta}) \exp\left\{-\frac{1}{2} (\boldsymbol{\theta} - \boldsymbol{\mu})^T \boldsymbol{\Sigma}^{-1} (\boldsymbol{\theta} - \boldsymbol{\mu})\right\} \|\boldsymbol{\Sigma}\|^{-1/2} \|\mathbf{D}_f(\boldsymbol{\theta})\|^{-1/2} \tag{5.62}$$

The probability density function of the eigenvalues is considered in the next section.

5.5 Probability Density Function of the Eigenvalues

Maximum Entropy Probability Density Function. Once the cumulants/moments of the eigenvalues are known, the pdf of the eigenvalues can be obtained using Maximum Entropy Method (MEM). Because equations (5.15), (5.16), (5.35), (5.36) and (5.56) can be used to calculate any arbitrary order cumulant and moment, the pdf can be obtained accurately by taking higher order terms. Here, following Kapur and Kesavan (1992), a general approach is presented.

Since $\mathbf{M}$ and $\mathbf{K}$ are symmetric and positive definite random matrices, all the eigenvalues are real and positive. Suppose the pdf of λ_j is given by $p_{\lambda_j}(u)$ where $u \in \mathbb{R}$ is positive, that is $u \in [0, \infty]$. Considering that only first n moments are used, the pdf of each eigenvalue must satisfy the following constraints:

$$\int_0^\infty p_{\lambda_j}(u) du = 1 \tag{5.63}$$

$$\text{and} \quad \int_0^\infty u^r p_{\lambda_j}(u) du = \mu_j^{(r)}, \quad r = 1, 2, 3, \cdots, n \tag{5.64}$$

Using Shannon's measure of entropy

$$\mathcal{S} = -\int_0^\infty p_{\lambda_j}(u) \ln p_{\lambda_j}(u) du \tag{5.65}$$

we construct the Lagrangian

$$\mathcal{L} = -\int_0^\infty p_{\lambda_j}(u)\ln p_{\lambda_j}(u)du - (\rho_0 - 1)\left[\int_0^\infty p_{\lambda_j}(u)du - 1\right] - \sum_{r=1}^{n}\rho_r\left[\int_0^\infty u^r p_{\lambda_j}(u)du - \mu_j^{(r)}\right] \tag{5.66}$$

where $\rho_r, r = 0, 1, 2, \cdots, n$ are Lagrange multipliers. The function $p_{\lambda_j}(u)$ which maximizes $\mathcal{L}$ can be obtained using the calculus of variations. Using the Euler-Lagrange equation the solution is given by

$$p_{\lambda_j}(u) = \exp\left\{-\rho_0 - \sum_{i=1}^{n}\rho_i u^i\right\} = \exp\{-\rho_0\}\exp\left\{-\sum_{i=1}^{n}\rho_i u^i\right\}, \quad u \geq 0 \tag{5.67}$$

The Lagrange multipliers can be obtained from the constraint equations (5.63) and (5.64) as

$$\exp\{\rho_0\} = \int_0^\infty \exp\left\{-\sum_{i=1}^{n}\rho_i u^i\right\} du \tag{5.68}$$

$$\text{and} \quad \exp\{\rho_0\}\mu_j^{(r)} = \int_0^\infty u^r \exp\left\{-\sum_{i=1}^{n}\rho_i u^i\right\} du, \quad \text{for} \quad r = 0, 1, 2, \cdots n \tag{5.69}$$

Closed-form expressions for ρ_r are in general not possible for all n. If we take $n = 2$, then the resulting pdf can be expressed by a truncated Gaussian density function

$$p_{\lambda_j}(u) = \frac{1}{\sqrt{2\pi}\sigma_j\,\Phi\left(\widehat{\lambda}_j/\sigma_j\right)}\exp\left\{-\frac{\left(u - \widehat{\lambda}_j\right)^2}{2\sigma_j^2}\right\}, \quad u \geq 0 \tag{5.70}$$

where σ_j is given by

$$\sigma_j^2 = \mu_j^{(2)} - \widehat{\lambda}_j^2 \tag{5.71}$$

The approach presented above can also be used in conjunction with the perturbation methods by transforming the cumulants obtained from equations (5.15), (5.16), (5.35) and (5.36) to moments. The truncated Gaussian density function derived here ensures that the probability of any eigenvalues becoming negative is zero.

Approximation by χ^2 Probability Density Function. We use an approximation analogous to Pearson's (Pearson, 1959) three moment central χ^2 approximation to the distribution of a noncentral χ^2 . The pdf of the eigenvalues are approximated as

$$p_{\lambda_j}(u) \approx \eta_j + \gamma_j \chi_{\nu_j}^2(u) \tag{5.72}$$

where $\chi_{\nu_j}^2(u)$ is a central χ^2 density function with ν_j degrees-of-freedom (see Abramowitz and Stegun, 1965, Chapter 26). The constants η_j, γ_j, and ν_j are obtained such that the first three moments of λ_j are equal to that of the approximated χ^2 pdf. The moment generating function of the approximated χ^2 pdf is given by

$$\mathrm{E}\left[\exp\left\{-s\left(\eta_j + \gamma_j \chi_{\nu_j}^2\right)\right\}\right] = \exp\{-s\eta_j\}(1 + 2s\gamma_j)^{-\nu_j/2} \tag{5.73}$$

Equating the first three moments we have

$$\eta_j + \nu_j\gamma_j = \mu_j^{(1)}, \tag{5.74}$$

$$\eta_j{}^2 + 2\eta_j\nu_j\gamma_j + \nu_j{}^2\gamma_j{}^2 + 2\nu_j\gamma_j{}^2 = \mu_j^{(2)} \tag{5.75}$$

and $$\eta_j{}^3 + 3\eta_j{}^2\nu_j\gamma_j + 3\eta_j\nu_j{}^2\gamma_j{}^2 + 6\eta_j\nu_j\gamma_j{}^2 + \nu_j{}^3\gamma_j{}^3 + 6\nu_j{}^2\gamma_j{}^3 + 8\nu_j\gamma_j{}^3 = \mu_j^{(3)} \tag{5.76}$$

This set of coupled non-linear equations can be solved exactly in closed-form to obtain η_j, γ_j, and ν_j:

$$\eta_j = \frac{\mu_j^{(1)^2}\mu_j^{(2)} - 2\,\mu_j^{(2)^2} + \mu_j^{(1)}\mu_j^{(3)}}{2\,\mu_j^{(1)^3} - 3\,\mu_j^{(1)}\mu_j^{(2)} + \mu_j^{(3)}} \tag{5.77}$$

$$\gamma_j = \frac{2\,\mu_j^{(1)^3} - 3\,\mu_j^{(1)}\mu_j^{(2)} + \mu_j^{(3)}}{4\left(\mu_j^{(2)} - \mu_j^{(1)^2}\right)}, \tag{5.78}$$

and $$\nu_j = 8\,\frac{\left(\mu_j^{(2)} - \mu_j^{(1)^2}\right)^3}{\left(2\,\mu_j^{(1)^3} - 3\,\mu_j^{(1)}\mu_j^{(2)} + \mu_3'\right)^2} \tag{5.79}$$

Moments of $\lambda_j(\mathbf{x})$ obtained in equation (5.56), can be used directly in the right-hand side of these equations. Alternatively, this approach can also be used in conjunction with the perturbation methods by transforming the cumulants obtained from equations (5.15), (5.16), (5.35) and (5.36) to moments. Using the transformation in equation (5.72) the approximate probability density function of $\lambda_j(\mathbf{x})$ is given by

$$p_{\lambda_j}(u) \approx \frac{1}{\gamma_j} p_{\chi^2_{\nu_j}}\left(\frac{u-\eta_j}{\gamma_j}\right) = \frac{(u-\eta_j)^{\nu_j/2-1}\exp\{-(u-\eta_j)/2\gamma_j\}}{(2\gamma_j)^{\nu_j/2}\Gamma(\nu_j/2)} \tag{5.80}$$

The two approximated pdfs proposed here have simple forms but it should be noted that they are not exhaustive. Given the moments/cumulants, different probability density functions can be fitted using different methods. Application of the approximate pdfs derived here is illustrated in the next section.

5.6 Numerical Examples

A two DOF system.

System model and computational methodology

A simple two-degree-of-freedom undamped system has been considered to illustrate a possible application of the expressions developed so far. The main purpose of this example is to understand how the proposed methods compare with the existing methods. Figure 5.1 shows the example, together with the numerical values of the masses and spring stiffnesses. The system matrices for the example are given by

$$\mathbf{M} = \begin{bmatrix} m_1 & 0 \\ 0 & m_1 \end{bmatrix} \quad \text{and} \quad \mathbf{K} = \begin{bmatrix} k_1 + k_3 & -k_3 \\ -k_3 & k_2 + k_3 \end{bmatrix} \tag{5.81}$$

It is assumed that only the stiffness parameters k_1 and k_2 are uncertain so that $k_i = \bar{k}_i(1 + \epsilon_i x_i)$, $i = 1, 2$ and $\bar{k}_i$ denote the deterministic values of the spring constants.

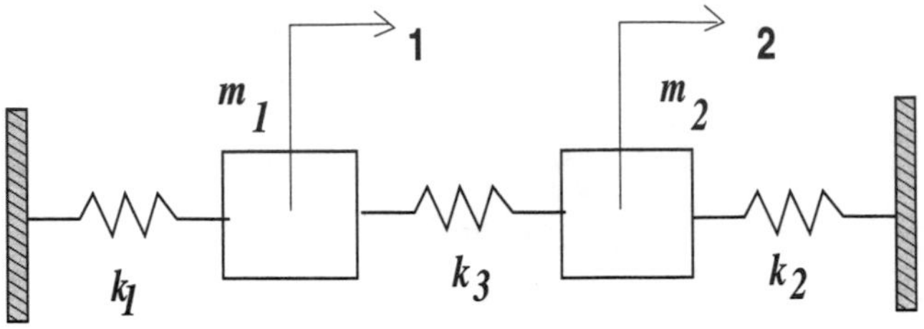

Figure 5.1. The undamped two degree-of-system system, $m_1 = 1$ Kg, $m_2 = 1.5$ Kg, $\bar{k}_1 = 1000$ N/m, $\bar{k}_2 = 1100$ N/m and $k_3 = 100$ N/m.

Here $\mathbf{x} = \{x_1, x_2\}^T \in \mathbb{R}^2$ is a vector of standard Gaussian random variables, that is $\boldsymbol{\mu} = \mathbf{0}$ and $\boldsymbol{\Sigma} = \mathbf{I}$. The numerical values of the 'strength parameters' are considered as $\epsilon_1 = \epsilon_2 = 0.25$. The strength parameters are selected so that the system matrices are almost surely positive definite. Noting that $\mathbf{M}$ is independent of $\mathbf{x}$ and $\mathbf{K}$ is a linear function of $\mathbf{x}$, the derivative of the system matrices with respect to the random vector $\mathbf{x}$ can be obtained as

$$\frac{\partial \mathbf{K}}{\partial x_1} = \epsilon_1 \begin{bmatrix} \bar{k}_1 & 0 \\ 0 & 0 \end{bmatrix}, \quad \frac{\partial \mathbf{K}}{\partial x_2} = \epsilon_2 \begin{bmatrix} 0 & 0 \\ 0 & \bar{k}_2 \end{bmatrix}, \tag{5.82}$$

$$\frac{\partial \mathbf{M}}{\partial x_i} = \mathbf{O} \quad \text{and} \quad \frac{\partial^2 \mathbf{K}}{\partial x_i \, \partial x_j} = \mathbf{O} \tag{5.83}$$

We calculate the raw moments and the probability density functions of the two eigenvalues of the system. Recall that the eigenvalues obtained from equation (5.1) are square of the natural frequencies ($\lambda_j = \omega_j^2$). The following six methods are used to obtain the moments and the pdfs.

1. *Mean-centered first-order perturbation:* This case arises when $\mathbf{D}_{\lambda_j}(\boldsymbol{\mu})$ in the Taylor series expansion (5.7) is assumed to be a null matrix so that only the first-order terms are retained. This is the simplest approximation, and as mentioned earlier, results in Gaussian distribution of the eigenvalues. Recalling that for this problem $\boldsymbol{\mu} = \mathbf{0}$ and $\boldsymbol{\Sigma} = \mathbf{I}$, the resulting statistics for this special case can be obtained from equations (5.17) and (5.18) as

$$\widehat{\lambda}_j = \overline{\lambda}_j \tag{5.84}$$

$$\text{and} \quad \mathrm{Var}\left[\lambda_j\right] = \mathbf{d}_{\lambda_j}^T(\mathbf{0}) \mathbf{d}_{\lambda_j}(\mathbf{0}) \tag{5.85}$$

   The gradient vector $\mathbf{d}_{\lambda_j}(\mathbf{0})$ can be obtained using the system derivative matrices (5.82) and (5.83).

2. *Mean-centered second-order perturbation:* In this case all the terms in equation (5.7) are retained. This approximation results in a quadratic form in Gaussian random variables. The resulting statistics can be obtained from equations (5.15) and (5.16) by substituting $\boldsymbol{\mu} = \mathbf{0}$ and $\boldsymbol{\Sigma} = \mathbf{I}$. The elements of the Hessian matrix $\mathbf{D}_{\lambda_j}(\mathbf{0})$ can be obtained using the system derivative matrices (5.82) and (5.83).

3. *Optimal point first-order perturbation:* This case arises when $\mathbf{D}_{\lambda_j}(\boldsymbol{\alpha})$ in the Taylor series expansion (5.21) is assumed to be a null matrix so that only the first-order terms are retained. Like its mean-centered counterpart, this approach also results

in a Gaussian distribution of the eigenvalues. From equation (5.29) we have

$$c_j = \lambda_j(\boldsymbol{\alpha}) - \mathbf{d}^T_{\lambda_j}(\boldsymbol{\alpha})\boldsymbol{\alpha}, \quad \mathbf{a}_j = \mathbf{d}_{\lambda_j}(\boldsymbol{\alpha}) \quad \text{and} \quad \mathbf{A}_j = \mathbf{O} \tag{5.86}$$

The equation to obtain the optimal point $\boldsymbol{\alpha}$ can be given from equation (5.28) as

$$\boldsymbol{\alpha} = \mathbf{d}_{\lambda_j}(\boldsymbol{\alpha})/\lambda_j(\boldsymbol{\alpha}) \quad \text{or} \quad \mathbf{d}_{\lambda_j}(\boldsymbol{\alpha}) = \lambda_j(\boldsymbol{\alpha})\boldsymbol{\alpha} \tag{5.87}$$

Using these equations, the mean and the variance can be obtained as special cases of equations (5.37) and (5.38)

$$\widehat{\lambda}_j = \lambda_j(\boldsymbol{\alpha}) - \mathbf{d}^T_{\lambda_j}(\boldsymbol{\alpha})\boldsymbol{\alpha} = \lambda_j(\boldsymbol{\alpha}) - \lambda_j(\boldsymbol{\alpha})\boldsymbol{\alpha}^T\boldsymbol{\alpha} \tag{5.88}$$

$$\text{or} \quad \widehat{\lambda}_j = \lambda_j(\boldsymbol{\alpha})\left(1 - |\boldsymbol{\alpha}|^2\right) \tag{5.89}$$

$$\text{and} \quad \mathrm{Var}\left[\lambda_j\right] = \mathbf{d}^T_{\lambda_j}(\boldsymbol{\alpha})\mathbf{d}_{\lambda_j}(\boldsymbol{\alpha}) = \lambda_j^2(\boldsymbol{\alpha})|\boldsymbol{\alpha}|^2 \tag{5.90}$$

4. *Optimal point second-order perturbation:* In this case all the terms in equation (5.21) are retained. Like the mean-centered approach, this approximation also results in a quadratic form in Gaussian random variables, but with different coefficients. The resulting statistics can be obtained from equations (5.35) and (5.36).
5. *Method based on the asymptotic integral:* In this case the moments can be obtained using equation (5.56). For the standardized Gaussian random vector substituting $\boldsymbol{\mu} = \mathbf{0}$ and $\boldsymbol{\Sigma} = \mathbf{I}$ in equation (5.62) the moment formula can be simplified to

$$\mu_j^{(r)} \approx \lambda_j^r(\boldsymbol{\theta}) \exp\left\{-\frac{1}{2}|\boldsymbol{\theta}|^2\right\} \left\|\mathbf{I} + \frac{1}{r}\boldsymbol{\theta}\boldsymbol{\theta}^T/r - \frac{r}{\lambda_j(\boldsymbol{\theta})}\mathbf{D}_{\lambda_j}(\boldsymbol{\theta})\right\|^{-1/2}, \tag{5.91}$$

$$\text{and} \quad \boldsymbol{\theta} = r\mathbf{d}_{\lambda_j}(\boldsymbol{\theta})/\lambda_j(\boldsymbol{\theta}), \quad r = 1, 2, 3, \cdots \tag{5.92}$$

The vector $\boldsymbol{\theta}$ needs to be calculated for each r and j from equation (5.92) using the iterative approach discussed before. The moments obtained from equation (5.91) can be used to obtain the pdf using the approach given in section 5.5.
6. *Monte Carlo Simulation:* The samples of two independent Gaussian random variables x_1 and x_2 are generated and the eigenvalues are computed directly from (5.1). A total of 15000 samples are used to obtain the statistical moments and pdf of both the eigenvalues. Results obtained from the Monte Carlo simulation are assumed to be the benchmark for the purpose of comparing the five analytical methods described above.

Numerical results

Figure 5.2 shows the percentage error for the first four raw moments of the first eigenvalue. The percentage error for an arbitrary kth moment of an eigenvalue obtained using any one of the five analytical methods is given by

$$\mathrm{Error}_{i\text{th method}} = \frac{\left|\{\mu_j^{(r)}\}_{i\text{th method}} - \{\mu_j^{(r)}\}_{\mathrm{MCS}}\right|}{\{\mu_j^{(r)}\}_{\mathrm{MCS}}} \times 100, \quad i = 1, \cdots 5 \tag{5.93}$$

Percentage error for the first four raw moments of the second eigenvalue is shown in figure 5.3. For both eigenvalues error corresponding to the mean-centered first-order perturbation method is more than the other four methods. Error corresponding to the optimal point first-order perturbation method follows next. Moments obtained from mean-centered and optimal point second-order perturbation methods are more accurate compared to their corresponding first-order counterparts. In general the moments ob-

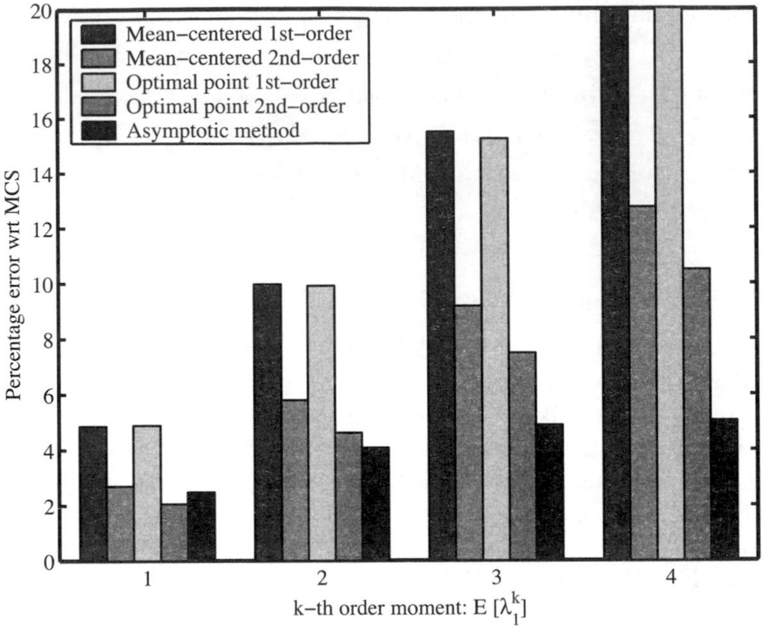

Figure 5.2. Percentage error for the first eigenvalue.

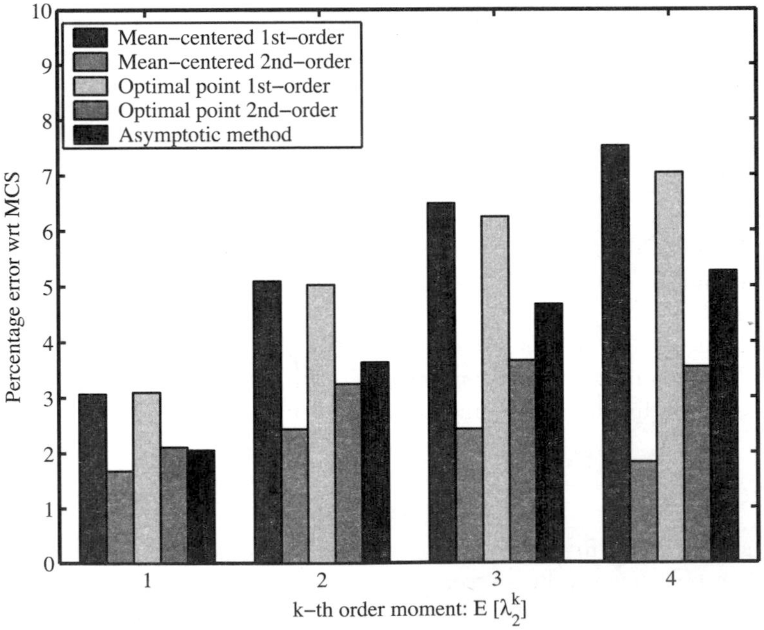

Figure 5.3. Percentage error for the second eigenvalue.

tained from the asymptotic formula (5.56) turns out to be quite accurate. Absolute error for the second eigenvalue is less compared to the first eigenvalue. For the first eigenvalue, the moments obtained from the asymptotic formula turns out to be the most accurate, while for the second eigenvalue, mean-centered second-order perturbation method yields most accurate results.

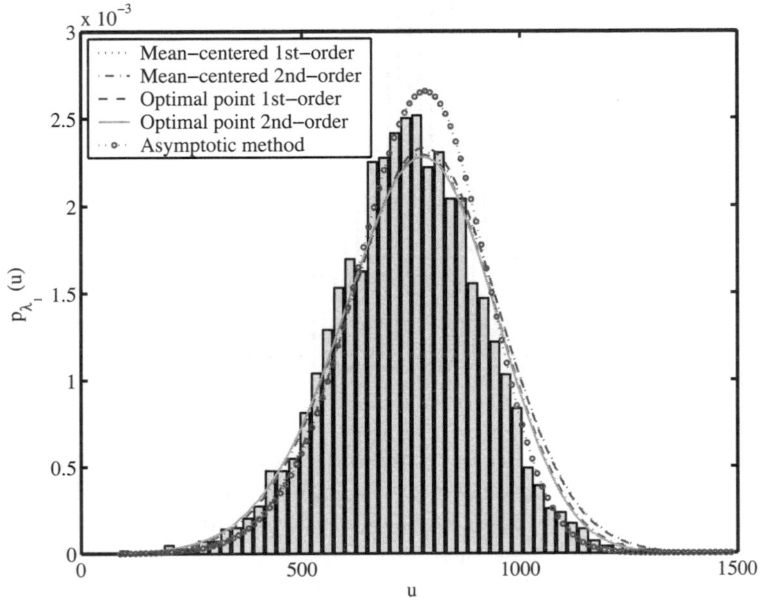

Figure 5.4. Probability density function of the first eigenvalue.

Now consider the probability density function of the eigenvalues. Figures 5.4 and 5.5 respectively show the pdf of the first and the second eigenvalue obtained from the five methods described earlier. The pdf corresponding to first five methods are obtained using the χ^2 distribution in equation (5.80). The constants appearing in this equation are calculated from the moments using equations (5.77)–(5.79). In the same plots, normalized histograms of the eigenvalues obtained from the Monte Carlo simulation are also plotted. For the first eigenvalue, pdf from the second-order perturbation methods are accurate in the lower and in the upper tail. For the second eigenvalue, pdf from the asymptotic moments is accurate over the whole curve.

5.7 Conclusions

The statistics of the eigenvalues of discrete linear dynamic systems with parameter uncertainties have been discussed in this chapter. It is assumed that the mass and stiffness matrices are smooth and at least twice differentiable functions of a set of random variables. The random variables are in general considered to be non-Gaussian. The usual assumption of small randomness employed in most mean-centered based perturbation analysis is not employed here. Two methods, namely (a) optimal point expansion

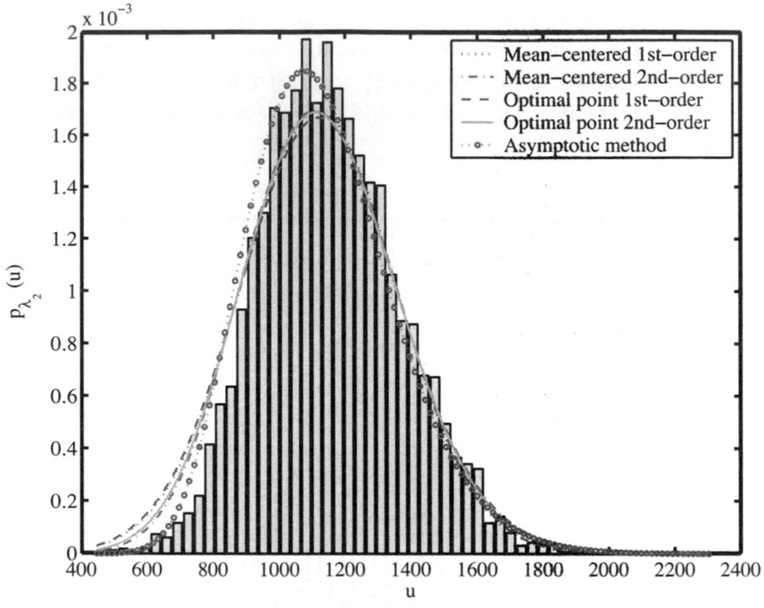

Figure 5.5. Probability density function of the second eigenvalue.

method, and (b) asymptotic moment method, have been outlined. The optimal point is obtained so that the mean of the eigenvalues are estimated most accurately. Both methods are based on an unconstrained optimization problem. Moments and cumulants of arbitrary orders are derived for both the approaches. Two simple approximations for the probability density function of the eigenvalues are derived. One is in terms of a truncated Gaussian random variable obtained using the maximum entropy principle. The other is a χ^2 random variable approximation based on matching the first three moments of the eigenvalues. Both formulations yield closed-form expressions of the pdf which can be computed easily. Proposed formulae are applied to a simple problems. The moments and the pdf match encouragingly well with the corresponding Monte Carlo simulation results.

Bibliography

M. Abramowitz and I. A. Stegun. *Handbook of Mathematical Functions, with Formulas, Graphs, and Mathematical Tables.* Dover Publications, Inc., New York, USA, 1965.

S. Adhikari. Dynamics of non-viscously damped linear systems. *ASCE Journal of Engineering Mechanics*, 128(3):328–339, March 2002.

S. Adhikari. Damping modelling using generalized proportional damping. *Journal of Sound and Vibration*, 293(1-2):156–170, May 2006.

S. Adhikari. Qualitative dynamic characteristics of a non-viscously damped oscillator. *Proceedings of the Royal Society of London, Series- A*, 461(2059):2269–2288, July 2005.

S. Adhikari. *Damping Models for Structural Vibration.* PhD thesis, Cambridge University Engineering Department, Cambridge, UK, September 2000.

S. Adhikari and J. Woodhouse. Identification of damping: part 2, non-viscous damping. *Journal of Sound and Vibration*, 243(1):63–88, May 2001.

R. J. Allemang and D. L. Brown. A correlation coefficient for modal vector analysis. In *Proceedings of the 1st International Modal Analysis Conference (IMAC)*, pages 110–116, Orlando, FL, 1982.

K. F. Alvin, L. D. Peterson, and K. C. Park. Extraction of normal modes and full modal damping from complex modal parameters. *AIAA Journal*, 35(7):1187–1194, July 1997.

R. L. Bagley and P. J. Torvik. Fractional calculus–a different approach to the analysis of viscoelastically damped structures. *AIAA Journal*, 21(5):741–748, May 1983.

M. Baruch. Identification of the damping matrix. Technical Report TAE No.803, Technion, Israel, Faculty of Aerospace Engineering, Israel Institute of Technology, Haifa, May 1997.

H. Benaroya. Random eigenvalues, algebraic methods and structural dynamic models. *Applied Mathematics and Computation*, 52:37–66, 1992.

M. A. Biot. Variational principles in irreversible thermodynamics with application to viscoelasticity. *Physical Review*, 97(6):1463–1469, 1955.

M. A. Biot. Linear thermodynamics and the mechanics of solids. In *Proceedings of the Third U. S. National Congress on Applied Mechanics*, pages 1–18, New York, 1958. ASME.

R. E. D. Bishop and W. G. Price. An investigation into the linear theory of ship response to waves. *Journal of Sound and Vibration*, 62(3):353–363, 1979.

D. R. Bland. *Theory of Linear Viscoelasticity.* Pergamon Press, London, 1960.

Norman Bleistein and Richard A. Handelsman. *Asymptotic Expansions of Integrals.* Holt, Rinehart and Winston, New York, USA, 1994.

W. E. Boyce. *Random Eigenvalue Problems.* Probabilistic methods in applied mathematics. Academic Press, New York, 1968.

T. K. Caughey. Classical normal modes in damped linear dynamic systems. *Transactions of ASME, Journal of Applied Mechanics*, 27:269–271, June 1960.

T. K. Caughey and M. E. J. O'Kelly. Classical normal modes in damped linear dynamic systems. *Transactions of ASME, Journal of Applied Mechanics*, 32:583–588, September 1965.

S. Y. Chen, M. S. Ju, and Y. G. Tsuei. Extraction of normal modes for highly coupled incomplete systems with general damping. *Mechanical System and Signal Processing*, 10(1):93–106, 1996.

J. D. Collins and W. T. Thomson. The eigenvalue problem for structural systems with statistical properties. *AIAA Journal*, 7(4):642–648, April 1969.

J. F. Doyle. *Wave Propagation in Structures.* Springer Verlag, New York, 1989.

I. Elishakoff and Y. J. Ren. *Large Variation Finite Element Method for Stochastic Problems.* Oxford University Press, Oxford, U.K., 2003.

Isaac Elishakoff. *Eigenvalues of Inhomogeneous Structures: Unusual Closed-Form.* CRC Press, Boca Raton, FL, USA, 2005.

D. J. Ewins. *Modal Testing: Theory and Practice.* Research Studies Press, Baldock, England, second edition, 2000.

N. J. Ferguson and W. D. Pilkey. Literature review of variants of dynamic stiffness method, Part 1 : The dynamic element method. *The Shock and Vibration Digest*, 25 (2):3–12, 1993a.

N. J. Ferguson and W. D. Pilkey. Literature review of variants of dynamic stiffness method, Part 1 : Frequency-dependent matrix and other. *The Shock and Vibration Digest*, 25(4):3–10, 1993b.

M. I. Friswell and J. E. Mottershead. *Finite Element Model Updating in Structural Dynamics.* Kluwer Academic Publishers, The Netherlands, 1995.

M. Géradin and D. Rixen. *Mechanical Vibrations.* John Wiely & Sons, New York, NY, second edition, 1997. Translation of: Théorie des Vibrations.

R. Ghanem and P.D. Spanos. *Stochastic Finite Elements: A Spectral Approach.* Springer-Verlag, New York, USA, 1991.

D. F. Golla and P. C. Hughes. Dynamics of viscoelastic structures - a time domain finite element formulation. *Transactions of ASME, Journal of Applied Mechanics*, 52: 897–906, December 1985.

T. K. Hasselsman. A method of constructing a full modal damping matrix from experimental measurements. *AIAA Journal*, 10(4):526–527, 1972.

Francois M. Hemez and Yakov Ben-Haim. The Good, The Bad, and The Ugly of Predictive Science. In Kenneth M. Hanson and François M. Hemez, editors, *Proceedings of the 4th International Conference on Sensitivity Analysis of Model Output*, pages 181–193, Santa Fe, New Mexico, USA, March 2004. see http://library.lanl.gov/.

R. A. Ibrahim. Structural dynamics with parameter uncertainties. *Applied Mechanics Reviews, ASME*, 40(3):309–328, 1987.

S. R. Ibrahim. Dynamic modeling of structures from measured complex modes. *AIAA Journal*, 21(6):898–901, June 1983.

Norman L. Johnson and Samuel Kotz. *Distributions in Statistics: Continuous Univariate Distributions - 2.* The Houghton Mifflin Series in Statistics. Houghton Mifflin Company, Boston, USA, 1970.

J. N. Kapur and H. K. Kesavan. *Entropy Optimization Principles With Applications.* Academic Press, San Diego, CA, 1992.

M. Kleiber and T. D. Hien. *The Stochastic Finite Element Method.* John Wiley, Chichester, 1992.

G. A. Lesieutre and D. L. Mingori. Finite element modeling of frequency-dependent material properties using augmented thermodynamic fields. *Journal of Guidance, Control and Dynamics*, 13:1040–1050, 1990.

R. H. Lyon and R. G. Dejong. *Theory and Application of Statistical Energy Analysis.* Butterworth-Heinmann, Boston, second edition, 1995.

N. M. M. Maia and J. M. M. Silva, editors. *Theoretical and Experimental Modal Analysis.* Engineering Dynamics Series. Research Studies Press, Taunton, England, 1997. Series Editor, J. B. Robetrs.

C. S. Manohar and S. Adhikari. Dynamic stiffness of randomly parametered beams. *Probabilistic Engineering Mechanics*, 13(1):39–51, January 1998.

C. S. Manohar and S. Gupta. Modeling and evaluation of structural reliability: Current status and future directions. In K. S. Jagadish and R. N. Iyengar, editors, *Research reviews in structural engineering, Golden Jubilee Publications of Department of Civil Engineering, Indian Institute of Science, Bangalore.* University Press, 2003.

C. S. Manohar and R. A. Ibrahim. Progress in structural dynamics with stochastic parameter variations: 1987 to 1998. *Applied Mechanics Reviews, ASME*, 52(5):177–197, May 1999.

A. M. Mathai and Serge B. Provost. *Quadratic Forms in Random Variables: Theory and Applications.* Marcel Dekker, Inc., 270 Madison Avenue, New York, NY 10016, USA, 1992.

D. J. McTavish and P. C. Hughes. Modeling of linear viscoelastic space structures. *Transactions of ASME, Journal of Vibration and Acoustics*, 115:103–110, January 1993.

L. Meirovitch. *Principles and Techniques of Vibrations.* Prentice-Hall International, Inc., New Jersey, 1997.

C. Minas and D. J. Inman. Identification of a nonproportional damping matrix from incomplete modal information. *Transactions of ASME, Journal of Vibration and Acoustics*, 113:219–224, April 1991.

J. E. Mottershead. Theory for the estimation of structural vibration parameters from incomplete data. *AIAA Journal*, 28(3):559–561, 1990.

A. Muravyov. Analytical solutions in the time domain for vibration problems of discrete viscoelastic systems. *Journal of Sound and Vibration*, 199(2):337–348, 1997.

N. C. Nigam. *Introduction to Random Vibration.* The MIT Press, Cambridge, Massachusetts, 1983.

Athanasios Papoulis and S. Unnikrishna Pillai. *Probability, Random Variables and Stochastic Processes.* McGraw-Hill, Boston, USA, fourth edition, 2002.

M. Paz. *Structural Dynamics.* CBS Publishers, New Delhi, 1985.

E. S. Pearson. Note on an approximation to the distribution of non-central χ^2. *Biometrica*, 46:364, 1959.

D. P. Pilkey and D. J. Inman. Survey of damping matrix identification. In *Proceedings of the 16th International Modal Analysis Conference (IMAC)*, volume 1, pages 104–110, 1998.

R. H. Plaut and K. Huseyin. Derivative of eigenvalues and eigenvectors in non-self adjoint systems. *AIAA Journal*, 11(2):250–251, February 1973.

W. H. Press, S. A. Teukolsky, W. T. Vetterling, and B. P. Flannery. *Numerical Recipes in C.* Cambridge University Press, Cambridge, 1992.

S. Anantha Ramu and R. Ganesan. A galerkin finite element technique for stochastic field problems. *Computer Methods in Applied Mechanics andEngineering*, 105:315–331, 1993.

Lord Rayleigh. *Theory of Sound (two volumes).* Dover Publications, New York, 1945 re-issue, second edition, 1877.

T. S. Sankar, S. A. Ramu, and R. Ganesan. Stochastic finite element analysis for high speed rotors. *Transactions of ASME, Journal of Vibration and Acoustics*, 115:59–64, 1993.

A. Sarkar and C. S. Manohar. Dynamic stiffness of a general cable element. *Archive of Applied Mechanics*, 66:315–325, 1996.

J. Vom Scheidt and Walter Purkert. *Random Eigenvalue Problems.* North Holland, New York, 1983.

Schlesinger. Terminology for model credibility. *Simulation,*, 32,(3):103104, 1979.

A. Sestieri and R.A. Ibrahim. Analysis of errors and approximations in the use of modal coordinates. *Journal of Sound and Vibration*, 177(2):145–157, 1994.

M. Shinozuka and F. Yamazaki. Stochastic finite element analysis: an introduction. In S. T. Ariaratnam, G. I. Schueller, and I. Elishakoff, editors, *Stochastic structural dynamics: Progress in theory and applications*, London, 1998. Elsevier Applied Science.

J. M. Montalvao E Silva and Nuno M. M. Maia, editors. *Modal Analysis and Testing: Proceedings of the NATO Advanced Study Institute*, NATO Science Series: E: Applied Science, Sesimbra, Portugal, 3-15 May 1998.

N. Wagner and S. Adhikari. Symmetric state-space formulation for a class of non-viscously damped systems. *AIAA Journal*, 41(5):951–956, 2003.

R. Wong. *Asymptotic Approximations of Integrals.* Society of Industrial and Applied Mathematics, Philadelphia, PA, USA, 2001. First published by Academic Press Inc. in 1989.

J. Woodhouse. Linear damping models for structural vibration. *Journal of Sound and Vibration*, 215(3):547–569, 1998.

Kemin Zhou, John C. Doyle, and Keith Glover. *Robust and Optimal Control.* Prentice-Hall Inc, Upper Saddle River, New Jersey 07458, 1995.

O. C. Zienkiewicz and R. L. Taylor. *The Finite Element Method.* McGraw-Hill, London, fourth edition, 1991.

Vibrations of Beams and Plates: Review of First Closed-Form Solutions in the Past 250 Years

Isaac Elishakoff

Department of Mechanical Engineering, Florida Atlantic University, Boca Raton, FL 33431, USA

Abstract. This paper is dedicated to derivation of eigenvalues of structures that possess modulus of elasticity and/or material density that vary from point to point. There is a large selection of methods that can deal with such a structures' vibration spectra. In very rare circumstances one has a possibility to obtain an exact solution, usually in terms of transcendental functions (hypergeometric, Bessel, Lommel, and other special functions).

In other cases, one resorts to powerful numerical methods, like the finite element method; finite difference method; Rayleight-Ritz or Galerkin methods; collocation method and others. In such circumstances, seeking to find a closed-form solution may appear to be a hopeless task.

In 1759, Leonhard Euler was able to derive some closed-form solutions for buckling of non-uniform columns. However, to the best of our knowledge, no closed-form solutions were available for the vibration problems of non-uniform and inhomogeneous structures, until very recently.

In this study some closed-form solutions are reported which we were fortunate to derive over the recent six years. Only a few solutions are discussed, in order to provide the scope of developments. It is hoped that this method, and its generalizations will be further developed, to advance the beautiful world of closed-form solutions, that retain their attractiveness even in the present era of numerical solutions.

Parts 1 and 2 are devoted to the beam vibrations, whereas Parts 3-5 deal with vibrations of circular plates.

1 Part 1: Beam Vibration

1.1 Motivation of Part 1

The aim of this study is to find some closed form solutions to the dynamic equation of a beam in which the Young's modulus and the density are both polynomial functions, with a deterministic or stochastic heterogeneities. The exact mode shape is searched also as a polynomial function, with attendant closed form expression for the natural frequencies. The considered case is the beam simply supported at both ends. For the bibliography of investigations on vibration and buckling of heterogeneous beams, one may consult with the papers by Eisenberger (1997), and Rollot and Elishakoff (1999).

The importance of the found solutions lies in the possibility of their use as benchmark solutions against which the efficacy of various approximate methods could be ascertained. Additionally, presently there is a considerable literature on so called stochastic finite element method (SFEM), that deals with heterogeneous structures involving random fields. The latter random functions can be represented as mean functions superimposed with deviation functions. Solution of the problem with properly chosen mean functions often constitutes an important part of the analysis (see *e.g.* Köylüoğlu *et al* 1994 and Elishakoff *et al* 1995). Thus, the closed form solutions, both in deterministic and stochastic settings possess attractive analytical advantages over approximate solutions where inherent approximations of various natures are needed. For alternative formulations of random eigenvalue problem, the reader may consult the papers by Shinozuka and Astill (1972) and Zhu and Wu (1991).

1.2 Formulation of the Problem

The dynamic behavior of a beam, is described by the following equation

$$\frac{d^2}{dx^2}\left[D(x)\frac{d^2w(x)}{dx^2}\right]-P(x)\omega^2w(x)=0 \tag{1.1}$$

where $D=E(x)I(x)$ is the flexural stiffness, $E=$ Young's modulus, $I=$ moment of inertia of the cross section, $P(x)=\rho A$ is the inertial coefficient, $\rho=$ density, $A=$ area of the cross section, $w(x)$ = displacement, and ω *is* the natural frequency.

In this study, it is assumed that the cross sectional area is constant, but both D and P are specified as polynomial functions, given by

$$P(\xi)=\sum_{i=0}^{m}a_i\xi^i \tag{1.2}$$

$$D(\xi)=\sum_{i=0}^{n}b_i\xi^i \tag{1.3}$$

where $\xi=x/L$ is a nondimensionnal axial coordinate.

We assume that $w(\xi)$ is also polynomial

$$w(\xi)=\sum_{i=0}^{p}w_i\xi^i \tag{1.4}$$

where w_i are coefficients to be determined. In these expressions, m, n and p are respectively the degree of the polynomials for $P(\xi)$, $D(\xi)$ and $w(\xi)$.

Eq (1.1) can be re-written as

$$\frac{d^2}{d\xi^2}\left[D(\xi)\frac{d^2w(\xi)}{d\xi^2}\right]-kL^4P(\xi)w(\xi)=0 \tag{1.5}$$

where

$$k = \omega^2 \tag{1.6}$$

As the involved functions are assumed to be polynomial ones, the degrees of each polynomial function must be linked, namely

$$n + (p-2) - 2 = m + p \tag{1.7}$$

or, simply

$$n - m = 4 \tag{1.8}$$

We observe that eq (1.8) does not depend on the degree p of the displacement $w(\xi)$. We arrive at the seemingly unexpected conclusion that any polynomial function for the displacement may be used in eq (1.5) if it also satisfies the boundary conditions. This fact will be used at a later stage. In view of eq (1.8) the expression for $D(\xi)$ can be written as follows

$$D(\xi) = \sum_{i=0}^{m+4} b_i \xi^i \tag{1.9}$$

1.3 Boundary Conditions

The case of the simply supported beam is associated with the following boundary conditions

$$w(0) = 0 \tag{1.10}$$

$$D(0)w''(0) = 0 \tag{1.11}$$

$$w(1) = 0 \tag{1.12}$$

$$D(1)w''(1) = 0 \tag{1.13}$$

where primes denote differentiation with respect to x.

Solution to eq (1.11) can be found with either $D(0)=0$ or $w''(0) = 0$. However, Young's modulus, which is zero on one point, has no physical sense, thus eq (1.11) is equivalent to $w''(0) = 0$. Hence we postulate $b_0>0$. The same reasoning can be applied to eq (1.13). So the displacement has to satisfy the following conditons

$$w(0) = 0 \tag{1.14}$$

$$w''(0) = 0 \tag{1.15}$$

$$w(1) = 0 \tag{1.16}$$

$$w''(1) = 0 \tag{1.17}$$

The satisfaction of the boundary conditions (1.14-1.17) requires that the degree of the displacement polynomial must at least be four. Assuming that $w(\xi)$ is a fourth order polynomial

$$w(\xi) = w_0 + w_1\xi + w_2\xi^2 + w_3\xi^3 + w_4\xi^4 \tag{1.18}$$

Then application of the boundary conditions yields

$$w(\xi) = w_1\left(\xi - 2\xi^3 + \xi^4\right) \tag{1.19}$$

1.4 Expansion of the Differential Equation

By substituting the different expressions of $D(\xi)$, $P(\xi)$, $w(\xi)$ in eq (1.5), we obtain

$$\begin{aligned} w_1\Bigg[\sum_{i=2}^{m+4} i(i-1)b_i\xi^{i-2}\left(-12\xi+12\xi^2\right)+\sum_{i=0}^{m+4}24b_i\xi^i+2\sum_{i=1}^{m+4}ib_i\xi^{i-1}(-12+24\xi) \\ -kL^4\sum_{i=0}^{m}a_i\xi^i\left(\xi-2\xi^3+\xi^4\right)\Bigg]=0 \end{aligned} \tag{1.20}$$

The latter expression can be re-written as follows

$$\begin{aligned} -12\sum_{i=1}^{m+3} i(i+1)b_{i+1}\xi^i + 12\sum_{i=2}^{m+4} i(i-1)b_i\xi^i + 24\sum_{i=0}^{m+4} b_i\xi^i - 24\sum_{i=0}^{m+3}(i+1)b_{i+1}\xi^i \\ +48\sum_{i=1}^{m+4} ib_i\xi^i - kL^4\sum_{i=1}^{m+1}a_{i-1}\xi^i + 2kL^4\sum_{i=3}^{m+3}a_{i-3}\xi^i - kL^4\sum_{i=4}^{m+4}a_{i-4}\xi^i = 0 \end{aligned} \tag{1.21}$$

Eq. (1.21) has to be satisfied for any ξ. This requirement yields the following relations

$$-24(b_1 - b_0) = 0, for\ \ i = 0$$

$$-kL^4 a_0 + 72(b_1 - b_2) = 0, for\ \ i = 1 \qquad (1.22)$$

$$-kL^4 a_1 + 144(b_2 - b_3) = 0, for\ \ i = 2 \qquad (1.23)$$

$$L^4 (2ka_0 - ka_2) + 240(b_3 - b_4) = 0, for\ \ i = 3 \qquad (1.24)$$

(1.25)

…

$$L^4 (2ka_{i-3} - ka_{i-4} - ka_{i-1}) + 12(i+1)(i+2)(b_i - b_{i+1}) = 0, for\ 4 \le i \le m+1 \qquad (1.26)$$

…

(1.27)

$$L^4 (2ka_{m-1} - ka_{m-2}) + 12(m+3)(m+4)(b_{m+2} - b_{m+3}) = 0, for\ \ i = m+2 \qquad (1.28)$$

$$L^4\ (2ka_m - ka_{m-1}) + 12(m+4)(m+5)(b_{m+3} - b_{m+4}) = 0, for\ \ i = m+3 \qquad (1.29)$$

$$-kL^4 a_m + 12(m^2 + 11m + 30)\, b_{m+4} = 0, for\ \ i = m+4$$

Note that eqs (1.22-1.29) are valid only if $m \geq 3$. For cases that satisfy the inequality $m < 3$, the reader is referred to the Appendix. Note also that eqs (1.22-1.29) have a recursive form with respect to coefficients b_j.

The sole unknown in eqs (1.22-1.29) is the natural frequency coefficient k, yet we observe that we have $m+5$ equations. We conclude that the parameters b_i and a_i have to satisfy some auxiliary conditions so that eqs (1.22-1.29) are compatible.

1.5 Compatibility Conditions

A first compatibility condition is given by the eq (1.22), leading to $b_0 = b_1$. From the other equations, several expressions for k can be found. Its values determined from eqs (1.22-1.29) respectively, are listed below

$$k = 72(b_1 - b_2)/L^4 a_0 \tag{1.30}$$

$$k = 144(b_2 - b_3)/L^4 a_1 \tag{1.31}$$

$$k = 240(a_2 - 2a_0)^{-1}(b_3 - b_4)/L^4 \tag{1.32}$$

...

$$k = 12(i+1)(i+2)(a_{i-1} + a_{i-4} - 2a_{i-3})^{-1}[b_i - b_{i+1}]/L^4 \ , \ for\ 4 \le i \le m+1 \tag{1.33}$$

...

$$k = 12(m+3)(m+4)(a_{m-2} - 2a_{m-1})^{-1}[b_{m+2} - b_{m+3}]/L^4 \tag{1.34}$$

$$k = 12(m+4)(m+5)(a_{m-1} - 2a_m)^{-1}[b_{m+3} - b_{m+4}]/L^4 \tag{1.35}$$

$$k = 12(m^2 + 11m + 30)b_{m+4}/L^4 a_m \tag{1.36}$$

To check the compatibility of these expressions, all expressions for k have to be equal to each other. We consider two separate problems: *(i)* material density coefficients a_i are specified; find coefficients b_i so that closed form solution holds; *(ii)* elastic modulus coefficients b_i are specified; find coefficients a_i so that closed form solution is obtainable.

1.6 Specified Inertial Coefficient Function

Let us assume that the function $P(\xi)$, of the inertial coefficient, and hence all a_i $(i=0,2...m)$ are given. Let us observe that if b_{m+4} is specified then the expression given in eq (36) is the final formula for the natural frequency coefficient k. Then eqs (1.30-1.36) allow an evaluation of remaining parameters b_i. Note that b_{m+4} and a_m have to have the same sign due to the positivity of k.

From eq (1.35), we get

$$b_{m+3} = \left\{ \left[\frac{m^2 + 11m + 30}{(m+4)(m+5)} \right] \left(\frac{a_{m-1}}{a_m} - 1 \right) + 1 \right\} b_{m+4} \tag{1.37}$$

Eq (1.34) yields

$$b_{m+2} = \left[\left(\frac{m+5}{m+3}\right)\frac{a_{m-2} - 2a_{m-1}}{a_{m-1} - 2a_m} + 1\right]b_{m+3} - \left(\frac{m+5}{m+3}\right)\frac{a_{m-2} - 2a_{m-1}}{a_{m-1} - 2a_m}b_{m+4} \tag{1.38}$$

Eq (1.33) results in

$$b_i = \left[\left(\frac{i+3}{i+1}\right)\frac{a_{i-1} - a_{i-4} - 2a_{i-3}}{a_i - a_{i-3} - 2a_{i-2}} + 1\right]b_{i+1} - \left(\frac{i+3}{i+1}\right)\frac{a_{i-1} - a_{i-4} - 2a_{i-3}}{a_i - a_{i-3} - 2a_{i-2}}b_{i+2} \tag{1.39}$$

where i belongs to the set $\{4,5,..,m+1\}$.

From eq (1.32) we obtain

$$b_3 = \left[\left(\frac{3}{2}\right)\frac{a_2 - 2a_0}{a_3 - a_0 - 2a_1} + 1\right]b_4 - \left(\frac{3}{2}\right)\frac{a_2 - 2a_0}{a_3 - a_0 - 2a_1}b_5 \tag{1.40}$$

Eq (1.31) leads to

$$b_2 = \left[\left(\frac{5}{3}\right)\frac{a_1}{a_2 - 2a_0} + 1\right]b_3 - \left(\frac{5}{3}\right)\frac{a_1}{a_2 - 2a_0}b_4 \tag{1.41}$$

From eq (1.29) we get

$$b_1 = \left(2\frac{a_0}{a_1} + 1\right)b_2 - 2\frac{a_0}{a_1}b_3 \tag{1.42}$$

And finally, eq (1.22) yields

$$b_0 = b_1 \tag{1.43}$$

Thus, for specified coefficients $a_0, a_1, \ldots a_m$ and b_{m+4}, eq (37-43) lead to the set of coefficients in elastic modulus such that the beam possesses mode shape given in eq (19). Note that if $a_i = a$, then coefficients b_i do not depend on the parameter a.

To sum up, if

$$P(\xi) = \sum_{i=0}^{m} a_i \xi^i \qquad D(\xi) = \sum_{i=0}^{m+4} b_i \xi^i \tag{1.44}$$

where b_i are computed via eqs (1.37-1.43), the fundamental mode shape of a beam is

$$w(\xi) = w_1\left(\xi - 2\xi^3 + \xi^4\right) \tag{1.45}$$

and the fundamental natural frequency squared reads

$$\omega^2 = 12\left(m^2 + 11m + 30\right)b_{m+4} / a_m L^4 \tag{1.46}$$

As we have seen, in order to obtain a closed solution it is sufficient that all a_i coefficients, as well as the coefficient b_{m+4} be specified. Yet, the requirements are not necessary ones. Indeed one can assume that a_i coefficient are given and instead of b_{m+4} any coefficients b_j ($j \neq m+4$) is specified. If this is the case, then from eq (1.33) one expresses b_{i+1} via b_i and k; substitution into subsequent equations allows us to express b_{m+2}, b_{m+3}, b_{m+4} via b_i; analogously, substitution of b_i into eqs (1.30-1.32) yields sought exact solutions.

In Tables 1 and 2, some sample specified function $D(x)$ and the attendant fundamental natural frequency coefficients are given. The polynomial functions $P(x)$ were specified as

$$P(\xi) = \sum_{i=0}^{m} \xi^i, \; D(\xi) = \sum_{i=0}^{m} (i+1)\xi^i \tag{1.47}$$

respectively, in Table 1 and Table 2.

1.7 Specified Flexural Stiffness Function

Consider now the case when the flexural stiffness function is specified, implying that all b_i (i=0,….,m+4) are given. The following question arises: Is it possible to determine the material density coefficients a_i (i=0,….,m), such that equations corresponding to eqs (1.22-1.29) are compatible? One immediately observes that there are (m+5) equations (1.22-1.29), while one has only m+1 unknowns, a_0, a_1,…,a_m. In actuality, however one has only m unknowns. In order for the process of determining of coefficient a_i to proceed, one of the a_j coefficients should be specified. The most convenient assumption is to fix either a_0 or a_1 or a_m, since in these cases only one equation, respectively eq (1.30) or eq (1.31) or eq (1.36) will be sufficient to determine the sought expression of the natural frequency coefficient. Let us assume that the coefficient a_0 is given thus, to check the compatibility of eqs (1.22-29), four b_i coefficients cannot be chosen arbitrarily.

Table 1.1

m	*D(ξ)*	*k*
0	$b\left(3+3\xi-2\xi^2-2\xi^3+\xi^4\right)$	$360\dfrac{b}{L^4}$
1	$(b/10)\left(59+59\xi-11\xi^2-46\xi^3-4\xi^4+10\xi^5\right)$	$504\dfrac{b}{L^4}$
2	$(b/15)\left(135+135\xi-5\xi^2-75\xi^3-33\xi^4-5\xi^5+15\xi^6\right)$	$672\dfrac{b}{L^4}$
3	$(b/35)\left(434+434\xi+14\xi^2-196\xi^3-70\xi^4-70\xi^5-10\xi^6+35\xi^7\right)$	$864\dfrac{b}{L^4}$
4	$(b/28)\left(452+452\xi+32\xi^2-178\xi^3-52\xi^4-52\xi^5-52\xi^6-7\xi^7+28\xi^8\right)$	$1080\dfrac{b}{L^4}$
5	$(b/36)\left(648+648\xi+69\xi^2-224\xi^3-56\xi^4-56\xi^5-56\xi^6-56\xi^7-8\xi^8+36\xi^9\right)$	$1320\dfrac{b}{L^4}$
6	$(b/15)\left(371+371\xi+41\xi^2-124\xi^3-25\xi^4-25\xi^5-25\xi^6-25\xi^7-25\xi^8-3\xi^9+15\xi^{10}\right)$	$1584\dfrac{b}{L^4}$
7	$(b/55)\left(1628+1628\xi+198\xi^2-517\xi^3-88\xi^4-88\xi^5-88\xi^6-88\xi^7-88\xi^8-88\xi^9-10\xi^{10}+55\xi^{11}\right)$	$1872\dfrac{b}{L^4}$
8	$(b/330)\left(5715+57151\xi+1492\xi^2-3513\xi^3-510\xi^4-510\xi^5-510\xi^6-510\xi^7-510\xi^8-510\xi^9-510\xi^{10}-55\xi^{11}+330\xi^{12}\right)$	$2184\dfrac{b}{L^4}$
9	$(b/26)\left(1053+1053\xi+143\xi^2-312\xi^3-39\xi^4-39\xi^5-39\xi^6-39\xi^7-39\xi^8-39\xi^9-39\xi^{10}-39\xi^{11}-4\xi^{12}+26\xi^{13}\right)$	$2520\dfrac{b}{L^4}$

Table 1.2

m	$D(\xi)$	k
0	$b\left(3+3\xi-2\xi^2-2\xi^3+\xi^4\right)$	$360\dfrac{b}{L^4}$
1	$(b/10)\left(38+38\xi+3\xi^2-32\xi^3-11\xi^4+10\xi^5\right)$	$252\dfrac{b}{L^4}$
2	$(b/45)\left(203+203\xi+63\xi^2-77\xi^3-119\xi^4-35\xi^5+45\xi^6\right)$	$224\dfrac{b}{L^4}$
3	$(b/140)\left(725+725\xi+305\xi^2-115\xi^3-241\xi^4-325\xi^5-85\xi^6+140\xi^7\right)$	$216\dfrac{b}{L^4}$
4	$(b/140)\left(815+815\xi+395\xi^2-25\xi^3-151\xi^4-235\xi^5-295\xi^6-70\xi^7+140\xi^8\right)$	$216\dfrac{b}{L^4}$
5	$(b/1512)\left(9751+9751\xi+5131\xi^2+511\xi^3-875\xi^4-1799\xi^5-2549\xi^6-2954\xi^7-644\xi^8+1512\xi^9\right)$	$220\dfrac{b}{L^4}$
6	$(b/1470)\left(10388+10388\xi+5768\xi^2+1148\xi^3-238\xi^4-1162\xi^5-1822\xi^6-1617\xi^7-2702\xi^8-546\xi^9+1470\xi^{10}\right)$	$\dfrac{1584}{7}\dfrac{b}{L^4}$

Note that the natural frequency coefficient k has to be positive thus, the difference b_1-b_2 and the coefficient a_0 have to have the same sign. Moreover, as the coefficient a_0 is positive, the difference b_1-b_2 should be positive. So, for $b_1>b_2$, one substitutes the value of k determined from eq (1.30) into eq (1.31); this allows to determine the coefficient a_1 so that the frequency coefficient k in eq (1.31) is positive, and so on.

First, eq (1.22) leads to

$$b_0 = b_1 \tag{1.48}$$

From eq (1.31), we get

$$a_1 = 2a_0 \frac{b_3 - b_2}{b_2 - b_1} \tag{1.49}$$

Eq (1.32) yields

$$a_2 = \frac{5a_1(b_4 - b_3) + 6a_0(b_3 - b_2)}{3(b_3 - b_2)} \tag{1.50}$$

Eq (1.33) where i=4, results in

$$a_3 = \frac{a_0(4b_4 - 6b_5 + 2b_3) + a_1(4b_4 - 4b_3) + a_2(3b_5 - 3b_4)}{2(b_4 - b_3)} \tag{1.51}$$

From eq (1.33), where $4 \le i \le m$, we obtain

$$\begin{aligned} a_i = \frac{1}{(i+1)(b_{i+1} - b_i)} \{ & a_{i-1}(i+3)(b_{i+2} - b_{i+1}) + 2a_{i-2}(i+1)(b_{i+1} - b_i) + \\ & a_{i-3}[b_{i+1}(i+5) - 2b_{i+2}(i+3) + b_i(i+1)] + a_{i-4}(i+3)(b_{i+2} - b_{i+1}) \} \end{aligned} \tag{1.52}$$

Then from eq (1.36) and eq (1.30), one can find an expression of b_{m+4}, so that, the compatibility of eq (1.22-1.29) is checked

$$b_{m+4} = \frac{6a_m(b_1 - b_2)}{a_0(m^2 + 11m + 30)} \tag{1.53}$$

From eq (1.35) and eq (1.36), a relation for b_{m+3} can be found

$$b_{m+3} = \frac{a_0 b_{m+4}(m^2 + 9m + 20) + 6a_{m-1}(b_1 - b_2) + 12a_m(b_2 - b_1)}{a_0(m+4)(m+5)} \tag{1.54}$$

Finally, eq (1.34) and eq (1.35) yield to an evaluation of b_{m+2}

$$b_{m+2} = \frac{a_0 b_{m+3}(m+3)(m+4) + 6a_{m-2}(b_1 - b_2) + 12a_{m-1}(b_2 - b_1)}{a_0(m+3)(m+4)} \quad (1.55)$$

To sum up, while specifying the elastic modulus function, only *m+1* coefficients b_i can be chosen arbitrarily; The other remaining four coefficients are connected with the arbitrary ones via eq (1.48) eqs (1.53-1.55).

Thus, if

$$P(\xi) = \sum_{i=0}^{m} a_i \xi^i \qquad D(\xi) = \sum_{i=0}^{m+4} b_i \xi^i \quad (1.56)$$

where a_i and four of b_i coefficients are computed via eqs (48-55), the fundamental mode shape of a beam is

$$w(\xi) = w_1\left(\xi - 2\xi^3 + \xi^4\right) \quad (1.57)$$

The fundamental natural frequency square reads

$$\omega^2 = 72(b_1 - b_2)/a_0 L^4 \quad (1.58)$$

The closed-form solutions could be utilized for comparison with approximate techniques. For example, utilization of the single term Boobnov-Galerkin method for the case

$$P(\xi) = 1 + 2\xi + 3\xi^2 + 4\xi^3 + 5\xi^4 \quad (1.59)$$

$$D(\xi) = b_8 L^8\left(\frac{163}{28} + \frac{163}{28}\xi + \frac{79}{28}\xi^2 - \frac{5}{28}\xi^3 - \frac{151}{140}\xi^4 - \frac{47}{28}\xi^5 - \frac{59}{28}\xi^6 - \frac{1}{2}\xi^7 + \xi^8\right) \quad (1.60)$$

yields, with *sin(πξ)* taken as a comparison function, the following expression

$$k = \frac{b_8 L^4}{22{,}050}\left(\frac{6{,}945{,}750 - 409{,}185\pi^4 + 391{,}612\pi^8 - 2{,}716{,}875\pi^2 + 173{,}56\pi^6}{10\pi^4 - 19\pi^2 + 15}\right) \quad (1.61)$$

or numerically $k=216.29697 b_8 L^4$, which differs from the exact solution $k=216 b_8 L^4$ by 0.13%.

1.8 Nature of Imposed Restrictions

In this chapter, in order to obtain the closed-form solutions for natural frequencies deterministically and/or stochastically heterogeneous simply-supported beam, the flexural stiffness and the inertial coefficient were assumed to be polynomial functions whose powers differ by four. One should stress that the a_i and b_i coefficients in eqs. (1.2) and (1.3) cannot be specified independently in order for a closed-form solution to exist. It is quite interesting to comment on the physical meaning of this restriction. Does it signify that a_i and b_i coefficients and therefore the inertial coefficient and flexural stiffness must depend of each other? To reply to this question consider a classical case of the closed-form solution, reported for nonlinear stochastic dynamics. Nigam (1983) studies the following set of equations

$$\ddot{Y}_j + \beta_j \dot{Y}_j + \partial V / \partial Y_j = Q_j(t) \quad , \quad \mathrm{j} = 1,2,...,\mathrm{n} \tag{1.62}$$

where Y_j = generalized forces, β_j = damping coefficients, V = potential function, $Q_j(t)$ = generalized forces, n = number of degree of freedom. Then the Fokker-Plank equation is constructed that is not reproduced here. Nigam (1983) notes:

"Assume that

$$\beta_j / \Phi_j = \gamma \text{ for every } j, \tag{1.63}$$

and define

$$H = \tfrac{1}{2} \sum_{j=1}^{n} z^2{}_{j+n} + V(z_1, z_2, ..., z_n), \tag{1.64}$$

then the solution can be expressed as

$$p(z_1, z_2, ..., z_{n+1}, ..., z_{2n}) = C \exp[-(\gamma / \pi) H]." \tag{1.65}$$

where p is the probability density function and Φ_j is the spectral density of $Q_j(t)$. As is seen the closed-form solution (1.65) is obtainable when the ratios between the *inner* characteristics – damping coefficients of the system on one hand and the power spectral densities Φ_j of *external* excitation on the other—satisfy the condition given in Eq. (1.63). This indicates that when these characteristics have a common numerical parameter then the closed-form solution is derivable. The condition (1.63) is necessary for the solution to be given by Eq. (1.64) which was the first derived by Ariaratnam (1960) for systems with two-degrees-of-freedom and was extended to multi-degree-of-freedom systems by Caughey (1963). For other closed-form solutions of this and other kinds, the reader may consult with works by Dimentberg (1982), Soize (1988), Scheurkogel and Elishakoff (1988), Lin and Cai (1995) and others.

One should stress that the importance of the derived closed-form solution is not diminished by the fact that certain conditions must be met. The appearance of conditions is natural too. Indeed, it can be expected that the solution of the *inverse* problem would depend upon the part or entire given data. Thus, if the heterogeneous beam has a polynomial inertial coefficient with given coefficients, it must be no surprize that the sought after flexural stiffness of the beam possesses the pre-selected mode shape, that is directly related to the specified inertial coefficients, in order to derive the closed-form solutions.

The following question arises: Is there any resemblence in the previous literature to the type of thinking adopted in this paper? The connection with the previous work was found via Saint-Venant's *semi-inverse* method. As Timoshenko (1953) writes:

> "In 1853, Saint-Venant presented his epochmaking memoir on torsion to the French Academy. The committee, composed of Cauchy, Poncelet, Piobert,

> and Lamé, were very impressed by the work and recommended its publication…
>
> In the introduction Saint-Venant states that the stresses at any point of an elastic body can be readily calculated if the functions representing the components u, v and w of the displacements are known… Saint-Venant then proposes the *semi-inverse method* by which he assumes only some features of the displacements and the forces and determines the remaining features of those quantities so as he ask by all the equations of elasticity. He remarks that an engineer, guided by the approximate solution of the elementary strength of materials, can obtain rigorous solutions of practical importance in this way."

Indeed, Saint-Venant in 1853 postulated the prior knowledge of the two displacement functions $u=\theta zy$ and $v=\theta zx$, and then determined the function $w=\theta\varphi(x,y)$ where $\varphi(x,y)$ is some function of x and y determined from basic equations. In the present paper we assumed the knowledge of the mode shape and derived the stiffness from the basic equations. This makes some (at least remote) connection with the Saint-Venant's semi-inverse method. It appears that the clarifying comments in this section further enhance the usefulness of this study.

1.9 Discussion of Part 1

The described class of deterministic and stochastic solutions contains infinite number of closed-form solutions. Indeed, the degree m of polynomial in the expression of the inertial coeficient in eq (44) can be chosen arbitrarily. Likewise, the coefficients a_i can be prescribed at the will of the engineer subject to a condition of positivity of both $P(\xi)$ and $D(\xi)$.

It should be noted that there is a connection between the present work with the subject of "inverse problems" of vibration (Gladwell, 1986, 1996, 2005). Indeed, whereas mathematical "direct problems" consist of finding solutions to equations with known *input* parameters, the mathematical "inverse problem" deals with the reconstruction of the parameters of the governing equations when the *output* quantities are known. According to Gladwell (1986), "inverse problems are concerned with the construction of a model of a given type; *e.g.*, a mass-spring system, a string, etc; which has given eigenvalues and/or eigenvectors or eigenfunctions; *i.e.* given *spectral* data. In general, if some such spectral data is given, there can be no system, a unique system, or many systems, having these properties." It is remarkable as the present study demonstrates, that there exist infinite beams, corresponding to $m=0,1,2,\ldots$, that possess the fundamental mode given in eq (1.19).

The natural question arises: *Is it possible to formulate the problem so as to obtain an unique solution?* The reply to this question is affirmative. Indeed, one can pre-select not only the fundamental mode shape, but also the fundamental natural frequency denoted by ω_1. Then the eq (1.46) yields the coefficient b_{m+4} that accomplishes this goal becomes

$$b_{m+4} = \omega_1^2 a_m L^4 / 12(m^2 + 11m + 30) \tag{1.66}$$

The polynomial expressions have been used prior to this study in deterministic analyses ; yet, to the best of authors' knowledge, this is the first collection of closed-form results in either deterministic setting for the natural frequencies and associated reliabilities of heterogeneous beams.

It is also notable that whereas in usual finite element stochastic method in setting, only small coefficients of variation can be allowed, the present formulation is not bound to small coefficients of variation. Therefore, the deterministic and probabilistic closed-form solutions that were uncovered in this study, can be utilized as benchmark solutions.

2 Part 2: Family of Other Analytical Polynomial Solutions

2.1 Motivation of Part 2

The differential equation for the vibrating beam is attributed to Daniel Bernouilli and Leonhard Euler. For the history of the beam's equation the reader is referred to the book by Todhunter and Pearson (1960). In over 250 years, that passed since then, apparently hundreds and even thousands of papers were devoted to free vibration of homogeneous and heterogeneous, uniform and non-uniform beams. Present author is unaware of closed-form solutions for heterogeneous and non-uniform beams' eigenvalues that were derived until the 21st century except only four investigations, those by Engesser (1893), Duncan (1937), Elishakoff and Rollot (1999) and Candan and Elishakoff (1999). The first three being devoted to buckling, whereas only the latter one deals with vibrations.

Duncan (1937) constructed an example with closed - form solution of the buckling load of a variable cross - section rod. The mode shape was taken as a polynomial of fifth degree. Elishakoff and Rollot (1999) found cases in which the simpler, fourth order polynomial serves as an exact buckling mode of the stiffness column. Later on, Candan and Elishakoff (1999) found numerous solutions for natural frequencies of vibrating beams with variable density and variable stiffness, apparently for the first time in the literature. They postulated, for the beam under various boundary conditions, the fourth order polynomial to serve as an exact mode shape. In present study we pose the following question: Is the simplest fourth order polynomial the only possible mode shape for the heterogeneous beam, or higher order polynomial mode shapes are also acceptable? Note that the fourth order polynomial represents an exact static deflection of the beam under the uniform load q_0. One can generate the higher order polynomials as the static deflections of the uniform beams to the loading represented as $q_0\xi^n$, where $q_0 = \text{constant}$, ξ is the non-dimensional coordinate, n is a positive integer. The so obtained static deflections satisfy all the boundary conditions. The question is then re-formulated as follows: Can such a static deflection serve as an exact mode shape of a vibrating heterogeneous beam? In case $n = 0$, the reply to this question was shown to be affirmative, by Candan and Elishakoff (1999). The present study addresses to cases when the power n in the loading $q_0\xi^n$ differs from zero, and obtain series of new closed-form solutions.

The present author is unaware of any other work in which the polynomial expressions serve as exact, closed-form vibration mode shapes, prior to the study by Candan and Elishakoff (1999).

Problem Formulation

Consider a uniform beam, simply supported at its both ends. It is subjected to the load $q(\xi) = q_0\xi^n$. The governing differential equation reads:

$$D_0 d^4 w / dx^4 = q_0 x^n \tag{2.1}$$

where D_0 is the stiffness of the beam, $w(x)$ = static displacement, x = axial coordinate, q_0 = load parameter. We introduce the non-dimensional coordinate

$$\xi = x / L \tag{2.2}$$

where L is the length. The equation becomes

$$D_0 d^4 w / d\xi^4 = \alpha \xi^n \tag{2.3}$$

The solution of this equation that satisfies the boundary conditions reads:

$$\begin{aligned} w(\xi) &= \frac{\alpha L^{n+4}}{(n+1)(n+2)(n+3)(n+4)D_0}\psi(\xi) \\ \psi(\xi) &= \xi^{n+4} - \frac{1}{6}\left(n^2+7n+12\right)\xi^3 + \frac{1}{6}\left(n^2+7n+6\right)\xi \end{aligned} \tag{2.4}$$

We address the following problem: Can the function $\psi(\xi)$ serve as an exact mode shape for the heterogeneous beam, *i.e.* beam with variable stiffness $D(\xi)$? We will explore cases of reconstruction of the stiffness function for various axial variations of the inertial coefficient $R(\xi) = \rho(\xi)A$.

2.2 Basic Equations

The differential equation that governs the mode shape $W(\xi)$ of the heterogeneous beam reads:

$$\frac{d^2}{dx^2}\left[D(x)\frac{d^2W}{dx^2}\right] - \omega^2 R(x) W(x) = 0 \tag{2.5}$$

where $W(x)$ is the mode shape in contrast to the function $w(x)$ denoting the static displacement in an auxiliary problem described in Section 2.2. The function $D(x)$ constitutes the stiffness that should be *reconstructed*, whereas $R(x)$ is the inertial coefficient.

With the non-dimensional coordinate ξ from equation (2.2), the governing equation becomes

$$\frac{d^2}{d\xi^2}\left[D(\xi)\frac{d^2\psi}{d\xi^2}\right]-\Omega^2R(\xi)\psi(\xi)=0 \tag{2.6}$$

where the substitution

$$W(\xi)=\psi(\xi) \tag{2.7}$$

has been made. Thus our objective is to construct the beam whose *vibration mode* in equation 2.5 *coincides* with the function $\psi(\xi)$ that is proportional to the static displacement in equation 2.4. We are confronted with determining $R(\xi)$ and $D(\xi)$ such that $\psi(\xi)$ serves as an exact mode shape. The variation of the inertial coefficient is taken as follows

$$R(\xi)=\sum_{i=0}^{m} r_i\xi^i \tag{2.8}$$

where r_i (i=0,1,...,m) are given coefficients. The second term of the governing equation (2.6) is a polynomial function with degree $n + m + 4$. In order to achieve the same polynomial order in the first term of this equation, the stiffness $D(\xi)$ has to be taken as a polynomial of degree $m + 4$:

$$D(\xi)=\sum_{i=0}^{m+4} b_i\xi^i \tag{2.9}$$

The left side of the governing equation (2.6) becomes a polynomial of ξ; in order for it to vanish for every ξ in the interval [0;1], we have to demand that its coefficients in front of ξ raised in any power, to be zero. This leads to a set of $n + m + 5$ linear algebraic equations. The problem is posed as the one of *reconstruction* of the stiffness, when the inertial coefficient and the mode shape are given. We have to determine the coefficients b_i of the stiffness as well as the natural frequency ω, which represent $m + 6$ unknowns, namely, $m + 5$ coefficients of the stiffness function, and the $(m + 6)$th unknown, which is natural frequency ω.

Note that when n is zero, we get a triangular system of $m + 5$ equations for $m + 6$ unknowns as studied by Candan and Elishakoff (1999). They succeeded to express the unknowns in terms of one of coefficients b_i, namely b_{m+4}, treated as an arbitrary constant. Yet, when n is larger than zero, the number of equations is greater than the number of unknowns. It is shown that in some cases new closed-form solutions are obtained. In other cases, only trivial solution can be established. The order of the polynomial in the right side of equation (2.8) will be specified and the determination of b_i coefficients in equation (2.9) will be pursued.

2.3 Constant Inertial Coefficient (m = 0)

Consider first the case $n = 0$; the pre-selected mode shape is given by

$$\psi(\xi)=\xi^4-2\xi^3+\xi \tag{2.10}$$

The inertial coefficient reduces to $R(\xi) = r_0$. The governing differential equation leads to the following set of linear algebraic equations

$$\begin{aligned}
&-24b_1 + 24b_0 = 0 \\
&-72b_2 + 72b_1 - r_0\Omega^2 = 0 \\
&-144b_3 + 144b_2 = 0 \\
&-240b_4 + 240b_3 + 2r_0\Omega^2 = 0 \\
&360b_4 - r_0\Omega^2 = 0
\end{aligned} \tag{2.11}$$

We have 5 equations for 6 unknowns: b_0, b_1, b_2, b_3, b_4 and Ω^2. We declare b_4 as an arbitrary parameter. We get Ω^2 from the last equation of the set (2.11)

$$\Omega^2 = 360b_4 / r_0 \tag{2.12}$$

Rest of equations lead to

$$b_0 = 3b_4, \quad b_1 = 3b_4, \quad b_2 = -2b_4, \quad b_3 = -2b_4 \tag{2.13}$$

The stiffness is then given by the following expression

$$D(\xi) = \left(3 + 3\xi - 2\xi^2 - 2\xi^3 + \xi^4\right)b_4 \tag{2.14}$$

which is positive within the interval [0;1] . This expression along with Eq (2.12) was derived previously by Candan and Elishakoff (1999). Consider now the case $n = 1$. Instead of (2.11) we obtain

$$\begin{aligned}
&b_1 = 0 \\
&-120b_2 + 120b_0 - \frac{7}{3}r_0\Omega^2 = 0 \\
&-b_3 + b_1 = 0 \\
&-400b_4 + 400b_2 + \frac{10}{3}r_0\Omega^2 = 0 \\
&b_3 = 0 \\
&840b_4 - r_0\Omega^2 = 0
\end{aligned} \tag{2.15}$$

We get a linear system of 6 equations for 6 unknowns. Since the determinant of the system (2.15) vanishes

$$\begin{vmatrix} 0 & 1 & 0 & 0 & 0 & 0 \\ 120 & 0 & -120 & 0 & 0 & -\frac{7}{3}r_0 \\ 0 & 1 & 0 & -1 & 0 & 0 \\ 0 & 0 & 400 & 0 & 400 & \frac{10}{3}r_0 \\ 0 & 0 & 0 & 1 & 0 & 0 \\ 0 & 0 & 0 & 0 & 840 & -r_0 \end{vmatrix} = 0 \tag{2.16}$$

we can solve the set, expressing the unknowns with b_4 taken as an arbitrary constant. Thus we get

$$\Omega^2 = 840 b_4 / r_0, \tag{2.17}$$

and the coefficients b_i are

$$b_0 = \tfrac{31}{3} b_4, \quad b_1 = 0 \quad b_2 = -6b_4, \quad b_3 = 0 \tag{2.18}$$

The stiffness,

$$D(\xi) = \left(\tfrac{31}{3} - 6\xi^2 + \xi^4\right) b_4 \tag{2.19}$$

which is positive within the interval [0;1]. One can demonstrate that for n larger than unity only trivial solutions are obtainable for simply supported beams with constant mass density.

2.4 Linearly Varying Inertial Coefficient (m=1)

Consider first the case $n = 0$, instead of (2.11) we have

$$\begin{aligned} & -24b_1 + 24b_0 = 0 \\ & -72b_2 + 72b_1 - r_0\Omega^2 = 0 \\ & -144b_3 + 144b_2 - r_1\Omega^2 = 0 \\ & -240b_4 + 240b_3 + 2r_0\Omega^2 = 0 \\ & -360b_5 + 360b_4 - (r_0 - 2r_1)\Omega^2 = 0 \\ & 504b_5 - r_1\Omega^2 = 0 \end{aligned} \tag{2.20}$$

This set represents a linear algebraic system of 6 equations for the 7 unknowns, b_0, b_1,…, b_5 and Ω^2. Hence, we express them in terms of b_5, treated as an arbitrary constant. We get Ω^2 from the last equation of the set (2.20)

$$\Omega^2 = 504b_5 / r_1 \tag{2.21}$$

while the remaining equations lead to

$$\begin{aligned} &b_0 = (17r_1 + 42r_0)b_5/10r_1, \quad b_1 = (17r_1 + 42r_0)b_5/10r_1, \quad b_2 = (17r_1 - 28r_0)b_5/10r_1 \\ &b_3 = -(9r_1 + 14r_0)b_5/5r_1, \quad b_4 = -(9r_1 - 7r_0)b_5/5r_1 \end{aligned} \tag{2.22}$$

It is easily shown that for simply supported beams with linearly varying mass density no polynomial solutions can be derived for n larger than zero.

2.5 Parabolically Varying Inertial Coefficient (m=2)

We consider first the case $n = 0$. We obtain the following set of 7 equations with 8 unknowns

$$\begin{aligned} &-b_1 + b_0 = 0 \\ &-72b_2 + 72b_1 - r_0\Omega^2 = 0 \\ &-144b_3 + 144b_2 - r_1\Omega^2 = 0 \\ &-240b_4 + 240b_3 + 2r_0\Omega^2 - r_2\Omega^2 = 0 \\ &-360b_5 + 360b_4 - r_0\Omega^2 + 2r_1\Omega^2 = 0 \\ &-504b_6 + 504b_5 - r_1\Omega^2 + 2r_2\Omega^2 = 0 \\ &672b_6 - r_2\Omega^2 = 0 \end{aligned} \tag{2.23}$$

We can therefore find a non-trivial solution when b_6 is regarded as a parameter. From the last equation of the set (2.23), we obtain the natural frequency squared

$$\Omega^2 = 672b_6 / r_2 \tag{2.24}$$

The remaining equations yield the coefficients in the stiffness as follows

$$b_0 = \frac{1}{15}\frac{\left(84r_0 + 34r_1 + 17r_2\right)b_6}{r_2}, \quad b_1 = \frac{1}{15}\frac{\left(84r_0 + 34r_1 + 17r_2\right)b_6}{r_2}$$
$$b_2 = -\frac{1}{15}\frac{\left(56r_0 + 34r_1 - 17r_2\right)b_6}{r_2}, \quad b_3 = -\frac{1}{15}\frac{\left(56r_0 + 36r_1 - 17r_2\right)b_6}{r_2} \tag{2.25}$$
$$b_4 = \frac{1}{15}\frac{\left(28r_0 - 36r_1 - 25r_2\right)b_6}{r_2}, \quad b_5 = \frac{1}{3}\frac{\left(4r_1 - 5r_2\right)b_6}{r_2}$$

When $n = 1$, the governing differential equation yields a set of 8 linear algebraic equations for 8 unknowns, namely the coefficients b_i in the stiffness and the natural frequency squared Ω^2:

$$\begin{aligned} &b_1 = 0 \\ &-120b_2 + 120b_0 - \tfrac{7}{3}r_0\Omega^2 = 0 \\ &-240b_3 + 240b_1 - \tfrac{7}{3}r_1\Omega^2 = 0 \\ &-400b_4 + 400b_2 + \tfrac{10}{3}r_0\Omega^2 - \tfrac{7}{3}r_2\Omega^2 = 0 \\ &-600b_5 + 600b_3 + \tfrac{10}{3}r_1\Omega^2 = 0 \\ &-840b_6 + 840b_4 - r_0\Omega^2 + \tfrac{10}{3}r_2\Omega^2 = 0 \\ &1120b_5 - r_1\Omega^2 = 0 \\ &1440b_6 - r_2\Omega^2 = 0 \end{aligned} \tag{2.26}$$

The determinant of system (2.26) equals $1895104512\text{x}10^9 r_1$. In order to find a non-trivial solution, the determinant must be zero, leading to the requirement $r_1 = 0$. We conclude that for an inertial coefficient equal to

$$R(\xi) = r_0 + r_2\xi^2 \tag{2.27}$$

where r_0 and r_2 are arbitrary, the set (2.26) yields

$$\Omega^2 = 1440b_6 / r_2 \tag{2.28}$$

and

$$b_0 = (620r_0 + 129r_2)b_6 / 35r_2, \quad b_1 = 0, \quad b_2 = -3(120r_0 - 43r_2)b_6 / 35r_2,$$
$$b_4 = 3(4r_0 - 11r_2)b_6 / 7r_2, \quad b_5 = 0 \tag{2.29}$$

Concerning the case $n = 2$, the governing equation yields a set of 8 equations for 9 unknowns, namely the coefficients b_i in the stiffness and the natural frequency squared Ω^2:

$$\begin{aligned}
&b_1 = 0 \\
&-180b_2 - 4r_0\Omega^2 = 0 \\
&-360b_3 + 360b_0 - 4r_1\Omega^2 = 0 \\
&-600b_4 + 400b_1 + 5r_0\Omega^2 - 4r_2\Omega^2 = 0 \\
&-900b_5 + 900b_2 + 5r_1\Omega^2 = 0 \\
&-1260b_6 + 900b_3 + 5r_2\Omega^2 = 0 \\
&1680b_4 - r_0\Omega^2 = 0 \\
&2160b_5 - r_1\Omega^2 = 0 \\
&2700b_6 - r_2\Omega^2 = 0
\end{aligned} \tag{2.30}$$

Since the number of equations is larger than the one of unknowns, in order to have a non-trivial solution, the rank of the set (30) must be less than 8. According to the definition of the rank, "a matrix is of rank p if it contains minors of order p different from 0, while all minors of order $p + 1$ (if there are such) are zero" (Upensky, 1948). Accordingly, the rank of the matrix of the system (30) is less than 8 if all minors of order 8 vanish. This last condition leads to a set of 9 equations out of which 3 are identically zero, the remaining equations can be reduced to a set of 2 equations as follows

$$\begin{aligned}
&-2199910648\ 32r_1 + 9599610101\ 76r_2 = 0 \\
&-1114240458\ 24r_0 + 9599610101\ 76r_2 = 0
\end{aligned} \tag{2.31}$$

The fulfillment of (2.31) yields

$$r_0 = \frac{56}{65}r_2, \quad r_1 = \frac{2688}{715}r_2, \tag{2.32}$$

where r_2 is treated as an arbitrary constant. We deduced that when the inertial coefficient is

$$R(\xi)=\left(\frac{56}{65}+\frac{2688}{715}\xi+\xi^2\right)r_2 \tag{2.33}$$

with an arbitrary parameter r_2, the set (2.30) yields

$$\Omega^2=2700b_6/r_2 \tag{2.34}$$

$$b_0=\frac{103172}{1001}b_6,\ b_1=0,\ b_2=-\frac{672}{13}b_6,\ b_3=-\frac{68}{7}b_6,\ b_4=\frac{18}{13}b_6,\ b_5=\frac{672}{143}b_6 \tag{2.35}$$

The stiffness is then

$$D(\xi)=\left(\frac{103172}{1001}-\frac{672}{13}\xi^2-\frac{68}{7}\xi^3+\frac{18}{13}\xi^4+\frac{673}{143}\xi^5+\xi^6\right)b_6 \tag{2.36}$$

which is positive within the interval [0;1].

For $n = 3$, we have to satisfy the following set of 10 equations for 8 unknowns

$$\begin{aligned}
&b_1=0\\
&-252b_2-6r_0\Omega^2=0\\
&-504b_3-6r_1\Omega^2=0\\
&-840b_4+840b_0+7r_0\Omega^2-6r_2\Omega^2=0\\
&-1260b_5+1260b_1+7r_1\Omega^2=0\\
&-1764b_6+1764b_2+7r_2\Omega^2=0\\
&b_3=0\\
&3024b_4-r_0\Omega^2=0\\
&3780b_5-r_1\Omega^2=0\\
&4620b_6-r_2\Omega^2=0
\end{aligned} \tag{2.37}$$

A non-trivial solution is obtainable, if the rank of the set (2.37) is less than 8. This demands the 45 minors of order 8 of the matrix of the linear system (2.37) to be zero. We obtain then 45 equations out of which 24 are identically zero, whereas 15 are proportional to r_1. Thus, the entire system reduces to

$$
\begin{aligned}
&r_1 = 0 \\
&-2192511832\,0306893619\,2r_0 + 1391401739\,5579374796\,80r_2 = 0
\end{aligned}
\tag{2.38}
$$

The solution of (2.38) is

$$r_0 = 26r_2/165, \quad r_1 = 0 \tag{2.39}$$

where r_0 represents an arbitrary constant. Therefore, when the inertial coefficient equals

$$R(\xi) = \left(\frac{26}{165} + \xi^2\right) r_2, \tag{2.40}$$

the set (2.37) yields

$$\Omega^2 = 4620 b_6 / r_2 \tag{2.41}$$

and

$$b_0 = \frac{7337}{270} b_6, \, b_1 = 0, \, b_2 = -\frac{52}{3} b_6, \, b_3 = 0, \, b_4 = \frac{13}{54} b_6, \, b_5 = 0 \tag{2.42}$$

The stiffness is then

$$D(\xi) = \left(\frac{7337}{270} - \frac{52}{3}\xi^2 + \frac{13}{54}\xi^4 + \xi^6\right) b_6 \tag{2.43}$$

One can show analytically that for n larger than 4 only trivial solution are obtained.

2.6 Cubic Inertial Coefficient (m = 3)

When $n = 0$, instead of (2.23) we obtain,

$$
\begin{aligned}
&-b_1 + b_0 = 0 \\
&-72b_2 + 72b_1 - r_0\Omega^2 = 0 \\
&-144b_3 + 144b_2 - r_1\Omega^2 = 0 \\
&-240b_4 + 240b_3 + 2r_0\Omega^2 - r_2\Omega^2 = 0 \\
&-360b_5 + 360b_4 - r_0\Omega^2 + 2r_1\Omega^2 - r_3\Omega^2 = 0 \\
&-504b_6 + 504b_5 - r_1\Omega^2 + 2r_2\Omega^2 = 0 \\
&-672b_7 + 672b_6 - r_2\Omega^2 + 2r_3\Omega^2 = 0 \\
&672b_7 - r_3\Omega^2 = 0
\end{aligned}
\tag{2.44}
$$

which constitutes 8 equations for 9 unknowns. We can, therefore, solve the system taking b_7 as a parameter. From the last equation of the set (44), we obtain the natural frequency squared

$$
\Omega^2 = 864b_7 / r_3 \tag{2.45}
$$

And the remaining equations yield the coefficients in the stiffness as follows

$$
\begin{aligned}
&b_0 = \frac{1}{35}\frac{(252r_0 + 102r_1 + 51r_2 + 29r_3)b_6}{r_3}, \quad b_1 = \frac{1}{35}\frac{(252r_0 + 102r_1 + 51r_2 + 29r_3)b_6}{r_3} \\
&b_2 = -\frac{1}{35}\frac{(168r_0 - 102r_1 - 51r_2 - 29r_3)b_7}{r_3}, \quad b_3 = -\frac{1}{35}\frac{(168r_0 + 108r_1 - 51r_2 - 29r_3)b_7}{r_3} \\
&b_4 = \frac{1}{35}\frac{(84r_0 - 108r_1 - 75r_2 + 29r_3)b_7}{r_3}, \quad b_5 = \frac{1}{7}\frac{(12r_1 - 15r_2 - 11r_3)b_7}{r_3} \\
&b_6 = \frac{1}{7}\frac{(9r_2 - 11r_3)b_7}{r_3}
\end{aligned}
\tag{2.46}
$$

Consider now the case $n = 1$; the governing differential equation gives a set of 9 linear algebraic equations for 9 unknowns:

$$
\begin{aligned}
& b_1 = 0 \\
& -120b_2 + 120b_0 - \frac{7}{3} r_0 \Omega^2 = 0 \\
& -240b_3 + 240b_1 - \frac{7}{3} r_1 \Omega^2 = 0 \\
& -400b_4 + 400b_2 + \frac{10}{3} r_0 \Omega^2 - \frac{7}{3} r_2 \Omega^2 = 0 \\
& -600b_5 + 600b_3 + \frac{10}{3} r_1 \Omega^2 - \frac{7}{3} r_3 \Omega^2 = 0 \\
& -840b_6 + 840b_4 - r_0 \Omega^2 + \frac{10}{3} r_2 \Omega^2 = 0 \\
& -1120b_7 + 1120b_5 - r_1 \Omega^2 + \frac{10}{3} r_3 \Omega^2 = 0 \\
& 1440b_6 - r_2 \Omega^2 = 0 \\
& 1800b_7 - r_3 \Omega^2 = 0
\end{aligned}
\tag{2.47}
$$

Again, we impose the determinant of the system to vanish, so that we get non-trivial solution. It produces the following equation,

$$
-9899134156\,8r_3 - 3411188121\,60r_1 = 0 \tag{2.48}
$$

whose solution reads

$$
r_1 = -74r_3 / 255 \tag{2.49}
$$

Thus, provided the inertial coefficient equals

$$
R(\xi) = r_0 - \frac{74}{255} r_3 \xi + r_2 \xi^2 + r_3 \xi^3 \tag{2.50}
$$

the set (2.47) yields

$$
\Omega^2 = 1800b_7 / r_3 \tag{2.51}
$$

$$
\begin{aligned}
b_0 &= \frac{1}{28}\frac{(129r_2 + 620r_0)b_7}{r_3}, \quad b_1 = 0 \\
b_2 &= -\frac{3}{28}\frac{(-43r_2 + 120r_0)b_7}{r_3}, \quad b_3 = -\frac{259}{51}b_7 \\
b_4 &= \frac{15}{28}\frac{(4r_0 - 11r_2)b_7}{r_3}, \quad b_5 = -\frac{82}{17}b_7, \quad b_6 = \frac{5}{4}\frac{r_2 b_6}{r_3}
\end{aligned}
\tag{2.52}
$$

For the case $n = 2$, the governing differential equation yields a set of 10 linear algebraic equations for 9 unknowns,

$$
\begin{aligned}
&b_1 = 0 \\
&-180b_2 - 4r_0\Omega^2 = 0 \\
&-360b_3 + 360b_0 - 4r_1\Omega^2 = 0 \\
&-600b_4 + 600b_1 + 5r_0\Omega^2 - 4r_2\Omega^2 = 0 \\
&-900b_5 + 900b_2 + 5r_1\Omega^2 - 4r_3\Omega^2 = 0 \\
&-1260b_6 + 1260b_3 - r_0\Omega^2 + 5r_2\Omega^2 = 0 \\
&-1680b_7 + 1680b_4 - r_0\Omega^2 + 5r_3\Omega^2 = 0 \\
&2160b_5 - r_1\Omega^2 = 0 \\
&2700b_6 - r_2\Omega^2 = 0 \\
&3300b_7 - r_3\Omega^2 = 0
\end{aligned}
\tag{2.53}
$$

A non-trivial solution is obtainable, if the rank of the set (2.53) is less than 9. This requires the 10 minors of order 9 of the matrix of the system (2.53) are zero. We get then 10 equations which reduce to the following two equations

$$
\begin{aligned}
&-316787133358080r_0 + 72597051394560r_1 - 63357426671616r_3 = 0 \\
&36769935121920r_0 - 31678713335808r_2 + 12702341223936r_3 = 0
\end{aligned}
\tag{2.54}
$$

The solution of (2.54) is then

$$
r_1 = \frac{48}{55}r_3 + \frac{48}{11}r_0, \quad r_2 = \frac{247}{616}r_3 + \frac{65}{56}r_0 \tag{2.55}
$$

where r_0 and r_3 are considered as arbitrary constants. We conclude for an inertial coefficient taken as follows

$$
R(\xi) = r_0 + \left(\frac{48}{55}r_3 + \frac{48}{11}r_0\right)\xi + \left(\frac{247}{616}r_3 + \frac{65}{56}r_0\right)\xi^2 + r_3\xi^3 \tag{2.56}
$$

The set (2.53) yields

$$\Omega^2 = 3300 b_7 / r_3 \tag{2.57}$$

$$\begin{aligned}
b_0 &= \frac{5}{882}\frac{(+25793r_0 + 4805r_3)b_7}{r_3}, \quad b_1 = 0 \\
b_2 &= -\frac{220}{3}\frac{r_0 b_7}{r_3}, \quad b_3 = -\frac{221}{882}\frac{(55r_0 + 19r_3)b_7}{r_3} \\
b_4 &= \frac{1}{28}\frac{(55r_0 - 247r_3)b_7}{r_3}, \quad b_5 = \frac{4}{3}\frac{(5r_0 + r_3)b_7}{r_3} \\
b_6 &= \frac{13}{504}\frac{(55r_0 + 19r_3)b_7}{r_3}
\end{aligned} \tag{2.58}$$

It turns out that for n larger than 3, the set of algebraic equations deduced from the governing equation does not possess a non-trivial solution.

2.7 Particular Case m = 4

We consider first the case $n = 0$. We get a set of 9 equations for 10 unknowns

$$\begin{aligned}
&-b_1 + b_0 = 0 \\
&-72b_2 + 72b_1 - r_0\Omega^2 = 0 \\
&-144b_3 + 144b_2 - r_1\Omega^2 = 0 \\
&-240b_4 + 240b_3 + 2r_0\Omega^2 - r_2\Omega^2 = 0 \\
&-360b_5 + 360b_4 - r_0\Omega^2 + 2r_1\Omega^2 - r_3\Omega^2 = 0 \\
&-504b_6 + 504b_5 - r_1\Omega^2 + 2r_2\Omega^2 - r_4\Omega^2 = 0 \\
&-672b_7 + 504b_6 - r_2\Omega^2 + 2r_3\Omega^2 = 0 \\
&-864b_8 + 864b_7 - r_3\Omega^2 + 2r_4\Omega^2 = 0 \\
&1080b_8 - r_4\Omega^2 = 0
\end{aligned} \tag{2.59}$$

We can, therefore, solve the system taking b_8 as a parameter. The last equation of the set (2.59) yields the natural frequency squared

$$\Omega^2 = 1080 b_8 / r_4 \tag{2.60}$$

and the remaining equations give the coefficients in the stiffness as follows

$$
\begin{aligned}
b_0 &= (252r_0 + 102r_1 + 51r_2 + 29r_3 + 18r_4)b_8 / 28r_4 \\
b_1 &= (252r_0 + 102r_1 + 51r_2 + 29r_3 + 18r_4)b_8 / 28r_4 \\
b_2 &= (-168r_0 + 102r_1 + 51r_2 + 29r_3 + 18r_4)b_8 / 28r_4 \\
b_3 &= (-168r_0 - 108r_1 + 51r_2 + 29r_3 + 18r_4)b_8 / 28r_4 \\
b_4 &= (84r_0 - 108r_1 - 75r_2 + 29r_3 + 18r_4)b_8 / 28r_4 \\
b_5 &= (60r_1 - 75r_2 - 55r_3 + 18r_4)b_8 / 28r_4 \\
b_6 &= (45r_2 - 55r_3 - 42r_4)b_8 / 28r_4, \quad b_7 = (5r_3 - 6r_4)b_8 / 4r_4
\end{aligned}
\tag{2.61}
$$

Note that equation (2.60) coincides with Candan and Elishakoff (1990) result, as part of a general case. Indeed substituting $m = 4$ in their equation (2.45) of the general natural frequency squared we get the same expression. When $n = 1$, the governing differential equation yields a set of 10 linear algebraic equations for 10 unknowns, specifically the coefficients b_i in the stiffness and the natural frequency squared Ω^2

$$
\begin{aligned}
& b_1 = 0 \\
& -120b_2 + 120b_0 - \tfrac{7}{3}r_0\Omega^2 = 0 \\
& -240b_3 + 240b_1 - \tfrac{7}{3}r_1\Omega^2 = 0 \\
& -400b_4 + 400b_2 + \tfrac{10}{3}r_0\Omega^2 - \tfrac{7}{3}r_2\Omega^2 = 0 \\
& -600b_5 + 600b_3 + \frac{10}{3}r_1\Omega^2 = 0 \\
& -840b_6 + 840b_4 - r_0\Omega^2 + \tfrac{10}{3}r_2\Omega^2 = 0 \\
& 1120b_5 - r_1\Omega^2 = 0 \\
& 1440b_6 - r_2\Omega^2 = 0
\end{aligned}
\tag{2.62}
$$

To find a non-trivial solution, the determinant must be zero. This leads to the following equation

$$
72597051394560r_1 - 63357426671616r_3 = 0 \tag{2.63}
$$

which is solved as follows

$$
r_3 = -255r_1 / 74 \tag{2.64}
$$

Hence, for the inertial coefficient

$$R(\xi) = r_0 + r_1\xi + r_2\xi^2 - \frac{255}{74} r_1\xi^3 + r_4\xi^4 \tag{2.65}$$

where r_0 and r_2 are arbitrary, the set (2.26) has the following non-trivial solution

$$\Omega^2 = 2200 b_8 / r_4 \tag{2.66}$$

$$\begin{aligned}
b_0 &= \frac{(20460r_0 + 4257r_2 + 1526r_4)b_8}{756r_4}, \quad b_1 = 0 \\
b_2 &= -\frac{(11880r_0 - 4257r_2 - 1526r_4)b_8}{756r_4}, \quad b_3 = -\frac{385r_1b_8}{18r_4} \\
b_4 &= \frac{(1980r_0 - 5445r_2 + 1526r_4)b_8}{756r_4}, \quad b_5 = \frac{2255r_1b_8}{111r_4} \\
b_6 &= \frac{(165r_2 - 442r_4)b_8}{108r_4}, \quad b_7 = -\frac{935r_1b_8}{222r_4}
\end{aligned} \tag{2.67}$$

Concerning the case $n = 2$, the governing equation yields the following set

$$\begin{aligned}
&b_1 = 0 \\
&-180b_2 - 4r_0\Omega^2 = 0 \\
&-360b_3 + 360b_0 - 4r_1\Omega^2 = 0 \\
&-600b_4 + 400b_1 + 5r_0\Omega^2 - 4r_2\Omega^2 = 0 \\
&-900b_5 + 900b_2 + 5r_1\Omega^2 - 4r_3\Omega^2 = 0 \\
&-1260b_6 + 1260b_3 + 5r_2\Omega^2 - 4r_4\Omega^2 = 0 \\
&-1680b_7 + 1680b_4 - r_0\Omega^2 + 5r_3\Omega^2 = 0 \\
&-2160b_8 + 2160b_5 - r_1\Omega^2 + 5r_4\Omega^2 = 0 \\
&2700b_6 - r_2\Omega^2 = 0 \\
&3300b_7 - r_3\Omega^2 = 0 \\
&3960b_8 - r_4\Omega^2 = 0
\end{aligned} \tag{2.68}$$

We have a set of 11 equations for 10 unknowns, namely the coefficients b_i in the stiffness and the natural frequency squared Ω^2. In order to have a non-trivial solution, the rank of the set (2.67) must be less than 10, or all 11 minors of order 10 of the matrix of the set (2.68) should vanish. This last condition leads to a set of 11 equations whose solution reads

$$r_2 = -\frac{65}{77}r_0 + \frac{1235}{2688}r_1 + \frac{8645}{46464}r_4, \quad r_3 = -5r_0 + \frac{55}{48}r_1 + \frac{245}{528}r_4 \tag{2.69}$$

where r_0, r_1 and r_4 are treated as arbitrary constants. Hence, we observe that for an inertial coefficient equal to

$$R(\xi) = r_0 + r_1\xi + \left(-\frac{65}{77} r_0 + \frac{1235}{2688} r_1 + \frac{8645}{46464} r_4 \right)\xi^2 + \left(-5r_0 + \frac{55}{48} r_1 + \frac{245}{528} r_4 \right)\xi^3 + r_4\xi^4 \tag{2.70}$$

the set (2.67) yields

$$\Omega^2 = 3960 b_8 / r_4 \tag{2.71}$$

$$\begin{aligned}
b_0 &= \frac{(933504\, r_0 + 2907025\, r_1 + 769993\, r_4) b_8}{77616\, r_4}, \quad b_1 = 0 \\
b_2 &= -\frac{88 r_0 b_8}{r_4}, \quad b_3 = \frac{(933504\, r_0 - 508079\, r_1 + 769993\, r_4) b_8}{77616\, r_4} \\
b_4 &= \frac{(136224\, r_0 - 29887\, r_1 - 12103\, r_4) b_8}{2464\, r_4}, \quad b_5 = \frac{(11 r_1 - 49 r_4) b_8}{6 r_4} \\
b_6 &= -\frac{(4224\, r_0 - 2299\, r_1 - 931 r_4) b_8}{44352\, r_4}, \quad b_7 = -\frac{(528 r_0 - 121 r_1 - 49 r_4) b_8}{88 r_4}
\end{aligned} \tag{2.72}$$

For $n = 3$, we have to satisfy the following set of equations

$$
\begin{aligned}
&b_1 = 0\\
&-252b_2 - 6r_0\Omega^2 = 0\\
&-504b_3 - 6r_1\Omega^2 = 0\\
&-840b_4 + 840b_0 + 7r_0\Omega^2 - 6r_2\Omega^2 = 0\\
&-1260b_5 + 1260b_1 + 7r_1\Omega^2 - 6r_3\Omega^2 = 0\\
&-1764b_6 + 1764b_2 + 7r_2\Omega^2 - 6r_4\Omega^2 = 0\\
&-2352b_7 + 2352b_3 + 7r_3\Omega^2 = 0\\
&-3024b_8 + 3024b_4 - r_0\Omega^2 + 7r_4\Omega^2 = 0\\
&3780b_5 - r_1\Omega^2 = 0\\
&4620b_6 - r_2\Omega^2 = 0\\
&5544b_7 - r_3\Omega^2 = 0\\
&6552b_8 - r_4\Omega^2 = 0
\end{aligned}
\tag{2.73}
$$

We get a set of 12 equations for 10 unknowns. A non-trivial solution is obtainable, if the rank of the set (2.73) is less than 10. This imposes all 66 minors of order 10 of the matrix of the linear system (2.37) to be zero. This condition is satisfied in this case, leading to the following inertial coefficient

$$
r_0 + \left(\frac{165}{26} r_0 + \frac{165}{182} r_4\right)\xi^2 + r_4\xi^4 \tag{2.74}
$$

where r_0 and r_4 are treated as arbitrary constants. The set (73) yields then

$$
\Omega^2 = 6552\frac{b_8}{r_4} \tag{2.75}
$$

$$
\begin{aligned}
&b_0 = \frac{(51359r_0 + 5935r_4)b_8}{210r_4}, \quad b_1 = 0\\
&b_2 = -\frac{156r_0b_8}{r_4}, \quad b_3 = 0\\
&b_4 = \frac{(13r_0 - 85r_4)b_8}{6r_4}, \quad b_5 = 0\\
&b_6 = \frac{(63r_0 + 9r_4)b_8}{7r_4}, \quad b_7 = 0
\end{aligned}
\tag{2.76}
$$

2.8 Discussion of Part 2

The procedure described in the body of this study can be applied for larger values of m. We discuss some general patterns obtainable for various values of m, representing the *maximum* power in the polynomial expression of the beam's material density. The solution depends on the value of integer n, with $n + 4$ defining the *maximum* degree of the polynomial in the postulated mode shape. We note that for $n = 0$, the number of equations is smaller than the number of unknowns and hence one of the coefficients b_i can be designated as an arbitrary one. For the particular case $n = 1$ the number of equations equals the number of unknowns. Since the set is a homogeneous one, the determinant must vanish for the non-trivial solution. This leads to a condition between the inertial coefficients, at which the solution is obtainable. For $n \geq 2$, the number of equations exceeds that of the unknowns. Hence, for the availability of the non-trivial solution the rank of obtained set of linear equations should be less than the number of unknowns. This results generally in large number of determinant equations which yield relationships between the inertial coefficients. The method expounded in this chapter appears to be extremely simple. It is hoped that it will trigger additional research and will find its way into future vibration textbooks and numerical studies as a benchmark solution.

3. Part 3: Vibration of Heterogeneous Clamped Circular Plates

3.1 Motivation of Part 3

The free vibrations of circular plates attracted investigators nearly two centuries ago (Chladni 1803, Poisson 1829). The vibrations of uniform plates have been studied quite extensively (Gontkevich 1964, Leissa 1969-1987). Although circular plates of variable thickness received much less attention than the uniform plates, still, they have been investigated quite intensively. An exact solution was derived by Conway *et al* 1964 by noting an analogy (Conway, 1957,1958) that exists between the free vibration of truncated-cone beams and linearly tapered plates for the special case when Poisson's ratio equals one third. Conway *et al* 1964 derived natural frequencies for clamped tapered circular plates. Series solutions have been provided by Soni (1972) for plates with quadratically varying thickness. Method of Frobenius was utilized by Jain (1972). Frequency parameters of clamped and simply supported plates were computed, for the first two modes, for various values of a taper parameter and inplane force, both for linear and parabolic variations of thickness. A perturbation method based on a small parameter was employed by Yang (1993). Zeroth and first-order asympototic solutions were obtained for the natural frequencies of a clamped plate with linearly varying thickness. Lenox and Conway (1980) obtained an exact expression for the buckling mode for an annular plate; at a later stage they performed numerical (computerized) calculations for natural frequencies.

Most of the reported solutions utilized approximate techniques. Laura *et al* (1982) used the Rayleigh-Ritz method with polynomial coordinate functions that identically satisfy the external boundary conditions to study the vibrations and elastic stability of polar orthotropic circular plates of linearly varying thickness. Lal and Gupta (1982) utilized Chebychev polynomials to obtain the frequencies of polar orthotropic annular plates of variable thickness. Gorman (1983) and Kanaka Raju (1977) employed the most universal technique, that of the finite element method for studying the dynamic behavior of polar orthotropic annular plates of variable thickness.

As far as exact solutions are concerned, one very important paper should be mentioned. Harris (1968) obtained a *closed-form* solution for both the mode shapes and the natural frequencies of the circular plate that is free at its edge. The stiffness was given as $D(r) = D_0\left(1 - r^2/R^2\right)^3$, where D_0 is the stiffness at the center, r is the polar coordinate, R is the radius.

The present part of this chapter shares with the paper by Harris (1968) the property of offering exact, closed-form solutions for the natural frequency. However, whereas Harris (1968) solves a *direct* vibration problem, we pose the *inverse* problem. Namely, we postulate the vibration mode and pursue the following objective : Find the variation of the stiffness that leads to the postulated vibration mode. In some circumstances such a problem has a unique solution with a remarkable by-product: a closed-form expression for the natural frequency. This paper complements that of Harris (1968)], who derived a solution for the circular plates that are free at the boundary, while present part deals with clamped plates.

In his note Conway (1958) derived the closed-form expression for the unsymetrical bending of a particular circular plate resting on a Winkler foundation. He mentioned: "What is remarkable about this solution is that it is in a closed-form which is far simpler than the constant thickness disk solution which involves Kelvin functions." In this chapter vibration case is considered and closed-form solution is derived. Yet, it appears that the presented solution is superior to that by Lenox and Conway (1980) for here we derive not only the closed-form expression for the displacement, but the *natural frequency* too. The unusual aspect characteristic in the paper by Conway (1980) is preserved in this study: While for the exact solution for homogeneous circular plates one has the Bessel functions involved, in the heterogeneous case the elementary functions turn out to suffice.

3.2 Basic Equations

The differential equation that governs the free nonaxisymmetric vibrations of the circular plate with variable thickness reads (Kovalenko, 1959)

$$D(r)r^3\Delta\Delta W + \frac{dD}{dr}\left(2r^3\frac{d^3W}{dr^3} + r^2(2+\nu)\frac{d^2W}{dr^2} - r\frac{dW}{dr}\right)$$
$$+\frac{d^2D}{dr^2}\left(r^3\frac{d^2W}{dr^2} + \nu r^2\frac{dW}{dr}\right) - \rho h\omega^2 r^3 W = 0 \tag{3.1}$$

where Δ is the Laplace operator in polar coordinates

$$\Delta = \frac{d^2}{dr^2} + \frac{1}{r}\frac{d}{dr} \tag{3.2}$$

D is the bending stiffness, assumed to vary along the radial coordinate r,

$$D = D(r) = \frac{Eh^3}{12(1-\nu^2)} \tag{3.3}$$

h is the thickness, ν is the Poisson ratio, ρ is the material density, r is the radial coordinate, φ is the circumferential coordinate, W the mode shape. The Poisson ratio ν is assumed to be constant. Note that the governing equation reported by Kovalenko (1959) is multiplied here by the term r^3, for further convenience. We are looking for the case when the inertial term

$$\delta(r) = \rho h \tag{3.4}$$

varies along r ; likewise, the stiffness is a function of r. we are interested in finding the *closed-form* solution for the natural frequency ω.

We pose the problem as an *inverse* vibration study. Note first that for the uniform circular plate that is under uniform load q_0 the displacement can be put in the following form (Timoshenko and Woinowsky - Krieger, 1959)

$$w = \frac{q_0}{64D}\left(R^2 - r^2\right)^2 \tag{3.5}$$

where R is the outer radius of the plate. We are interested in determining such a variation of $D(r)$ in Eq. (3.1) that the function

$$W(r) = \left(R^2 - r^2\right)^2 \tag{3.6}$$

serves as an exact mode shape, obtained from the static displacement (3.5) by formally putting $q_0 = 64D$. We confine our interest to circular plate that is clamped along the boundary $r = R$.

3.3 Method of Solution

We assume that the inertial term is represented as a polynomial

$$\delta(r) = \sum_{i=0}^{m} a_i r^i \tag{3.7}$$

Since $W(r)$ is the fourth order polynomial expression in terms of r, in view of Eq. (3.7) the last term in Eq. (3.1) is the polynomial expression of order $m + 7$. Since the operator $\Delta\Delta$ in Eq. (3.1) involves the four-fold differentiation with respect to r, in order for the highest degree of the first term's polynomial expression in $Dr^3\Delta\Delta W$ to be of order $m + 7$, it is necessary and sufficient the stiffness to be represented as a polynomial of degree $m + 4$. Thus, the sought after stiffness can be put in the following form

$$D(r) = \sum_{i=0}^{m+4} b_i r^i \tag{3.8}$$

Furhter steps involve the substitution of Eqs. (3.6)-(3.8) into the governing differential equation (3.1) and demanding that the resulting polynomial expression vanish. This implies that all the coefficients in front of powers r^i must be zero, leading, in turn, to the set of algebraic equations in terms of b_i and ω^2. Various expressions for the inertial term $\delta(r)$ in Eq. (3.4) will be considered.

3.4 Constant Inertial Term (m = 0)

In this case the stiffness is sought as a fourth order polynomial

$$D(r) = b_0 + b_1 r + b_2 r^2 + b_3 r^3 + b_4 r^4 \tag{3.9}$$

We get instead the differential equation (1) the following equation

$$\sum_{i=0}^{7} c_i r^i = 0 \tag{3.10}$$

where

$$\begin{aligned} &c_0 = 0, \quad c_1 = 0, \quad c_2 = -4(1+\nu)R^2 b_1, \quad c_3 = 64 b_0 - 16(1+\nu)R^2 b_2 - a_0\omega^2 R^4 \\ &c_4 = 12(11+\nu)b_1 - 36(1+\nu)R^2 b_3, \quad c_5 = 32(7+\nu)b_2 - 64(1+\nu)R^2 b_4 + 2a_0\omega^2 R^2 \\ &c_6 = (340+60\nu)b_3, \quad c_7 = 96(5+\nu)b_4 - a_0\omega^2 \end{aligned} \tag{3.11}$$

Since the left side of the differential Eq. (3.10) must vanish for any r within $[0;R]$, all the coefficients c_i must be zero. This leads to a homogeneous set of 6 linear algebraic equations for 6 unknowns. This is equivalent to the determinant of the matrix of the set derived from Eq. (3.11) being identically zero. Therefore, a non-trivial solution is obtainable. From the requirement $c_7 = 0$, the natural frequency squared is obtained as follows

$$\omega^2 = 96(5+\nu)b_4 / a_0 \tag{3.12}$$

Upon substitution of Eq. (3.12) into (3.11), the remaining equations yield the coefficients in the stiffness

$$b_0 = \frac{13+\nu}{2} R^4 b_4, \quad b_1 = 0, \quad b_2 = -4R^2 b_4, \quad b_3 = 0 \tag{3.13}$$

Hence, the stiffness reads

$$D(r) = \left(\frac{13+\nu}{2} R^4 - 4R^2 r^2 + r^4 \right) b_4 \tag{3.14}$$

3.5 Linearly Varying Inertial Term (m = 1)

Instead of the set (3.11) we get here, 7 linear algebraic equations with 7 unknowns

$$\begin{aligned}
&-4(1+\nu)R^2 b_1 = 0 \\
&64b_0 - 16(1+\nu)R^2 b_2 - a_0\omega^2 R^4 = 0 \\
&12(11+\nu)b_1 - 36(1+\nu)R^2 b_3 - a_1\omega^2 R^4 = 0 \\
&32(7+\nu)b_2 - 64(1+\nu)R^2 b_4 + 2a_0\omega^2 R^2 = 0 \\
&(340+60\nu)b_3 - 100(1+\nu)R^2 b_5 + 2a_1\omega^2 R^2 = 0 \\
&96(5+\nu)b_6 - a_0\omega^2 = 0 \\
&(644+140\nu)b_5 - a_1\omega^2 = 0
\end{aligned} \tag{3.15}$$

In order to have a non-trivial solution the determinant of the set of equations (3.15)

$$(1+\nu)(7+\nu)\left(13765 + 5643\nu + 728\nu^2 + 30\nu^3\right)a_1 = 0 \tag{3.16}$$

must vanish, leading to $a_1 = 0$. Yet this result signifies that the linear inertial coefficient is a constant. Thus, the problem solved in the previous section is obtained.

3.6 Parabolically Varying Inertial Term (m = 2)

For $m = 2$, i.e. the plate whose material density varies parabolically

$$\delta(r) = a_0 + a_1 r + a_2 r^2 \tag{3.17}$$

the bending stiffness has to be sought as the sixth order polynomial

$$D(r) = b_0 + b_1 r + b_2 r^2 + b_3 r^3 + b_4 r^4 + b_5 r^5 + b_6 r^6 \tag{3.18}$$

Substitution of Eq. (3.6) in conjunction with Eqs. (3.17) and (3.18) into the governing differential equation (3.1) yields

$$\sum_{i=0}^{9} d_i r^i = 0 \tag{3.19}$$

where

$$d_0=0,\quad ,d_1=0,\quad d_2=-4R^2(1+\nu)b_1,\quad d_3=64b_0-16R^2(1+\nu)b_2-a_0\omega^2R^4$$
$$d_4=12(11+\nu)b_1-36R^2(1+\nu)b_3-a_1\omega^2R^4$$
$$d_5=32(7+\nu)b_2-64R^2(1+\nu)b_4+\omega^2\left(2a_0R^2-a_2R^4\right)$$
$$d_6=20(17+3\nu)b_3-100R^2(1+\nu)b_5+2a_1\omega^2R^2 \tag{3.20}$$
$$d_7=96(5+\nu)b_4-144R^2(1+\nu)b_6-\omega^2\left(a_0-2a_2R^2\right)$$
$$d_8=(644+140\nu)b_5-a_1\omega^2,\quad d_9=64(13+3\nu)b_6-a_2\omega^2$$

As in the case of the constant inertial term, we demand that all $d_i=0$. Thus, we get a set of 8 equations with 8 unknowns (7 coefficients b_i and ω^2). The resulting determinant equation is

$$(1+\nu)(7+\nu)a_1\left(178945+114654\nu+26393\nu^2+2574\nu^3+90\nu^4\right)=0 \tag{3.21}$$

In order the homogeneous system to possess a non-trivial solution the coefficient a_1 to vanish. We substitute $a_1=0$ into the set (3.20). For the natural frequency we arrive at the following expression, obtainable from the requirement $d_9=0$:

$$\omega^2=64(13+3\nu)b_6/a_2 \tag{3.22}$$

Then, the coefficients in the stiffness are obtained as follows

$$b_0=\frac{R^4b_6}{12}\frac{\left(295+353\nu+61\nu^2+3\nu^3\right)R^2a_2+\left(4732+2132\nu+292\nu^2+12\nu^3\right)a_0}{\left(35+12\nu+\nu^2\right)a_2}$$
$$b_1=0,\quad b_2=\frac{R^2b_6}{3}\frac{\left(295+58\nu+3\nu^2\right)R^2a_2-\left(728+272\nu+24\nu^2\right)a_0}{\left(35+12\nu+\nu^2\right)a_2} \tag{3.23}$$
$$b_3=0,\quad b_4=-\frac{b_6}{6}\frac{(95+15\nu)R^2a_2-(52+12\nu)a_0}{(5+\nu)a_2},\quad b_5=0$$

where b_6 is an arbitrary constant. In order the natural frequency squared to be a positive quantity we demand that the ratio b_6/a_2 be positive. We have two sub-cases : (i) both b_6 and a_2 are positive, (ii) both are negative. In the former case (i) the necessary condition $b_0\geq 0$ for positivity of the stiffness *D(r)* is identically satisfied. In the latter case the above inequality reduces to

$$\left(295+353\nu+61\nu^2+3\nu^3\right)R^2a_2+\left(4732+2132\nu+292\nu^2+12\nu^3\right)a_0\geq 0 \tag{3.24}$$

leading to the following inequality

$$\frac{a_0}{|a_2|R^2} \le \frac{295+353\nu+61\nu^2+3\nu^3}{4732+2132\nu+292\nu^2+12\nu^3} \tag{3.25}$$

One can immediately see that the ratio

$$\frac{a_0}{|a_2|R^2} < 1 \tag{3.26}$$

Hence, the associated variation of the inertial coefficient

$$\delta(r) = a_0 + a_2 r^2 \tag{3.27}$$

Takes a negative value at $r = R$. Thus, the possibility that both a_2 and b_6 be negative should be discarded as physically unrealizable one. We conclude, therefore, that in Eq. (3.22) both a_2 and b_6 must constitute positive quantities.

3.7 Cubic Inertial Term (m = 3)

For $m = 3$, the following set of 9 linear algebraic equations with 9 unknowns is obtained

$$\begin{aligned}
&-4R^2(1+\nu)b_1 = 0\\
&64b_0 - 16R^2(1+\nu)b_2 - a_0\omega^2R^4 = 0\\
&12(11+\nu)b_1 - 36R^2(1+\nu)b_3 - a_1\omega^2R^4 = 0\\
&32(7+\nu)b_2 - 64R^2(1+\nu)b_4 + \omega^2\left(2a_0R^2 - a_2R^4\right) = 0\\
&20(17+3\nu)b_3 - 100R^2(1+\nu)b_5 + \omega^2\left(2a_1R^2 - a_3R^4\right) = 0\\
&96(5+\nu)b_4 - 144R^2(1+\nu)b_6 - \omega^2\left(a_0 - 2a_2R^2\right) = 0\\
&(644+140\nu)b_5 - 196(1+\nu)R^2b_7 - \omega^2\left(a_1 - 2a_3R^2\right) = 0\\
&(832+192\nu)b_6 - a_2\omega^2 = 0\\
&(1044+252\nu)b_7 - a_3\omega^2 = 0
\end{aligned} \tag{3.28}$$

The resulting determinant is

$$(1+\nu)(7+\nu)M = 0 \tag{3.29}$$

where

$$M = \left(490685R^2 + 791887\nu + 367095\nu^2 + 71999\nu^3 + 6316\nu^4 + 210\nu^5\right)R^2 a_3$$
$$\left(5189405 + 4577581\nu + 1567975\nu^2 + 259397\nu^3 + 20628\nu^4 + 630\nu^5\right)a_1 \quad (3.30)$$

The solution of Eq. (3.29) is

$$a_1 = -\frac{\left(7549 + 8931\nu + 1452\nu^2 + 70\nu^3\right)a_3}{79837 + 36033\nu + 4916\nu^2 + 210\nu^3} \quad (3.31)$$

Upon substitution of Eq. (3.28) into all equations of the set (3.25), the natural frequency squared equals

$$\omega^2 = \frac{36(29 + 7\nu)b_7}{a_3} \quad (3.32)$$

and we get the following solution for the coefficients in the stiffness

$$\begin{aligned}
b_0 &= 3\left[\left(8555 + 12302\nu + 4240\nu^2 + 514\nu^3 + 21\nu^4\right)R^2 a_2 + \left(137228 + 94952\nu \right.\right.\\
&\quad \left.\left. + 23392\nu^2 + 2392\nu^3 + 84\nu^4\right)a_0\right]R^4 b_7 / 64a_3\left(455 + 261\nu + 49\nu^2 + 3\nu^3\right)\\
b_1 &= 0\\
b_2 &= 3\left[\left(8555 + 3747\nu + 493\nu^2 + 21\nu^3\right)R^2 a_2 - \left(21112 + 12984\nu + 2600\nu^2 \right.\right.\\
&\quad \left.\left. + 168\nu^3\right)a_0\right]R^2 b_7 / 16a_3\left(455 + 261\nu + 49\nu^2 + 3\nu^3\right)\\
b_3 &= \frac{\left(7549 + 1382\nu + 70\nu^2\right)R^4 b_7}{2753 + 578\nu + 30\nu^2}\\
b_4 &= \frac{2b_7\left[\left(1508 - 712\nu + 84\nu^2\right)a_0 - \left(2755 + 1100\nu + 105\nu^2\right)R^2 a_2\right]}{32a_3\left(65 + 28\nu + 3\nu^2\right)}\\
b_5 &= -\frac{2R^2\left(4255 + 832\nu + 42\nu^2\right)b_7}{2753 + 578\nu + 30\nu^2}, \quad b_6 = \frac{9(29 + 7\nu)a_2 b_7}{16a_3(13 + 3\nu)}
\end{aligned} \quad (3.33)$$

where b_7 is an arbitrary constant. For the particular case $\nu = 1/3$, the stiffness equals

$$D(r)=\left[\frac{2773}{2464}\frac{R^6a_2}{a_3}+\frac{235}{16}\frac{R^4a_0}{a_3}+\left(\frac{8319}{2464}\frac{R^4a_2}{a_3}-\frac{141}{112}\frac{R^2a_0}{a_3}\right)r^2+\frac{72157}{26541}R^4r^3\right.$$
$$\left.+\left(-\frac{3525}{896}\frac{R^2a_2}{a_3}+\frac{141}{64}\frac{a_0}{a_3}\right)r^4-\frac{9074}{2949}R^2r^5+\frac{141}{112}\frac{a_2}{a_3}r^6+r^7\right]b_7 \tag{3.34}$$

3.8 General Inertial Term (m ≥ 4)

Consider now the general expression of the inertial term given in Eq. (3.7), and the stiffness in Eq. (3.8), for $m \geq 4$. Substitution of Eqs. (3.6), (3.7) and (3.8) into the terms of the differential equation yields

$$r^3D(r)\Delta\Delta W=-64r^3\sum_{i=0}^{m+4}b_ir^i \tag{3.35}$$

$$\frac{dD}{dr}\left(2r^3\frac{d^3W}{dr^3}+r^2(2+\nu)\frac{d^2W}{dr^2}-r\frac{dW}{dr}\right)=4\left[(17+3\nu)r^2-R^2(1+\nu)\right]r^2\sum_{i=1}^{m+4}ib_ir^{i-1} \tag{3.36}$$

$$\frac{d^2D}{dr^2}\left(r^3\frac{d^2W}{dr^2}+\nu r^2\frac{dW}{dr}\right)=4\left[(3+\nu)r^2-R^2(1+\nu)\right]r^3\sum_{i=2}^{m+4}i(i-1)b_ir^{i-2} \tag{3.37}$$

and

$$\rho h\omega^2r^3W=\omega^2r^3\left(R^2-r^2\right)^2\sum_{i=0}^{m}a_ir^i \tag{3.38}$$

Demanding the sum of (3.41)-(3.44) to be zero, we obtain the following equation

$$\sum_{i=0}^{m+7}g_ir^i=0 \tag{3.39}$$

where the coefficients g_i are

$$g_0=0,\quad g_1=0 \tag{3.40}$$

$$g_2=-4(1+\nu)b_1 \tag{3.41}$$

$$g_3 = 64b_0 - 16R^2(1+\nu)b_2 - a_0R^4\omega^2 \tag{3.42}$$

$$g_4 = 12(11+\nu)b_1 - 36R^2(1+\nu)b_3 - a_1R^4\omega^2 \tag{3.43}$$

$$g_5 = 32(7+\nu)b_2 - 64R^2(1+\nu)b_4 - \omega^2(a_2R^4 - 2a_0R^2) \tag{3.44}$$

$$g_6 = 20(17+3\nu)b_3 - 100R^2(1+\nu)b_5 - \omega^2(a_3R^4 - 2a_1R^2) \tag{3.45}$$

$$\vdots$$

for $7 \le i \le m+3$

$$g_i = [64 + 4(i-1)(17+3\nu) + 4(i-2)(i-1)(3+\nu)]b_{i-1} - 4(i+1)^2R^2(1+\nu)b_{i+1} - \omega^2(a_{i-1}R^4 - 2a_{i-3}R^2 + a_{i-5}) \tag{3.46}$$

$$\vdots$$

$$g_{m+4} = [64 + 4(m+1)(17+3\nu) + 4m(m+1)(3+\nu)]b_{m+1} - 4(m+3)^2R^2b_{m+3} - \omega^2(-2a_{m-1}R^2 + a_{m-3}) \tag{3.47}$$

$$g_{m+5} = [64 + 4(m+2)(17+3\nu) + 4(m+1)(m+2)(3+\nu)]b_{m+2} - 4(m+4)^2R^2b_{m+4} - \omega^2(-2a_mR^2 + a_{m-2}) \tag{3.48}$$

$$g_{m+6} = [64 + 4(m+3)(17+3\nu) + 4(m+2)(m+3)(3+\nu)]b_{m+3} - \omega^2a_{m-1} \tag{3.49}$$

$$g_{m+7} = [64 + 4(m+4)(17+3\nu) + 4(m+3)(m+4)(3+\nu)]b_{m+4} - \omega^2a_m \tag{3.50}$$

We demand all coefficients g_i to be zero, thus, we get a set of m + 6 homogeneous linear algebraic equations for m + 6 unknowns. In order to find a non-trivial solution the determinant of the set (3.40)-(3.50) must vanish. We expand the determinant along the last column of the matrix of the set, getting a linear algebraic expression with the coefficients a_i as coefficients. The determinant yields a condition for which the non-trivial solution is obtainable. In this case the general expression of the natural frequency squared is obtained from the equation $g_{m+7} = 0$, resulting in

$$\omega^2 = [64 + 4(m+4)(17+3\nu) + 4(m+3)(m+4)(3+\nu)]b_{m+4}/a_m \tag{3.51}$$

Note that the formulas pertaining the cases $m = 0$, $m = 2$ and $m = 3$ are formally obtainable from Eq. (3.51) by appropriate substitution.

3.9 Alternative Mode Shapes

Let us pose now the following question. In previous sections we postulated the expression given in Eq. (3.6), which is proportional to the deflection of uniform circular plates under distributed loading. Eq. (3.6) represents a fourth order polynomial. A natural question arises: Can an heterogeneous circular plate possess a simpler expression? A simplest polynomial expression, that satisfies the boundary conditions is

$$\psi(r) = (R - r)^2 \tag{3.52}$$

which represents a second order polynomial. Third order polynomial

$$\psi(r) = (R - r)^3 \tag{3.53}$$

as well as the fourth order polynomial

$$\psi(r) = (R - r)^4 \tag{3.54}$$

also satisfy the boundary conditions. Note that Eq. (3.54) is also a fourth order polynomial, as Eq. (3.6) although they are different. Expressions (3.52)-(3.54) are the candidate functions for both Rayleigh-Ritz or Boobnov-Galerkin methods. Thus, in essence we ask if the coordinate functions utilizable for *approximate* evaluation of natural frequencies of either homogeneous or heterogeneous plates, can serve as *exact* buckling modes. We consider the candidate mode shape given in Eq. (3.52).

3.10 Parabolic Mode Shape

Substitution of Eq. (3.52) into the differential equation (3.1) in conjunction with Eq. (3.9) for constant inertia term ($m = 0$) yields the following equation

$$\sum_{i=0}^{7} e_i r^i = 0 \tag{3.55}$$

where

$$\begin{aligned}
&e_0 = -2b_0 R, \quad e_1 = 0, \quad e_2 = 2(1+\nu)b_1 + 2R(1-2\nu)b_2 \\
&e_3 = 8(1+\nu)b_2 + 4R(1-3\nu)b_3 - a_0 R^2\omega^2 \\
&e_4 = 18(1+\nu)b_3 - 6R(1-4\nu)b_4 + 2a_0 R^2\omega^2 \\
&e_5 = 32(1+\nu)b_4 - a_0\omega^2, \quad e_6 = 0, \quad e_7 = 0
\end{aligned} \tag{3.56}$$

In order Eq. (3.55) to be valid for every r, we require $e_i = 0$, for i taking values 0, 2, 3, 4, 5, for the remaining requirements are identically satisfied. We get 5 equations for 6 unknowns. Taking b_4 to be an arbitrary constant we get the expression for the natural frequency squared

$$\omega^2 = 32(1+\nu)b_4/a_0 \tag{3.57}$$

With attendant stiffness coefficients

$$\begin{aligned} b_0 &= 0, \quad b_1 = \frac{R^3(-107+155\nu+106\nu^2+24\nu^3)b_4}{18(1+3\nu+3\nu^2+\nu^3)} \\ b_2 &= \frac{R^2(107+59\nu+12\nu^2)b_4}{18(1+2\nu+\nu^2)}, \quad b_3 = -\frac{5R(7+4\nu)b_4}{9(1+\nu)} \end{aligned} \tag{3.58}$$

The necessary condition for non-negativity of the stiffness is $b_1 \geq 0$. The roots of the equation $b_1 = 0$ are

$$\nu_1 = 1/2, \quad \nu_2 = \left(-59+i\sqrt{1655}\right)/24, \quad \nu_3 = \left(-59-i\sqrt{1655}\right)/24 \tag{3.59}$$

Last two roots, as complex numbers, have no physical meaning. Only the first root, corresponding to incompressible material, is acceptable. The associated expression for the stiffness is

$$D(r) = \left(\frac{31R^2}{9}r^2 - \frac{10R}{3}r^3\right)b_4 \tag{3.60}$$

The candidate mode shapes given in Eq. (3.53) and (3.54) should be investigated separately. It is conjectured that the polynomial function, that satisfies the boundary conditions, may not correspond to physically realizable material density and / or stiffness distribution. For example, for the Poisson's ratio that differs from 1/2, the plate does not possess the parabolic shape given in Eq. (3.52).

3.11 Discussion of Part 3

Exact, closed-form solutions have been obtained here, for the heterogeneous circular plates clamped along their boundary.

Part 4: Vibration of Heterogeneous Free Circular Plates

4.1. Motivation of Part 4

In the previous section of this chapter the free vibrations of the circular plates, that were clamped at their edges, were investigated. Harris (1968) obtained closed-form solutions for the circular plate with free edge. His study appears to be very interesting. He considered a specific case of the variation of the thickness as a function of the radial coordinate r

$$h(r) = h_0\left[1 - (r/R)^2\right] \tag{4.1}$$

Leading to the following bending stiffness

$$D(r) = D_0\left[1 - (r/R)^2\right]^3 \tag{4.2}$$

where R is the radius of the plate, $D_0 = Eh_0^3/12(1-\nu^2)$, $\nu =$ Poisson's ratio. It observed that due to the fact that $D(r)$ satisfies the conditions

$$D(R) = dD(R)/dr = 0 \tag{4.3}$$

The boundary conditions, demanding that the bending moment

$$M_r = -D(r)\left(\frac{d^2W}{dr^2} + \frac{\nu}{r}\frac{dW}{dr}\right) \tag{4.4}$$

and the shearing force

$$V_r = -D(r)\frac{d}{dr}\nabla^2 W - \frac{dD(r)}{dr}\left(\frac{d^2W}{dr^2} + \frac{\nu}{r}\frac{dW}{dr}\right) \tag{4.5}$$

to vanish at the outer boundary are identically satisfied. This led Harris (1968) to the possibility to choose, de facto any expression of the mode shape. He chose the function

$$W(\rho) = \sum_{i=0}^{n} \rho^{2i}, \quad \rho = r/R \tag{4.6}$$

and obtained closed-form expression for the natural frequency. The work by Harris (1968) appears the *only one* both for plates with constant thickness, as well as of variable thickness on which both the mode shape and the natural frequency are derived in the closed-form. Contrary to the paper by Lennox and Conway (1980) not only did he obtain the expression for the mode shape, but also for the eingenvalue. Somewhat different avenue with attendant closed-form solution will be pursued here.

4.2. Formulation of the Problem

Consider an heterogeneous circular plate that is free at its ends. We pose the following inverse problem: Find a distribution of the modulus of elasticity $E(\rho)$ and the material density $\rho(r)$, so that the free plate will possess the postulated following mode shape

$$W(\rho) = 1 + \alpha\rho^2 + \beta\rho^4 \tag{4.7}$$

where α and β are parameters yet to be determined. If such a plate will be found, it is natural to visualize that its stiffness will be dependent upon the parameters α and β, thus

$$D(\rho) = D(\rho, \alpha, \beta) \tag{4.8}$$

The free parameters α and β must be chosen in such a form that the conditions

$$D(\rho, \alpha, \beta) = 0, \quad \text{at} \quad \rho = 1 \tag{4.9}$$

$$\frac{dD(\rho, \alpha, \beta)}{d\rho} = 0 \quad \text{at} \quad \rho = 1 \tag{4.10}$$

This leads to two equations for α and β, and thus, to the closed-form solution for the first non-zero natural frequency.

4.3. Basic Equations

The governing differential equation for the free vibration of the heterogeneous plate is

$$\begin{aligned} D(r)r^3\Delta\Delta W &+ \frac{dD}{dr}\left(2r^3\frac{d^3W}{dr^3} + r^2(2+\nu)\frac{d^2W}{dr^2} - r\frac{dW}{dr}\right) \\ &+ \frac{d^2D}{dr^2}\left(r^3\frac{d^2W}{dr^2} + \nu r^2\frac{dW}{dr}\right) - \rho h\omega^2 r^3 W = 0 \end{aligned} \tag{4.11}$$

We represent the mass density and the stiffness in the following forms respectively

$$\rho(r) = \sum_{i=0}^{m} c_i r^i \tag{4.12}$$

$$D(r) = \sum_{i=0}^{m+4} b_i r^i \tag{4.13}$$

Substitution of Eqs. (4.12) and (4.13) into the governing differential equation leads to the following result

$$\sum_{i=0}^{m+7} d_i r^i = 0 \tag{4.14}$$

where

$$\begin{aligned} &d_0 = 0, \quad d_1 = 0, \quad d_2 = 2b_1\alpha(1+\nu), \quad d_3 = 64b_0\alpha + 8b_2\beta - a_0\omega^2 \\ &d_4 = 132b_1\beta(1+\nu) + 18b_3\alpha(1+\nu), \quad d_5 = 32b_2\beta(7+\nu) + 32b_4\alpha(1+\nu) - a_0\alpha\omega^2 \\ &d_6 = 20(17+3\nu)b_3\beta, \quad d_7 = 96(5+\nu)b_4\beta - a_0\beta\omega^2 \end{aligned} \tag{4.15}$$

In order Eq. (4.14) to be valid, all d_i's must vanish. We get six non-trivial equations for 5 coefficients b_0, b_1, b_2, b_3, b_4 and ω^2, i.e. total of 6 unknowns. for non-triviality of the solution, the determinant should vanish. It turns out to be identically zero. Taking b_4 to be an undetermined coefficient, the solution is written as

$$\begin{aligned} &b_0 = -\left[\alpha^2(1+\nu) - 6\beta(5+\nu)\right]b_4/4\beta^2, \quad b_1 = 0, \\ &b_2 = 2\alpha b_4/\beta, \quad b_3 = 0 \end{aligned} \tag{4.16}$$

with attendant natural frequency squared

$$\omega^2 = 96(5+\nu)b_4/a_0 \tag{4.17}$$

Eq. (4.16) yields the following stiffness

$$D(\rho) = -\frac{\left[\alpha^2(1+\nu) - 6\beta(5+\nu)\right]b_4}{4\beta^2} + \frac{2\alpha b_4}{\beta}\rho^2 + b_4\rho^4 \tag{4.18}$$

We require that $D(1) = 0$, along with $D'(1) = 0$,

$$\begin{aligned} &-\frac{\left[\alpha^2(1+\nu) - 6\beta(5+\nu)\right]b_4}{4\beta^2} + \frac{2\alpha b_4}{\beta} + b_4 = 0 \\ &\frac{4\alpha b_4}{\beta} + 4b_4 = 0 \end{aligned} \tag{4.19}$$

Solution of Eq. (4.19) and (4.20) yields

$$\alpha = -6, \quad \beta = 6 \tag{4.20}$$

Thus the mode shape of the plate reads

$$W(\rho) = 1 - 6\rho^2 + 6\rho^4 \qquad (4.21)$$

whereas the stiffness is

$$D(\rho) = b_4\left(1 - \rho^2\right)^2 \qquad (4.22)$$

4.4. Discussion of Part 4

This study presents a new closed form solution for natural frequency of the heterogeneous circular plate, that is free at its boundary. While Harris (1968) dealt with the stiffness in the form $D(\rho) = D_0\left(1 - \rho^2\right)^3$, here we arrive, by using inverse vibration analysis, to the simpler expression for it, namely $D(\rho) = D_0\left(1 - \rho^2\right)^2$.

Part 5: Vibration of Heterogeneous Simply Supported Circular Plates

5.1. Motivation of Part 5

Conway (1980) found an unusual closed-form solution for a variable-thickness plate on an elastic foundation, in a static setting. Whereas constant thickness plate involved Kelvin function, Conway's (1980) solution was derived in closed-form. Analogous, closed-form solutions were derived by Harris (1968) for plates of variable thickness, free on their boundary and by Lenox and Conway(1980) who studied plates with arbitrary conditions, and with parabolic thickness variation. In recent papers, Elishakoff (2000a, 2000b) derived closed-form solutions for heterogeneous circular plates that are either simply supported (Elishakoff, 2000a) or free (Elishakoff 2000b) at the boundary. Here closed-form solutions for the heterogeneous plates that are simply supported at their boundary are derived.

There appears to be a single monograph, that of Kovalenko (1959) solely devoted to plates of variable thickness. There are several papers dedicated to vibrations of plates with thickness variations.

Axisymetric variations of circular plates of linearly varying thickness was studied by Prasad, Jain and Soni (1983), whereas plates with double linear thickness were studied by Sing and Sascena (1995).

Various analytical and approximate techniques have been studied (Barakst and Baumann, 1998; Chen and Ren, 1998; Gupta and Ansari, 1998; Liew and Yang, 1999; Singh and Hassan, 1998; Singh and Chaakraverty, 1991, 1992; Yang and Xie, 1984; Yeh, 1994).

In this section, a semi-inverse method will be utilized to report closed-form solutions, that can serve as benchmark solutions for validations of various numerical techniques.

5.2. Basic Equations

Timoshenko and Woinowsky-Krieger (1959) report following static displacement of the uniform circular simply supported plate

$$W = \left(R^2 - r^2\right)\left(\theta R^2 - r^2\right) \tag{5.1}$$

where the parameter θ depends solely of the Poisson ratio

$$\theta = (5+\nu)/(1+\nu) \tag{5.2}$$

We pose the following problem : Find the variation of the bending stiffness as a function of the polar coordinate r that leads to the following postulated mode shape

$$\psi(r) = \left(R^2 - r^2\right)\left(\theta R^2 - r^2\right) \tag{5.3}$$

that is proportional to the static displacement in Eq. (5.52) as given by Timoshenko and Woinowsky- Krieger (1959).

5.3. Constant Inertial Term (m = 0)

In this case the stiffness is sought as a fourth order polynomial

$$D(r) = b_0 + b_1 r + b_2 r^2 + b_3 r^3 + b_4 r^4 \tag{5.4}$$

We get then the differential equation as follows

$$\sum_{i=0}^{7} c_i r^i = 0 \tag{5.5}$$

where

$$\begin{aligned}
&c_0 = 0, \quad c_1 = 0, \quad c_2 = -4(3+\nu)R^2 b_1, \quad c_3 = 64 b_0 - 16(3+\nu)R^2 b_2 - a_0\omega^2 R^4 \frac{5+\nu}{1+\nu} \\
&c_4 = 12(11+\nu)b_1 - 36(3+\nu)R^2 b_3 \\
&c_5 = 32(7+\nu)b_2 - 64(3+\nu)R^2 b_4 + 2a_0\omega^2 R^2 \frac{3+\nu}{1+\nu} \\
&c_6 = (340+60\nu)b_3, \quad c_7 = 96(5+\nu)b_4 - a_0\omega^2
\end{aligned} \tag{5.6}$$

Since the left side of the differential equation must vanish for any r within [0;R], we demand that all the coefficients c_i to be zero. This leads to a homogeneous set of 6 linear algebraic equations for 6 unknowns. It turns out that the determinant of the matrix of the derived set of equations is zero. Therefore, a non-trivial solution is obtainable; the natural frequency is obtained from the coefficient c_7

$$\omega^2 = 96(5+\nu)b_4/a_0 \tag{5.7}$$

The remaining equations yield the coefficients in the stiffness

$$b_0 = \frac{57+18\nu+\nu^2}{2(1+\nu)}R^4 b_4, \quad b_1 = 0, \quad b_2 = -4\frac{3+\nu}{1+\nu}R^2 b_4, \quad b_3 = 0 \tag{5.8}$$

Hence, the stiffness reads

$$D(r) = \left(\frac{57+18\nu+\nu^2}{2(1+\nu)}R^4 - 4\frac{3+\nu}{1+\nu}R^2 r^2 + r^4\right)b_4 \tag{5.9}$$

5.4. Linearly Varying Inertial Term (m = 1)

Instead of the set (5.11) we get here, 7 linear algebraic equations with 7 unknowns

$$\begin{aligned}
&-4(3+\nu)R^2 b_1 = 0\\
&64b_0 - 16(3+\nu)R^2 b_2 - a_0\omega^2 R^4\frac{5+\nu}{1+\nu} = 0\\
&12(11+\nu)b_1 - 36(3+\nu)R^2 b_3 - a_1\omega^2 R^4\frac{5+\nu}{1+\nu} = 0\\
&32(7+\nu)b_2 - 64(3+\nu)R^2 b_4 + 2a_0\omega^2 R^2\frac{3+\nu}{1+\nu} = 0\\
&(340+60\nu)b_3 - 100(3+\nu)R^2 b_5 + 2a_1\omega^2 R^2\frac{3+\nu}{1+\nu} = 0\\
&96(5+\nu)b_6 - a_0\omega^2 = 0\\
&(644+140\nu)b_5 - a_1\omega^2 = 0
\end{aligned} \tag{5.10}$$

In order to have a non-trivial solution the determinant of the set (5.10)

$$-100663296R^6\frac{(3+\nu)(7+\nu)(27730+18641\nu+4439\nu^2+439\nu^3+15\nu^4)}{1+\nu}a_1 = 0 \tag{5.11}$$

must vanish, leading to $a_1 = 0$. Yet this result signifies that the linear inertial coefficient is a constant. Thus, the problem solved in the previous section is obtained.

5.5. Parabolically Varying Inertial Term (m = 2)

For $m = 2$, i.e. the plate whose material density varies parabolically

$$\delta(r) = a_0 + a_1 r + a_2 r^2 \tag{5.12}$$

the bending stiffness has to be sixth order polynomial

$$D(r) = b_0 + b_1 r + b_2 r^2 + b_3 r^3 + b_4 r^4 + b_5 r^5 + b_6 r^6 \tag{5.13}$$

Substitution of Eq. (5.6) in conjunction with Eqs. (5.12) and (5.13) into the governing differential equation yields

$$\sum_{i=0}^{9} d_i r^i = 0 \tag{5.14}$$

where

$$\begin{aligned}
&d_0 = 0, \quad ,d_1 = 0, \quad d_2 = -4R^2(3+\nu)b_1\\
&d_3 = 64b_0 - 16R^2(3+\nu)b_2 - a_0\omega^2 R^4 \frac{5+\nu}{1+\nu}\\
&d_4 = 12(11+\nu)b_1 - 36R^2(3+\nu)b_3 - a_1\omega^2 R^4 \frac{5+\nu}{1+\nu}\\
&d_5 = 32(7+\nu)b_2 - 64R^2(3+\nu)b_4 + \omega^2\left(2a_0R^2\frac{3+\nu}{1+\nu} - a_2R^4\frac{5+\nu}{1+\nu}\right)\\
&d_6 = 20(17+3\nu)b_3 - 100R^2(3+\nu)b_5 + 2a_1\omega^2R^2\frac{3+\nu}{1+\nu}\\
&d_7 = 96(5+\nu)b_4 - 144R^2(3+\nu)b_6 - \omega^2\left(a_0 - 2a_2R^2\frac{3+\nu}{1+\nu}\right)\\
&d_8 = (644+140\nu)b_5 - a_1\omega^2, \quad d_9 = (832+192\nu)b_6 - a_2\omega^2
\end{aligned} \tag{5.15}$$

As in the case of the constant inertial term, we demand that all $d_i = 0$, thus, we get a set of 8 equations with 8 unknowns (7 coefficients b_i and ω^2). The resulting determinantal equation is

$$\begin{aligned}
&(3+\nu)(7+\nu)a_1(360490 + 325523\,\nu + 113630\,\nu^2\\
&+ 19024\,\nu^3 + 1512\,\nu^4 + 45\nu^5)/1+\nu = 0
\end{aligned} \tag{5.16}$$

In order the homogeneous system to have a non-trivial solution we must demand the coefficient a_1 to vanish. For the natural frequency we arrive at the following expression

$$\omega^2 = 64(13+3\nu)b_6/a_2 \tag{5.17}$$

Then, the coefficients in the stiffness are given by

$$\begin{aligned}
b_0 &= R^4 b_6 \left[\left(3285+2670\nu+744\nu^2+82\nu^3+3\nu^4\right)R^2 a_2, \right. \\
&\quad \left. +\left(20748+14304\nu+3496\nu^2+352\nu^3+12\nu^4\right)a_0 \right] / 12\left(35+47\nu+13\nu^2+\nu^3\right)a_2, \\
b_1 &= 0, \\
b_2 &= \frac{R^2 b_6}{3} \frac{\left(1095+525\nu+73\nu^2+3\nu^4\right)R^2 a_2 - \left(2184+1544\nu+344\nu^2+24\nu^4\right)a_0}{\left(35+47\nu+13\nu^2+\nu^3\right)a_2}, \\
b_3 &= 0, \quad b_4 = -\frac{b_6}{6} \frac{\left(285+140\nu 15\nu^2\right)R^2 a_2 - \left(52+64\nu+12\nu^2\right)a_0}{\left(5+6\nu+\nu^2\right)a_2}, \quad b_5 = 0,
\end{aligned} \tag{5.18}$$

where b_6 is an arbitrary constant. In order for the natural frequency squared to be a positive quantity we demand that the ratio b_6/a_2 be positive. We have two sub-cases : (1) both b_6 and a_2 are positive, or (2) both are negative. In the former case (1) the necessary condition for positivity of the stiffness $b_0 \geq 0$ is identically satisfied. In the latter case (2) the above inequality reduces to

$$\begin{aligned}
&\left(3285+2670\nu+744\nu^2+82\nu^3+3\nu^4\right)R^2 \\
&+\left(20748+14304\nu+3496\nu^2+352\nu^3+12\nu^4\right)a_0/a_2 \geq 0
\end{aligned} \tag{5.19}$$

Leading to the following inequality

$$\frac{a_0}{|a_2|R^2} \leq \frac{3285+2670\nu+744\nu^2+82\nu^3+3\nu^4}{20748+14304\nu+3496\nu^2+352\nu^3+12\nu^4} \tag{5.20}$$

For example, for $\nu = 1/3$, the following inequality must hold

$$\frac{a_0}{|a_2|R^2} \leq \frac{3595}{21868} \tag{5.21}$$

and the stiffness reads

$$D(r) = \left[\left(\frac{3595R^6}{528} + \frac{497R^4 a_0}{12a_2} \right) + \left(\frac{719R^4}{88} - \frac{35R^2 a_0}{2a_2} \right) r^2 \right.$$
$$\left. + \left(-\frac{125R^2}{16} + \frac{7a_0}{4a_2} \right) r^4 + r^6 \right] b_6 \tag{5.22}$$

5.6. Cubic Inertial Term (m = 3)

For $m = 3$, the following set of 9 linear algebraic equations with 9 unknowns is obtained

$$-4R^2(3+\nu)b_1 = 0$$
$$64b_0 - 16R^2(3+\nu)b_2 - a_0\omega^2 R^4 \frac{5+\nu}{1+\nu} = 0 \tag{5.23}$$
$$12(11+\nu)b_1 - 36R^2(3+\nu)b_3 - a_1\omega^2 R^4 \frac{5+\nu}{1+\nu} = 0$$

$$32(7+\nu)b_2 - 64R^2(3+\nu)b_4 + \omega^2\left(2a_0R^2\frac{3+\nu}{1+\nu} - a_2R^4\frac{5+\nu}{1+\nu}\right) = 0$$
$$20(17+3\nu)b_3 - 100R^2(3+\nu)b_5 + \omega^2\left(2a_1R^2\frac{3+\nu}{1+\nu} - a_3R^4\frac{5+\nu}{1+\nu}\right) = 0 \tag{5.24}$$
$$96(5+\nu)b_4 - 144R^2(1+\nu)b_6 - \omega^2\left(a_0 - 2a_2R^2\frac{3+\nu}{1+\nu}\right) = 0$$

$$(644+140\nu)b_5 - 196(3+\nu)R^2b_7 - \omega^2\left(a_1 - 2a_3R^2\frac{3+\nu}{1+\nu}\right) = 0$$
$$(832+192\nu)b_6 - a_2\omega^2 = 0 \tag{5.25}$$
$$(1044+252\nu)b_7 - a_3\omega^2 = 0$$

The determinant equation stemming from it reads

$$\left(2527200+3153105\nu+1582173\nu^2+406618\nu^3+56058\nu^4 +3893\nu^5+105\nu^6\right)R^2a_3+\left(10454210+11963597\nu+5573931\nu^2 +1347106\nu^3+177016\nu^4+11889\nu^5+315\nu^6\right)a_1=0 \tag{5.26}$$

The solution of Eq. (5.26) is

$$a_1=-\frac{\left(38880+31761\nu+8865\nu^2+971\nu^3+35\nu^4\right)R^2a_3}{160834+114773\nu+28889\nu^2+2983\nu^3+105\nu^4} \tag{5.27}$$

Upon substitution of Eq. (5.24) into all equations of the set (5.20), the natural frequency squared equals

$$\omega^2=\frac{36(29+7\nu)b_7}{a_3} \tag{5.28}$$

and we get the following solution for the coefficients in the stiffness

$$\begin{aligned}
b_0&=3\left[\left(95265+100425\nu+40266\nu^2+7586\nu^3+661\nu^4+21\nu^5\right)R^2a_2\right.\\
&\quad+\left(601692+560052\nu+201512\nu^2+34680\nu^3+2812\nu^4\right.\\
&\quad\left.\left.+84\nu^5\right)a_0\right]R^4b_7/64a_3M\\
b_1&=0\\
b_2&=3\left[\left(31755+22890\nu+5792\nu^2+598\nu^3+21\nu^4\right)R^2a_2-\left(63336+60064\nu\right.\right.\\
&\quad\left.\left.+20784\nu^2+3104\nu^3+168\nu^4\right)a_0\right]R^2b_7/16a_3M\\
b_3&=\left(64800+44295\nu+105597\nu^2+1041\nu^3+21\nu^4\right)R^4b_7/N\\
b_4&=3b_7\left[\left(1508-2220\nu+796\nu^2+84\nu^3\right)a_0\right.\\
&\quad\left.-\left(8265+6055\nu+1415\nu^2+105\nu^3\right)R^2a_2\right]/a_3\left(65+93\nu+31\nu^2+3\nu^3\right)\\
b_5&=-2R^2\left(25527+20092\nu+5424\nu^2+584\nu^3+21\nu^4\right)b_7/N\\
b_6&=9(29+7\nu)a_2b_7/16a_3(13+3\nu)
\end{aligned} \tag{5.29}$$

where $M=455+716\nu+310\nu^2+52\nu^3+3\nu^4$

$N=5546+8165\nu+2983\nu^2+379\nu^3+15\nu^4$

where b_7 is an arbitrary constant. For the particular case $\nu=1/3$, the stiffness equals

$$D(r) = \left[\frac{1689965}{19712}\frac{R^6 a_2}{a_3} + \frac{3337}{64}\frac{R^4 a_0}{a_3} + \left(\frac{101379}{9856}\frac{R^4 a_2}{a_3} - \frac{705}{32}\frac{R^2 a_0}{a_3}\right)r^2 \right.$$
$$\left. + \frac{21524}{2295}R^4 r^3 + \left(-\frac{17625}{1792}\frac{R^2 a_2}{a_3} + \frac{141}{64}\frac{a_0}{a_3}\right)r^4 - \frac{389}{51}R^2 r^5 + \frac{141}{112}\frac{a_2}{a_3}r^6 + r^7\right]b_7 \quad (5.30)$$

5.7. General Inertial Term (m ≥ 4)

Consider now the general expression of the inertial term given in Eq. (5.7), and the stiffness in Eq. (5.8), for $m \geq 4$. Substitution of Eqs. (5.6), (5.7) and (5.8) into the terms of the differential equation yields

$$r^3 D(r)\Delta\Delta W = -64 r^3 \sum_{i=0}^{m+4} b_i r^i \quad (5.31)$$

$$\frac{dD}{dr}\left(2r^3\frac{d^3W}{dr^3} + r^2(2+\nu)\frac{d^2W}{dr^2} - r\frac{dW}{dr}\right) = 4\left[(17+3\nu)r^2 - R^2(3+\nu)\right]r^2 \sum_{i=1}^{m+4} i b_i r^{i-1} \quad (5.32)$$

$$\frac{d^2D}{dr^2}\left(r^3\frac{d^2W}{dr^2} + \nu r^2\frac{dW}{dr}\right) = 4\left[(3+\nu)r^2 - R^2(3+\nu)\right]r^3 \sum_{i=2}^{m+4} i(i-1) b_i r^{i-2} \quad (5.33)$$

and

$$\rho h \omega^2 r^3 W = \omega^2 r^3 \left(R^2 - r^2\right)\left(\theta R^2 - r^2\right)\sum_{i=0}^{m} a_i r^i \quad (5.34)$$

Demanding the sum of (5.31)-(5.44) to be zero, we obtain the following equation

$$\sum_{i=0}^{m+7} g_i r^i = 0 \quad (5.35)$$

where the coefficients g_i are

$$g_0 = 0, \quad g_1 = 0 \quad (5.36)$$

$$g_2 = -4(3+\nu)b_1 \tag{5.37}$$

$$g_3 = 64b_0 - 16R^2(3+\nu)b_2 - \theta a_0 R^4\omega^2 \tag{5.38}$$

$$g_4 = 12(11+\nu)b_1 - 36R^2(3+\nu)b_3 - \theta a_1 R^4\omega^2 \tag{5.39}$$

$$g_5 = 32(7+\nu)b_2 - 64R^2(3+\nu)b_4 - \omega^2\left[\theta a_2 R^4 - (1+\theta)a_0 R^2\right] \tag{5.40}$$

$$g_6 = 20(17+3\nu)b_3 - 100R^2(3+\nu)b_5 - \omega^2\left[\theta a_3 R^4 - (1+\theta)a_1 R^2\right] \tag{5.41}$$

$$\vdots$$

for $7 \le i \le m+3$

$$\begin{aligned} g_i = {} & \left[64 + 4(i-3)(17+3\nu) + 4(i-3)(i-4)(3+\nu)\right]b_{i-3} - 4(i-1)^2 R^2(3+\nu)b_{i-1} \\ & - \omega^2\left[\theta a_{i-3}R^4 - (1+\theta)a_{i-5}R^2 + a_{i-7}\right] \end{aligned} \tag{5.42}$$

$$\vdots$$

$$\begin{aligned} g_{m+4} = {} & \left[64 + 4(m+1)(17+3\nu) + 4m(m+1)(3+\nu)\right]b_{m+1} - 4R^2(m+3)^2(3+\nu)b_{m+3} \\ & - \omega^2\left[a_{m-3} - (1+\theta)a_{m-1}R^2\right] \end{aligned} \tag{5.43}$$

$$\begin{aligned} g_{m+5} = {} & \left[64 + 4(m+2)(17+3\nu) + 4(m+1)(m+2)(3+\nu)\right]b_{m+2} - 4R^2(m+4)^2(3+\nu)b_{m+4} \\ & - \omega^2\left[a_{m-2} - (1+\theta)a_m R^2\right] \end{aligned} \tag{5.44}$$

$$g_{m+6} = \left[64 + 4(m+3)(17+3\nu) + 4(m+2)(m+3)(3+\nu)\right]b_{m+3} - \omega^2 a_{m-1} \tag{5.45}$$

$$g_{m+7} = \left[64 + 4(m+4)(17+3\nu) + 4(m+3)(m+4)(3+\nu)\right]b_{m+4} - \omega^2 a_m \tag{5.46}$$

One gets a set of $m + 6$ homogeneous linear algebraic equations for $m + 6$ unknowns. In order to find a non-trivial solution the determinant of the set (5.36)-(5.46) must vanish. The determinant along the last column of the matrix of the set is expanded, getting a linear algebraic expression with the coefficients a_i as coefficients. Determinant equation yields a condition for which the non-trivial solution is obtainable. In this case the general expression of the natural frequency squared is obtained from the equation $g_{m+7} = 0$, resulting in

$$\omega^2 = [64 + 4(m+4)(17+3\nu) + 4(m+3)(m+4)(3+\nu)]b_{m+4}/a_m \tag{5.47}$$

Note that the formulas pertaining the cases $m = 0$, $m = 2$ and $m = 3$ are formally obtainable from Eq. (5.47) by appropriate substitution.

Conclusion

Numerous closed-form solutions have been reported in this study based upon the polynomial representation of mode shapes. Immediately, the following question arises: *Are there some non-polynomial closed-form solutions?* The answer to this question is affirmative. Caliò and Elishakoff (2002, 2004, 2005) were able to derive several *trigonometric* solutions for structures. Elishakoff (2000), Baruch, Elishakoff and Catellani (2203), Catellani and Elishakoff (2004) obtained solutions for the mode shape in terms of *rational* functions, i.e. ratios of polynomial functions.

Another question that immediately pops up is as follows: *Is it possible to obtain closed-form solutions in buckling?* The reply is again affirmative. Interested reader can consult with the monograph by Elishakoff (2005), papers by Elishakoff and Endres (2005) as well as the forthcoming article by Elishakoff, Ruta and Stavsky (2005).

Yet another question with a positive reply is as follows: *Are there structures beyond the beams or isotropic circular plates that allow for the closed-form solutions?* Several works are now in the pipeline in cooperative efforts with several scientists of Israel, namely, Professors Menachem Baruch and Yehuda Stavsky; of USA: Professor Charles W. Bert, Dr. Joel Storch, Mr. Demetris Pentaras, Mr. Achilles Perez, Mr. David Chandra, Mr. Jason Yost; and of Italy: Professor Ivo Caliò of the University of Catania, Professor Giuseppe Ruta of the University of Rome "La Sapienza," Dr. Giulia Catellani of University of Modena, Dr. Roberta Santoro of the University of Palermo, Dr. Christina Gentelini of the University of Bologna on vibrating heterogeneous bars and polar orthotropic circular plates, and will be published elsewhere.

Acknowledgement

This review is based on papers by the author, as well as those co-authored with Mr. Syleyman Candan and Mr. Zakoua Guédé of IFMA (Aubière, France). Thanks are to Ms. Barbara Ferracuti of University of Bologna and Mr. Demetris Pentaras of the FAU for dedicated editorial assistance. Partial support of the J.M. Rubin Foundation of the FAU is greatly appreciated. The partial support by the National Science Foundation (Program Director: Dr. K.P. Chong) is acknowledged through the grant 99-10195. Opinions, findings and conclusions expressed are those by the writer and do not necessarily reflect the views of the sponsor.

References

S. T. Ariaratnam. Random vibration of nonlinear suspensions. *Journal of Mechanical Engineering Sciences*, **2** (**3**), 195-201, 1960.

M. Baruch, I. Elishakoff and G. Catellani. Solution of semi-inverse buckling problems yield a flexural rigidity as a rational function. *International Journal of Structural Stability and Dynamics,* Vol. (3), 207-334, 1960.

I. Caliò and I. Elishakoff. Can a harmonic function constitute a closed-form buckling mode of an inhomogeneous column? *AIAA Journal*, Vol. 40, 2532-2537, 2002.

I. Caliò and I. Elishakoff. Closed-form trigonometric solution for inhomogeneous beam columns in elastic foundation. *International Journal of Structural Stability and Dynamics,* Vol. 4(1), 139-141, 2004.

I. Caliò and I. Elishakoff. Can a trigonometric function serve both as the vibration and buckling mode of an axially graded structures? *Mechanics Based Design of Structures and Machines*, Vol. 32(4), 401-421, 2004.

I. Caliò and I. Elishakoff. Closed-form solutions for axially graded beam-columns. *Journal of Sound and Vibration,* Vol. 280, 1083-1094, 2005.

S. Candan and I. Elishakoff. Infinite number of closed-form solutions exist for frequencies and reliabilities of stochastically nonhomogeneous beams. *Applications of Probability and Statistics*, (R. Melchers and M. G. Stewart, eds.), Balkema Publishers, Rotterdam, pp. 1059-1067, 1999.

G. Catellani and I. Elishakoff. Apparently first closed-form solutions of semi-inverse buckling problems involving distributed and concentrated loads. *Thin-Walled Structures*, Vol. 42, 1719-1733, 2004.

T. K. Caughey. Derivation and application of the Fokker-Planck equation to discrete nonlinear dynamic systems subjected to white random excitation. *Journal of Acoustical Society of America*, **35** (**11**), 1683-1692, 1963.

D.Y. Chen and B.S. Ren. Finite element analysis of the lateral vibration of thin annular and circular plates with variable thickness. *Journal of Vibration and Acoustics*, 120 (3), 747-752, 1998.

E.F.F. Chladni. *Die Akustik*, Leipzig, 1803.

H.D. Conway. An analogy between the flexural vibrations of a cone and a disc of linearly varying thickness. *ZAMM*, 37 (9, 10), 406-407, 1957.

H.D. Conway. Some special solutions for flexural vibrations of discs of varying thickness. *Ingenieur-Archiv*, 26, 408-410, 1958.

H.D. Conway. An unusual closed-form solution for a variable thickness plate on an elastic foundation. *Journal of Applied Mechanics*, **47**, 204, 1980.

H.D. Conway, E.C.H. Becker and J.F. Dubil. Vibration frequencies of tapered bars and circular plates. *Journal of Applied Mechanics*, 31,329-331, 1964.

W. J. Duncan. Galerkin's method in mechanics and differential equations. *Aeronautical Research Committee, Reports and Memoranda* **1798**, 1937.

M. Eisenberger. Dynamic stiffness vibration analysis of non-uniform members. *International Symoposium on Vibrations on Continuous Systems* , (Leissa A. W., organizer), Estates Park, Colorado, 11-15 Aug. 1997, pp. 13-15, 1997.

I. Elishakoff. Axisymetric vibration of inhomogeneous clamped circular plates: an unusual closed-form solution. *Journal of Sound and Vibration*, **233**, 727-738, 2000a.

I. Elishakoff. Axisymetric vibration of inhomogeneous free circular plates: an unusual exact closed-form solution. *Journal of Sound and Vibration*, **234**, 167-170, 2000b.

I. Elishakoff. *Eigenvalues of Inhomogeneous Structures: Unusual Closed Form Solutions*, CRC Press, Boca Raton, 2005.

I. Elishakoff. Resurrection of the method of successive approximation to yield closed-form solutions for vibrating inhomogeneous beams, *Journal of Sound and Vibration*, **234 (2)**, 349-362, 2000.

I. Elishakoff and S. Candan. Apparently first closed-form solutions for vibrating inhomogeneous beams, *International Journal of Solids and Structures*, Vol. 38, 3411-3441, 2001.

I. Elishakoff and J. Endres. Extension of Euler's problem to axially graded columns : 260 years later, *Journal of Intelligent Material Systems and Structures,* Vol. 16(1), 77-83, 2005.

I. Elishakoff and Z. Guédé. A remarkable nature of the effect of boundary conditions on closed-form solutions for vibrating inhomogeneous Bernoulli-Euler beams. *Chaos, Soliform and Fractals*, Vol.12, 659-704, 2001.

I. Elishakoff, Y.J Ren. and M. Shinozuka. Some exact solutions for bending of beams with spatially stochastic stiffness. *International Journal of Solids and Structure*, Vol. 32, pp. 2315-2327 (Corrigendum: Ve. 33, p.3491, 1996) , 1995.

I. Elishakoff and O. Rollot. New closed-form solutions for buckling of a variable stiffness column by Mathematica®. *Journal of Sound and Vibration*, **224 (1)**, 172-182, 1999.

I. Elishakoff, G. Ruta and Y. Stavsky. A folmulation leading to closed-form solutions for the buckling of circular plates. *Acta Mechanica*, to appear, 2006.

F. Engesser. Über die Berechnung auf Knickfestigkeit beanspruchten Stäbe aus Schweiβ- und Guβ. *Z. Österr. Ing. Arch. Ver.,* No 45, 506-508 (in German) , 1893.

G.M.L. Gladwell. *Inverse Problems in Vibration.* Martinus Nijhoff Publishers, Dordrecht (Second Enlarged Edition 2005 Kluwer Academic Publishers), 1986.

G.M.L. Gladwell. Inverse problem in vibration II. *Applied Mechanics Reviews,* Vol. 49 (10), Part2, pp. 13-15, 1996 .

V.S.Gontkevich. *Free Vibrations of Plates and Shells*. "Naukova Dumka" Publishing, Kiev (in Russian), 1964.

D.G. Gorman. Natural frequencies of transverse vibration of polar orthotropic variable thickness annular plate. *Journal of Sound and Vibration*, 86, 47-60, 1983.

U.S. Gupta and A.H. Ansari. Free vibration of polar orthotropic circular plates of variable thickness with elastically restrained edge. *Journal of Sound and Vibration*, 213 (3), 429-445, 1998.

C. Z. Harris. The normal modes of a circular plate of variable thickness. *Quarterly Journal of Mechanics and Applied Mathematics*, **21**, 320-327, 1968.

R.K. Jain. Vibrations of circular plates of variable thickness under an inplane force. *Journal of Sound and Vibration*, 23 (4), 406-414, 1972.

K. Kanaka Raju. Large amplitude vibrations of circular plates with varying thickness. *Journal of Sound and Vibration*, 50, 399-403, 1977.

A. D. Kovalenko. *Circular Plates of Variable Thickness*. "Naukova Dumka" Publishing, Kiev (in Russian), 1959.

H. Köylüoğlu, A. S. Cakmak, and S. A. K. Nielsen. Response of stochastically loaded Bernoulli-Euler beams with randomly varying bending stiffness. i*n Structural Safety and Reliability,*G. I. Schuëller, M. Shinozuka, J. T. P. Yao, eds., pp. 267-274, 1994.

R. Lal and U.S. Gupta. Axisymmetric vibrations of polar orthotropic annular plates of variable thickness. *Journal of Sound and Vibration*, 83, 229-240, 1982.

P.A.A. Laura, D.R. Avalos and C.D. Galles. Vibrations and elastic stability of polar orthotropic circular plates of linearly varying thickness, *Journal of Sound and Vibration*, 82, 151-156, 1982.

A.W. Leissa. *Vibration of Plates*, NASA SP 160, 1969.

A.W. Leissa. Plate vibration research: 1976-1980, *The Shock and Vibration Digest*, 10, 19-36, 1981.

A.W. Leissa. Recent studies in plate vibrations: 1981-85, Part II, Complicating effects. *The Shock and Vibration Digest*, 19, 10-24, 1987.

T. A. Lenox and H. D. Conway. An exact closed-form solution for the flexural vibration of a thin annular plate having a parabolic thickness variation. *Journal of Sound and Vibration*, 68, 231-239, 1980.

K.M. Liew and B. Yang. Three-dimensional solutions for free vibrations of circular plates by a polynomials-Ritz analysis. *Computer Methods in Applied Mechanics and Engineering*, 175 (1-2), 189-201, 1999.

Y. K. Lin and G. Q. Cai. *Probabilistic Structural Dynamics*. McGraw Hill, New York, 1995.

N. C. Nigam. *Introduction to Random Vibrations*. MIT Press, Cambridge, pp. 258-259, 1983.

M. K. Ochi. *Applied Probability and Stochastic Processes*. Wiley-Interscience, New York, pp. 150, 1989.

S.D. Poisson. *Memoires de l'Academie Royales des Sciences de l'Institut de la France*, L'Équilibre et le mouvement des corps élastiques, Ser. 2, 8, 357, 1829.

C. Prasad, R.K. Jain and S.R. Soni. Axisymmetric vibrations of circular plates of linearly varying thickness, *ZAMP*, 23, 941-948, 1983 .

A. Scheurkogel and I. Elishakoff. Nonlinear random vibration of a two degree-of-freedom systems. *Non-Linear Stochastic Engineering Systems*, (F. Ziegler and G. I. Schuëller, eds.), Springer, Berlin, 285-299, 1988.

M. Shinozuka and C. J. Astill. Random eigenvalue problems in structural analysis. *AIAA Journal*, Vol. 10, pp. 456-462, 1972.

R. Sing and V. Sascena. Axisymmetric vibration of circular plate with double linear variable thickness. *Journal of Sound and Vibration*, 179 (1), 879-897, 1995.

B. Singh and S.M. Hassan. Transverse vibration of a circular plate with arbitrary thickness variation. *International Journal of Mechanical Sciences*, 40(11), 1089-1104, 1998.

S.R. Singh and S. Chakraverty. Transverse vibration of circular and elliptic plates with variable thickness. *Indian Journal of Pure and Applied Mathematics*, 22(9), 787-803, 1991.

S.R. Singh. and S. Chakraverty, Transverse vibration of circular and elliptical plates with quadratically varying thickness. *Applied Mathematics Modeling*, 16, 269-274, 1992.

C. Soize. Steady-state solution of Fokker-Planck equation in high dimension. *Probabilistic Engineering Mechanics*, **3**, 196-206, 1988.

S.R. Soni. Vibrations of elastic plates and shells of variable thickness. Ph.D. Thesis, University of Roorkee, 1972.

S. P. Timoshenko. *History of Strength of Materials*. McGraw-Hill, New York, 1953.

S. Timoshenko and S. Woinowsky – Krieger. *Theory of Plates and Shells*. McGraw Hill Book Company, New York, 1959.

I. Todhunter and K. Pearson. *A History of the Theory of Elasticity and of the Strength of Materials*. Dover, New York, Vol. 1, 1960.

J. V. Upensky. *Theory of Equations*. McGraw-Hill Book Company, 1948.

J.S. Yang. The vibration of a circular plate with varying thickness. *Journal of Sound and Vibration*, 165 (1), 178-184, 1993.

J.S. Yang and Z. Xie. Perturbation method in the problem of large deflections of circular plates with nonuniform thickness. *Applied Mathematics and Mechanics*, 5, 1237-1242, 1984.

K-Y. Yeh. Analysis of high-speed rotating discs with variable thickness and inhomogeneity. *Journal of Applied Mechanics*, 61 (1), 186-192, 1994.

W. Q. Zhu and W. Q. Wu. A stochastic finite element method for real eigenvalue problem. In *Stochastic Structural Dynamics,* I. Elishakoff and Y. K. Lin, eds., Springer Verlag, Berlin, Vol. 2, pp. 337-351, 1991.

Appendix

Case 1 : m=0

In this sub-case, the expressions of $P(\xi)$ and $D(\xi)$ read

$$P(\xi) = a_0 \, , D(\xi) = \sum_{i=0}^{4} b_i \xi^i \tag{A.1}$$

By substituting the latter expressions in eq (5), we obtain

$$\begin{aligned} &-12\sum_{i=1}^{3} i(i+1)b_{i+1}\xi^i + 12\sum_{i=2}^{4} i(i-1)b_i\xi^i + 24\sum_{i=0}^{4} b_i\xi^i - 24\sum_{i=0}^{3}(i+1)b_{i+1}\xi^i \\ &+ 48\sum_{i=1}^{4} ib_i\xi^i - kL^4 a_0\left(\xi - 2\xi^3 + \xi^4\right) = 0 \end{aligned} \tag{A.2}$$

The eq (A.2) has to be satisfied for any ξ. This requirement yields

$$-24b_1 + 24b_0 = 0 \tag{A.3}$$

$$-72b_2 + 72b_1 - ka_0 = 0 \tag{A.4}$$

$$-144b_3 + 144b_2 = 0 \tag{A.5}$$

$$-240b_4 + 240b_3 + 2kL^4 a_0 = 0 \tag{A.6}$$

$$360b_4 - kL^4 a_0 = 0 \tag{A.7}$$

To satisfy the compatibility equations, b_i, where $i = \{0,1,2,3\}$, has to be

$$b_1 = 3b_4 \tag{A.8}$$
$$b_0 = 3b_4 \tag{A.9}$$
$$b_3 = -2b_4 \tag{A.10}$$
$$b_2 = -2b_4 \tag{A.11}$$

To sum up, if conditions (A.1) are satisfied, where b_i are given by eq (A.8-A.11), then the fundamental mode shape is expressed by eq (44), where the fundamental natural frequency reads

$$\omega^2 = 360\, b_4 / a_0 L^4 \tag{A.12}$$

Case 2 : m=1

In this sub-case, the expressions of $P(\xi)$ and $D(\xi)$ read

$$P(\xi) = a_0 + a_1\xi \;,\; D(\xi) = \sum_{i=0}^{5} b_i \xi^i \tag{A.13}$$

By substituting the latter expressions in eq (5), we obtain

$$\begin{aligned} &-12\sum_{i=1}^{4} i(i+1)b_{i+1}\xi^i + 12\sum_{i=2}^{5} i(i-1)b_i\xi^i + 24\sum_{i=0}^{5} b_i\xi^i - 24\sum_{i=0}^{4}(i+1)b_{i+1}\xi^i \\ &+ 48\sum_{i=1}^{5} ib_i\xi^i - kL^4 a_0(\xi - 2\xi^3 + \xi^4) - kL^4 a_1\xi(\xi - 2\xi^3 + \xi^4) = 0 \end{aligned} \tag{A.14}$$

The eq (A.14) has to be satisfied for any ξ. This requirement yields

$$-24b_1 + 24b_0 = 0 \tag{A.15}$$
$$-72b_2 + 72b_1 - kL^4 a_0 = 0 \tag{A.16}$$
$$-144b_3 + 144b_2 - kL^4 a_1 = 0 \tag{A.17}$$
$$-240b_4 + 240b_3 + 2kL^4 a_0 = 0 \tag{A.18}$$
$$-360b_5 + 360b_4 + 2kL^4 a_1 - kL^4 a_0 = 0 \tag{A.18}$$
$$-504b_5 + kL^4 a_0 = 0 \tag{A.20}$$

To satisfy the compatibility equations, b_i, where $i = \{0,1,2,3,4\}$, has to be

$$b_4 = -\frac{9a_1 - 7a_0}{5a_1} b_5 \tag{A.15}$$

$$b_3 = -\frac{9a_1 + 14a_0}{5a_1} b_5 \tag{A.16}$$

$$b_2 = \frac{17a_1 - 28a_0}{10a_1} b_5 \tag{A.17}$$

$$b_1 = \frac{17a_1 + 42a_0}{10a_1} b_5 \tag{A.18}$$

$$b_0 = \frac{17a_1 + 42a_0}{10a_1} b_5 \tag{A.25}$$

To sum up, if conditions (A.13) are satisfied, where b_i are given by eq (A.21-A.25), then the fundamental mode shape is expressed by eq (45), where the fundamental natural frequency reads

$$\omega^2 = 504\, b_5 / a_1 L^4 \tag{A.26}$$

Case 3 : m=2

In this sub-case, the expressions of $P(\xi)$ and $D(\xi)$ reads

$$P(\xi) = a_0 + a_1\xi + a_2\xi^2 \quad , \quad D(\xi) = \sum_{i=0}^{6} b_i \xi^i \tag{A.27}$$

By substituting the latter expressions in eq (5), we obtain

$$\begin{aligned} &-12\sum_{i=1}^{5} i(i+1)b_{i+1}\xi^i + 12\sum_{i=2}^{6} i(i-1)b_i\xi^i + 24\sum_{i=0}^{6} b_i\xi^i - 24\sum_{i=0}^{5}(i+1)b_{i+1}\xi^i \\ &+ 48\sum_{i=1}^{6} ib_i\xi^i - kL^4\left(a_0 + a_1\xi + a_2\xi^2\right)\left(\xi - 2\xi^3 + \xi^4\right) = 0 \end{aligned} \tag{A.28}$$

The eq (A.14) has to be satisfied for any ξ. This requirement yields

$$-24b_1 + 24b_0 = 0 \quad \text{(A.29)}$$
$$-72b_2 + 72b_1 - kL^4 a_0 = 0 \quad \text{(A.30)}$$
$$-144b_3 + 144b_2 - kL^4 a_1 = 0 \quad \text{(A.31)}$$
$$-240b_4 + 240b_3 + 2kL^4 a_0 - kL^4 a_2 = 0 \quad \text{(A.32)}$$
$$-360b_5 + 360b_4 + 2kL^4 a_1 - kL^4 a_0 = 0 \quad \text{(A.33)}$$
$$-504b_6 + 504b_5 + 2kL^4 a_2 - kL^4 a_1 = 0 \quad \text{(A.34)}$$
$$672b_6 - kL^4 a_2 = 0 \quad \text{(A.35)}$$

To satisfy the compatibility equations, b_i, where $i = \{0,1,2,3,4,5\}$ has to be

$$b_5 = -\frac{5a_2 - 4a_1}{3a_2} b_6 \tag{A.36}$$

$$b_4 = -\frac{25a_2 + 36a_1 - 28a_0}{15a_2} b_6 \tag{A.37}$$

$$b_3 = \frac{17a_2 - 36a_1 - 56a_0}{15a_2} b_6 \tag{A.38}$$

$$b_2 = \frac{17a_2 + 34a_1 - 56a_0}{15a_2} b_6 \tag{A.39}$$

$$b_1 = \frac{17a_2 - 36a_1 + 84a_0}{15a_2} b_6 \tag{A.40}$$

$$b_0 = \frac{17a_2 - 36a_1 + 84a_0}{15a_2} b_6 \tag{A.41}$$

To sum up, if conditions (A.27) are satisfied, where b_i are givne by eq (A.36-A.41), then the fundamental mode shape is expressed by eq (45), where the fundamental natural frequency reads

$$\omega^2 = 672\, b_6 / a_2 L^4 \tag{A.42}$$

Closed Form Trigonometric Solution of Inhomogeneous Beam-Columns

Buckling Problem

Ivo Caliò[1] and Isaac Elishakoff[2]

[1] Dipartimento di Ingegneria Civile e Ambientale, Università di Catania, Catania, Italy
[2] Department of Mechanical Engineering, Florida Atlantic University, Boca Raton, Florida, USA

Abstract. In this study four cases of harmonically varying buckling modes are postulated and semi-inverse problems are solved that result in the distributions of the flexural rigidity compatible to the pre-selected modes and to specified axial load distributions. In all cases the closed-form solutions are obtained for the eigenvalue parameter. For comparison the obtained closed form solution is contrasted with an approximate solution based on an appropriate polynomial shape, serving as trial function in an energy method

1 Introduction

In thc previous chapter and in several previous works (Elishakoff et al. 2001-2005) it has been shown that the vibration modes of a special class of inhomogeneous beams can assume simplc polynomial forms. Now we pose the legitimate question if other closed form solutions can be derived, by postulating mode shapes differing from polynomials. It will be shown that the reply to this query is affirmative, as reported by Caliò and Elishakoff 2002-2005 in some previous papers. The natural candidate for such a function is a trigonometric function. Obviously, the trigonometric function cannot be abandoned ab initio as candidate mode shapes. Indeed, it is found to be the mode shape of free vibrations of an uniform beam, or buckling mode of an uniform column. Therefore the following seemingly provocative question is posed: “Can a trigonometric functions serve as buckling or vibration mode shape also for an inhomogeneous beam-column”? It is shown, in the following, that the reply to this question is “yes”. We also demonstrate that an inhomogeneous beam-column that is elastically attached at its ends can possess a trigonometric mode shape that serves both as the vibration and the buckling mode.

This work is organized in two parts, in the first part the buckling semi-inverse problem of an inhomogeneous column subjected to variable external compressive loading is considered therefore in the second part the problem is widen to the dynamic context and it is demonstrated that a harmonic function can serve both as the vibration and the buckling mode of an axially graded beam-column.

2 Buckling Trigonometric Closed Form solution for Inhomogeneous Columns

In the following we demonstrate that the same buckling mode of the homogeneous column can be used in order to obtain closed form solutions for inhomogeneous columns with trigonometric variability of axial load and flexural stiffness.

The differential equation that governs the buckling of the non-uniform column under a prescribed distributed axial load is

$$\frac{d^2}{dx^2}\left[D(x)\frac{d^2w}{dx^2}\right]+\frac{d}{dx}\left[q\,N(x)\frac{dw}{dx}\right]=0 \tag{1}$$

where x is the axial coordinate, $w(x)$ is the transverse displacement, $E(x)$ is the modulus of elasticity, $I(x)$ is the moment of inertia and $N(x)$ is the axial compressive load associated of a prescribed distribution of external axial load $p(x) = dN/dx$. The product $D(x)=E(x)I(x)$ is the axially varying flexural rigidity and q is the sought eigenvalue.

In the following the differential equation (1) will be solved for four sets of boundary conditions corresponding to the following columns: (a) simply supported at its both ends, (b) clamped at one end and free at the other, (c) simply supported at one end and guided at the other, (d) clamped at both ends. For simplicity, the non-dimensional coordinate $\xi = x/L$ is introduced where L is the length of the column. Then the differential equation becomes

$$\frac{d^2}{d\xi^2}\left[D(\xi)\frac{d^2w}{d\xi^2}\right]+\frac{d}{d\xi}\left[qL^3N(\xi)\frac{dw}{d\xi}\right]=0 \tag{2}$$

The semi-inverse problem is posed as follows: Find an inhomogeneous column with a specified trigonometric buckling mode $w(\xi)=\psi(\xi)$, that satisfies the boundary conditions and the differential equation (2). This problem requires the determination of the distribution of flexural stiffness $D(\xi)$ that together with the specific distribution of axial load and the prescribed buckling mode satisfy the governing eigenvalue problem.

The flexural rigidity $D(\xi)$ is represented as follows

$$D(\xi)= A_o + A_1\sin(\alpha\pi\xi) + A_2\cos(\alpha\pi\xi) \tag{3}$$

where A_o, A_1, A_2 are constants and α is a real number. The semi-inverse problem may have no solution or multiple solutions or unique solution. It will be shown that for a specified distribution of axial load the solution turns out to be unique; moreover, a single constant will be needed only.

2.1 Column That is Simply Supported at its Both Ends

Let us consider a simply supported column subjected to a distributed axial load $p(\xi)$ which leads to a variable distribution of axial force $N(\xi)$, Figure 1.

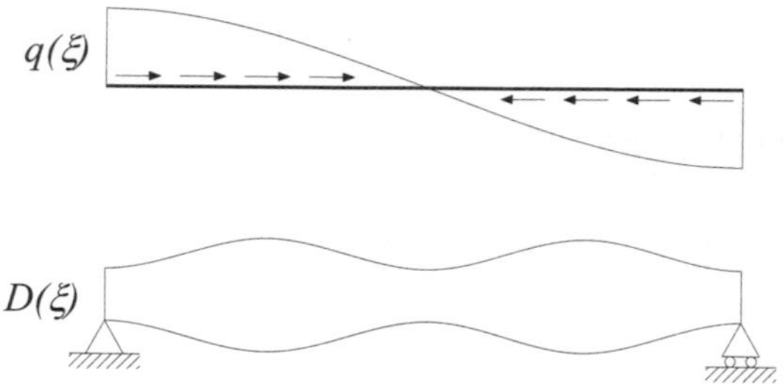

Figure 1. Inhomogeneous column that is simply supported at both ends subjected to a variable axial load.

The following harmonic representation of the stiffness is introduced

$$D(\xi) = A_o + A_1 \sin(\pi\xi) \tag{4}$$

In addition the well known buckling mode of the homogeneous column under constant axial force is postulated

$$\psi(\xi) = \sin(\pi\xi) \tag{5}$$

Bearing in mind the expressions (4) and (5) the first term in the differential equation (2) can be re-written as

$$\frac{d^2}{d\xi^2}\left[D(\xi)\frac{d^2\psi}{d\xi^2}\right] = -\pi^2\frac{d^2}{d\xi^2}\left[A_o \sin(\pi\xi) + A_1 \sin^2(\pi\xi)\right] =$$

$$= -\pi^3\frac{d}{d\xi}\left[A_o \cos(\pi\xi) + 2A_1 \sin(\pi\xi)\cos(\pi\xi)\right] \tag{6}$$

while the second term assumes the expression

$$\frac{d}{d\xi}\left[qL^3 N(\xi)\pi\cos(\pi\xi)\right] \tag{7}$$

In view of the expressions (6) and (7) the differential equation (2) takes the following simple multiplicative form

$$\frac{d}{d\xi}\left\{\left[\frac{qL^3}{\pi^2}N(\xi) - A_o - 2A_1\sin(\pi\xi)\right]\cos(\pi\xi)\right\} = 0 \tag{8}$$

that is naturally satisfied for all the distributions of axial force N(ξ) that are proportional to

$$N(\xi) \propto A_o + 2A_1 \sin(\pi\xi) \tag{9}$$

Now, we consider the following particular cases:

Case 1: ($A_o = 0$; $A_1 \neq 0$). In this case the stiffness distribution is

$$D(\xi) = A_1 \sin(\pi\xi) \tag{10}$$

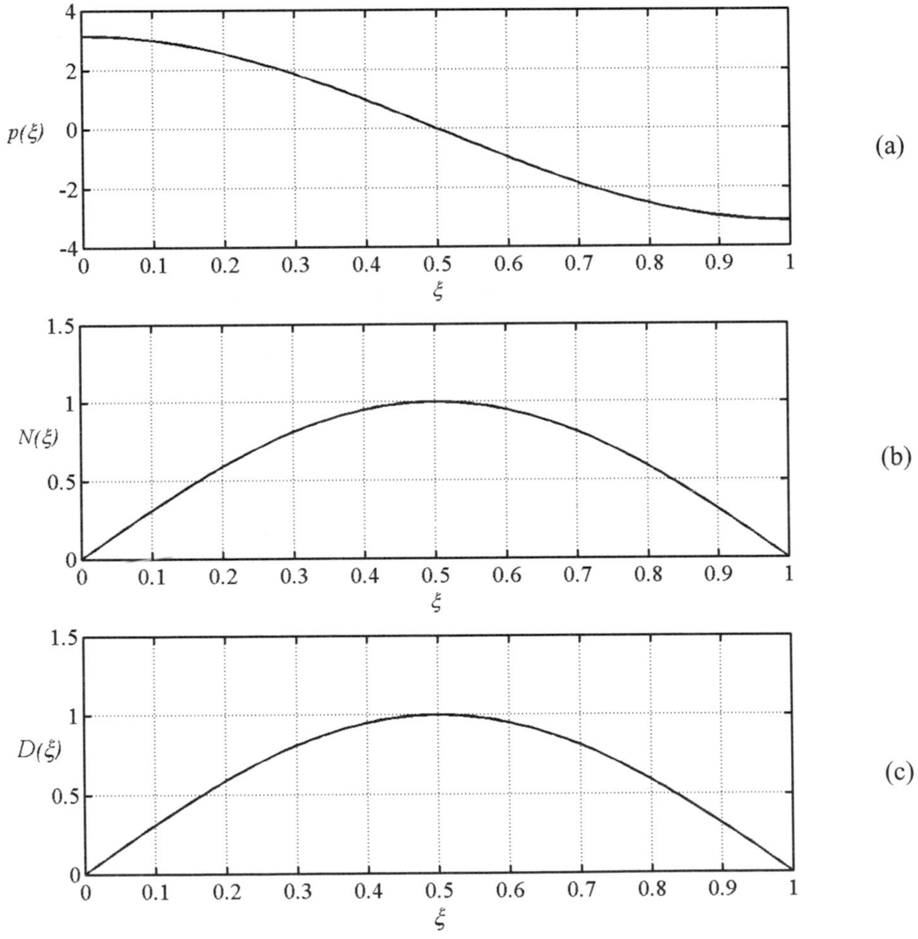

Figure 2. Column that is simply supported at both ends ($A_o = 0$; $A_1 = 1$).
(a) External loading distribution. (b) Axial load distribution. (c) Stiffness distribution.

Then assuming $N(\xi) = \sin(\pi\xi)$ the following eigenvalue is obtained

$$q = 2A_1 \pi^2 / L^3 \tag{11}$$

In Figure 2a and 2b are reported the axial load distribution and the corresponding axial force for the values of the parameters $A_o = 0; A_1 = 1$, while in Figure 3b is reported the corresponding stiffness distribution.

In order to obtain a simple geometric representation of the inhomogeneous column, the variability in the stiffness distribution can be associated to a variability of the cross section of the column. In Figure 3 is reported the qualitative shape of a column with circular cross section to which correspond a stiffness distribution given by expression (10).

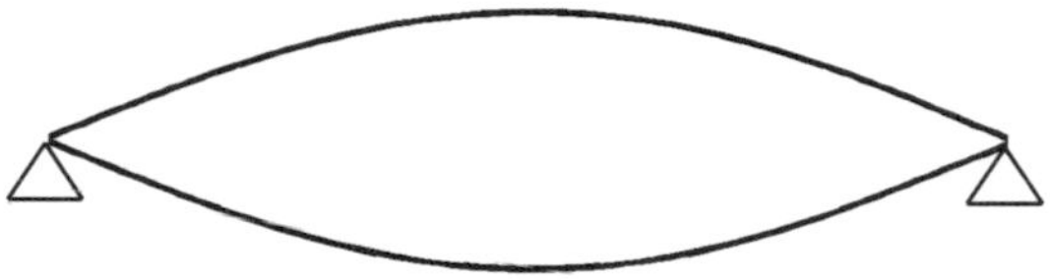

Figure 3. Qualitative shape of an inhomogeneous column, with circular cross section, that is simply supported at its both ends, ($A_o = 0; A_1 \neq 0$).

Case 2: **(** $A_o \neq 0$ **;** $A_1 = 0$ **).** This case corresponds to the homogeneous column

$$D(\xi) = A_0 \tag{12}$$

Then, assuming $N(\xi) = P = constant$, the non-dimensional equilibrium differential equation may be written in the form

$$A_o \frac{d^4\psi}{d\xi^4} + PL^2 \frac{d^2\psi}{d\xi^2} = 0 \tag{13}$$

from which the well known critical load of the homogeneous column under constant axial load is obtained

$$P = A_o \pi^2 / L^2 \tag{14}$$

General Case: **(** $A_o \neq 0; A_1 \neq 0$ **).** This case corresponds to the inhomogeneous column with stiffness distribution

$$D(\xi) = A_o[1 + \beta \sin(\pi\xi)] \tag{15}$$

where for convenience the ratio $\beta = A_1 / A_o$ has been introduced.

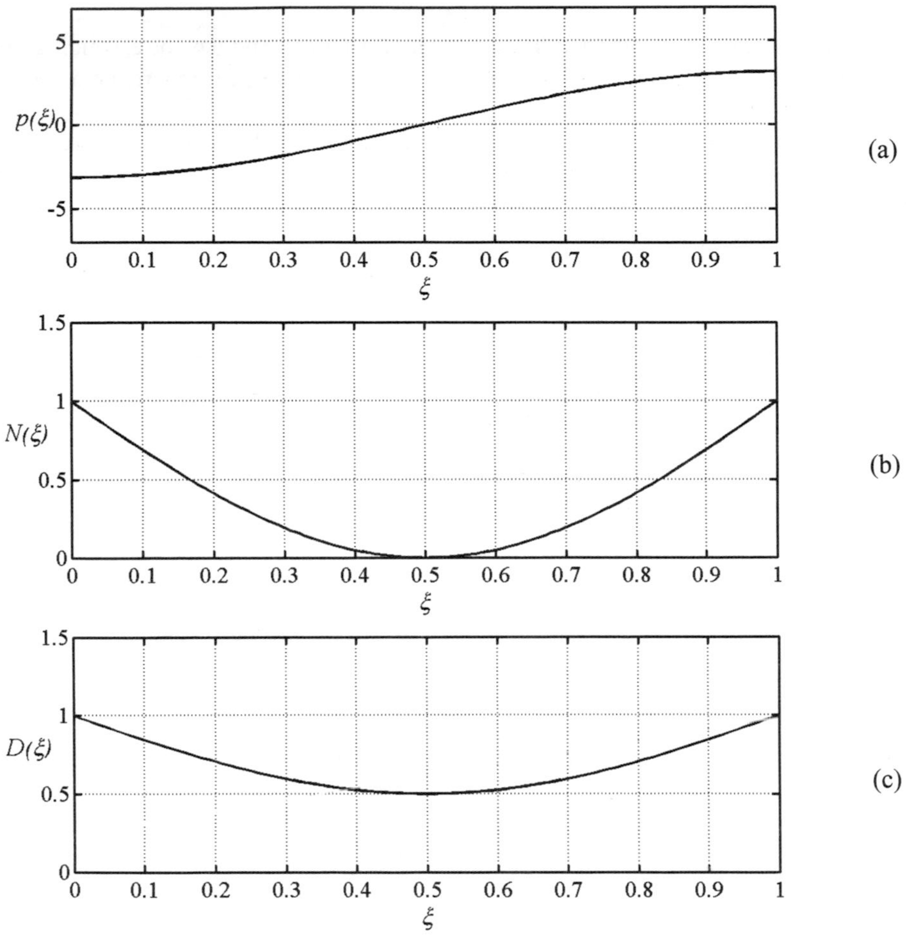

Figure 4. Column that is simply supported at both ends ($A_o = 1; \beta = -0.5$).
(a) External loading distribution. (b) Axial load distribution. (c) Stiffness distribution.

Considering that the stiffness must be positive the following inequality must be verified:

$$A_o > 0 \text{ and } \beta > -1$$

Furthermore observing that the axial load must be positive and proportional to $N(\xi) \propto 1 + 2\beta \sin(\pi\xi)$ the ratio β must also satisfy the inequality

$$\beta \geq -1/2$$

Assuming $N(\xi) = 1 + 2\beta \sin(\pi\xi)$, that may be associated to a unit concentrated axial load plus a distributed axial load $p(\xi) = -2\beta\pi L \cos(\pi\xi)$, the following eingenvalue is obtained

$$q = A_o \pi^2 / L^3 \tag{16}$$

Figures 4 report respectively the variability of the external loading distribution, of the internal axial load and of the flexural stiffness corresponding to the values of the parameters A_o=1 and β = -0.5.

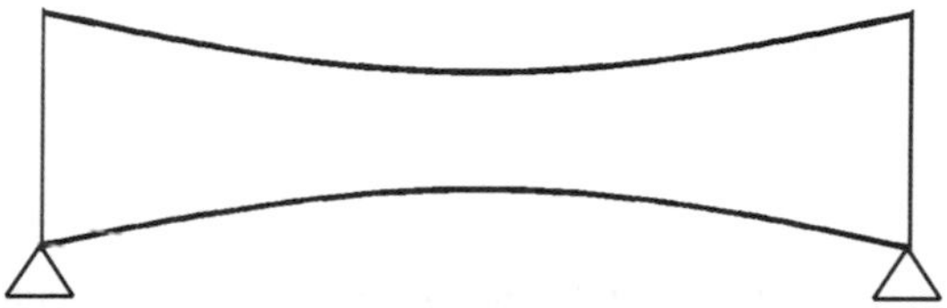

Figure 5. Qualitative shape of an inhomogeneous column, with circular cross section, that is simply supported at its both ends, ($A_o \neq 0; A_1 \neq 0$).

Figure 5 reports the qualitative shape of an inhomogeneous column with circular cross section representative of this general case.

2.2 Column That is Clamped at One End and Free at the Other

For a cantilever column, a simple harmonic function that satisfies the boundary conditions is

$$\psi(\xi) = 1 - \cos\left(\frac{\pi}{2}\xi\right) \tag{17}$$

Employing a trigonometric representation of the flexural rigidity ,as follows

$$D(\xi) = A_o + A_1 \cos\left(\frac{\pi}{2}\xi\right) \tag{18}$$

the equilibrium differential equation may be written as

$$\frac{d}{d\xi}\left\{\left[\frac{4}{\pi^2}qL^3N(\xi) - A_o - 2A_1\cos(\frac{\pi}{2}\xi)\right]\sin\left(\frac{\pi}{2}\xi\right)\right\} = 0 \tag{19}$$

It is easy to recognize that all the distributions of axial force $N(\xi)$ that are proportional to

$$N(\xi) \propto A_o + 2A_1 \cos\left(\frac{\pi}{2}\xi\right) \tag{20}$$

satisfy the equilibrium differential equation. We consider the following particular cases:

Case 1: ($A_o = 0$; $A_1 \neq 0$). In this case the stiffness distribution is

$$D(\xi) = A_1 \cos\left(\frac{\pi}{2}\xi\right) \tag{21}$$

then assuming $N(\xi) = \cos\left(\frac{\pi}{2}\xi\right)$ the following eigenvalue is obtained

$$q = A_1 \pi^2 / 2L^3 \tag{22}$$

The distribution of external load and the corresponding distribution of axial force are reported in Figure 6a and 6b while the stiffness distribution is reported in Figure 6c.

The qualitative shape of a column with variable circular cross section corresponding to this particular case is represented in Figure 7.

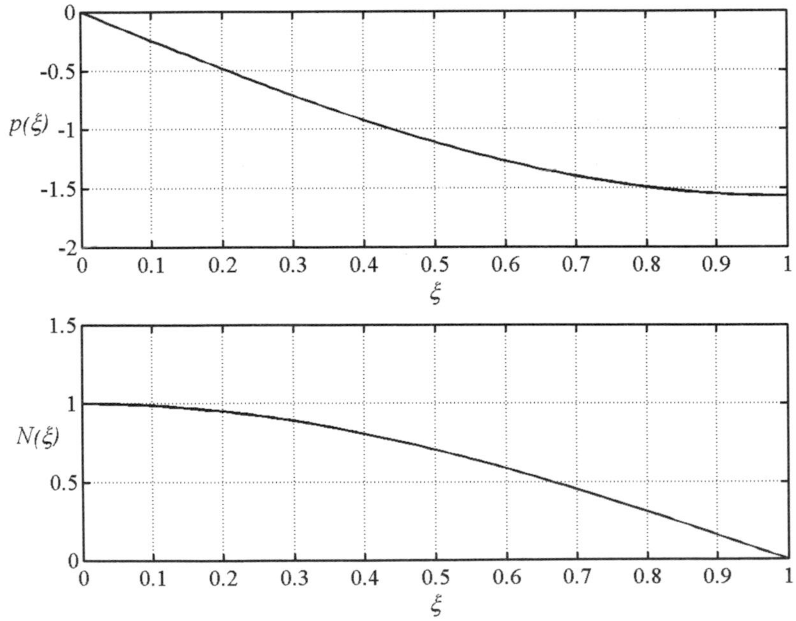

(a)

(b)

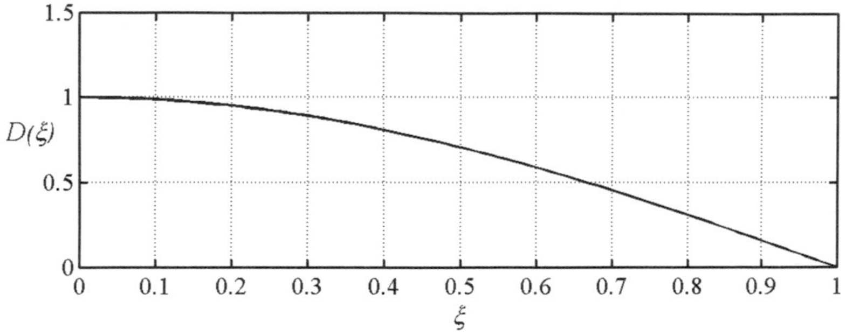

(c)

Figure 6. Inhomogeneous cantilever column ($A_o = 0; A_1 = 1$).
(a) External loading distribution. (b) Axial load distribution. (c) Stiffness distribution.

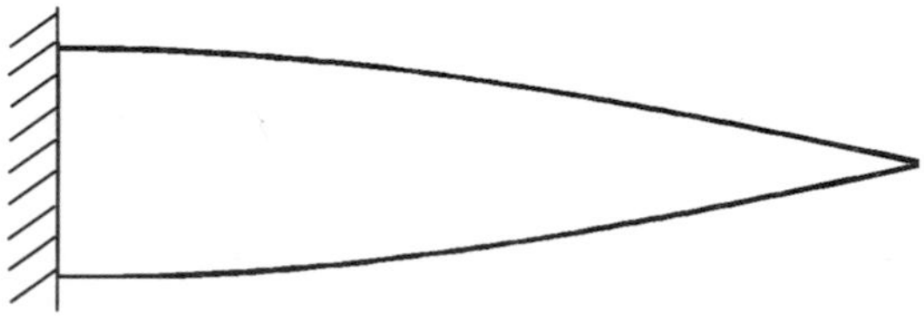

Figure 7. Qualitative shape of an inhomogeneous column, with circular cross section, that is clamped at one end and free at the other, **(** $A_o = 0; A_1 \neq 0$ **).**

Case 2: ($A_o \neq 0$ **;** $A_1 = 0$ **).** This case corresponds to the homogeneous column

$$D(\xi) = A_0 \tag{23}$$

then, assuming $N(\xi) = P$ and $q = 1$, the critical load of the homogeneous cantilever beam is obtained

$$P = A_o \pi^2 / 4L^2 \tag{24}$$

General Case: ($A_o \neq 0$ **;** $A_1 \neq 0$ **).** This case corresponds to the inhomogeneous column with stiffness distribution

$$D(\xi) = A_o \left[1 + \beta \cos(\frac{\pi}{2}\xi) \right] \tag{25}$$

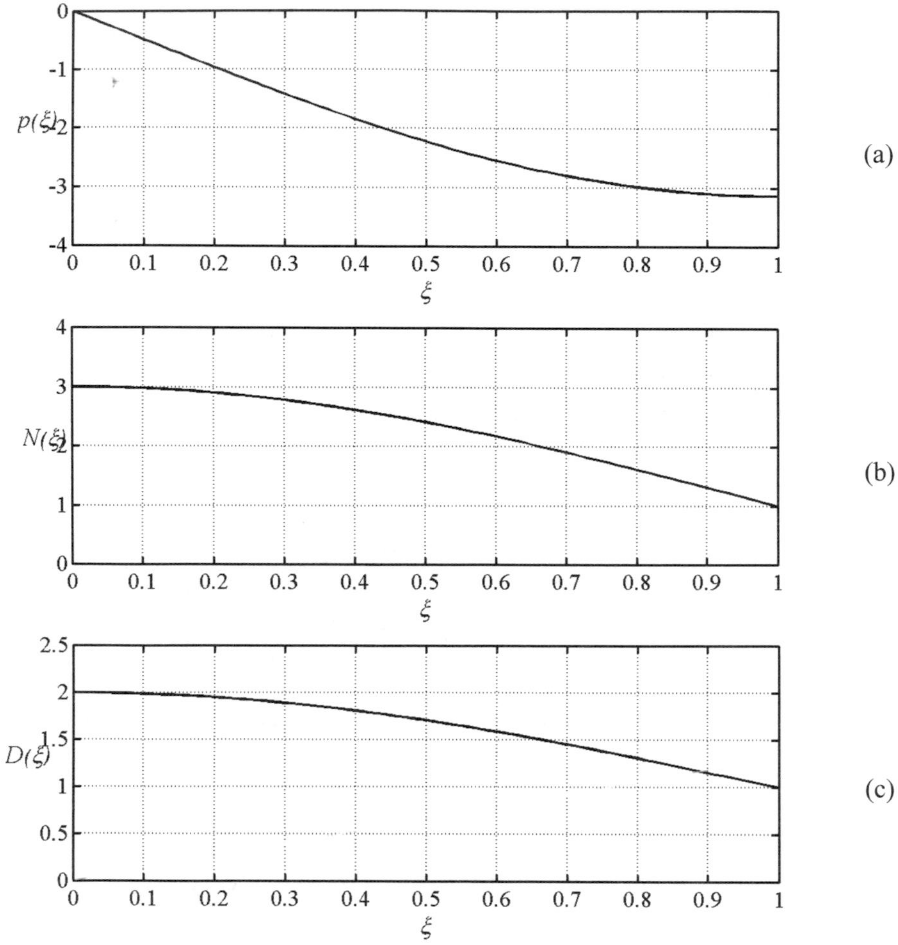

Figure 8. Inhomogeneous cantilever column ($A_o = 1;\ \beta = 1$).
(a) External loading distribution. (b) Axial load distribution. (c) Stiffness distribution.

Assuming $N(\xi) = \left[1 + 2\beta\cos\left(\frac{\pi}{2}\xi\right)\right]$ the following eingenvalue is obtained

$$q = \frac{A_o}{4}\frac{\pi^2}{L^3} \tag{26}$$

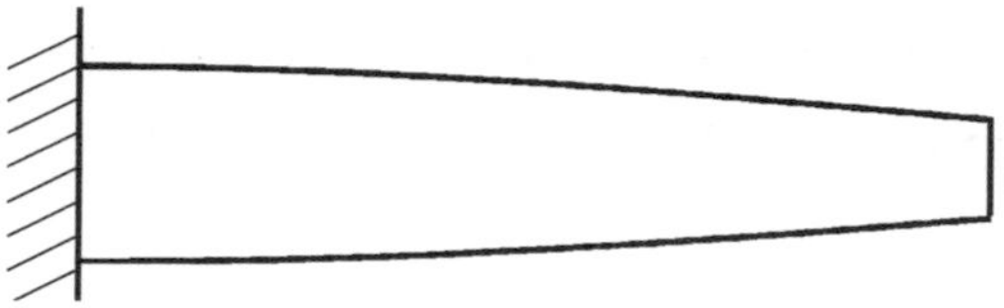

Figure 9. Qualitative shape of an inhomogeneous column, with circular cross section, that is clamped at one end and free at the other, ($A_o \neq 0; A_1 \neq 0$).

Also in this case the stiffness constants A_o and β stiffness must satisfy the following inequalities:

$$A_o > 0 \quad \text{and} \quad \beta \geq -1/2$$

In Figures 8 the results corresponding to the values of the stiffness coefficients $A_o = \beta = 1$ are reported, while in Figure 9 the qualitative shape of a cantilever column with variable circular cross section is schematically represented.

Verification of the Natural Boundary Conditions. For this particular case the natural boundary conditions at the free end need to be verified, since they are not obvious. Namely the following boundary conditions at the free end must be verified:

$$D(\xi)\psi''(\xi) = 0 \tag{27}$$

$$[D(\xi)\psi''(\xi)]' + q_{cr}N(\xi)\psi'(\xi) = 0 \tag{28}$$

Condition (27) is met because $\psi''(\xi)$ is zero at the free end. In the following the condition (28) will be verified for the *case* 1 relative to the distributed load, the verification relative to the general case is similar. The critical load is $q = A_1\pi^2/2L^3$. Therefore the boundary condition (28) becames

$$\left[A_1 \cos\left(\frac{\pi}{2}\xi\right)\frac{\pi^2}{4}\cos\left(\frac{\pi}{2}\xi\right)\right]' + \frac{A_1}{2}\frac{\pi^2}{L^3}\cos\left(\frac{\pi}{2}\xi\right)\frac{\pi}{2}\sin\left(\frac{\pi}{2}\xi\right) = \tag{29}$$

$$= A_1\frac{\pi^3}{4}\cos\left(\frac{\pi}{2}\xi\right)\sin\left(\frac{\pi}{2}\xi\right) + \frac{A_1}{2}\frac{\pi^2}{L^3}\cos\left(\frac{\pi}{2}\xi\right)\frac{\pi}{2}\sin\left(\frac{\pi}{2}\xi\right) = 0 \quad \forall \xi \tag{30}$$

2.3 Column That is Simply Supported at One End and Guided at the Other

For a column that is simply supported at one end and guided at the other the desired buckling mode is taken as

$$\psi(\xi)=\sin\left(\frac{\pi}{2}\xi\right) \tag{31}$$

It should be noted that this mode shape coincides with the exact buckling mode for the uniform column under axial constant compression.

Considering a simple trigonometric representation of the flexural rigidity as follows

$$D(\xi)=A_o+A_1\sin\left(\frac{\pi}{2}\xi\right) \tag{32}$$

the first term of the equilibrium differential equation may be written in the following form

$$\frac{d^2}{d\xi^2}\left[D(\xi)\frac{d^2\psi}{d\xi^2}\right]=-\frac{\pi^3}{8}\frac{d}{d\xi}\left\{\left[-A_o-2A_1\sin\left(\frac{\pi}{2}\xi\right)\right]\cos\left(\frac{\pi}{2}\xi\right)\right\} \tag{33}$$

while the second term assumes the expression

$$\frac{\pi}{2}\frac{d}{d\xi}\left[qL^3N(\xi)\cos\left(\frac{\pi}{2}\xi\right)\right] \tag{34}$$

and the equilibrium differential equation (2) may be written as

$$\frac{d}{d\xi}\left\{\left[\frac{4}{\pi^2}qL^3N(\xi)-A_o-2A_1\sin\left(\frac{\pi}{2}\xi\right)\right]\cos\left(\frac{\pi}{2}\xi\right)\right\}=0 \tag{35}$$

as a consequence the following class of distributions of axial force

$$N(\xi)\propto A_o+2A_1\sin\left(\frac{\pi}{2}\xi\right) \tag{36}$$

satisfy the equilibrium differential equation. The following particular cases are considered:

Case 1: **(** $A_o=0;\ A_1\neq 0$ **).** In this case the stiffness distribution becomes

$$D(\xi)=A_1\sin\left(\frac{\pi}{2}\xi\right) \tag{37}$$

then assuming $N(\xi) = \sin\left(\frac{\pi}{2}\xi\right)$ the following eigenvalue is obtained

$$q = A_1\pi^2 / 2L^3 \tag{38}$$

The external axial loading, the corresponding internal axial load and the flexural stiffness distribution are reported in the Figures 10a, 10b and 10c for the values of the parameters A_o=0, A_1 =1. In Figure 11 is represented the qualitative shape of an inhomogeneous column in which the elastic modulus is assumed constant while the variability of the flexural stiffness is associated to the variation of the circular cross section.

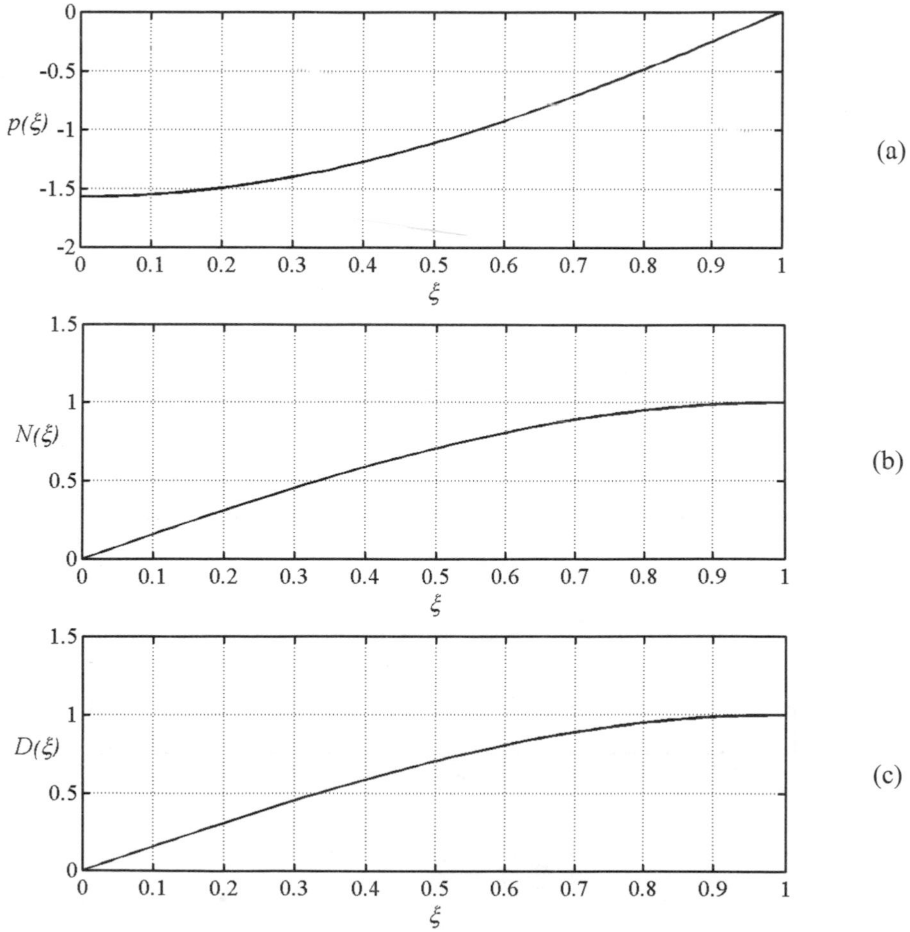

Figure 10. Inhomogeneous column that is simply supported at one end and guided at the other ($A_o = 0; A_1 = 1$). (a) External loading distribution. (b) Axial load distribution. (c) Stiffness distribution.

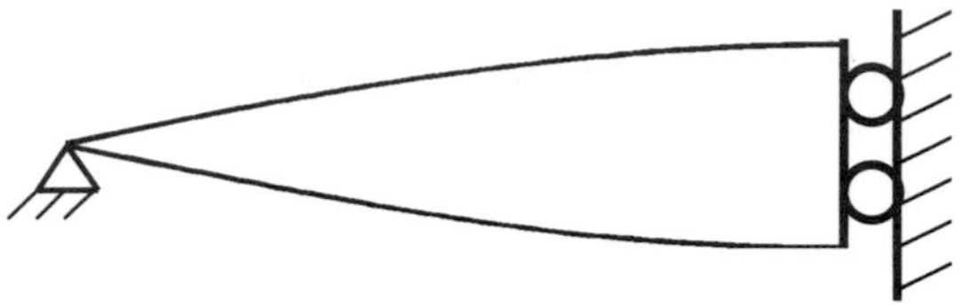

Figure 11. Qualitative shape of an inhomogeneous column, with circular cross section, that is simply supported at one end and guided at the other, ($A_o = 0; A_1 \neq 0$).

Case 2: ($A_o \neq 0; A_1 = 0$). This case corresponds to the homogeneous column

$$D(\xi) = A_0 \tag{39}$$

Then, assuming $N(\xi) = P$ and $q = 1$, the critical load of the homogeneous simply supported-guided column is obtained

$$P = A_o \pi^2 / 4L^2 \tag{40}$$

General Case: ($A_o \neq 0$; $A_1 \neq 0$). This case corresponds to the inhomogeneous column with stiffness distribution

$$D(\xi) = A_o \left[1 + \beta \sin(\frac{\pi}{2}\xi) \right] \tag{41}$$

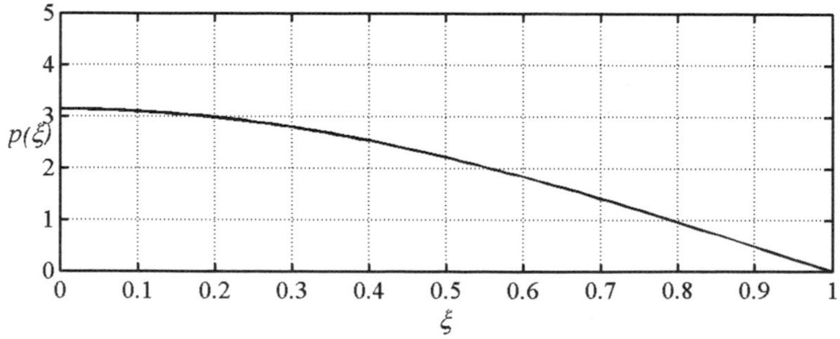

(a)

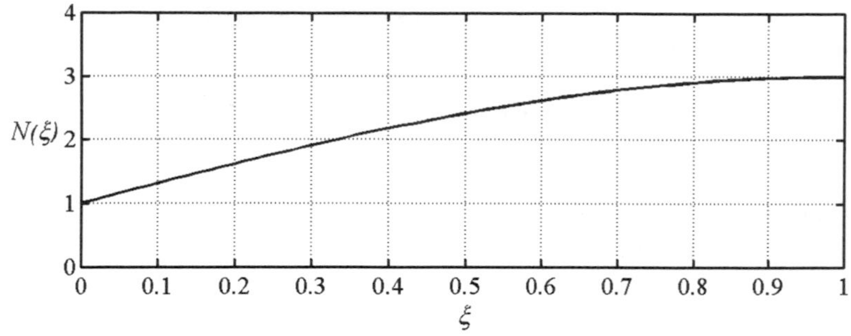

(b)

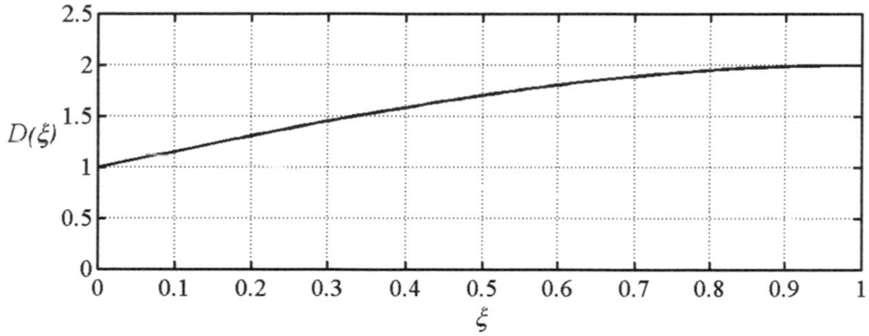

(c)

Figure 12. Inhomogeneous column that is simply supported at one end and guided at the other ($A_o = 1; \beta = 1$). (a) External loading distribution. (b) Axial load distribution. (c) Stiffness distribution.

Assuming $N(\xi) = [1 + 2\beta \sin \pi / 2\xi)]$ the following eingenvalue is obtained

$$q = A_o \pi^2 / 4L^3 \tag{42}$$

The stiffness and the axial load must be non negative therefore the following inequalities must be verified:

$$A_o > 0 \quad \text{and} \quad \beta > -1/2$$

It is remarkable that the expressions (38) and (42) coincide formally with its counterparts (22) and (26) for the cantilever column.

Similarly to the previously considered cases, the results corresponding to this case for the values of the parameters A_0=0, A_1 =1 are reported in figures 12 in terms of external loading, axial load distribution and stiffness distribution. The qualitative shape of an inhomogeneous column in which the stiffness variability is due the variation of the circular cross section is sketched in Figure 13.

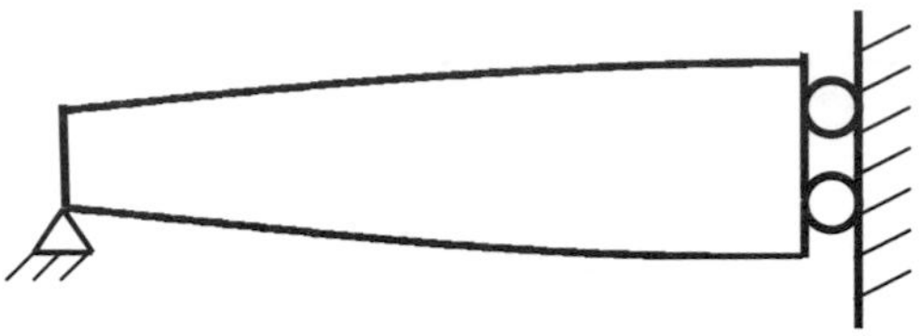

Figure 13. Qualitative shape of an inhomogeneous column, with circular cross section, that is simply supported at one end and guided at the other, ($A_o \neq 0; A_1 \neq 0$).

2.4 Column Clamped at its Both Ends

Let us consider a column clamped at its both ends. The boundary conditions are satisfied for the following buckling mode

$$\psi(\xi) = (1 - \cos(2\pi\xi)) \tag{43}$$

Assuming the following trigonometric representation of the flexural rigidity

$$D(\xi) = A_o + A_1(1 - \cos(2\pi\xi)) \tag{44}$$

after algebraic manipulation the first term of the equilibrium differential equation may be written as

$$\frac{d^2}{d\xi^2}\left[D(\xi)\frac{d^2\psi}{d\xi^2}\right] = 8\pi^3 \frac{d}{d\xi}\{[-A_o - A_1(1 - 2\cos(2\pi\xi))]\sin(2\pi\xi)\} \tag{45}$$

and the second term assumes the expression

$$2\pi \frac{d}{d\xi}\left[qL^3 N(\xi)\sin(2\pi\xi)\right] \tag{46}$$

Therefore, considering the expressions (45) and (46) the differential equation (2) may be put as

$$\frac{d}{d\xi}\left\{\left[\frac{1}{4\pi^2}qL^3 N(\xi) - A_o - A_1(1 - 2\cos(2\pi\xi))\right]\sin(2\pi\xi)\right\} = 0 \tag{47}$$

from which it is apparent that all the distributions of axial force $N(\xi)$ that are proportional to

$$N(\xi) \propto A_o + A_1[1 - 2\cos(2\pi\xi)] \tag{48}$$

satisfy the equilibrium differential equation for a specified value of the buckling load.

Bearing in mind that $N(\xi)$ must be positive the case $A_o = 0;\ A_1 \neq 0$ must be discarded while the following particular cases are considered:

Case 1: ($A_o \neq 0; A_1 = 0$). This case corresponds to the homogeneous column

$$D(\xi) = A_0 \tag{49}$$

Then, assuming a constant axial force P and $q = 1$, the critical load of the homogeneous column is obtained

$$P = 4A_o\pi^2 / L^2 \tag{50}$$

General Case: ($A_o \neq 0; A_1 \neq 0$). This case corresponds to the inhomogeneous column with stiffness distribution

$$D(\xi) = A_o[1 + \beta(1 - \cos(2\pi\xi))] \tag{51}$$

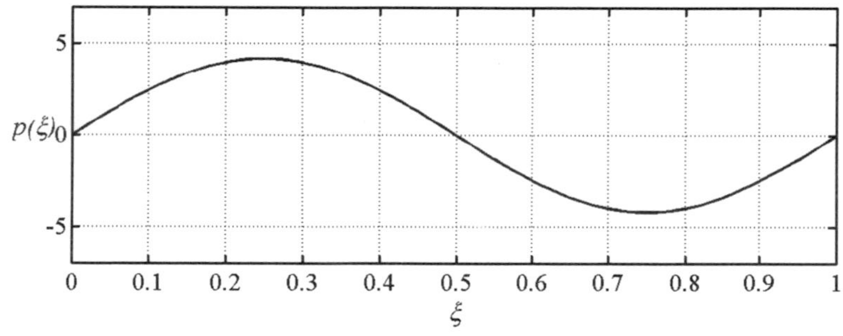

(a)

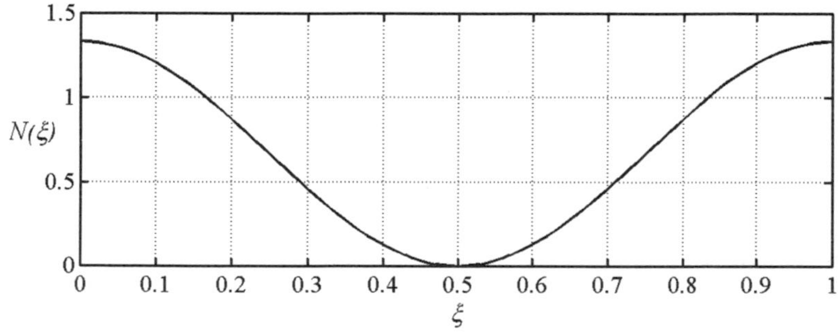

(b)

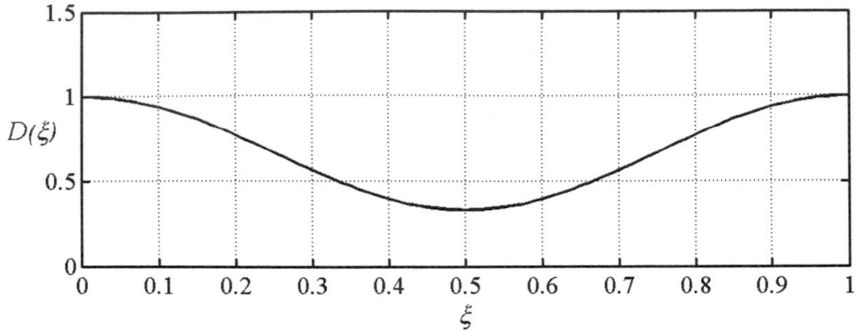

(c)

Figure 14. Inhomogeneous column that is clamped at its both ends ($A_o = 1; \beta = 1$).
(a) External loading distribution. (b) Axial load distribution. (c) Stiffness distribution.

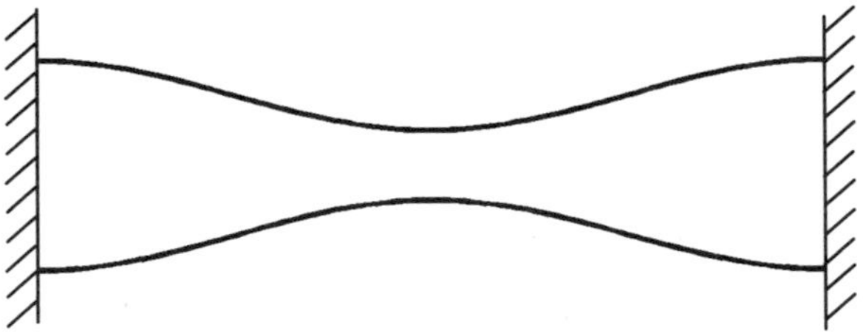

Figure 15. Qualitative shape of an inhomogeneous column, with circular cross section, that is clamped at its both ends, ($A_o \neq 0; A_1 \neq 0$).

Assuming $N(\xi) = 1 + \beta[1 - 2\cos(2\pi\xi)]$ the following eingenvalue is obtained

$$q = 4A_o\pi^2 / L^3 \tag{52}$$

Observing that the stiffness and the axial load cannot be negative, the following inequalities must be verified:

$$A_o > 0 \text{ and } -1/3 < \beta \leq 1$$

Similarly to the previously considered case the results corresponding to the values A_o=1 and β=-1/3 are reported in figures 14 and 15 respectively.

2.5 Comparison with approximate results

It appears to be instructive to list also the buckling loads obtained by some other methods. In order to compare the obtained analytical results with those obtained considering approximate solutions, we also conducted a calculation by the energy method. For the trial function polynomial expression were used for simplicity. The results are reported in Table 1.

Table 1. Verification of the analytical results

Boundary Condition	Closed Form Solution	Polynomial Trial Function	Rayleigh's Quotient	Percentages Difference with Closed Form Solution	Stiffness Distribution	Shape of a Column with Circular Cross Section
S-S	$2\pi^2 \frac{A_1}{L^3}$	$\xi - 2\xi^3 + \xi^4$	$19.754 \frac{A_1}{L^3}$	0.076	$A_o = 0;$ $A_1 \neq 0$	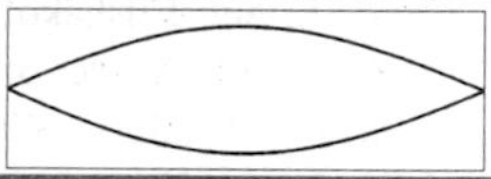
	$\pi^2 \frac{A_o}{L^3}$		$9.886 \frac{A_o}{L^3}$	0.172	$A_o \neq 0;$ $A_1 = -1/2 \cdot A_o$	
C-F	$\pi^2/2 \frac{A_1}{L^3}$	$6\xi^2 - 4\xi^3 + \xi^4$	$5.406 \frac{A_1}{L^3}$	9.566	$A_o = 0;$ $A_1 \neq 0$	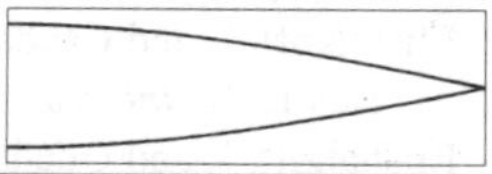
	$\pi^2/4 \frac{A_o}{L^3}$		$2.7521 \frac{A_o}{L^3}$	10.442	$A_o \neq 0;$ $A_1 = A_o$	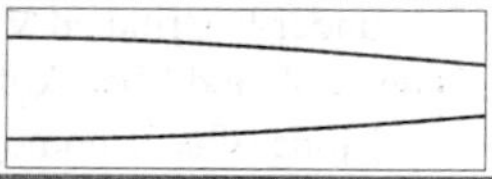
S-G	$\pi^2/2 \frac{A_1}{L^3}$	$\frac{3}{2}\xi - \frac{1}{2}\xi^3$	$4.988 \frac{A_1}{L^3}$	1.094	$A_o = 0;$ $A_1 \neq 0$	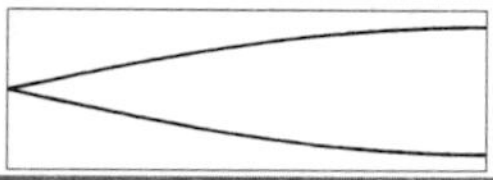
	$\pi^2/4 \frac{A_o}{L^3}$		$2.497 \frac{A_o}{L^3}$	1.216	$A_o \neq 0;$ $A_1 = A_o$	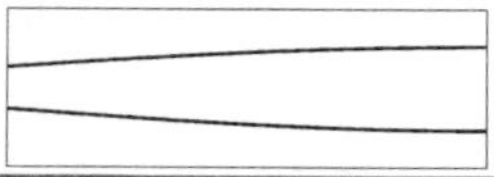
C-C	$\pi^2/4 \frac{A_1}{L^3}$	$\xi^2 - 2\xi^3 + \xi^4$	$2.846 \frac{A_1}{L^3}$	15.363	$A_o \neq 0;$ $A_1 = -1/3 \cdot A_o$	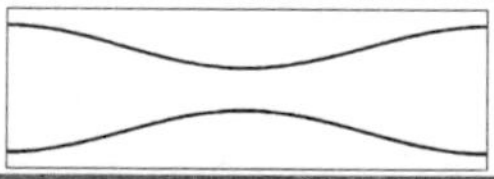
	$\pi^2/4 \frac{A_o}{L^3}$		$2.752 \frac{A_o}{L^3}$	11.552	$A_o \neq 0;$ $A_1 = A_o$	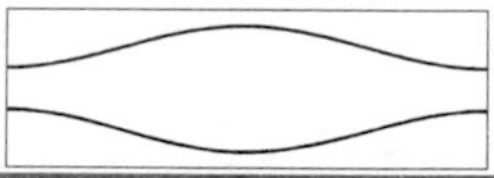

3 Conclusion

In this work trigonometric closed form solutions have been derived for the buckling mode of inhomogeneous columns subjected to a distribution of axial compressive load. In this study four cases of harmonically varying buckling modes have been postulated and semi-inverse problems have been solved that result in the distributions of the flexural rigidity compatible to the pre-selected modes and to specified axial load distributions. In all cases the closed-form solutions have been obtained for the eigenvalue parameter. The presented methodology of solving semi-inverse problems represents in actuality a design buckling problem within the class of inhomogeneous columns.

References

Caliò, I., and Elishakoff, I. (2002). Can a Harmonic Function Constitute a Closed-Form Buckling Mode of an Inhomogeneous Column?. In *AIAA Journal*, **40**(12): 2532-2537.

Caliò, I., and Elishakoff, I. (2004). Closed-form Trigonometric Solutions for Inhomogeneous Beam Columns on Elastic Foundation. In *International Journal of Structural Stability and Dynamics*, **4**(1): 139-146.

Caliò, I., and Elishakoff, I. (2004). Can a Trigonometric Function Serve Both as the Vibration and the Buckling Mode of Axially Graded Structures?. In *Mechanics Based Design of Structures and Machines*, **32**(4): 401-421.

Caliò, I., and Elishakoff, I. (2005). Closed-form Solutions for Axially Graded Beam-Columns. In *Journal of Sound and Vibration*, **280**: 1083-1094.

Elishakoff, I. (2005) Eigenvalues of Inhomogeneous Structures: *Unusual Closed-Form Solutions*, CRC Press, Boca Raton, XVII + pp. 720.

Elishakoff, I. and Candan, S. (2001) Apparently First Closed-Form Solution for Vibrating: Inhomogeneous Beams. In *Inter. J. of Solids and Structures*, **38**(19): 3411-3441.

Elishakoff, I. and Guédé, (2001) Z. Novel Closed-Form Solutions in Buckling of Inhomogeneous Columns under Distributed Variable Loading. In *Chaos, Solitons and Fractals*, **12**: 1075 – 1089.

Guédé, Z. and Elishakoff, (2002) I. A Fifth-Order Polynomial That Serves as Both Buckling and Vibratioin Mode of an Inhomogeneous Structure. In *Chaos, Solitons and Fractals*, **12**: 1267-1298.

Closed Form Trigonometric Solution of Inhomogeneous Beam-Columns

Vibration Problem

Isaac Elishakoff[1] and Ivo Caliò[2]

[1] Department of Mechanical Engineering, Florida Atlantic University, Boca Raton, Florida, USA
[2] Dipartimento di Ingegneria Civile e Ambientale, Università di Catania, Catania, Italy

Abstract. In the previous work buckling modes has been postulated and semi-inverse problems has been solved that result in the distributions of the flexural rigidity compatible to the pre-selected modes and to specified axial load distributions. In this part II the semi-inverse problem is widen to the dynamic context and it is demonstrated that a harmonic function can serve both as the vibration and the buckling mode of an axially graded beam-column.

1 Introduction

In the previous work relative to the buckling problem it has been shown that, for some specified boundary conditions, inhomogeneous columns subjected to a variable distribution of axial load can possess the same trigonometric buckling modes of the homogeneous column subjected to constant axial load.

Now it will be demonstrated that a harmonic function can also serve both as the vibration and the buckling mode of a particular class of axially graded beam-column in which the inhomogeneity is associated to the flexural stiffness and to the mass density of the beam.

2 Vibration Trigonometric Closed Form solution for Inhomogeneous Beam-Columns

Let us consider an inhomogeneous, axially loaded beam-column of length L, with constant cross-sectional area A and moment of inertia I, varying material density $\rho(x)$ and varying modulus of elasticity $E(x)$. The governing differential equation of the dynamic behavior of such inhomogeneous Bernoulli-Euler beam is given by

$$\frac{\partial^2}{\partial x^2}\left[E(x)I\frac{\partial^2 w(x,t)}{\partial x^2}\right]+\frac{\partial}{\partial x}\left[N(x)\frac{\partial w(x,t)}{\partial x}\right]+\rho(x)A\frac{\partial^2 w(x,t)}{\partial t^2}=0 \tag{1}$$

where $w(x,t)$ is the transverse displacement, $N(x)$ is the axial compressive load distribution, x is the axial coordinate, and t is the time.

In the following the differential equation (1) will be solved in a closed form for three sets of boundary conditions corresponding to the following beam-columns: (a) simply supported at its both ends, (b) simply supported at one end and guided at the other, (c) guided at its both ends. For simplicity, the non-dimensional coordinate $\xi = x/L$ is introduced. Harmonic vibration is studied so that the displacement $w(x,t)$ is represented as follows:

$$w(\xi,t) = W(\xi)e^{i\omega t}, \tag{2}$$

where $W(\xi)$ is the postulated mode shape, ω is the corresponding natural frequency, that should be determined. Upon substitution of Eq. (2) into (1), the latter becomes

$$\frac{d^2}{d\xi^2}\left[E(\xi)I\frac{d^2W(\xi)}{d\xi^2}\right] + L^2\frac{d}{d\xi}\left[N(\xi)\frac{dW(\xi)}{d\xi}\right] - AL^4\omega^2\rho(\xi)W(\xi) = 0\,. \tag{3}$$

The following semi-inverse problem is posed: Find an inhomogeneous beam with a specified harmonic mode $W(\xi)$, which satisfies the boundary conditions and the governing dynamic equilibrium differential equation. This semi-inverse problem requires the determination of the distribution of elastic modulus, $E(\xi)$, that together with a specified law of material density, $\rho(\xi)$, and axial load distribution, $N(\xi)$, satisfy the governing eigenvalue problem.

In the previous chapters the polynomial modes have been postulated to solve analogous problems. Here the objective is to widen the class of the possible mode shape by including trigonometric functions.

The elastic modulus, $E(\xi)$, and the axial force, $N(\xi)$, are represented as follows

$$E(\xi) = A_o + A_1\sin(\varphi\pi\xi) + A_2\cos(\varphi\pi\xi) \tag{4}$$

$$N(\xi) = B_o + B_1\sin(\gamma\pi\xi) + B_2\cos(\gamma\pi\xi) \tag{5}$$

where A_o, A_1, A_2, B_o, B_1, B_2 are constants, while φ and γ are real numbers. In the following it will be shown that for a specified distribution of elastic modulus and material density the posed semi-inverse problem possesses a unique solution.

2.1 Beam That Is Simply Supported At Both Ends

Let us consider a simply supported inhomogeneous beam subjected to a variable distribution of axial load.

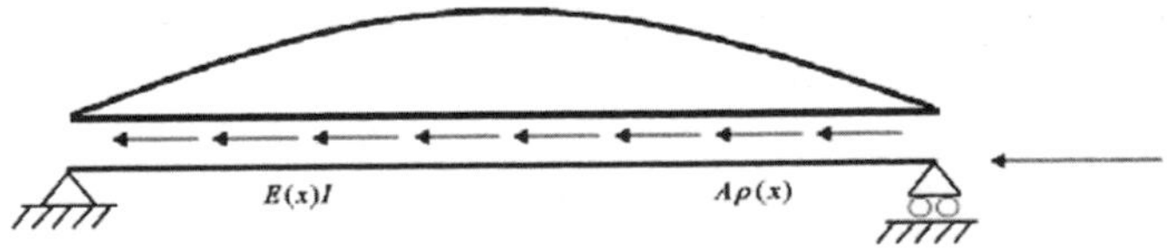

Figure 1. Inhomogeneous beam-column that is simply supported at both end.

The following harmonic representation of the elastic modulus is introduced

$$E(\xi) = A_o + A_2\cos(\pi\xi) = A_o[1+\alpha\cos(\pi\xi)] \tag{6}$$

where, in agreement with the equation (4), $\alpha = A_2 / A_o$, $A_1 = 0$ and $\varphi = 1$. Furthermore, the following harmonic axial load, compatible to the general expression (5), is considered

$$N(\xi) = \lambda[1+\beta\cos(\pi\xi)] \tag{7}$$

where β is a real number and λ is a load parameter. In addition the vibration mode of the homogeneous beam, $W(\xi) = \sin(\pi\xi)$, is postulated. Therefore, considering equations (6) and (7) the first two terms in the differential equation (3) can be re-written as

$$\pi^2\left[\pi^2 A_o I(1+4\alpha\cos(\pi\xi)) - \lambda L^2(1+2\beta\cos(\pi\xi))\right]\sin(\pi\xi) \tag{8}$$

With the purpose of obtaining a closed form solution, we set $\beta = 2\alpha$. Hence, bearing in mind the postulated vibration mode the differential equation (3) reduces to

$$\left[\pi^2\left(\pi^2 A_o I - \lambda L^2\right)(1+4\alpha\cos(\pi\xi)) - AL^4\omega^2\rho(\xi)\right]\sin(\pi\xi) = 0 \tag{9}$$

It is easy to recognize that equation (9) is naturally satisfied for all the distributions of material density $\rho(\xi)$ that are proportional to

$$\rho(\xi) \propto 1+4\alpha\cos(\pi\xi) \tag{10}$$

Since the distribution of density must be positive, we arrive at the conclusion that α must satisfy the inequality $-1/4 < \alpha < 1/4$. It is worth noticing that if this inequality is valid, both the clastic modulus and the distribution of axial load, given respectively by expressions (6) and (7), are positive functions too, as they should be.

Now, we consider the following particular cases:

Case 1: Homogeneous Beam ($\alpha = 0$). This case corresponds to the homogeneous beam, with $E(\xi) = A_0$, subjected to constant axial loading $N(\xi) = \lambda = P$. Then, for $\rho(\xi) = \rho = const$, the natural frequency of the homogeneous beam under constant axial load is recovered

$$\omega^2 = \frac{\pi^2\left(\pi^2 A_o I - PL^2\right)}{\rho A L^4} \tag{11}$$

In this classical case the natural frequency vanishes if the load P equals the critical buckling value, $P_{cr} = \pi^2 A_o I / L^2$.

Case 2: Inhomogeneous Beam under Distributed Axial Load ($\alpha \neq 0$). This case corresponds to the inhomogeneous beam under distributed axial load. The beam with the material density and axial load distribution, respectively

$$\rho(\xi) = \rho_o\left(1 + 4\alpha\cos(\pi\xi)\right); \quad N(\xi) = \lambda\left[1 + 2\alpha\cos(\pi\xi)\right], \tag{12}$$

has the following natural frequency

$$\omega^2 = \frac{\pi^2\left(\pi^2 A_o I - \lambda L^2\right)}{\rho_o A L^4} \tag{13}$$

This expression vanishes at $\lambda = \lambda_{cr} = \pi^2 A_o I / L^2$, that represent the critical value of the load distribution. It is remarkable that the expression (13) formally coincides with its counterpart, in equation (11), that is valid for the homogeneous beam under constant axial load. Thus, for the inhomogeneous beam-column, we derive, in exact terms, the natural frequency in the same form as for the homogeneous one. This mathematical affinity appears to be remarkable. Furthermore, the obtained eigenvalue does not depend on the parameter α, it depends only on A_0 and ρ_o. In Figures 2a, 2b, and 2c the elastic modulus, the axial load distribution and the variability of the material density corresponding to the case $A_0 = 1$, $\rho_o = 1$ and $\alpha = 0.1$ are reported.

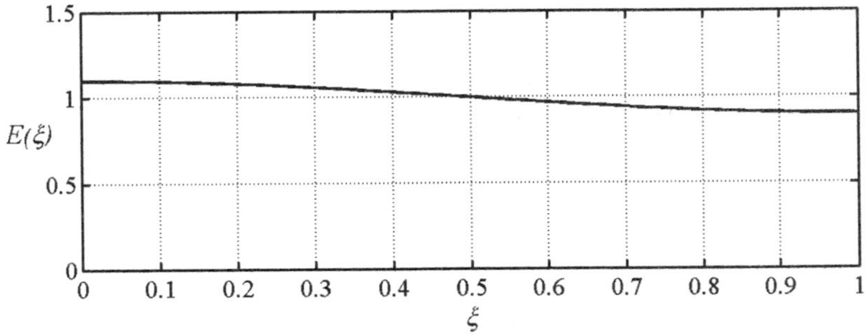

(a)

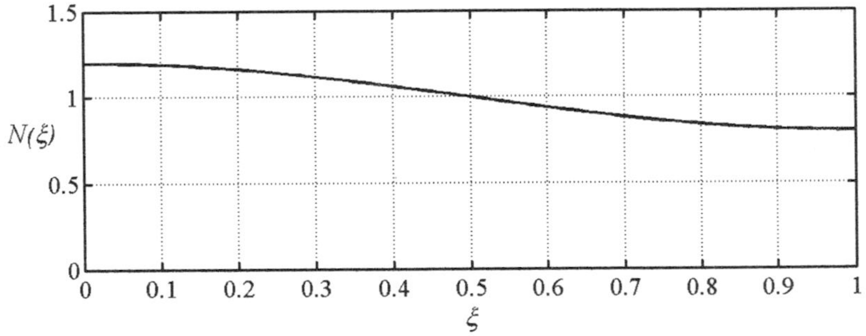

(b)

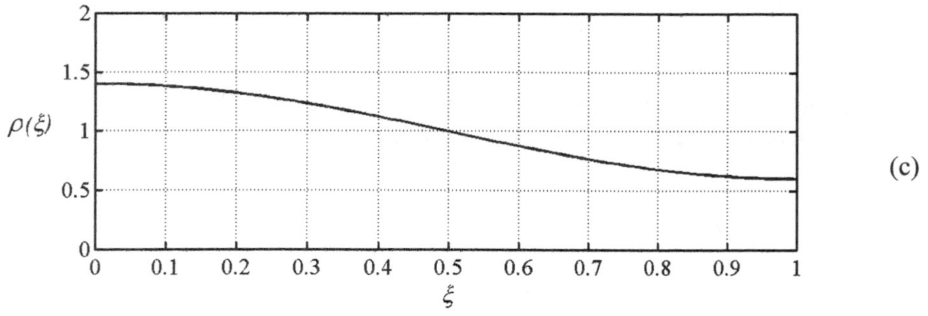

Figure 2. Beam that is simply supported at both ends ($\alpha = 0.1$).
(a) Elastic modulus distribution. (b) Axial load distribution. (c) Density distribution.

2.2 Beam That Is Simply Supported At One End and Elastically Guided At the Other

For a beam that is simply supported at one end and guided, with an elastic spring, at the other, subjected to a variable distribution of axial load, Figure 3, the boundary conditions read

$$\text{(a):} \quad \psi(0)=0\,; \qquad \text{(b):} \quad \psi''(0)=0\,;$$

$$\text{(c):} \quad \psi'(1)=0\,; \qquad \text{(d):} \quad \left[E(\xi)I\psi''(\xi)\right]' - kL^3\psi(\xi)\Big|_1 = 0\,. \tag{14}$$

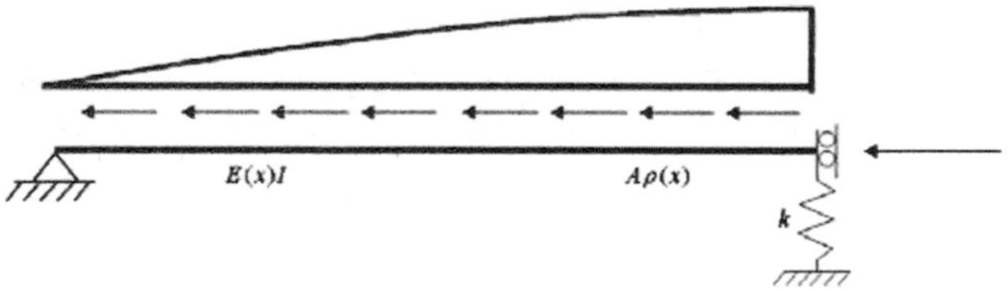

Figure 3. Inhomogeneous beam-column that is simply supported at one end and guided at the other.

The demanded vibration mode is taken as $\psi(\xi)=\sin(\pi/2\cdot\xi)$ that is equal to the buckling mode of the homogeneous column. Furthermore, the following trigonometric representation of the elastic modulus is considered

$$E(\xi)=\left[1+\alpha\cos\left(\frac{\pi}{2}\xi\right)\right] \tag{15}$$

in conjunction with the harmonic distribution of axial load

$$N(\xi)=\lambda\left[1+\beta\cos\left(\frac{\pi}{2}\xi\right)\right] \tag{16}$$

By considering the assumed distributions of elastic modulus and axial load together with the postulated vibration mode the first two terms of the differential equation (3) may be written as

$$\frac{1}{16}\pi^2\left[\pi^2A_oI\left(1+4\alpha\cos\left(\frac{\pi}{2}\xi\right)\right)-4L^2\lambda\left(1+2\beta\cos\left(\frac{\pi}{2}\xi\right)\right)\right]\sin\left(\frac{\pi}{2}\xi\right) \tag{17}$$

In order to obtain a multiplicative representation it is sufficient to fix β at 2α. Therefore, bearing in mind the postulated vibration mode the differential equation (3) takes the following final form

$$\left[\frac{1}{16}\pi^2\left(\pi^2A_oI-4L^2\lambda\right)\left[1+4\alpha\cos\left(\frac{\pi}{2}\xi\right)\right]-AL^4\omega^2\rho(\xi)\right]\sin\left(\frac{\pi}{2}\xi\right)=0 \tag{18}$$

It easy to recognize that equation (18) is automatically satisfied for all the distributions of material density $\rho(\xi)$ that are proportional to

$$\rho(\xi)\propto 1+4\alpha\left(\cos\frac{\pi}{2}\xi\right) \tag{19}$$

In order to obtain a positive density distribution the inequality $\alpha>-1/4$ must be imposed. We consider the following particular cases:

Case 1: Homogeneous Beam ($\alpha=0$). This case corresponds to homogeneous beam with $E(\xi)=A_0$. The beam is subjected to a constant axial loading, $N(\xi)=\lambda=P$. Then, for $\rho(\xi)=\rho=const$, the natural frequency of the homogeneous beam under constant axial load is obtained

$$\omega^2=\frac{\pi^2}{16}\frac{\left(\pi^2A_oI-4PL^2\right)}{\rho AL^4} \tag{20}$$

The frequency vanishes for the critical value of the load $P=P_{cr}=\pi^2A_oI/4L^2$.

Case 2: Inhomogeneous Beam under Distributed Axial Load ($\alpha\neq0$). By considering a distribution of material density $\rho(\xi)$ as follows

$$\rho(\xi)=\rho_o\left[1+4\alpha\cos\left(\frac{\pi\xi}{2}\right)\right] \tag{21}$$

the following frequency is obtained

$$\omega^2 = \frac{\pi^2}{16} \frac{\left(\pi^2 A_o I - 4\lambda L^2\right)}{\rho_o A L^4} \tag{22}$$

This expression vanishes for $\lambda = \lambda_{cr} = \pi^2 A_o I / 4L^2$, that represents the critical value of the load distribution. It is worth noticing that the expression (22) formally coincides with its counterpart (20) for the homogeneous beam under constant axial load.

Now we pose the much needed question whether the boundary condition (14d) is satisfied. Naturally, for arbitrary value of the stiffness k the boundary condition will not be satisfied. The special value of the stiffness of the linear elastic spring at the guided end is evaluated imposing the satisfaction of the natural boundary condition, which involves the reaction of the spring (14d).

By substituting the postulated vibration mode and the expressions (14d) and (15) we get

$$\frac{1}{8} A_o I \pi^3 \left\{ \alpha \sin\left(\frac{\pi\xi}{2}\right)^2 - \cos\left(\frac{\pi\xi}{2}\right)\left[1 + \alpha\cos\left(\frac{\pi\xi}{2}\right)\right]\right\} - kL^3 \sin\left(\frac{\pi\xi}{2}\right)\Bigg|_1 = 0 \tag{23}$$

from which the sought value of the elastic spring at the guided end is derived

$$k = \alpha A_o I \pi^3 / 8L^3 \tag{24}$$

For any fixed value of the parameter A_o the stiffness of the elastic spring varies linearly with the parameter α that identifies the particular distribution of elastic modulus and material density. It is worth noticing that in the limiting case of the homogeneous beam, $\alpha = 0$, the boundary condition (14d) is satisfied for k=0, which correspond to the guided beam. Furthermore by considering that the stiffness k must be positive the more stringent inequality $\alpha \geq 0$ is derived than the one, $\alpha > -1/4$, derived before above.

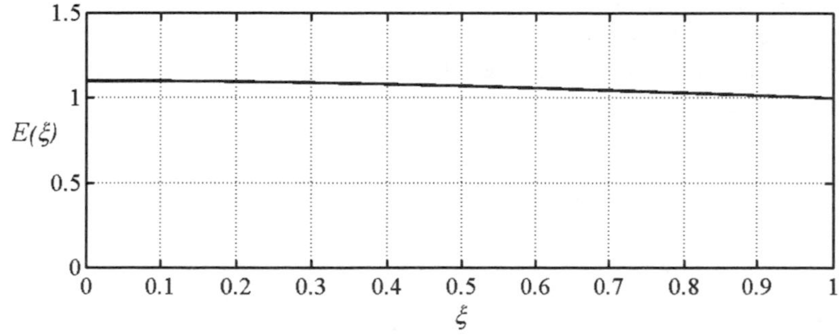

(a)

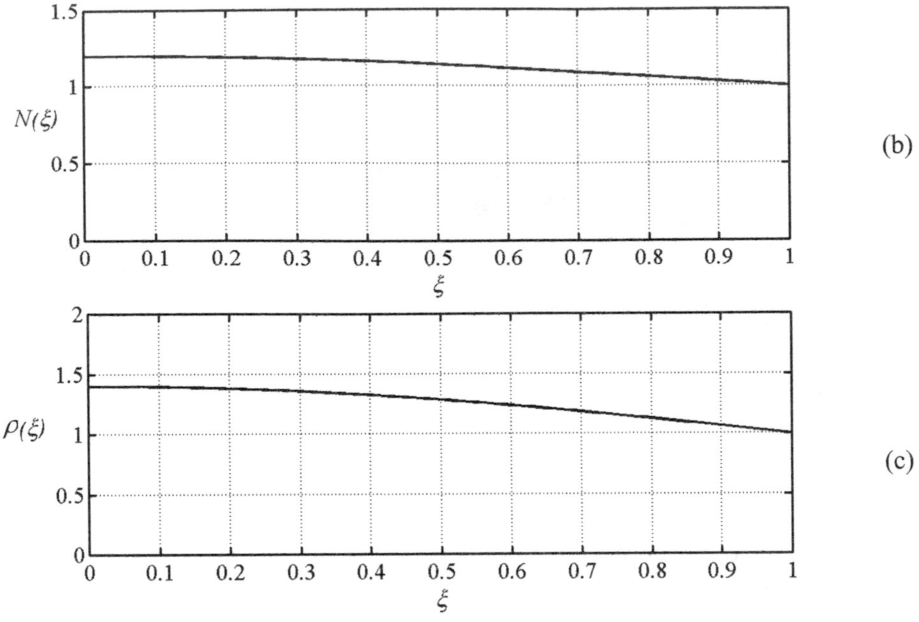

Figure 4. Beam simply supported at one end and guided at the other ($\alpha = 0.1$). (a) Elastic modulus distribution. (b) Axial load distribution. (c) Density distribution.

With reference to inhomogeneous beams it is important to emphasize the remarkable and unanticipated independence of the eigenvalue, given by expression (22), from the coefficient α. Hence different beams characterized by the same values of A_o and ρ_o but different value of α have the same eigenvalues and mode shape. For example, with reference to a vibration problem without axial load distribution, this means that if we consider two beams with different values of α (corresponding to different distributions of elastic modulus and material density) but with springs characterized by the corresponding values of $k(\alpha)$, given by equation (24), this two beams will have the same frequency and share the same vibration mode.

For example, two beams characterized by:

$$\text{(I)}\ \ k = \varepsilon A_o I\pi^3/8L^3; \quad E(\xi) = \left[1 + \varepsilon\cos\left(\frac{\pi}{2}\xi\right)\right]; \quad \rho(\xi) = \rho_o\left[1 + 4\varepsilon\cos\left(\frac{\pi\xi}{2}\right)\right] \quad \text{with } \varepsilon \geq 0$$

$$\text{(II)}\ \ k = \gamma A_o I\pi^3/8L^3; \quad E(\xi) = \left[1 + \gamma\cos\left(\frac{\pi}{2}\xi\right)\right]; \quad \rho(\xi) = \rho_o\left[1 + 4\gamma\cos\left(\frac{\pi\xi}{2}\right)\right] \quad \text{with } \gamma \geq 0$$

have the same vibration mode, $\psi(\xi) = \sin(\pi/2 \cdot \xi)$, and the same frequency, expressed by equation (22).

In Figures 4a, 4b, and 4c the distributions of elastic modulus, axial load and material density corresponding to the case $\alpha = 0.1$, $A_0 = 1$ and $\rho_o = 1$ are reported; it may be observed that these distributions, relative to the supported-guided beam, correspond to the distributions of the simply supported beam with reference to the first half beam, this seem to indicate that the spring with its rigidity simulates the contribution of the remaining half beam both for the dynamic and the instability problem.

2.3 Beam That Is Elastically Guided At Both Ends

For a beam that is elastically guided at both ends subjected to a variable axial load, Figure 5, the boundary conditions, are

$$\text{(a):}\quad \psi'(0)=0; \qquad \text{(b):}\quad \left.[E(\xi)I\psi''(\xi)]' + kL^3\psi(\xi)\right|_0 = 0;$$
$$\text{(c):}\quad \psi'(1)=0; \qquad \text{(d):}\quad \left.[E(\xi)I\psi''(\xi)]' - kL^3\psi(\xi)\right|_1 = 0. \tag{25}$$

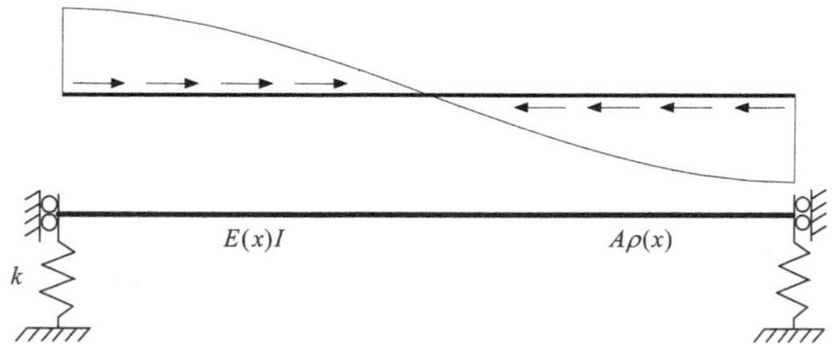

Figure 5. Inhomogeneous beam-column that is elastically guided at its both ends.

Furthermore a cosinusoidal vibration mode, $\psi(\xi)=\cos(\pi\xi)$, is postulated and the following representation of the elastic modulus is considered

$$E(\xi)=A_o[1+\alpha\sin(\pi\xi)], \tag{26}$$

while the distribution of axial load is taken as

$$N(\xi)=\lambda[1+\beta\sin(\pi\xi)] \tag{27}$$

Therefore the first two terms of the differential equation (3) becomes

$$\pi^2\left[\pi^2 A_o I(1+4\alpha\sin(\pi\xi)) - \lambda L^2(1+2\beta\sin(\pi\xi))\right]\cos(\pi\xi) \tag{28}$$

By setting $\beta = 2\alpha$ and considering the postulated vibration mode the differential equation (3) takes the following final form

$$\left[\pi^2\left(\pi^2 A_o I - \lambda L^2\right)\left[1+4\alpha\sin(\pi\xi)\right] - AL^4\omega^2\rho(\xi)\right]\cos(\pi\xi) = 0 \tag{29}$$

Equation (29) is satisfied for all the distributions of material density $\rho(\xi)$ that is proportional to

$$\rho(\xi) \propto 1 + 4\alpha(\sin\pi\xi) \tag{30}$$

Furthermore, the inequality $\alpha > -1/4$ must be introduced in order to obtain a positive material density distribution. We consider the following particular cases:

Case 1: Homogeneous Beam ($\alpha = 0$). This first case corresponds to the homogeneous beam, $E(\xi) = A_0$, subjected to a constant axial load, $N(\xi) = \lambda = P$. Assuming $\rho(\xi) = \rho = const$, the natural frequency is obtained

$$\omega^2 = \pi^2 \frac{\left(\pi^2 A_o I - PL^2\right)}{\rho A L^4} \tag{31}$$

The natural frequency vanishes for the critical value of the load $P = P_{cr} = \pi^2 A_o I / L^2$.

Case 2: Inhomogeneous Beam under Distributed Axial Load ($\alpha \neq 0$). According to the expression (30) the distributions of material density $\rho(\xi)$ is taken as follows

$$\rho(\xi) = \rho_o\left[1 + 4\alpha\cos\left(\frac{\pi\xi}{2}\right)\right] \tag{32}$$

The resulting natural frequency reads

$$\omega^2 = \pi^2 \frac{\left(\pi^2 A_o I - \lambda L^2\right)}{\rho_o A L^4} \tag{33}$$

The natural frequency vanishes if the distributed load, λ, attains the critical value $\lambda_{cr} = \pi^2 A_o I / L^2$.

It must be noted that the frequencies and the buckling load obtained for the beam guided at its both ends have the same analytical expressions of those obtained for the simply supported beam but refer to different distributions of elastic modulus, axial load and material density.

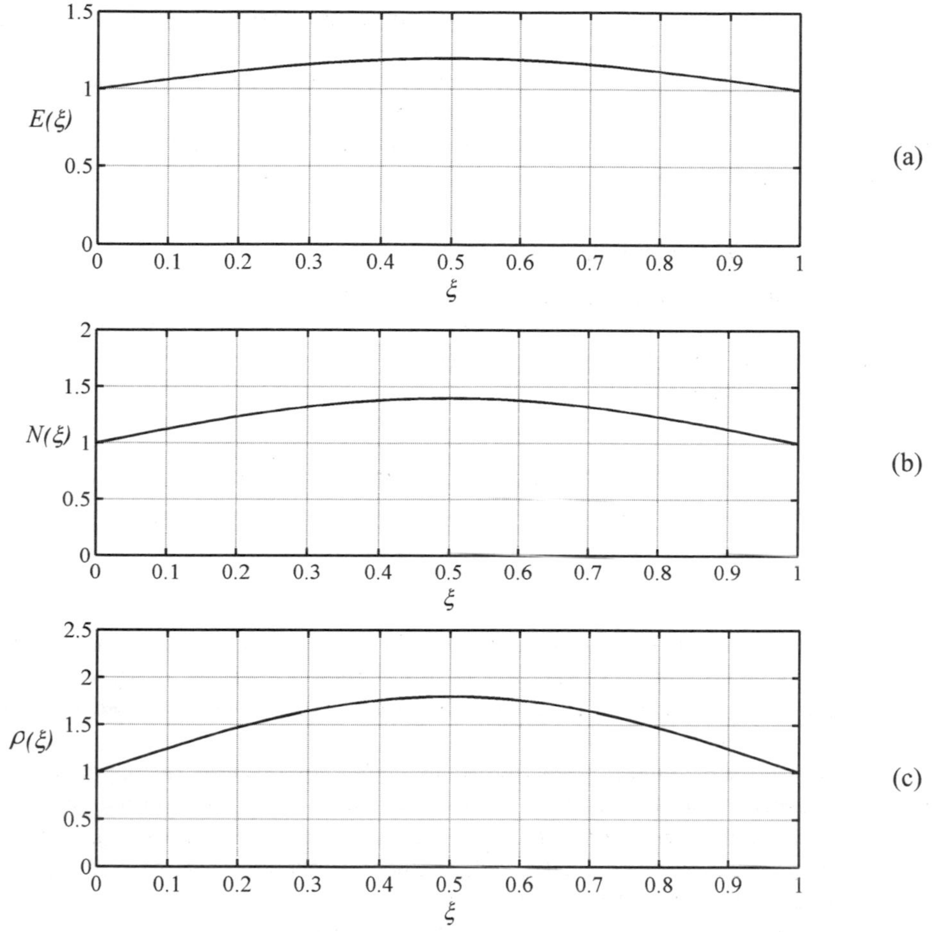

Figure 6. Beam that is elastically guided at both ends ($\alpha = 0.2$).
(a) Elastic modulus distribution. (b) Axial load distribution. (c) Density distribution.

The values of the stiffness of the elastic supports, corresponding to the reported solution, are derived hereinafter by imposing the fulfillment of the natural boundary conditions at both ends. By substituting the postulated mode and the distribution of the elastic modulus in the expressions (25b) and (25d) the value of the elastic spring, equal at both ends, is derived

$$k = \alpha A_o I \pi^3 / L^3 \tag{34}$$

It is worth to notice that guided inhomogeneous beams for which the stiffness of the springs at the ends are specified by equation (34) and have elastic modulus and density distribution given by Eqs. (26) and (32) and are eventually subjected to an axial load distribution, specified by Eq. (27), are all characterized by the same mode, $\psi(\xi) = \cos(\pi\xi)$, and the same fundamental frequency,

given by expression (33), independently by the particular value of the parameter α that characterizes the particular distributions of elastic modulus, density and axial load. Bearing in mind that the stiffness k must be positive, also in this case the more stringent inequality $\alpha \geq 0$ must be considered than the one, $\alpha > -1/4$, derived before above.
The distributions of elastic modulus, axial load and material density are reported in Figures 6a, 6b, and 6c for the case $A_0 = 1$, $\rho_o = 1$ and $\alpha = 0.2$.

2.4 Comparison with the Rayleigh quotient

Since all the solutions reported above are written in the closed form, they do not need an additional corroboration. Still, it is of interest to contrast them with approximate methods in order to gain some additional insight. Therefore, we also conducted a calculation by the energy method using the Rayleigh quotient

$$\omega^2 = \frac{\int_0^L E(x)I[\phi''(x)]^2 dx - \int_0^L N(x)[\phi'(x)]^2 dx + k[\phi(0)]^2 + k[\phi(L)]^2}{\int_0^L A\rho(x)[\phi(x)]^2 d\xi} \tag{35}$$

By substituting in the Rayleigh quotient the exact shape of vibration, the exact eigenvalues, expressed by the equations (13), (22) and (33), has been re-obtained. With reference to the approximate evaluation, for the trial function polynomial expressions were used for simplicity. The comparison is reported in Table 1.
For the simply-supported-guided beam and for the guided-guided beam the approximate results are slightly dependent on the parameter α, while the closed form solution does not have this dependence. The results reported in the table are referred to a value of α=0.5. It is important to recognize that the true vibration shapes yield frequencies slightly lower than the approximate ones for any admissible value of the parameter α.
The solutions reported herein can be used as benchmark problems. Also, at the time when the technology will be available to produce a desired arbitrary distribution of elastic modulus along the axis of the beam and specific variation of material density ρ (ξ) one will be able to design inhomogeneous beams with pre-selected frequency or buckling load values. Thus, the presented methodology of solving semi-inverse problems may represent a design tool for buckling and vibration problems within the class of inhomogeneous beam.

3 Conclusion

In this work trigonometric closed form solutions have been derived for the natural frequencies of beam-columns with simply supported or elastically guided end conditions. The trigonometric function was postulated for the mode shape and conditions were established for which this postulate holds. The distributions of the material density as well as the elastic modulus were sought in terms of trigonometric functions. This seemingly transparent approach uncovers a series of new closed form solutions.

Table 1. Verification of the analytical results

Boundary Condition	Closed Form Solution Frequency and corresponding critical load distribution	Polynomial Trial Function	Rayleigh's Quotient	Percentages Difference with Closed Form Solution
S-S	$\omega^2 = \frac{\pi^4 A_o I - \pi^2 \lambda L^2}{\rho_o A L^4}$	$\xi - 2\xi^3 + \xi^4$	$\omega^2 = \frac{97.548 A_o I - 9.871 \lambda L^2}{\rho_o A L^4}$	
	$\omega^2 = \pi^4 \frac{A_o I}{\rho_o A L^4}$		$\omega^2 = 97.548 \frac{A_o I}{\rho_o A L^4}$	0.14%
	$\lambda_{cr} = \pi^2 \frac{A_o I}{L^2}$		$\lambda_{cr} = 9.882 \frac{A_o I}{L^2}$	0.12%
S-G	$\omega^2 = \frac{\pi^4 A_o I - 4\pi^2 \lambda L^2}{16 \rho_o A L^4}$	$8\xi - 4\xi^3 + \xi^4$	$\omega^2 = \frac{\pi^4 A_o I - \pi^2 \lambda L^2}{16 \rho_o A L^4}$	
	$\omega^2 = \frac{\pi^4 A_o I}{16 \rho_o A L^4}$		$\omega^2 = 6.0944 \frac{A_o I}{A L^4}$	0.10%
	$\lambda_{cr} = \frac{\pi^2}{4} \frac{A_o I}{L^2}$		$\lambda_{cr} = 2.469 \frac{A_o I}{L^2}$	0.09%
G-G	$\omega^2 = \frac{\pi^4 A_o I - \pi^2 \lambda L^2}{\rho_o A L^4}$	$1 - 6\xi^2 + 4\xi^3$	$\omega^2 = \frac{98.339 A_o I - 9.879 \lambda L^2}{\rho_o A L^4}$	
	$\omega^2 = \pi^4 \frac{A_o I}{\rho_o A L^4}$		$\omega^2 = 98.339 \frac{A_o I}{\rho_o A L^4}$	0.95%
	$\lambda_{cr} = \pi^2 \frac{A_o I}{L^2}$		$\lambda_{cr} = 9.953 \frac{A_o I}{L^2}$	0.84%

References

Caliò, I., and Elishakoff, I. (2002). Can a Harmonic Function Constitute a Closed-Form Buckling Mode of an Inhomogeneous Column?. In *AIAA Journal*, **40**(12): 2532-2537.

Caliò, I., and Elishakoff, I. (2004). Closed-form Trigonometric Solutions for Inhomogeneous Beam Columns on Elastic Foundation. In *International Journal of Structural Stability and Dynamics*, **4(**1): 139-146.

Caliò, I., and Elishakoff, I. (2004). Can a Trigonometric Function Serve Both as the Vibration and the Buckling Mode of Axially Graded Structures?. In *Mechanics Based Design of Structures and Machines*, **32**(4): 401-421.

Caliò, I., and Elishakoff, I. (2005). Closed-form Solutions for Axially Graded Beam-Columns. In *Journal of Sound and Vibration*, **280**: 1083-1094.

Elishakoff, I. (2005) Eigenvalues of Inhomogeneous Structures: *Unusual Closed-Form Solutions*, CRC Press, Boca Raton, XVII + pp. 720.

Elishakoff, I. and Candan, S. (2001) Apparently First Closed-Form Solution for Vibrating: Inhomogeneous Beams. In *Inter. J. of Solids and Structures*, **38**(19): 3411-3441.

Elishakoff, I. and Guédé, (2001) Z. Novel Closed-Form Solutions in Buckling of Inhomogeneous Columns under Distributed Variable Loading. In *Chaos, Solitons and Fractals*, **12**: 1075 – 1089.

Guédé, Z. and Elishakoff, (2002) I. A Fifth-Order Polynomial That Serves as Both Buckling and Vibratioin Mode of an Inhomogeneous Structure. In *Chaos, Solitons and Fractals*, **12**: 1267-1298.